S0-EIH-158

Temperature Measurement

WILEY SERIES IN MEASUREMENT SCIENCE AND TECHNOLOGY

Instruments and Experiences: Papers on Measurement and Instrument Design
R. V. Jones

Introduction to Measurement Science and Engineering
P. H. Sydenham, N. H. Hancock and R. Thorn

Temperature Measurement
L. Michalski, K. Eckersdorf and J. McGhee

Temperature Measurement

L. Michalski, K. Eckersdorf
Politecnika Lódźa, Katedra Elektrotermii, Lodz, Poland

and

J. McGhee
Department of Electronic and Electrical Engineering,
University of Strathclyde, Glasgow, Scotland

JOHN WILEY & SONS
Chichester · New York · Brisbane · Toronto · Singapore

Updated, extended and thoroughly revised edition of
Pomiary Temperatury, originally published in 1986 by
Wydawnictwa Naukowo-Techniczne, Warsaw, Poland.

Other Wiley Editorial Offices

John Wiley & Sons, Inc., 605 Third Avenue,
New York, NY 10158-0012, USA

Jacaranda Wiley Ltd, G.P.O. Box 859, Brisbane,
Queensland 4001, Australia

John Wiley & Sons (Canada) Ltd, 22 Worcester Road,
Rexdale, Ontario M9W 1L1, Canada

John Wiley & Sons (SEA) Pte Ltd, 37 Jalan Pemimpin 05-04,
Block B, Union Industrial Building, Singapore 2057

Library of Congress Cataloging-in-Publication Data:

Michalski, L.
Temperature measurement / L. Michalski, K. Eckersdorf, J. McGhee.
p. cm.—(Wiley series in measurement science and
technology)
Includes bibliographical references and index.
ISBN 0 471 92229 3
1. Temperature measurements. I. Eckersdorf, K. II. McGhee, J.
III. Title. IV. Series.
QC271.M483 1991
536′.5′0287—dc20 90-25320
CIP

A catalogue record for this book is available from the British Library

Typeset by Dobbie Typesetting Limited, Tavistock, Devon
Printed in Great Britain by Courier International, East Kilbride

Contents

Series Editor's Preface

Accounts of knowledge on measurement science and technology have been provided, on an ad-hoc basis, by a variety of books from numerous sources. Few publishers have concentrated on its specific scholarship. Such books have often been designed to capture markets by broad coverage—with subsequently shallow depth treatment.

This Series commissions works that cover core knowledge of measurement science and its technology. Books already published, or planned, deal with aspects such as presentation of the breadth of knowledge involved, professionalism of its practice, how to teach this material at undergraduate level and provision of accounts of measured theory and practice.

Temperature Measurement is the first of several books in production that give book length accounts about a single measurand. As well as being an authoritative account about temperature measurement the treatment readily shows how quality measurement asks of the measurer an understanding that is deeper than generally regarded as necessary. In this way it contributes to better scholarship in measurement science.

Professor Peter H Sydenham, Chief Editor
Measurement and Instrumentation Systems Centre
University of South Australia

Preface

Temperature influences all natural physical phenomena as well as all physiological, technological and thermal processes. It is one of the most important parameters in any kind of research.

From the vast and ever growing number of possible methods of temperature measurement, that best suited for each application should be chosen in order to attain readings which are as precise as possible. As the errors of the method are more important in most cases than those of the instrumentation, this book concentrates on the analysis of the different temperature measurement methods and sources of errors.

The scope of the book covers all of the principal temperature measurement methods and instruments. Special stress is placed on such problems as temperature measurement of solids and gases as well as temperature measurement in industrial heating appliances. Other important aspects considered include logging and recording instruments as well as conditioners and transmitters of temperature signals.

As the International Temperature Scale of 1990 (ITS-90) has been introduced, it is described in detail. The importance of maintaining and disseminating this temperature scale, which is closer to the thermodynamic scale than any previous one, is given significant treatment through the detailed consideration of calibration and testing techniques for temperature measuring instruments.

A large number of numerical examples, tables and diagrams are given to provide assistance in choosing and implementing the temperature measuring system best suited for a particular application. Many references enable the reader to find supplementary information regarding those aspects which could not be treated in detail in the book.

Interpretation of temperature scales in the context of contemporary measurement theory is given in an appendix.

This book is conceived for engineers, applied scientists, pure scientists and student readers who wish to master the beautiful art of temperature measurement.

June 1990

THE AUTHORS

List of Principal Symbols

A	amplitude, area
a	thermal diffusivity
C	radiation constant, electrical capacitance
c	specific heat
D, d	diameter
E	thermal e.m.f.
e	thermal e.m.f. in a junction
f	frequency
$G(s)$, $F(s)$	transfer function
$G(j\omega)$, $F(j\omega)$	frequency response
I	electric current
K	gain
k	general coefficient
L	time lag also called dead time
l	length
N	time constant
P	power
Q	energy
q	heat flux density
R	resistance
r	radius
s	Laplace operator
T	temperature in K
t	time
V	voltage, volume
v	velocity
W	thermal resistance, radiant intensity
α	heat transfer coefficient, coefficient of linear thermal expansion, temperature coefficient of resistance, absorptivity
β	coefficient of cubic thermal expansion
Δ	error, difference, amplitude
δ	relative error, penetration depth
ϵ	emissivity
ϑ	temperature in °C or °F
Θ	excess temperature over a reference temperature such as ambient or original value

λ	wavelength, thermal conductivity
ρ	density, reflectivity, resistivity
Φ	heat flux or rate of heat flow
ϕ	phase angle
ω	angular frequency
τ	transmissivity

Subscripts

a	adjustable, ambient, average
C	correction, corrector
c	compensating, conduction
d	disturbance
e	effective, end-value, equivalent
gr	grey body
i	indicated, input value
k	convection
l	leads, limit value, loop
M	measuring instrument
n	nominal value
o	black body, output value
r	radiation, reference, reflection, relative
s	set-point value, shield, solid
T	temperature sensor
t	true value
w	wall
λ	spectral

1 Temperature and Temperature Scales

1.1 HISTORICAL BACKGROUND

The concept of temperature makes one think of physiological experiences whilst touching or approaching some solid. Some of them may be described as cold, cool or tepid, others as hot or warm. Warmer bodies transfer heat to other cooler bodies. Both bodies tend to equalize their temperatures, approaching a new common intermediate temperature. Thus the correctness of the definition, given to temperature by the Scotsman James Clerk Maxwell, may be seen. He stated that the temperature of a body is its thermal state, regarded as a measure of its ability to transfer heat to other bodies. At the present time this definition compels the attribution of larger numerical values to those bodies which have a higher ability to transfer heat to other bodies. This assumption forms the basis of all temperature scales in use today. Science took a long, difficult and tortuous route, full of errors, to this contemporary definition of temperature.

In ancient Rome, during the second century BC, the physician C. Galen introduced four degrees of coldness regarding the effects of different medicines upon human organisms. These medicines were supposed either to warm or to cool them. Galen also introduced a neutral temperature, attributing to it a value of zero degrees. He claimed that this neutral temperature depended upon geographical latitude. Building upon Galen's basic assumption, later scientists tried to attribute the effects of these medicines to different bodies by ascribing to them different degrees of warmth or coldness.

The first device, which was used to measure the degree of warmth or coldness, seems to have been invented by Galileo Galilei some time between the years 1592 and 1603. This instrument, which is shown in Figure 1.1, consisted of a glass bulb connected to a long tube immersed in a coloured liquid. After a preliminary heating of the contained gas, its subsequent cooling caused a certain amount of the liquid to be sucked in. The liquid column rose or fell as a function of the ambient temperature, thus indicating the degree of warmth or coldness. In the absence of any evidence that the instrument had any graduation, it is better to call it a thermoscope. As the indicated values were also a function of the atmospheric pressure its precision must have been quite poor. Subsequently, about the year 1650 the members of the Florentine Academy of Sciences made the first thermometer which is represented in Figure 1.2. This consisted of a spiral-shaped tube with a closed end and a graduation. However, no numbers were ascribed to the graduation marks (Lindsay, 1962).

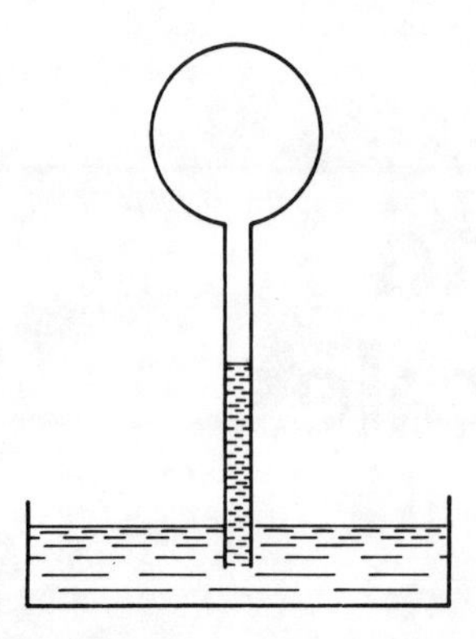

Figure 1.1
Thermoscope of Galileo (1592).

In the course of time the need arose to define temperature fixed points, to standardize those thermometers which existed at that time. One of the first proposals came, in 1669, from H. Fabri from Leida. His proposal was for two fixed points. The lower should be the temperature of snow and the higher the temperature of the hottest summer day. A later proposal, which was made by C. Rinaldi from Padua in 1693, suggested that the fixed points should be the temperatures corresponding to the melting point of ice and the boiling point of water. Between these two points twelve divisions should be introduced. In the same year, and for the first time, the British scientist E. Halley applied mercury as a thermometric liquid.

Further development in thermometry is due to D. G. Fahrenheit from Danzig (now the city of Gdansk in Poland). In 1724 he described the mercury-in-glass thermometer. There were three temperature fixed points associated with the device as follows:

- A mixture of ice, water and ammonium chloride was taken as the zero point.
- A mixture of ice and water was taken as 32°.
- A human body temperature was taken as 96°.

Even yet there is no clear reason why Fahrenheit chose such a scale division based upon these assumed temperature fixed points.

As Newton Friend (1937) indicated, the reasons for choosing such a scale division by Fahrenheit might have been that in the eighteenth century the majority of thermometers were intended for meteorological purposes. Taking the freezing point of water as zero would have involved the repeated use of negative values for winter temperatures. To avoid this, Fahrenheit proposed to use the lowest attainable temperature of those days as zero. In the case of the upper fixed point, the temperature of boiling water was rejected as being unnecessarily high for meteorological purposes. In his decision to assume 96° for the temperature of blood, Fahrenheit was influenced by the then existing Roemer's scale. He merely changed Roemer's 90° to 96°, probably

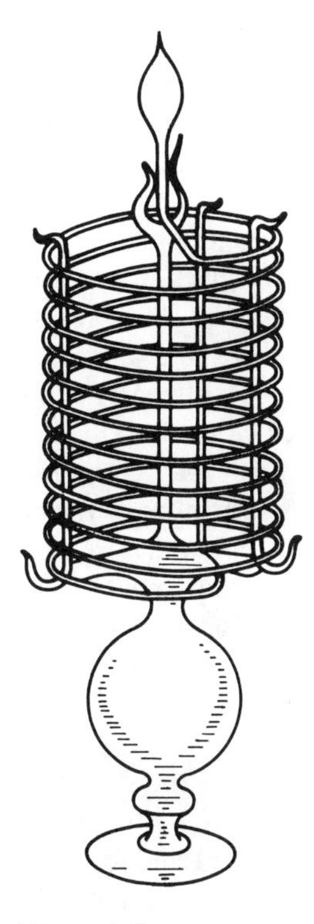

Figure 1.2
Thermometer of the Florentine Academy of Sciences (1650).

because 96 is divisible not only by 2 but also by multiples of 3 and hence 12. (The decimal system was not in general use at that time.)

Further development of the mercury-in-glass thermometer, in 1742, was due to the Swedish astronomer and physicist A. Celsius. He assigned 0° to the temperature of boiling water and 100° to the temperature of melting ice. The region between these two points was divided into 100 equal steps. Subsequently M. Stromer reversed these values.

A true thermodynamic temperature scale, described below, had been the unconscious aim of all of the previous efforts. Such a scale was not possible until 1854 when its foundations were laid by the Belfast born William Thomson, who later became Professor of Natural Philosophy in the University of Glasgow, Scotland, and assumed the title Lord Kelvin of Largs. Of course, as shown in the Appendix, the aim of any scale of temperature, but especially the thermodynamic scale, is the representation of the hotness relations between objects and events in the real physical world by numbers.

1.2 THERMODYNAMIC TEMPERATURE SCALE

A temperature scale is applied to correlate numerical values to some defined temperatures. Temperature fixed points, to which certain temperature values are attributed, are needed.

To enable the definition of temperatures between these temperature fixed points, a thermometric working substance, one of its properties and a correlating function must be assumed. The chosen function provides the means of attributing the chosen property of the working substance with a certain temperature. Because of the diversity of materials and their properties there is an unlimited number of these temperature scales. Properties which may be relevant are, for example, the length of a rod, the pressure of saturated steam, the resistance of a wire and so on. In the given temperature range the property must be consistently repeatable and reproducible. In normal conditions, corresponding to 101.325 kPa, let the ice-point temperature be 0° and the temperature of boiling water 100°. Assuming that the chosen property is linearly dependent upon the temperature it is apparent that any temperature scale based upon say the thermal expansion of copper rod, will not coincide with a scale based upon the thermal expansion of another metal or on any change of its resistance with temperature.

The difficulties associated with the measurement of temperature are evident as there are no direct methods for it, in contrast say to length measurement. As only indirect temperature measurements are possible, any temperature scale depends upon the chosen thermometric working substance and its chosen property. In principle, any working substance may be applied. However, its use will be restricted to some finite range determined by its thermal resistance. As an example, the application of a mercury-in-glass thermometer is limited on the low temperature side by the solidification of the mercury as it freezes and on the high temperature side by the glass resistance.

An ideal solution to the problem of proposing a suitable temperature scale would be to find one valid in any temperature range and totally independent of the working substance. The thermodynamic Kelvin scale, based upon the efficiency of the ideal reversible Carnot cycle, is such a scale (Herzfeld, 1962; McGee, 1988). A reversible Carnot cycle consists of a reversible heat engine operating between two isotherms at the temperatures T_2 and T_1, with $T_2 > T_1$, and of two adiabatic processes. A reversible heat engine absorbs the heat, Q_2, from the high temperature source, at the temperature T_2, and discharges the heat Q_1 to the low temperature source, at the temperature T_1. The difference between the absorbed heat Q_2 and the discharged heat Q_1, which is the external work, A, performed by the engine, may be written as:

$$A = Q_2 - Q_1 \tag{1.1}$$

Reversing the engine action, indicates that it may be driven by a second identical engine, working between the same two heat sources. The effect of such action might be the heat flow from the lower to the higher temperature source. Using the properties of reversible processes it may be proven that the ratio Q_2/Q_1 is a function only of the two source temperatures, so that:

$$\frac{Q_2}{Q_1} = f(T_1, T_2) \tag{1.2}$$

Following Kelvin's proposal it may be assumed that the functional relation in Equation (1.2) is:

$$\frac{Q_2}{Q_1}=\frac{T_2}{T_1} \tag{1.3}$$

Equation (1.3) is the basis of the thermodynamic temperature scale and thus the efficiency of a reversible heat engine is defined as:

$$\eta=\frac{(Q_2-Q_1)}{Q_2}=\frac{(T_2-T_1)}{T_2}=1-\frac{T_1}{T_2} \tag{1.4}$$

This efficiency and the definition of temperature, which is based upon it, may be shown to be independent of the working substance. As a result it may be used to define the thermodynamic temperature scale:

$$T_1=T_2(1-\eta) \tag{1.5}$$

By means of this scale any chosen thermal state, such as the melting point of ice, may be assigned a certain value of thermodynamic temperature. The thermodynamic temperature scale may be founded upon a defined temperature difference between two temperature fixed points or on a defined value of one temperature fixed point.

In the course of the development of technology the manner of defining the thermodynamic temperature scale has changed. Until 1954, it was assumed that 100° represented the difference between the boiling point of water and the melting point of ice. Since then, there has been a return to the original and older proposals of Kelvin, in 1848, Mendeleyev, in 1874, and Giauque in 1939. Thus, since 1954, the thermodynamic temperature scale is based upon one temperature fixed point, which is the *triple point of water*. A temperature of 273.16 has been assigned to this temperature fixed point.

In 1967 the Thirteenth General Conference on Weights and Measures (CGPM) introduced a new definition for the scale and a new symbol for the unit of thermodynamic temperature. This unit is called the *kelvin* denoted by the symbol K. It is defined as 1/273.16 part of the thermodynamic temperature of the triple point of water.

However, an ideal Carnot cycle is impossible to realize in practice. Nevertheless, it can be demonstrated that the thermodynamic scale may be reproduced by a gas thermometer with an ideal gas as the working substance. Here again, although the ideal gas is quite fictitious, it could be replaced by a noble gas at very low pressure. Either pressure difference at constant volume or volume difference at constant pressure can be chosen as the measure of temperature. When the readings of temperature at constant volume, T_v, and the similar readings at constant pressure, T_p, are extrapolated to zero they tend to the same value, $T_v=T_p=T$, independently of the properties of the gas. Thus, the thermodynamic temperature scale may be reproduced using gas thermometers which have an application range up to about 1350 K.

Another simple method of reproducing the scale at thermodynamic temperatures above 1337 K is allowed by means of thermal radiation from heated bodies. When this radiation is in thermodynamic equilibrium with the radiating body, some properties of this radiation are directly linked to the temperature of the body (Herzfeld, 1962).

The concepts of black body radiation are essential for proper utilization of the method. For thermal radiation to possess similar properties to that from black body radiators it should be emitted from an aperture which is sufficiently deep and narrow with a uniform temperature distribution. When these conditions are complied with, it may be shown that the radiation intensity and its spectral distribution only depend upon the temperature of the body and not upon its material. Take, as a reference system, a heated body, which is radiating heat with some radiation intensity and whose temperature, T_1, is within the measurement range of a gas thermometer. The radiant intensity of the body provides a means of extending the thermodynamic temperature scale to higher temperatures. A relation between the ratio of spectral radiant intensities of a black body at two different temperatures, T_1 and T_2, at one wavelength, λ, exists. This relation is obtained from Planck's law (given later in Equation (7.7)) which is:

$$\frac{W_{\lambda T_1}}{W_{\lambda T_2}} = \frac{e^{c_2/\lambda T_2} - 1}{e^{c_2/\lambda T_1} - 1} \tag{1.6}$$

where $W_{\lambda T_1}$, $W_{\lambda T_2}$ are the spectral radiant intensities of a black body at the temperatures T_1 and T_2 respectively, c_2 = Planck's constant = 0.014 388 m K, and λ = wavelength in metre.

Equation (1.6) presents the ratio of the spectral radiant intensities of a black body at two temperatures T_1 and T_2 at the same single wavelength, λ. The temperature T_2 is to be determined, whereas T_1 is the temperature of a fixed point measured by a gas thermometer.

1.3 INTERNATIONAL TEMPERATURE SCALES

1.3.1 *Introduction*

It has already been stated that one of the ways in which the thermodynamic temperature scale may be realized is to apply a gas thermometer. Extending the scale is accomplished by means of radiant intensity measurements. A gas thermometer is a complex piece of apparatus which is only appropriate for use as a primary standard in fundamental laboratory measurements. This renders the thermodynamic temperature scale of no practical use as the gas thermometer needs to be replaced by some other, more practically convenient types. To this aim, in 1911, Germany, Great Britain and USA agreed to accept one common, practical temperature scale. In 1927, when this scale was accepted by the Seventh General Conference on Weights and Measures, it was called the International Temperature Scale. Subsequent minor revisions of this scale in 1948, 1954 and 1960 finally led to the establishment of what is called the International Practical Temperature Scale of 1968 (IPTS-68). This was later modified in 1975 and 1976. Finally the International Temperature Scale of 1990 (ITS-90) was adopted by the International Committee of Weights and Measures at its meeting in September 1989. This scale supersedes IPTS-68, amended edition of

1975 and the Echelle Provisoire de Température of 1976 (EPT-76) between 0.5 K and 30 K (Rusby, 1987; NPL, 1989; Preston-Thomas, 1990).

As IPTS-68, which has important underlying principles, has been used since 1968 and still is in practical use, both it and ITS-90 will be described. Both scales are based on assigned values of the temperatures of a number of highly reproducible equilibrium states called Defining Fixed Points, on Standard Instruments calibrated at those temperatures and on Interpolation Formulae used for interpolation between fixed point temperatures.

1.3.2 The International Practical Temperature Scale of 1968: IPTS-68

The scale was based upon the thermodynamic temperature, T, which has the unit kelvin and symbol K. Celsius temperature, t, was defined by the relation $t = T - T_0$, where $T_0 = 273.15$ K. The unit of Celsius temperature, which is equal to 1 K, is indicated by the symbol °C. A difference in temperature may be expressed in either kelvins or degrees Celsius.

Choice of IPTS-68 was made in such a way that every temperature measurement based upon it closely approximated the true thermodynamic temperature. Any difference was due to instrument inadequacies which limited the achievable accuracy at the time of its definition.

In IPTS-68 a distinction exists between the International Practical Kelvin Temperature, symbolized by T_{68}, and the International Practical Celsius Temperature, symbolized by t_{68}. This is expressed as

$$t_{68}(°\mathrm{C}) = T_{68}(\mathrm{K}) - 273.15 \tag{1.7}$$

The unit of T_{68} is the kelvin, symbol K, as in the thermodynamic temperature, T. In the same way as the Celsius temperature, t, the temperature, t_{68}, has the unit degree Celsius and the symbol °C.

Temperature values, which define fixed points and standard instruments calibrated at these temperatures constitute the foundation of IPTS-68. These defining fixed points of IPTS-68 are given in Table 1.1. The standard instruments, which are essential for IPTS-68, are defined as follows.

1. *In the range from 13.81 K to 630.74 °C* the standard instrument is the platinum resistance thermometer. (Kelvin temperatures are used below 0 °C to avoid the use of negative temperature values.) Only pure platinum, which is annealed and strain-free, may be used in the manufacture of the resistors. Its resistance ratio, $W(T_{68})$, is defined by the relation:

$$W(T_{68}) = \frac{R(T_{68})}{R(273.15\ \mathrm{K})} \tag{1.8}$$

where $R(T_{68})$ is the resistance at the given temperature, T_{68}. At the temperature, $T_{68} = 373.15$ K, this ratio must be such that $W(T_{68}) \geqslant 1.392\,50$.

1(a). *Below 0 °C* the relative resistance is given by the relation:

$$W(T_{68}) = W_{CCT\text{-}68}(T_{68}) + \Delta W_i(T_{68}) \tag{1.9}$$

where $W_{CCT\text{-}68}(T_{68})$, is the resistance ratio defined by the reference function set out by Comité Consultatif de Thermometrie (CCT), $\Delta W_i(T_{68})$ is the deviation calculated at the Defining Fixed Points as the difference between the measured values of $W(T_{68})$ and the values of $W_{CCT\text{-}68}(T_{68})$.

The values of these deviations are given in a table which is a supplement to the text of IPTS-68.

1(b). *In the range 0 °C to 630.74 °C*, t_{68} is defined by:

$$t_{68} = t' + 0.045\left(\frac{t'}{100\ °C}\right)\left(\frac{t'}{100\ °C} - 1\right)\left(\frac{t'}{419.58\ °C} - 1\right)\left(\frac{t'}{630.74\ °C} - 1\right) \tag{1.10}$$

Table 1.1
The temperature fixed points which define IPTS-68 and ITS-90.

	Scale			
	IPTS-68		ITS-90	
Equilbrium state	T_{68} K	t_{68} °C	T_{90} K	t_{90} °C
Vapour pressure point of helium	Not defined		3 to 5	270.15 to 268.19
Triple point of equilibrium hydrogen	13.81	−259.34	13.8033*	−259.346*
Boiling point of hydrogen at a pressure 33 330.6 Pa	17.042	−256.108	17*	−256.15*
Boiling point of equilibrium hydrogen (see IPTS-68) or gas thermometer point of He	20.28	−252.87	20.3*	−252.85*
Triple point of neon	Not defined		24.5561	−248.5939
Boiling point of neon	27.102	−246.048	Not defined	
Triple point of oxygen	54.361	−218.789	54.3584	−218.7916
Triple point of argon	83.798	−189.352	83.8058	−189.3442
Condensation point of oxygen	90.188	−182.362	Not defined	
Triple point of mercury	Not defined		234.3156	−38.8344
Triple point of water	273.16	0.01	273.16	0.01
Melting point of gallium	Not defined		302.9146	29.7646
Boiling point of water	373.15	100	Not defined	
Freezing point of indium	Not defined		429.7485	156.5985
Freezing point of tin	505.1181	231.9681	505.078	231.928
Freezing point of zinc	692.73	419.58	692.677	419.527
Freezing point of aluminium	Not defined		933.473	660.323
Freezing point of silver	1235.08	961.93	1234.93	961.78
Freezing point of gold	1337.58	1064.43	1337.33	1064.18
Freezing point of copper	Not defined		1357.77	1084.62

The values of the temperature fixed points with the exception of the triple points and the 17.042 K point are given at a pressure $p_o = 101\ 325$ Pa.
*Given for $e-H_2$, which is hydrogen at the equilibrium concentration of the *ortho* and *para* molecular forms.

where t' is defined by the equation:

$$t' = \frac{1}{\alpha}[W(t') - 1] + \delta\left(\frac{t'}{100\,°\mathrm{C}}\right)\left(\frac{t'}{100\,°\mathrm{C}} - 1\right)\,°\mathrm{C} \quad (1.11)$$

$$\text{with } W(t') = R(t')/R_0$$

The constants R_0, α and δ are determined by measurement of the resistance $R(t')$ at the triple point of water, the boiling point of water (or freezing point of tin) and the freezing point of zinc.

2. *From 630.74 °C to 1064.43 °C* the temperature is determined by a standard Pt10Rh–Pt thermocouple. Whilst the platinum electrode should be of high purity platinum such that $R_{100}/R_0 \geqslant 1.3920$, the platinum–rhodium one should contain 90% platinum and 10% rhodium by weight. In reference conditions corresponding to the temperature, t_{68}, of the measuring junction and 0 °C of the reference junction, the thermoelectric force is given by the equation:

$$E(t_{68}) = a + bt_{68} + ct_{68}^2 \quad (1.12)$$

Calculation of the constants a, b and c is performed from the values of E at 630.74 °C ± 0.2 °C as determined by a platinum resistance thermometer and at the freezing points of silver and gold.

3. *In the range above 1064.43 °C* the temperature to be measured, T_{68}, is defined in terms of the spectral concentrations $L(T_{68})$ and $L(T_{68,\mathrm{Au}})$ of a black body radiator at wavelength λ. The temperature $T_{68,\mathrm{Au}}$ is the freezing point of gold. In the notation of the original text (IPTS-68) and taking the value of $c_2 = 0.014\,388$ m K, the relation is expressed by the equation:

$$\frac{L_\lambda(T_{68})}{L_\lambda(T_{68,\mathrm{Au}})} = \frac{e^{c_2/\lambda T_{68,\mathrm{Au}}} - 1}{e^{c_2/\lambda T_{68}} - 1} \quad (1.13)$$

The original text of IPTS-68 (1976) also gives a list of secondary reference points shown in Table 1.2. Some of these points are now defining fixed points of ITS-90 with slightly modified values.

1.3.3 The International Temperature Scale of 1990: ITS-90

This new temperature scale has the following advantages over IPTS-68:

- it is in better agreement with the corresponding thermodynamic temperatures;
- it has improved continuity, precision and reproducibility throughout its range;

- it extends to lower temperatures down to 0.65 K so superseding EPT-76;
- its possession of sub-ranges with alternative definitions in certain ranges greatly facilitates its use;
- it eliminates most boiling points as defining fixed points.

Although the differences, existing between values of T_{90} and the corresponding values of T_{68}, are significant to those working to high levels of accuracy, they exert no practical influence on everyday industrial measurement. The changes in some of the adopted Defining Fixed Points of ITS-90 are shown in Table 1.1.

A noteworthy change is the elimination of the Pt10Rh–Pt thermocouple as an interpolating instrument. The range of the platinum resistance thermometer as an interpolating instrument is extended from 13.8033 K (the triple point of equilibrium hydrogen) to 961.78 °C (the freezing point of silver).

The scale uses the kelvin, symbol K, as the unit of thermodynamic temperature, symbol T. One kelvin is defined as 1/273.16 of the thermodynamic temperature of the triple point of water. Celsius temperature, t, is defined by the relation:

$$t(^{\circ}\mathrm{C}) = T(\mathrm{K}) - 273.15 \tag{1.14}$$

The unit of Celsius temperature is the degree Celsius, symbol °C, which is equal to one kelvin. A difference in temperature may be expressed in either kelvins or degrees Celsius.

In ITS-90, similarly as in IPTS-68, a distinction exists between the International Kelvin Temperature, symbolized by T_{90}, and the International Celsius Temperature, symbolized by t_{90}, as expressed by:

$$t_{90}(^{\circ}\mathrm{C}) = T_{90}(\mathrm{K}) - 273.15 \tag{1.15}$$

In this book the Celsius temperature will be indicated by ϑ to avoid confusion with the unit of time, which is indicated by t.

Interpolation between the Defining Fixed points of ITS-90, which are listed in Table 1.1, are as follows.

1. *From 0.65 K to 5.0 K:* T_{90} is defined in terms of the vapour pressure temperature relations of ^{3}He and ^{4}He.

2. *From 3.0 K to 24.5561 K* (the triple point of neon): the constant volume type of ^{3}He or ^{4}He gas thermometer is used. It is calibrated at three experimentally realizable temperatures of defining fixed points using specified interpolation procedures.

3. *From 13.8033 K* (the triple point of equilibrium hydrogen) to 961.78 °C (the freezing point of silver): the standard instrument is a platinum resistance thermometer calibrated at specified sets of defining fixed points and using specified interpolation procedures. Pt thermometers of different construction are used in different

Table 1.2
Secondary reference points of IPTS-68

Equilibrium state	T_{68} K	t_{68} °C
Triple point of normal hydrogen	13.956	−259.194
Boiling point of normal hydrogen	20.397	−252.753
Triple point of neon	24.561	−248.589
Triple point of nitrogen	63.146	−210.004
Boiling point of nitrogen	77.349	−195.806
Boiling point of argon	87.294	−185.856
Sublimation point of CO	194.674	−78.476
Freezing point of mercury	234.314	−38.836
Ice point	273.15	0
Triple point of phenoxybenzene	300.02	26.87
Triple point of benzoic acid	395.52	122.37
Freezing point of indium	429.784	156.634
Freezing point of bismuth	544.592	271.442
Freezing point of cadmium	594.258	321.108
Freezing point of lead	600.652	327.502
Boiling point of mercury	629.81	356.66
Boiling point of sulphur	717.824	449.674
Freezing point of CuAl-eutectic	821.41	548.26
Freezing point of antimony	903.905	630.755
Freezing point of aluminium	933.61	660.46
Freezing point of copper	1358.03	1089.88
Freezing point of nickel	1728	1455
Freezing point of cobalt	1768	1495
Freezing point of palladium	1827	1554
Freezing point of platinum	2042	1769
Freezing point of rhodium	2236	1963
Freezing point of Al_2O_3	2327	2054
Freezing point of iridium	2720	2447
Melting point of niobium	2750	2477
Melting point of molybdenum	2896	2623
Melting point of tungsten	3695	3422

The values of the temperature secondary reference points beyond the triple points are given at a pressure of $p_0 = 101\ 325$ Pa.

temperature ranges. The temperatures are determined from the thermometer resistance ratio defined by the relation:

$$W(T_{90}) = \frac{R(T_{90})}{R(273.16\,\mathrm{K})} \tag{1.16}$$

where $R(273.16\,\mathrm{K})$ is the thermometer resistance at the triple point of water. This definition differs from the previous one used in IPTS-68, where the resistance at the ice point (0 °C) was assumed as a reference value.

An acceptable platinum resistance thermometer must be made from pure, strain free, annealed platinum, satisfying at least one of the following relations:

at the gallium melting point,

$$W(29.7646\ ^\circ\mathrm{C}) \geqslant 1.118\,07 \tag{1.17}$$

at the triple point of mercury,

$$W(-38.8344\ ^\circ\mathrm{C}) \leqslant 0.844\,235 \tag{1.18}$$

If used up to the freezing point of silver it must also satisfy the relation:

$$W(961.78\ ^\circ\mathrm{C}) \geqslant 4.2844 \tag{1.19}$$

In each of the resistance thermometer ranges, T_{90} is obtained from $W_r(T_{90})$ as given by an appropriate reference function and the deviations $W(T_{90}) - W_r(T_{90})$. At the Defining Fixed Points this deviation is obtained directly from the calibration of the thermometer. At intermediate temperatures it is obtained by means of the appropriate deviation functions.

3(a). *In the range from 13.8033 K* (the triple point of equilibrium hydrogen) *to 273.16 K* (the triple point of water), the thermometer is calibrated at the triple points of equilibrium hydrogen (13.8033 K), neon (24.5561 K), oxygen (54.3584 K), argon (83.8058 K), mercury (234.3156 K) and water (273.16 K) and at two additional temperatures close to 17.0 K and 20.3 K, using a gas thermometer.

3(b). *In the range from 0 °C to 961.78 °C* (the freezing point of silver) the thermometer is calibrated at the triple point of water (0.01 °C) and at the freezing points of tin (231.928 °C), zinc (419.527 °C), aluminium (660.323 °C) and silver (961.78 °C). In both of the ranges described above at 3(a) and 3(b), for sub-ranges with limited upper temperatures, fewer calibration points may be used, as precisely specified in ITS-90.

4. *Above 961.78 °C* (the freezing point of silver) the Planck radiation law is to be used. The temperature T_{90} is defined by the equation:

$$\frac{L_\lambda(T_{90})}{L_\lambda[T_{90}(x)]} = \frac{\exp[c_2/\lambda T_{90}(x)] - 1}{\exp[c_2/\lambda T_{90}] - 1} \tag{1.20}$$

where $T_{90}(x)$ refers to any of the freezing points of silver, gold, or copper, $L(T_{90})$ and $L[T_{90}(x)]$ are the spectral concentrations of the radiance of a black body at wavelength λ at T_{90} and $T_{90}(x)$ respectively, and c_2 is a constant with a value of 0.014 388 m K.

The full text of ITS-90 will be accompanied by two further documents, namely 'Supplementary Information for ITS-90' and 'Techniques for Approximating the ITS-90', describing practical details and current good practice of ITS-90.

The scale ITS-90 seems to represent thermodynamic temperatures with an accuracy of ± 2 mK from 1 to 273 K increasing to about ± 7 mK at 900 K. Differences between

Table 1.3
Conversion of temperature scales.

Scale	Given	To be determined in °C	°F
Celsius	X °C	X	$1.8X+32$
Fahrenheit	X °F	$0.5556(X-32)$	X
Kelvin	X K	$X-273.15$	$1.8(X-273.15)+32$

ITS-90 and IPTS-68 are between one and several mK in the platinum resistance thermometer range and not smaller than 0.2 K in the thermocouple range. These differences are smaller very close to the fixed points (McGee, 1988).

Besides those temperature scales described above, the Fahrenheit Scale is still widely used. The conversion between the temperature scales is specified in Table 1.3 and in Table I at the end of the book.

REFERENCES

Hall, A. A. and Barber, C. R. (1964) Calibration of temperature measuring instruments. Ed. 3. *Notes on Applied Science No. 12*. National Physical Laboratory, London.

Herzfeld, C. M. (1962) The thermodynamic temperature scale, its definition and realisation. *Temperature: Its Measurement and Control in Science and Industry*, Vol. 3, Part 1, Reinhold Publ. Co., New York, pp. 41–50.

IPTS-48 (1960) *Echelle Internationale Pratique de Température de 1948. Edition amendée de 1960*. Comité International des Poids et Mésures.

IPTS-68 (1968) Echelle Internationale Pratique de Temperature de 1968. *Comptes rendues de la 13e Conférence Générale des Poids et Mésure et Comité Consultatif de Thermométrie*, 8e Session.

IPTS-68 (1976) The International Practical Temperature Scale of 1968. Amended Edition of 1975. *Metrologia*, **12** (12), 7–17.

Lindsay, R. B. (1962) The temperature concept for systems in equilibrium. *Temperature: Its Measurement and Control in Science and Industry*, Vol. 3, Part 1, Reinhold Publ. Co., New York, pp. 3–13.

McGee, T. (1988) *Principles and Methods of Temperature Measurement*. John Wiley and Sons, New York.

Newton Friend, J. (1937) The origin of Fahrenheit's thermometric scale. *Nature*, **139**, 395–398.

NPL (1989) *Adoption of international temperature scale of 1990, ITS-90*.

Preston-Thomas, H. (1990) The International Temperature Scale of 1990 (ITS-90), *Metrologia*, **27** (1), 4–10.

Rusby, R. L. (1987) The basis of temperature measurement. *Meas. and Cont.*, **20** (6), 7–10.

2 Classification of Temperature Measuring Instruments

Temperature measuring instruments and methods applied in laboratories and in industry will be described in this book. No highly specialized or rarely used instruments will be considered. There are a number of ways in which temperature measurement may be classified. As the instruments belong to the general class of information machine (McGhee and Henderson, 1989) the division or grouping should be based upon the functions they perform and the structures they possess. This criterion of classification is the essential technique used extensively in this book. Comparing the similarities of various types without in any way diminishing the important differences which exist between them, a fundamental ingredient of taxonomy (McGhee and Henderson, 1989), leads to incisive understanding.

From the functional point of view temperature measuring instruments may be grouped as self-generators or modulators (Middlehoek and Noorlag, 1981) on the basis of the ranges of temperature over which they may perform their measuring function. The instruments to be considered have a temperature measuring range which mainly extends from −50 °C to approximately +2000 °C. As there is no single instrument which provides total coverage of this range, it is important to clarify the specific range appropriate to each type. Figure 2.1 illustrates the division of temperature measuring instruments by temperature range. From this illustration it is clear that the diversity of possibilities demands that the theoretical and descriptive material for each type should be well organized. Such is the nature and purpose of this book.

At the structural level there is also diversity. An important criterion of structural classification is the method of heat transfer between the thermometer and the body under measurement. As heat may be transferred by conduction, convection and radiation the structural classification must take these into account. Consequently the grouping given in Figure 2.2 becomes apparent. Those instruments which depend upon heat transfer by contact are known as *contact thermometers*. In contact thermometers the heat transfer between the temperature sensor and the medium to be measured, occurs by conduction and convection. Both the system under measurement and the temperature sensor are in a state of thermal equilibrium. Where radiation heat transfer is the means of energy exchange the instruments may be called *non-contact thermometers* or simply *pyrometers*.

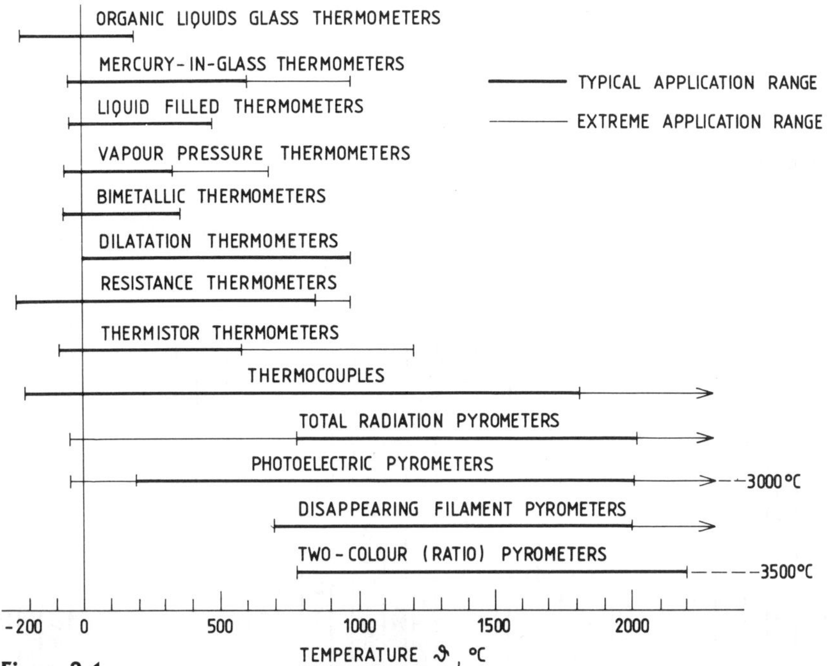

Figure 2.1
Measuring ranges of temperature measuring instruments.

Pyrometers, in a literal translation from the ancient Greek roots of the word, are fire measurers. However, this original meaning has been somewhat distorted since the lower temperature limit is well below 0 °C, far below that of 'fire'. In these instruments the sensitive element, which responds to the incident thermal radiation emitted by the body whose temperature is to be measured, transforms a certain part of the incident radiation into another energy form. This other energy form is most conveniently electrical as its signals are more easily manipulated.

Since there are generally six energy forms (Middlehoek and Noorlag, 1981) of which the thermal energy form is one, it is possible in principle to transduce heat energy into any one of the other five. Further division of temperature measuring instruments, which is given in Figure 2.2, is based upon this idea. Consequently, the class of contact thermometers may be grouped upon the energy form of the output signal. Before the electrical signals became dominant, thermometers were based upon physical expansion effects. Such instruments based upon solid expansion, liquid expansion and gas/vapour expansion were and still are of significance. As electrical signals are easily manipulated and processed using both analogue and digital techniques, the class of contact thermometers which transduce to the electrical energy form is important. Thermoelectric thermometers, as their name indicates, are self-generating thermometers. The other classes of electrical thermometers are based upon thermal modulation of the thermometric property of the applied material.

The science of classification, also called taxonomy, whose methods and purposes are not described in detail in this book, is an essential aid in the analysis design and

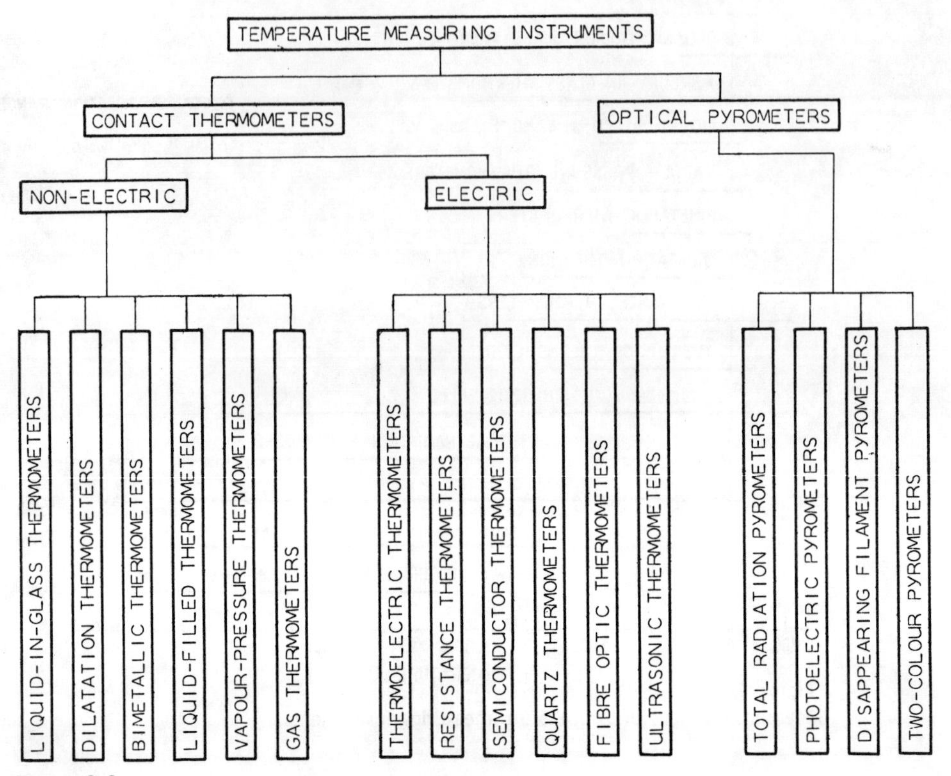

Figure 2.2
Classification of temperature measuring instruments.

use of temperature measuring instruments. Application of this science to engineering systems, which has recently been proposed by McGhee and Henderson (1989), requires further development. Henderson and McGhee (1990) have used the basic techniques of taxonomy to provide a more detailed grouping of temperature measuring instruments and thermal systems identification. Another classification is given by Hamidi and Swithenbank (1987).

REFERENCES

Hamidi, A. A. and Swithenbank, J. (1987) Temperature measurement in a hostile environment. *Thermal and Temperature Measurement in Science and Industry*, Proc. 3rd IMEKO Symposium TEMPMEKO 87, Sheffield, Institute of Measurement and Control, London, pp. 99–113.

Henderson, I. A. and McGhee, J. (1990) A taxonomy of temperature measuring instruments. *Temperature and Thermal Measurement in Industry and Science*, Proc. 4th IMEKO Symposium TEMPMEKO 90, Helsinki, pp. 400–405.

McGhee, J. and Henderson, I. A. (1989) Holistic perception in measurement and control: applying keys adapted from classical taxonomy, in Linkens, D. A. and Atherton D. P. (eds), *Trends in Control and Measurement Education*, IFAC Proc. Series 1989, No 5, pp. 85–90.

Middlehoek, S. and Noorlag, D. J. W. (1981) Three dimensional representation of input and output transducers. *Sensors and Actuators*, **2**, 29–41.

3 Non-electric Thermometers

3.1 LIQUID-IN-GLASS THERMOMETERS

3.1.1 Physical principles

Liquid-in-glass thermometers are based upon the temperature dependent variation of the volume of the liquid which is used. The thermometer consists of a liquid filled bulb connected to a thin capillary with a temperature scale as shown in Figure 3.1.

Assuming that the bulk volume, V_b, is much greater than that of the liquid contained in the capillary, the volume variation, ΔV, of the liquid corresponding to the measured temperature variation, $\Delta\vartheta$, is:

$$\Delta V = V_b \beta_a \Delta\vartheta \tag{3.1}$$

where β_a is the average apparent coefficient of cubic thermal expansion of the thermometric liquid in the given glass. This coefficient, which also covers small changes of the bulb volume as a function of the measured temperature, has an average value for a given application range of the thermometer. It is expressed as the difference of the respective coefficients of cubic expansion, β_l, of the liquid and, β_g, of the glass so that:

$$\beta_a = \beta_l - \beta_g \tag{3.2}$$

Assume that the inner capillary has a diameter, d, and that the temperature difference, $\Delta\vartheta$, corresponds to a change of length, Δl, of the liquid column. Using Equation (3.1) the thermometer sensitivity is:

$$\frac{\Delta l}{\Delta\vartheta} = \frac{\Delta V}{\pi d^2 \Delta\vartheta/4} = \frac{4 V_b \beta_a \Delta\vartheta}{\pi d^2 \Delta\vartheta} = \frac{4 V_b(\beta_l - \beta_g)}{\pi d^2} \tag{3.3}$$

Equation (3.3) indicates that the sensitivity increases in direct proportion with growing bulb volume, V_b, and coefficient, β_a, but as the inverse square of capillary diameter, d. There are some practical limits to increasing this sensitivity. Firstly, an increase in bulb volume increases the thermal inertia of the thermometer. Secondly, if the bore of the capillary is too small, the liquid column may break easily under the influence of surface tension. In mercury-in-glass thermometers, for the Celsius scale, the bulb volume is about 6000 times the capillary volume, corresponding to the length of one degree Celsius of the thermometer scale.

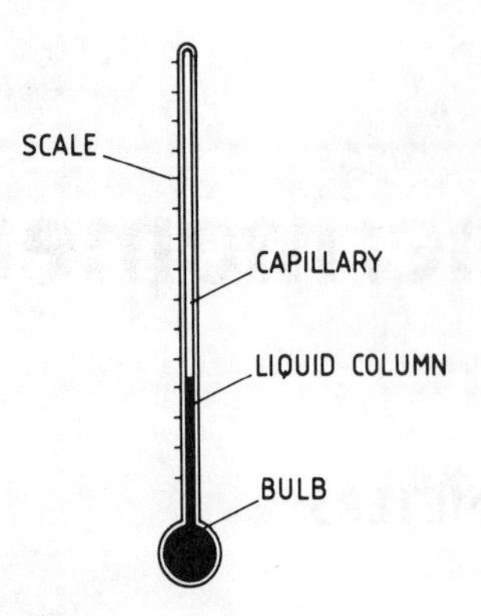

Figure 3.1
Liquid-in-glass thermometer.

Laboratory thermometers are standardized for use with a liquid column which is totally immersed in the heating medium. If applying a thermometer, standardized in such a way, in conditions when the liquid column is not totally immersed, some portion of the liquid column will be at a different temperature from that in the bulb. To compensate for any systematic error due to the partial immersion, a correction should be applied to the indicated value. This correction can be calculated from the formula:

$$\Delta\vartheta = \beta_a n(\vartheta_i - \vartheta_m) \qquad (3.4)$$

The average apparent coefficient of cubic thermal expansion of the thermometric liquid in the given glass is equal to β_a, n is the length of the emergent liquid column, given in degrees of the thermometer scale, ϑ_i is the indicated temperature and ϑ_m is the average value of the temperature of the emergent liquid column. In the case when ϑ_m is higher than the indicated temperature the correction, of course, is negative.

Under ordinary industrial conditions, it is not generally possible to arrange that the whole liquid column of the thermometer is immersed in the medium to be measured. Special thermometers, standardized with a partially immersed liquid column, are then applied. Normally, the nominal immersion depth and the average value of the temperature of the emergent liquid column are stated on the thermometer scale. If such a thermometer is used at the correct immersion depth, with the emergent column temperature, ϑ_m, different from the nominal value, ϑ'_m, a corresponding correction must be applied. Such a correction is given by:

$$\Delta\vartheta = \beta_a n(\vartheta'_m - \vartheta_m) \qquad (3.5)$$

In Equation (3.5), ϑ'_m is the nominal value of the average temperature of the emergent liquid column whilst the other symbols are the same as in Equation (3.4). In both cases, the mean temperature of the emergent liquid column is calculated by measuring its temperature at some points along its length. Alternatively, this average emergent temperature may be directly estimated by applying a second auxiliary thermometer with an elongated bulb placed close to the emergent column.

Numerical example

A mercury-in-glass laboratory thermometer has been standardized by total immersion. When immersed up to the scale division +50 °C in hot water it indicated a water temperature of +95 °C. If the average value of the emergent column temperature is +35 °C calculate the correction which is required. Assume that the effective coefficient of cubic thermal expansion, β_a, is 0.000 16 1/°C.

Using Equation (3.4) the calculated correction is:

$$\Delta\vartheta = \beta_a n(\vartheta_i - \vartheta_m) = 0.000\ 16 \times (95-50) \times (95-35) = 0.43\ °C$$

3.1.2 Materials and structures

Thermometer liquids for use in liquid-in-glass thermometers should have the following properties:

- constancy in time of physical and chemical properties,
- constant value of coefficient of cubic thermal expansion in the measuring temperature range,
- low freezing temperature,
- high boiling temperature,
- easily obtained in pure form.

Commonly applied thermometric liquids and thermometric glasses are displayed in Table 3.1 and given in BS 1041.

Mercury-in-glass thermometers, used up to about 200 °C, have a vacuumized capillary. For measuring temperatures in excess of 200 °C a suitable compressed inert gas is used. This gas prevents both boiling of the mercury and condensation of its vapours in the upper part of the capillary. Nitrogen and hydrogen at pressures up to 2 MPa (about 20 atmospheres) at 600 °C, are commonly used.

In *thermometers with organic liquids*, which wet the glass, the capillaries are always gas filled to prevent the liquid column from breaking. In addition, problems due to breaking of the liquid column also occur at low temperatures, where the viscosity of the thermometric liquid increases. For this reason, it is always advisable, but especially at low temperatures, to cool the thermometer bulb very slowly. If the bulb is cooled too quickly there is a real danger that the liquid column may break under the influence of the adhesive forces between the liquid and the capillary walls. Glass exhibits hysteresis under changing temperatures. For example, when heated and then allowed to cool to its original temperature, it does not return to quite its original dimensions immediately. This phenomenon of hysteresis causes a so-called depression

Table 3.1
Liquids and glasses for liquid-in-glass thermometers. (Reproduced with permission from BS 1041: Section 2.1. 1985.)

Glass type	Borasilicate glass	Other normal glasses			
Liquid type	**Mercury**	**Pentane**	**Toluene**	**Ethanol**	**Mercury**
Temperature (°C)	**Apparent coefficient of cubic thermal expansion β_a (l/°C)**				
−180	—	0.9×10^{-3}	—	—	
−120	—	1.0×10^{-3}	—	—	
−80	—	1.0×10^{-4}	0.9×10^{-3}	1.04×10^{-3}	
−40	—	1.2×10^{-3}	1.0×10^{-3}	1.04×10^{-3}	
0	1.64×10^{-4}	1.4×10^{-3}	1.0×10^{-3}	1.04×10^{-3}	1.58×10^{-4}
20	—	1.5×10^{-3}	1.1×10^{-3}	1.04×10^{-3}	
100	1.64×10^{-4}				1.58×10^{-4}
200	1.67×10^{-4}				1.59×10^{-4}
300	1.74×10^{-4}				1.64×10^{-4}
400	1.82×10^{-4}				
500	1.95×10^{-4}				

of zero of a glass thermometer. It may take several hours or even days to recover. The value of the lowering of the zero point and the recovery time depend upon the type of glass.

Laboratory and industrial thermometers are available in two main forms described by Busse (1941). *Etched-stem thermometers*, more popular in the UK, are made from a glass rod of about 4 to 6 mm diameter with an axial capillary, as shown in Figure 3.2. The scale is etched on the rod surface, the curvature of which acts as a magnifying lens for the liquid column. To facilitate reading of the thermometer, a white enamel backing is placed behind the capillary. In the *enclosed scale type*, which is shown in Figure 3.3, a thin wall capillary and a milk glass scale are contained in a thin-walled glass tube. As in the former type the capillary curvature acts as a magnifying

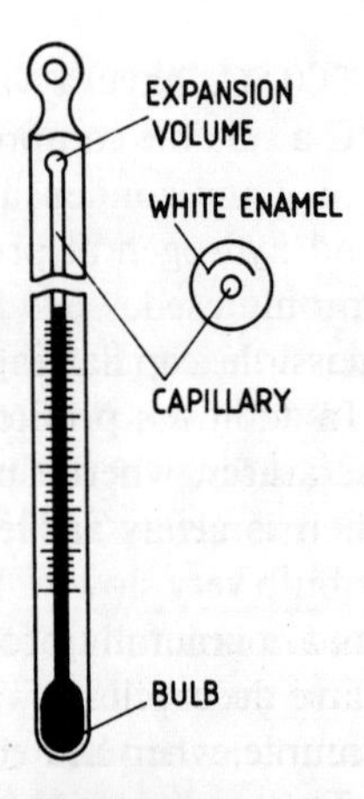

Figure 3.2
Etched-stem thermometer.

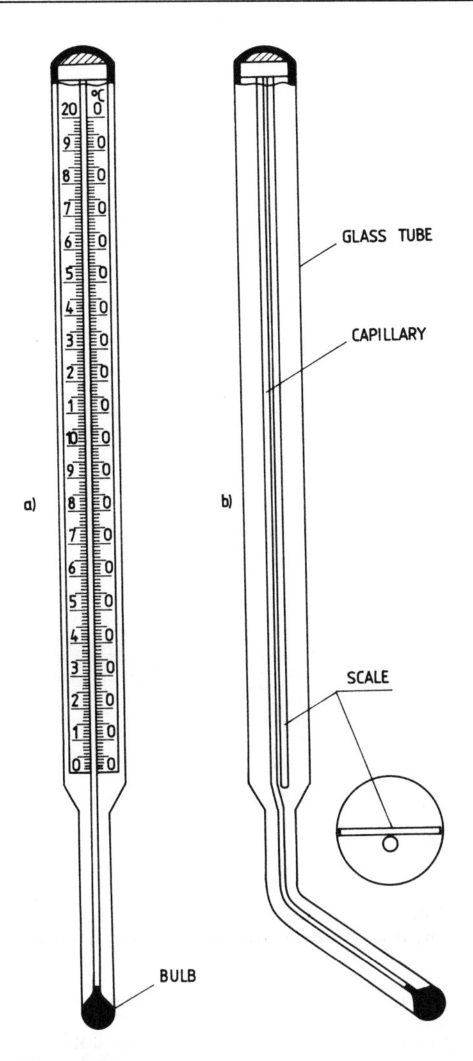

Figure 3.3
Enclosed scale thermometer: a) straight, b) angled.

lens. The bore enlargement at the capillary end (expansion volume) shown in Figure 3.2 protects the thermometer from damage in the case of possible overheating. One of the advantages of the enclosed scale type thermometers is their mechanical robustness to breaking although they suffer from a drawback owing to the possibility of a scale shift relative to the capillary. Etched-stem thermometers are less robust but do not suffer from this scale shift.

As shown in Figure 3.4, industrial glass thermometers are generally protected by a steel sheath. Whereas industrial glass thermometers have accuracies of ±0.02 °C to ±10 °C, laboratory thermometers may even have accuracies around ±0.01 °C.

A great diversity of types of glass thermometers exists. They include such categories as maximum thermometers, max–min thermometers, domestic thermometers and

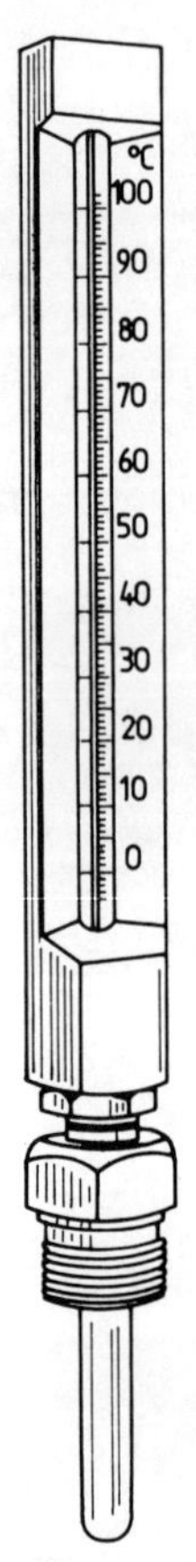

Figure 3.4
Liquid-in-glass thermometer in a steel sheath.

others. Most types have been standardized to ISO 386 adopted as BS 1041. Detailed information on liquid-in-glass thermometers may be found in the works by Thomson (1962), Busse (1941) and also in BS 1041. A selection of commercially available liquid-in-glass thermometers is shown in Figure 3.5.

3.2 DILATATION THERMOMETERS

Differences between the coefficients of linear thermal expansion of two different materials may be used in the construction of thermometers. This difference is given by the equation

$$\Delta l = l(1+\alpha_1 \Delta \vartheta) - l(1+\alpha_2 \Delta \vartheta) = l(\alpha_1 - \alpha_2)\Delta \vartheta \qquad (3.6)$$

where l is the sensor length, α_1 and α_2 are the coefficients of linear thermal expansion of the two applied materials and $\Delta \vartheta$ is the temperature difference.

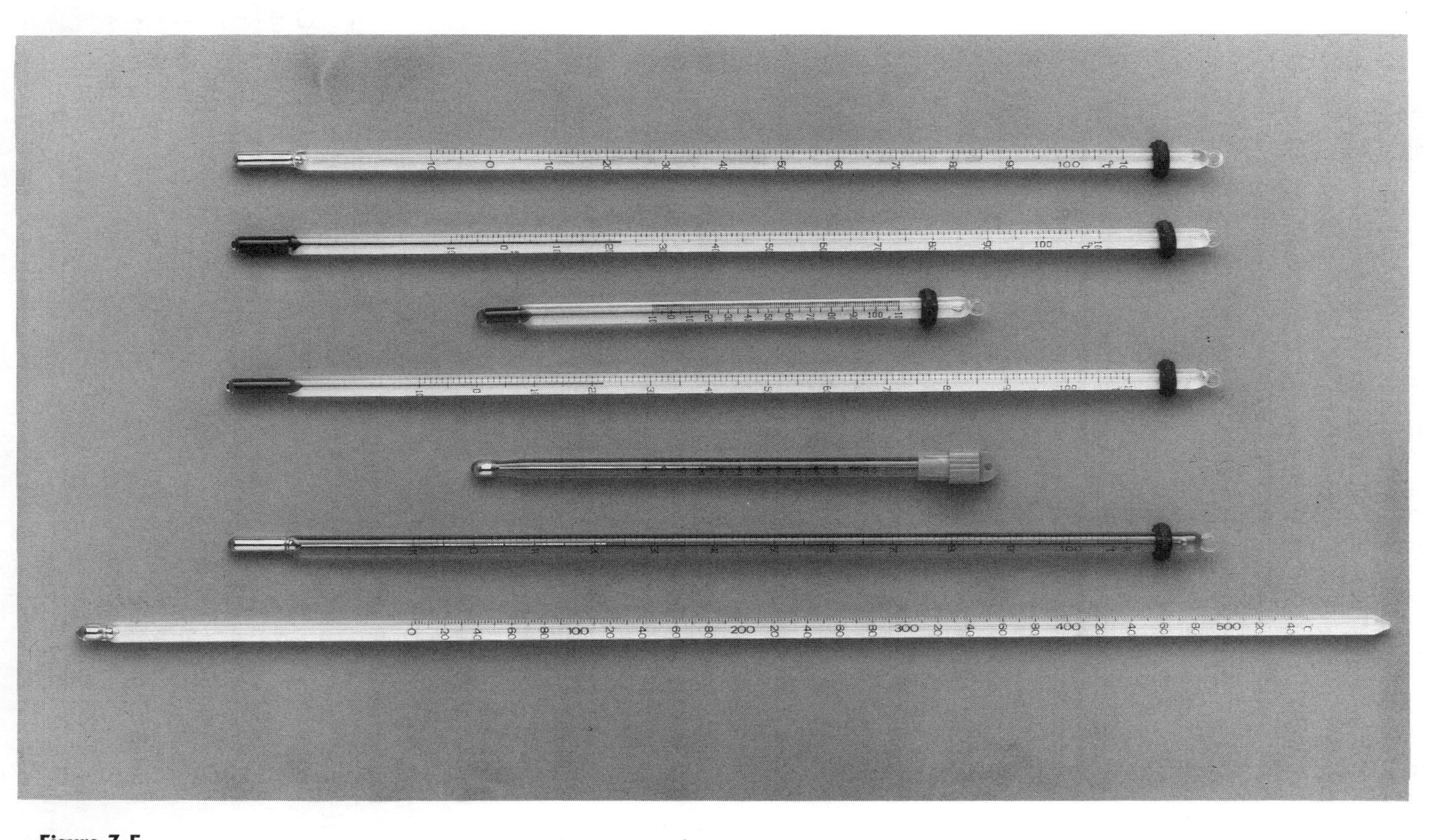

Figure 3.5
A selection of liquid-in-glass thermometers. THL-210: 75 mm immersion; mercury. THL-320: 75 mm immersion; red spirit. THL-333-010X: total immersion; blue spirit. THL-333-020K: 75 mm immersion; blue spirit. Stirring thermometers with permanent amber-stained graduations. THL-340: total immersion; mercury. THL-360: 76 mm immersion; mercury. (Courtesy of Griffin and George.)

In most cases the sensors are constructed as a tube of material having a bigger expansion coefficient, α_1, with a coaxial rod made of the material of smaller coefficient, α_2. They are called respectively the active and passive materials as indicated in Figure 3.6. The pairs of applied materials should have as big a difference as possible of the coefficients α_1 and α_2, high permissible working temperature and good resistance against corrosion and oxidation (Figure 3.7 and Table 3.2).

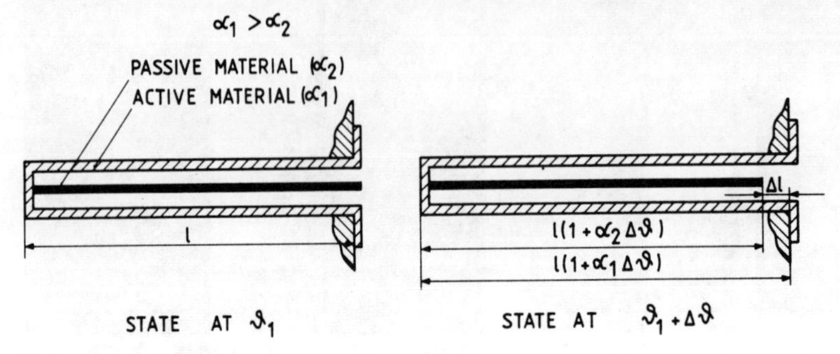

Figure 3.6
Principle of dilatation thermometer.

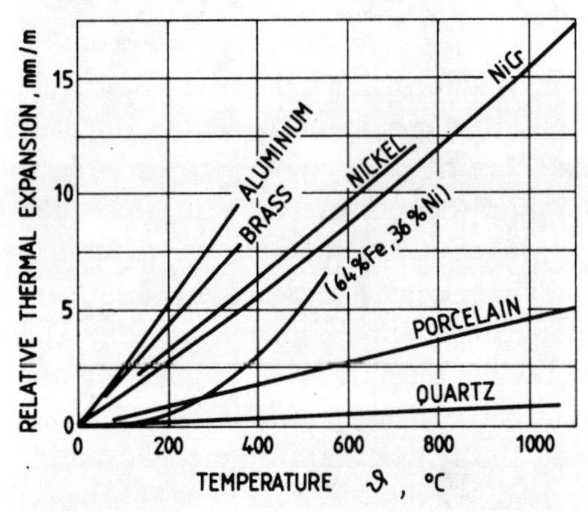

Figure 3.7
Materials applied for dilatation thermometers.

Table 3.2
Materials used in dilatation thermometers.

Materials		Temperature range (°C)	Coefficient of linear thermal expansion, α (l/°C) (mean value in application range)
Active	Aluminium	0–600	23×10^{-6}
	Brass	0–300	18×10^{-6}*
	Nickel	0–600	13×10^{-6}
	Chromium-Nickel alloy	0–1000	16×10^{-6}*
Passive	Porcelain	0–1000	4×10^{-6}
	Invar (64%Fe, 36%Ni)	0–200	3×10^{-6}
	Quartz	0–1000	0.54×10^{-6}

*Approximate values depending upon the exact material composition.

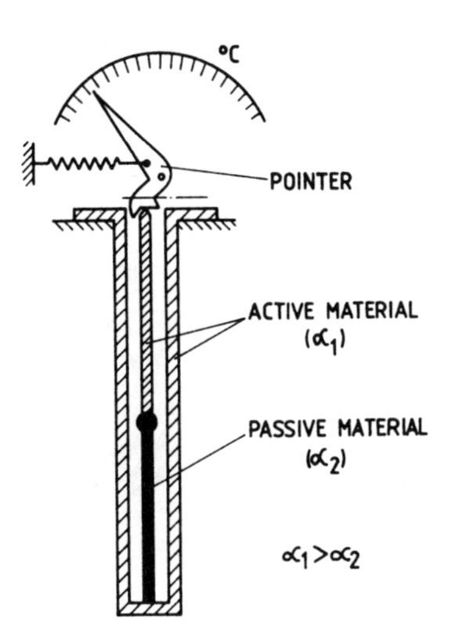

Figure 3.8
Dilatation thermometer cross-section.

As the expansion difference of two materials of reasonable length is usually too small to give a direct indication of temperature, it needs to be amplified by a mechanical transmission. A dilatation thermometer indicates the average value of temperature along its length. They are rarely used and only produced by very few firms. Indication errors for dilatation thermometers are around ±1% to ±2% of the temperature range. The highest measured temperatures are about 1000 °C. The cross-section of a dilatation thermometer is given in Figure 3.8. That part of the sensor inner rod which is outside the nominal immersion length is made from a material having the same expansion coefficient as the outer tube. In this way the variations of the ambient temperature and the possible heating of the emergent part of the sensor have no influence on the reading. When measuring the temperature of metallic parts, the thermal elongation of these parts may be used as a direct replacement for the active material.

3.3 BIMETALLIC THERMOMETERS

Two metal strips of different coefficients of linear thermal expansion, welded or hot-rolled together, form a bimetallic strip similar to that shown in Figure 3.9. As in

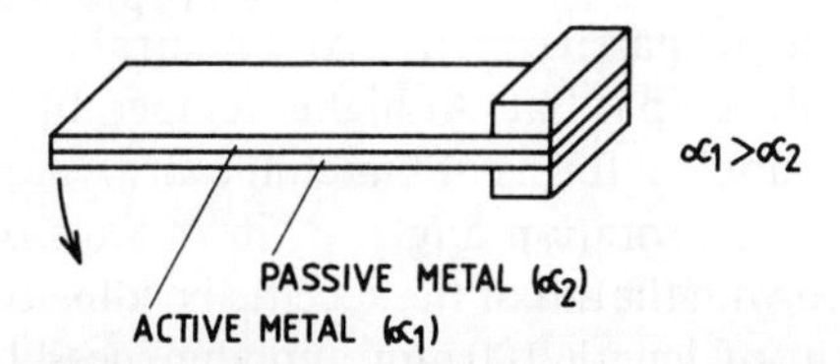

Figure 3.9
Bimetallic strip.

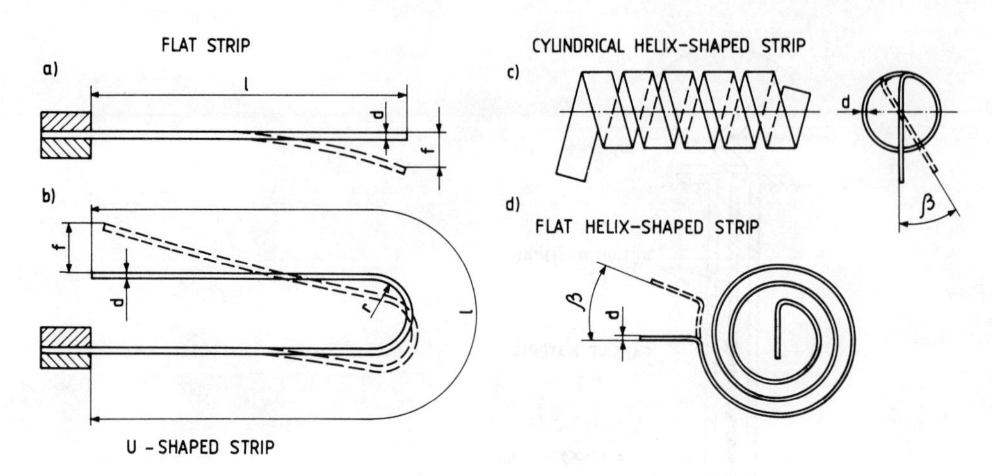

Figure 3.10
Forms of bimetallic strips.

Table 3.3
Bimetallic thermometers.

Passive metal	Active metal	Temperature range (°C)	Specific bending (l/°C) (mean value in range)
Invar (64%Fe, 36%Ni)	Alloy (27% Ni, 68%Fe, 5%Na)	0–200	0.16
	Brass	0–150	0.16*
	Copper	0–150	0.16
	Constantan	0–200	0.14*
	Nickel	0–150	0.12
	Iron	0–150	0.11
	Non-magnetic steel	0–120	0.18*
Alloy (58%Fe, 42%Ni)	Alloy (27%Ni, 68%Fe, 5%Na)	0–500	0.12
	Constantan	0–350	0.11*
	Nickel	0–400	0.09
	Alloy (42%Ni, 53%Fe, 5%Na)	0–350	0.09

*Approximate values depending upon the exact material composition.

dilatation thermometers, the metal with the high value coefficient is called the active metal and the other with the low value the passive metal. At a neutral temperature, which is most often 20 °C, the bimetallic strip is flat. At higher temperatures it bends towards the passive metal. Commonly applied forms of bimetallic strips are as shown in Figure 3.10. The shift f, in mm or the rotation angle, β, in radians of the end of the bimetallic strip may be expressed with the aid of the specific bending coefficient, k. This is the bending of a flat strip of length 100 mm and thickness 1 mm at a temperature 1 °C over neutral. The values of the coefficients k for different metals are given in Table 3.3. They are the mean values for the working range of a bimetallic strip.

Huston (1962) gives formulae for the movement, f, of the end of the strip and rotation angle, β, as follows:

1. Flat strip as in Figure 3.10(a)

$$f = k\frac{\Delta\vartheta l^2}{d\times 10^4} \tag{3.7}$$

2. U-shaped strip as in Figure 3.10(b)

$$f = k\frac{\Delta\vartheta l^2}{2d\times 10^4}\left\{\begin{array}{c}\text{gilts for } r\ll l\\ \text{where } r \text{ is the radius as in Figure 3.10(b)}\end{array}\right\} \tag{3.8}$$

3. Helix-shaped strip as in Figure 3.10(c), (d)

$$\beta = k\frac{2\Delta\vartheta l}{d\times 10^4} \tag{3.9}$$

where $\Delta\vartheta$ is the temperature difference above neutral temperature, l is the strip length in mm, k is the specific bending in 1/°C, and d is the strip thickness in mm.

In Equation (3.9) the expression for the rotation angle in radians has been derived with l given, neglecting the bent endings.

Numerical example

A bimetallic helical thermometer of the type illustrated in Figure 3.10(c) has $k = 0.156$ 1/°C and thickness $d = 0.2$ mm. How long should the strip be to give a rotation angle $\beta = \pi$ rad over the temperature change 0 °C to 200 °C?

From Equation (3.9) it follows that:

$$l = \frac{\beta d\times 10^4}{2k\Delta\vartheta} = \frac{\pi\times 0.2\times 10^4}{2\times 0.156\times 200} = 101 \text{ mm}$$

Typical materials used for bimetallic strips, with their parameters, are given in Table 3.3. Overheating of a bimetallic strip may cause the elastic limit of the materials used to be exceeded. In that case permanent deformation of the bimetallic element renders it useless.

Cylindrical helical bimetallic strips are predominantly used in the manufacture of bimetallic thermometers. They are inserted in a stainless steel protective tube of length around 250 mm or occasionally as long as 1 metre (Figure 3.11). The tube diameter may vary between 6 mm to 10 mm. Useful temperature ranges covered by these thermometers may be from −40 °C to as high as 500 °C with errors between ±1% to ±2% of full scale.

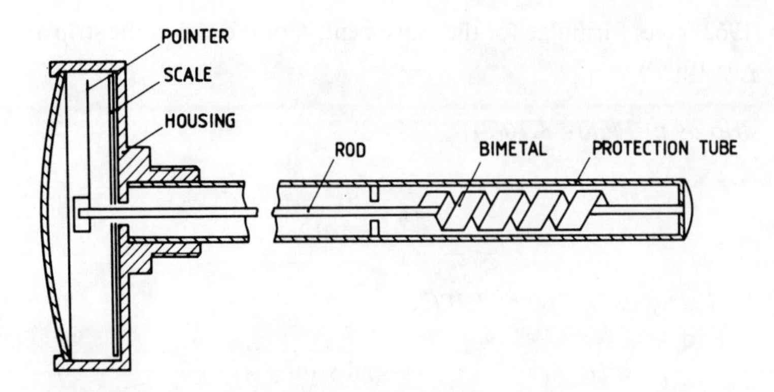

Figure 3.11
Bimetallic thermometer.

Bimetallic thermometers are simple and robust in structure. They possess accuracies and physical sizes which are compatible with most industrial applications. Their settling time, of less than 1 min, is accompanied by low sensitivity to both vibration and electrical disturbances.

Bimetallic thermometers are especially suitable for the measurement of the temperatures in live electrical equipment and also when only local, non-remote, readings are required. Other typical applications may be encountered in measuring the temperature of liquids and gases in containers, boilers and baths and also the temperature of the oil in power transformers. It is also possible to apply thermometers, made from flat helical strip, for surface temperature measurement. More detailed information is available in the work reported in Huston (1962) and in the *Temperature Measurement Handbook* (Omega Engineering Inc., USA).

3.4 MANOMETRIC THERMOMETERS

Although the physical principles for these thermometers depend upon the particular type, they have similar physical structure. They may be considered under the headings of variable volume or variable pressure types. Variable volume thermometers are liquid-filled units while the variable pressure class depends upon the thermometric behaviour of vapours and gases.

3.4.1 Liquid-filled thermometers

This type of manometric thermometer is illustrated in Figure 3.12. Its whole system, which comprises a steel tube, connecting capillary and elastic element, is filled with thermometric liquid. An increase in bulb temperature causes the liquid to expand and to dilate the elastic element. Subsequently this dilatation moves the pointer through the transmission element. As the liquid may be regarded as incompressible, the deformation of the elastic element is proportional to the increase in temperature so

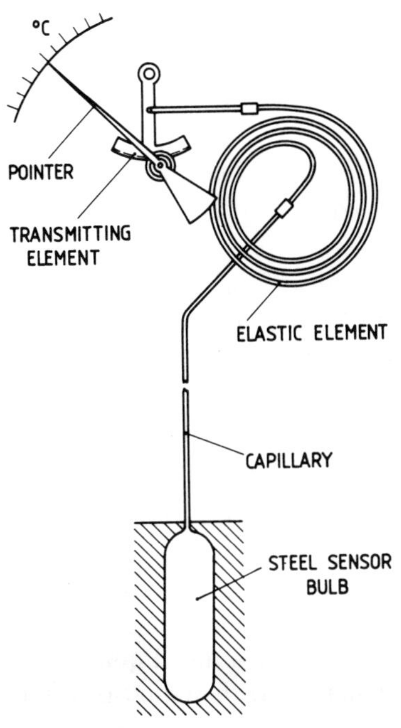

Figure 3.12
Liquid-filled thermometer.

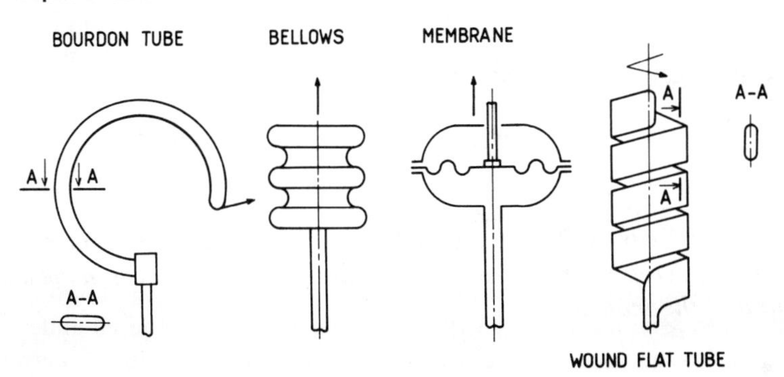

Figure 3.13
Elastic elements of liquid-filled thermometers.

that the scale is practically linear. Diverse forms of elastic element may be applied depending upon the internal pressure in the system (Figure 3.13). For example, membranes are suitable for pressures around 0.5 MPa (or 5 atm), Bourdon tubes and bellows between 0.2 MPa and 2.5 MPa. Flat helical tubes or cylindrically wound flat tube are suitable for the range 2.5 MPa to 60 MPa.

The temperature induced increase in the liquid volume, ΔV, resulting in a deformation of the elastic element, is given by:

$$\Delta V = V_b(\beta_l - 3\alpha)\Delta\vartheta \tag{3.10}$$

where V_b is the bulb volume, β_l is the coefficient of cubic thermal expansion of the liquid, α is the coefficient of linear thermal expansion of the bulb material, and $\Delta\vartheta$ is the temperature difference.

The sensitivity of the thermometer is:

$$\frac{\Delta V}{\Delta\vartheta} = V_b(\beta_l - 3\alpha) \tag{3.11}$$

Normally these thermometers are produced with the same elastic element and the same mechanical system for different temperature ranges. Hence, the volume of the bulb, which is necessary, may be calculated as:

$$V_b = \frac{\Delta V_{max}}{R(\beta_l - 3\alpha)} \tag{3.12}$$

where ΔV_{max} is the increase of the total system volume for the maximum deformation of the elastic element and R is the measured temperature range.

As a result of relation (3.12) it can be seen that the bigger the temperature range of the thermometer, the smaller the sensor bulb volume has to be.

Inserting the corresponding values from Table 3.4 into Equation (3.12) allows a calculation of the commonly used bulb volumes. If the liquid is mercury these are:

$$V_b = \frac{0.04\ .\ .\ .\ 0.10}{1.3\times 10^{-3}\times R} \approx \frac{300\ .\ .\ .\ 800\ \mathrm{cm}^3}{R}$$

whereas for organic liquids they are:

$$V_b = \frac{0.04\ .\ .\ .\ 0.15}{1.1\times 10^{-3}\times R} \approx \frac{35\ .\ .\ .\ 140\ \mathrm{cm}^{-3}}{R}$$

A number of factors influence the readings of the thermometer. These are *variations in ambient temperature and atmospheric pressure* and *differences in the level between the bulb and the indicator.*

Table 3.4
Manometric liquid-filled thermometers.

Liquid	Bulb and capillary	Temperature range (°C)	$(\beta - 3\alpha)$ (l/°C) (Equation (3.10))	V_{max} (cm³) (Equation (3.12))	Minimum temperature span (°C)	Liquid pressure (MPa)
Mercury or mercury thallium	Chromium or nickel steel	−30– +600	1.3×10^{-4}*	0.04–0.10	25	10–15
Organic liquids	Copper, brass, bronze	−35– +350	1.1×10^{-4}*	0.04–0.15	15–25	0.5–5

*Approximate values.

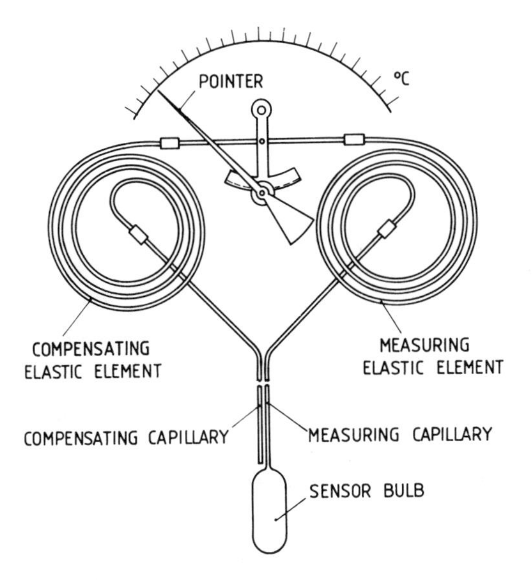

Figure 3.14
Liquid-filled thermometer with full compensation of ambient temperature influence.

Ambient temperature variations cause a change in the volume of the liquid in the capillary and the elastic element, giving rise to reading errors. A reference temperature of 20 °C is usually assumed. In mercury-in-steel thermometers this error is of the order 0.1% to about 0.2% of the full scale range for 1 metre of the capillary length and 10 °C of temperature variation. Errors at capillary lengths up to about 20 metres do not need compensation. Automatic compensation of ambient temperature influence may be achieved by a number of different methods.

Full compensation may be arranged with the assistance of a second identical capillary and elastic element system but without a sensor bulb as shown in Figure 3.14. Both capillaries are installed in parallel along their whole length. The driving moment of the pointer is the difference between the moments due to each elastic element. For variations in temperature of ±30 °C negligibly small errors arise.

Partial compensation may be achieved in one of two ways. The first compensates for errors due to temperature variations which arise in the indicator. As in Figure 3.15, a bimetallic insert between the elastic element and the pointer provides the compensation. Consequently the resulting error is some 0.05% of the full scale range for each 10 °C variation in ambient temperature. The second method (Considine *et al.*, 1957; Lieneweg, 1975), which is only applied in the mercury thermometer, is based on a capillary with an invar core to compensate for the errors which arise in the capillary itself. The volume of this core is independent of temperature. A correct choice of core and capillary diameters ensures that the increase of volume between them equals the increase which occurs in the volume of the thermometric liquid.

There are two other methods of full compensation. A suitable combination of the above two partial compensation schemes may be applied to the mercury thermometer. Such an arrangement compensates for inaccuracies arising from either the mechanical

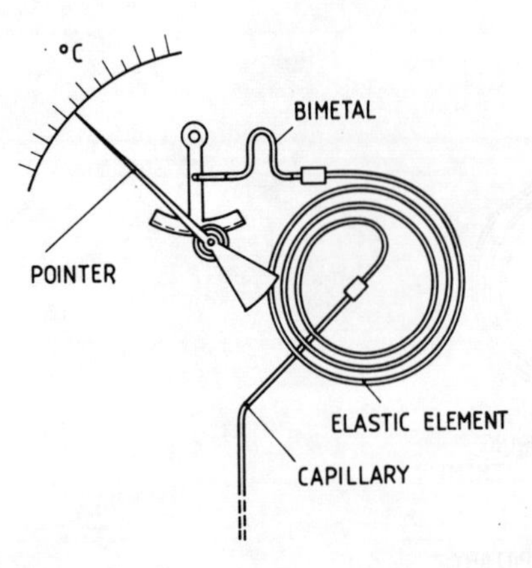

Figure 3.15
Liquid-filled thermometer with partial compensation of ambient temperature influence.

imperfections of the indicator or from temperature dependent liquid effects. When the bimetallic insert for indicator compensation is calculated in the appropriate manner it can compensate for the mechanical and liquid inadequacies simultaneously. However, it only works when the capillary and the indicator have the same temperature and provided that this temperature does not change too rapidly. When the volume of the elastic element is a function of the measured temperature, these last two full compensation methods provide full compensation only at about half the measuring range. At the lower and upper limit of the measuring range and for variations of ambient temperature in the range ± 20 °C, the errors may be up to $\pm 1.5\%$.

Level difference between the bulb and the indicating instrument causes errors owing to the hydrostatic pressure which then exists between them. Readings are too high or too low as the sensor is respectively installed above or below the indicating instrument. As these level difference errors are a function of the density of the thermometric liquid, they are greatest in the case of mercury-in-steel thermometers. In that case the value of the error may be anything between 1% and 1.5% of the full scale range at a level difference of around 20 metres. If the thermometric liquid is organic this error is considerably smaller. Compensation of these errors is achieved by adjusting the zero position of the pointer after the thermometer has been installed.

Atmospheric pressure variations only give rise to insignificant errors.

3.4.2 Vapour-pressure thermometers

Manometric thermometers of this type are based upon the temperature dependence of the saturated vapour pressure of the thermometric liquid. The thermometer system is filled by a mixture composed of the thermometric liquid and its saturated vapour. Saturated vapour pressure is a function of the temperature of the liquid–vapour interface. Although it is essential that both phases of the thermometric fluid exist in the sensor bulb, only one phase need be present in the capillary and the elastic element. The liquid phase is always present in the coolest part of the system.

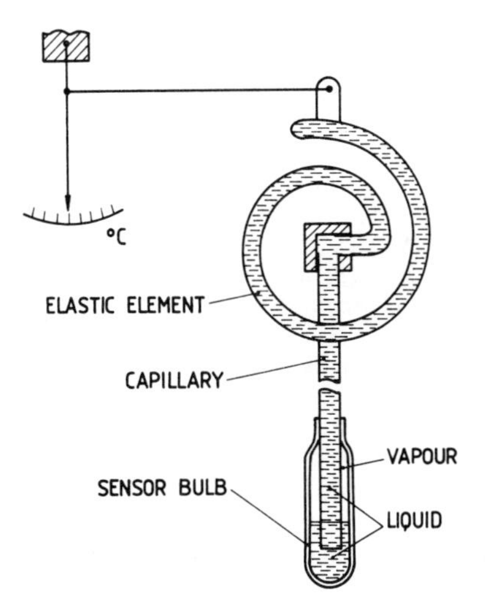

Figure 3.16
Vapour-pressure thermometer for measured temperatures above ambient.

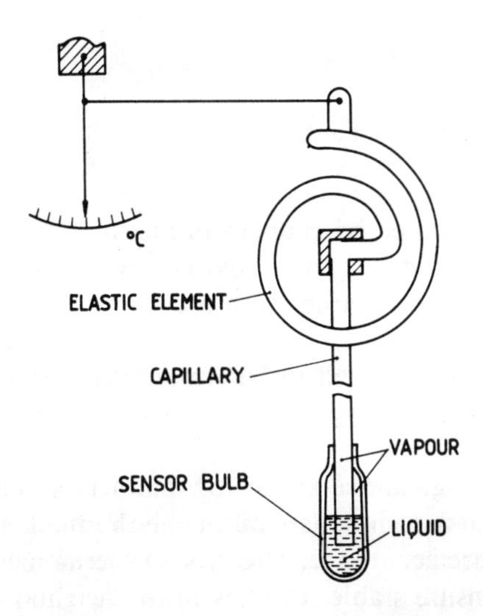

Figure 3.17
Vapour-pressure thermometer for measured temperatures below ambient.

Although the vapour-pressure thermometer has the same general form as the liquid-filled thermometer, shown in Figure 3.12, it can have four different structures depending upon the range of temperature to be measured.

For measured temperatures above ambient, vapour-pressure manometric thermometers have a design similar to that illustrated in Figure 3.16. This represents the most commonly applied structure. Both the capillary and the elastic element are liquid filled. The bulb must be large enough to accommodate any volume change in the contents of the capillary and elastic element. These volume changes may be

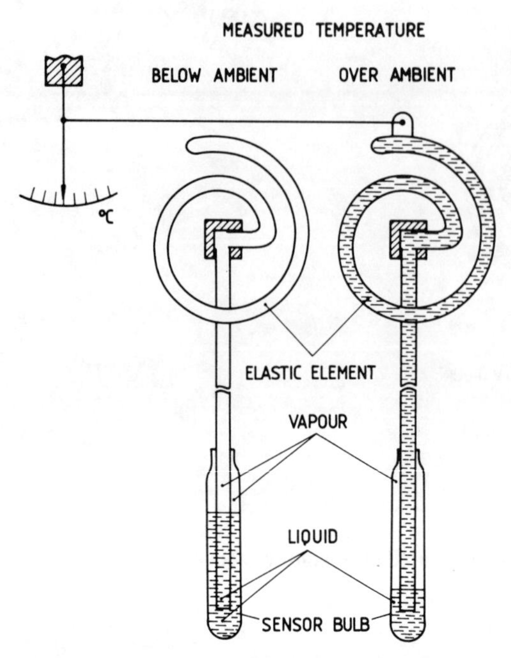

Figure 3.18
Vapour-pressure thermometer for measured temperatures below and above ambient.

due to ambient temperature changes caused by changes in measuring temperature or to changes in the volume of the elastic element across the range. At the lowest measured temperature the bulb should not be filled to more than about 60% of its volume.

If the measured temperature is below ambient the thermometer may use a bulb with a smaller volume, as shown in Figure 3.17, since the capillary and elastic element are vapour filled.

Thermometers with a straddling range above and below ambient should have a bulb volume which is large enough to accommodate all of the thermometric liquid at the lowest temperature of its measurement range. This type of thermometer, which is shown in Figure 3.18, does not ensure stable readings in the neighbourhood of the ambient temperature as the thermometric liquid may flow from or to the sensor bulb.

Manometric thermometers exist which will allow the measurement of temperatures below, about and above ambient. In this type, shown in Figure 3.19, the liquid and its saturated vapour only exist in the sensor bulb. Pressure variations are transmitted to the elastic element by an additional closed, liquid-filled system. This consists of an elastic bellows, a capillary and an elastic element.

A number of factors are important in the choice and design of an appropriate manometric thermometer. Firstly, a comprehensive range of liquids are suitable as thermometric liquids. Examples are ether, ethanol, methanol, pentane, benzene, toluene, xylene, chloroethylene, chloromethylene and chloroform. Other factors are

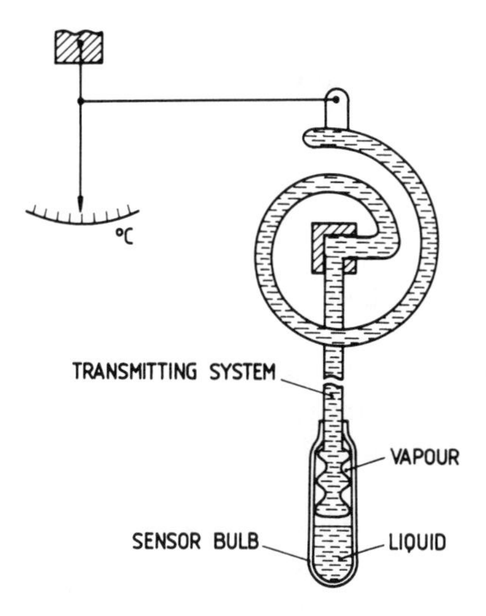

Figure 3.19
Vapour–pressure thermometer for measured temperatures below, about and above ambient.

concerned with the functional dependence of the pressure upon temperature, the useful measurement range and the sources of error. Although the pressure of a saturated vapour is an exponential function of the measured temperature, it should not be lower than 0.7 MPa. Provided this pressure is exceeded it is possible to guarantee that the mechanical forces at the elastic element are high enough. As the exponential dependence makes the scale non-linear, with its division being wider apart toward one end of the scale, it is advisable to apply the thermometers at the upper part of the measuring range. Vapour-pressure manometric thermometers may be applied in the range between about -50 °C and $+350$ °C, where achievable accuracies of the order of $\pm 1\%$ to $\pm 2\%$ of its full scale range are possible. Capillary length may be as large as 50 metres while the saturated vapour pressure usually lies between 0.7 MPa and 2.5 MPa.

Error in saturated vapour manometric thermometers may arise from the same sources as for liquid-filled manometric thermometers. However, there are some differences in emphasis. The principle of operation of vapour-pressure thermometers does not depend upon the temperature of the elastic element or upon that of either the liquid or vapour in the capillary. Hence, in principle, ambient temperature variations do not influence the readings obtained from them. However, if the capillary is liquid filled, correct functioning with temperature can only be ensured if the temperature of the capillary does not exceed the bulb temperature. Conversely, if it is vapour filled its temperature must not be allowed to drop below the bulb temperature. A possible source of error may arise from the temperature dependence of the modulus of elasticity of the elastic element. However, they are well below 1% in the ambient temperature range from 0 °C to 50 °C. These comments do not apply to those thermometers with an additional transmitting system similar to that shown in Figure 3.19.

Level difference between the bulb and the indicating instrument causes some errors in the case of the liquid-filled capillary. Vapour-pressure thermometers, of the type shown in Figure 3.16 or with additional transmitting systems as in Figure 3.19, which are employed in measurements above ambient temperature, should be graduated for a given level difference. A thermometer, intended for use in ranges below or above ambient temperature as illustrated in Figure 3.18, has a capillary filled with either liquid or vapour. Thus, as units of this type do not permit the introduction of any correction they should be operated with very little or no difference between the level of the bulb and indicator.

Atmospheric pressure variations influence the elastic element, so causing error. At atmospheric pressure differences of 4 kPa the errors introduced may amount to 2 °C at the lower end of the range decreasing to about 0.2 °C at the upper end of the scale.

If there are likely to be big changes in ambient temperature, and especially when precise measurements are needed at the upper part of the temperature range, then the cheaper vapour-pressure thermometer is to be preferred over the liquid-filled type.

3.4.3 Gas thermometers

In constant volume gas thermometers, the gas pressure is a function of temperature. Their structure is analogous with that of the vapour-pressure type but with a larger bulb. Nitrogen or helium are the most commonly used gases which allows their application over the range −200 °C to +500 °C in a linear scale. Ambient temperature variations cause some error owing to the temperature difference between the capillary and the elastic element. Moreover these errors increase with increases in the volume of both capillary and the elastic element. Compensating mechanisms are rarely used in gas thermometers. As gas thermometers only produce a small force to drive the pointer, they are infrequently used. No description of highly specialized laboratory gas thermometers will be given in this book.

3.4.4 Summary of the properties of manometric thermometers

As manometric thermometers are mechanically robust with excellent resistance to vibration and also with a long effective life their use in industry is widespread. They are the only type of non-electric thermometer which allows the reading of temperature remote from the point of measurement. High values of driving moment in the pointer mechanism allow its replacement by an inking system to provide hard copy from a simple low cost temperature recorder. It is also possible to build explosion-proof manometric thermometers because of the nature of their functional principle. A significant disadvantage is the difficulty involved in their repair. A summary of the properties of manometric thermometers appears in Table 3.5.

Table 3.5
Properties of manometric thermometers.

	Liquid-filled thermometers		Vapour-pressure thermometers
	Mercury	**Organic liquids**	
Application range	−30–+600 °C	−35–+350 °C	−50–+350 °C
Minimum temperature span	25 °C	15 °C	25 °C
Maximum capillary length	60 m	60 m	50 m
Scale	Linear	Linear	Non-linear
Errors			
Ambient temperature variations	Existing in both types compensation needed in both		Non-existent
Level difference between bulb and instrument	Exists	Exists small	Exists only when capillary is liquid filled
Atmospheric pressure variations	Negligibly small	Negligibly small	Small
Limits	±1–±2%	±1–±2%	±1–±2%

3.5 TEMPERATURE INDICATORS

Temperature indicators only allow estimation of temperature. It is not possible to use them as precision measuring instruments.

The commonly used temperature indicators are the *Harrison pyrometric cones*, also known as *Seger cones* on the European continent or *Orton cones* in the USA. They are shaped like asymmetrical triangular pyramids, with one face vertical, and about 60 mm in height. Variation of the composition of the cones provides cones with different melting or fusion points, which are characterized by giving each type of cone a particular distinguishing number. As the cones soften under the influence of temperature, the pyramid bends until eventually its tip touches the base as represented in Figure 3.20. Each cone number is correlated with a specific bending temperature. A total of 42 different numbers in the temperature range from 600 °C to 2000 °C, stepped at 10 °C to 25 °C, is defined in BS 1041: Part 7. Numbered cones, consecutively increasing in number, bend at progressively higher temperatures as indicated in Figure 3.20. The standard also defines the permissible half-range uncertainty of tolerance upon the nominal temperature. Industrial cones may have ±15 °C uncertainty whilst laboratory cones have ±10 °C. Indicated temperature values also depend upon the rate at which the temperature increases. Pyrometric cones are widely used in the ceramics industry to determine the firing temperatures of ceramic charges. Bars and rings are also used, as specified in BS 1041: Part 7.

Temperature indicating or thermochromic paints are very popular in different fields of application. They change colour as a function of temperature in either a continuous or discontinuous way. Some of them change colour only once, whilst others may change from twice to as many as even four times at well defined temperatures.

Reversible paints regain their former colour when cooled down whilst irreversible paints undergo permanent colour change (BS 1041: Part 7). A thin paint layer is administered to the object whose temperature is to be measured.

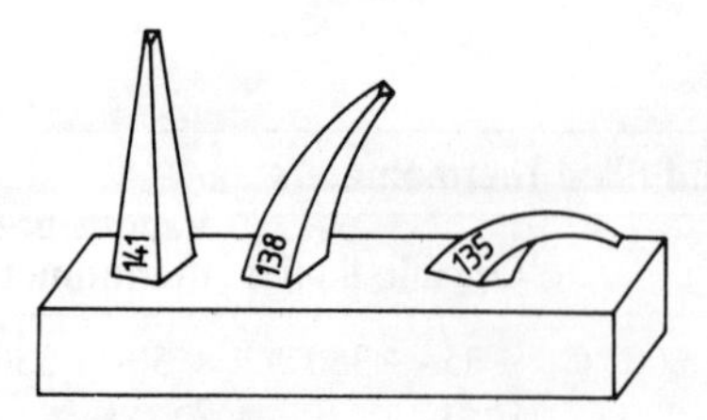

Figure 3.20
Pyrometric cones.

Paints, whose colour changes continuously, are produced for the range from 120 °C to 400 °C, whereas discontinuous ones are available for the range from 40 °C to 1350 °C. The colour transition is also a function of heating time. Normally manufacturers define a nominal paint temperature as that at which the colour transition occurs after a heating time of 30 minutes. It is also normal to specify the colours for shorter and longer heating times. Lower colour transition temperatures correspond to longer heating times. For heating times between some seconds to some hours the temperatures may differ significantly.

If a temperature is to be determined a first application of a multiple changing paint allows a rough estimate. The investigated surface is then covered by strips or dots of paints for the indication of some neighbouring temperatures. After a heating time of 30 minutes the achievable precision is about ±5 °C. As most of the paints are of the irreversible kind the temperature may be read even after the measured object has cooled down.

A typical example of a multiple temperature indicating paint is the paint No. 69 manufactured by Synthetic and Industrial Finishes Ltd. It has the following colour transition temperatures:

Transition temperature (°C)	*Colour* *(original colour light tan)*
150	Bronze green
240	Deep purple brown
310	Pale indian red

These temperature indicating paints are usually employed in the estimation of the surface temperatures of motors, pipe-lines, cooling ribs and high voltage apparatus among many others. They also find application in estimating the temperature of moving elements such as rotating wheels, ventilator fan blades, transmission gearing or charges which are continuously heated or dried. Frequently they are employed as warning indicators of over-heating of some stored or transported goods such as chemicals, films, foodstuffs and so on.

For rapid temperature determination temperature indicating crayons, with a colour change response time of 1 s to 2 s, may be used. A range of from 65 °C to 670 °C, in steps of from 10 °C to 100 °C, is produced. They are useful for rapid measurement of the temperature of previously heated objects. Melting temperature indicators, which leave a dry chalky mark at temperatures below their rating, are also produced as

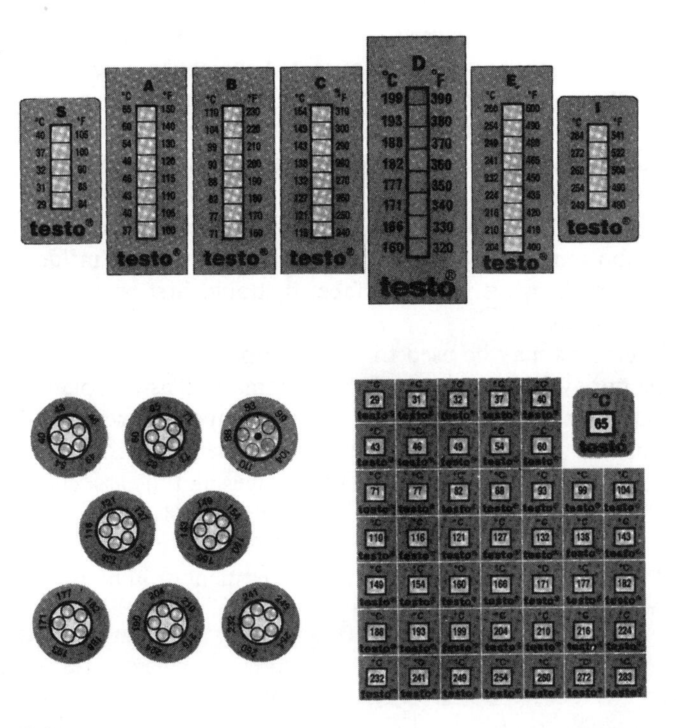

Figure 3.21
Self-adhesive miniature temperature indicators. (Courtesy of Testoterm GmbH, Germany.)

crayons or rapidly drying liquids. When their rated temperature is exceeded this chalky mark becomes glossy. Such crayons and paints are produced for the temperature range from +50 °C to 1400 °C, in steps of from 5 °C to 30 °C, with typical accuracies of ±1% (Kasanof and Kimmel, 1962).

A similar idea to temperature indicating, or thermochromic paints is embodied in miniature self-adhesive temperature indicators, shown in Figure 3.21, which mostly turn black when the temperature rating, shown on the label, is reached. As they belong to the irreversible group of indicators they register a temperature history of the object of interest. For example, circular indicators have diameters ranging from 4 mm to 10 mm, and their application range covers from 30 °C to 280 °C in 3 °C to 10 °C steps. A single label may have four indicators as for example the 'Tempilabel' of Tempil Big Three Industries (USA). Such a particular label consecutively indicates, by way of example, 79/93/107/121 °C or perhaps 107/121/135/149 °C with an accuracy of ±1 °C.

As self-adhesive indicators are of small size they can be used to determine the temperature of small components such as transistors, integrated circuits and other components used in electronic equipment. In the temperature range from 260 °C to 600 °C these indicators are produced on refractory steel labels which are glued to the surface by a special cement supplied by the producer.

Recently, the use of liquid crystals to image thermal fields has been described by Kaiser (1979), Meier *et al.* (1975), Stegmeyer (1973) and Stephens and Sinnadurai (1974). These crystals are compounds which change their reflection factor in the visible radiation range. In this manner they appear to change colour. They are applied by spreading a thin layer over the investigated surface in a 10 μ to 20 μm thick surface skin. Suitably chosen mixtures of cholesteryl ester or carboxylic acids, depending upon temperature range, are used for the production of such crystals. An experimental determination of their colour-temperature dependence allows the achievement of good thermal field imaging with resolution between 0.01 °C and 10 °C over a temperature range from −40 °C to +283 °C (Weichert *et al.* 1976). During investigations, but only in a laboratory environment, it is possible to avoid the influence of any radiation reflected from the investigated surface. It should first be covered by a thin layer of black matt varnish.

Liquid crystals may be used to build temperature indicating instruments as well as for the purely labelling function. A thermally sensitive plate, or perhaps a tape, is fabricated with several numbers. These numbers correspond to the temperature of the sensitive crystal to which they refer. When a specific temperature is reached the reflectivity of the respective crystal changes and the associated number becomes visible. The numbers at the adjacent higher and lower temperatures are also distinguishable but with a lower contrast. They have a precision of about ±1 °C and are used mainly in room temperature measurement or in body temperature measurement for medical purposes.

REFERENCES

Busse, J. (1941) Liquid-in-glass thermometers. *Temperature: Its Measurement and Control in Science and Industry*, Reinhold Publ. Co., New York, pp. 228–255.

Baldinger, O. (1974) Die Sichtbarmachung von Temperaturfeldern. *Elektrizitätsverwertung*, **49**(3), 122–124.

Considine, D. *et al.* (1957) *Process Instruments and Controls Handbook*, McGraw-Hill, New York.

Das Kanthal Bimetall Handbuch (1966) A. B. Kanthal, Hallstahammar, Sweden.

Huston, W. D. (1962) The accuracy and reliability of bimetallic temperature measuring elements. *Temperature: Its Measurement and Control in Science and Industry*. Vol. 3, Part 2, Reinhold Publ. Co., New York, pp. 949–957.

Kaiser, E. (1979) Temperaturfeldmessung mit Flüssigkristallen in der Umgebung von Berührungsthermometern. *m.s.r.*, **22**(3), 122–124.

Kasanof, D. R. and Kimmel, E. (1962) Recent developments in fusible temperature indicators. *Temperature: Its Measurement and Control in Science and Industry*. Vol. 3, Part 2, Reinhold Publ. Co., New York, pp. 1005–1007.

Lieneweg, F. (1975) *Handbuch, Technische Temperaturmessung*. F. Vieweg, Braunschweig.

Meier, G., Sackmann, E. and Grab-Meier, J. G. (1975) *Application of Liquid Crystals*, Springer-Verlag, Berlin.

Stegmeyer, H. (1973) Anwendung von cholesterischen, flüssigen Kristallen zur Darstellung von Temperaturefelder. *VDI-Berichte, 198, Technische Temperaturmessung*. Düsseldorf, VDI-Verlag GmbH.

Stephens, C. E. and Sinnadurai, F. N. (1974) A surface temperature limit detector using nematic liquid crystals with an application to microcircuits. *J. Phys. E: Sc. Instr.* **7**(8), 641–643.

Temperature Measurement Handbook Omega Engineering Inc., USA.

Thomson, D. (1962) Recent developments in liquid-in-glass thermometry, *Temperature: Its Measurement and Control in Science and Industry*, Vol. 3, Part 1, Reinhold Publ. Co., New York. pp. 201–208.

Weichert, L. *et al.* (1976) *Temperaturmessung in der Technik-Grundlagen und Praxis*. Lexika-Verlag, Grafenau.

4 Thermoelectric Thermometers

4.1 PHYSICAL PRINCIPLES

4.1.1 Thermoelectric forces

It was T. Seebeck who discovered, in 1821, the effect, which now bears his name. He observed that a current flows in a closed loop of two dissimilar metals when their junctions are at two different temperatures. As Ohm's law was not formulated by G. S. Ohm until 1826, the quantitative description of this phenomenon was not possible at the time of its discovery.

In 1834, I. C. A. Peltier, discovered that a junction of two dissimilar metals is respectively heated or cooled when a current is passed through it either in one direction or the other. This phenomenon is quite distinct from and in addition to the I^2R Joule heat, which is generated by the current flowing in the lead resistance.

Another themoelectric force, distinguishably different from the previous, was discovered by Lord Kelvin (W. Thomson) in 1854. He concluded that a homogeneous current carrying conductor, lying in a temperature gradient, will absorb or generate heat, in addition to and independently of Joule heating. This absorption or generation of heat depends upon the metal properties and the respective direction of the current.

Whereas the *Peltier effect* concerns the generation of a certain thermoelectric force in the junction of two dissimilar metals, the *Thomson effect* describes the generation of a thermoelectric force along the conductors which form a closed circuit and lie in a temperature gradient. In this book the thermoelectric force will be called *thermal electromotive force*, which may also be abbreviated to *thermal e.m.f.* or simply *e.m.f.*

As the Peltier effect results from the difference in the number of free electrons on both sides of a junction of two dissimilar metals at the given temperature, the junction temperature determines the generated thermal e.m.f. The Thomson effect is accounted for by the different free electron density existing at points along a conductor which lies in a temperature gradient. In such a conductor, the Thomson thermal e.m.f., E_T, is related to the Thomson coefficient, σ_T, of a given metal conductor and the temperatures, ϑ_1 and ϑ_2, of its ends by the relation:

$$E_T = \int_{\vartheta_1}^{\vartheta_2} \sigma_T \mathrm{d}\vartheta = \sigma_T(\vartheta - \vartheta_1) \qquad (4.1)$$

Equation (4.1) indicates that the Thomson thermal e.m.f., E_T, is only a function of the difference in temperature between the ends of the conductor. It is independent of the length of the conductor and of the temperature distribution along it. Laboratory experiments have shown that the Thomson coefficient of lead is nearly zero. Thus, the coefficients of other metals may be referred to lead. Their values at 0 °C are:

Constantan	$-23\ \mu v/°C$
Pt10Rh	$-10\ \mu V/°C$
Pt	$-9\ \mu V/°C$
Fe	$-8\ \mu V/°C$
Cu	$+2\ \mu V/°C$

The positive sign indicates that the heat in the conductor due to the Thomson effect is generated when the electric current flows in the direction from the higher to the lower temperature.

A summary of the foregoing, which appears in Figure 4.1, applies to two homogeneous metals A and B. When they are joined at both ends, with the junctions at the different temperatures ϑ_1 and ϑ_2, the following four, separate and distinct thermoelectric forces, exist in the circuit:

$E_P(\vartheta_1)$ is the Peltier e.m.f. in junction 1.
$E_P(\vartheta_2)$ is the Peltier e.m.f. in junction 2.
E_{TA} is the Thomson e.m.f. in conductor A.
E_{TB} is the Thomson e.m.f. in conductor B.

It is important to adopt a convenient convention when summing thermal e.m.fs. In this book thermal e.m.fs are summed in a clockwise direction. Those e.m.fs which are compatible with this direction, possess positive polarity. Summing all of the partial e.m.fs in the circuit of Figure 4.1 gives the resultant e.m.f., E, as:

$$E = E_P(\vartheta_1) - E_P(\vartheta_2) + \sigma_B(\vartheta_1 - \vartheta_2) - \sigma_A(\vartheta_1 - \vartheta_2) \tag{4.2}$$

As it is difficult to precisely identify the particular e.m.fs, they are assumed to be lumped in the two circuit junctions. When the e.m.fs in these two junctions of Figure 4.1 are added in a clockwise direction the resultant e.m.f. of the circuit for the two metals, A and B, at the temperatures ϑ_1 and ϑ_2 yields:

$$E_{AB}(\vartheta_1, \vartheta_2) = e_{AB}(\vartheta_1) + e_{BA}(\vartheta_2) \tag{4.3}$$

It is apparent, and deserves emphasis, that the resultant e.m.f. of the circuit of Figure 4.1 is only a function of the types of metal A and B and of the temperatures ϑ_1 and ϑ_2 of both junctions. From the principles of the double subscript notation it is known that:

$$e_{BA}(\vartheta_2) = -e_{AB}(\vartheta_2) \tag{4.4}$$

As a result Equation (4.3) may be written as:

$$E_{AB}(\vartheta_1, \vartheta_2) = e_{AB}(\vartheta_1) - e_{AB}(\vartheta_2) = f_1(\vartheta_1, \vartheta_2) \tag{4.5}$$

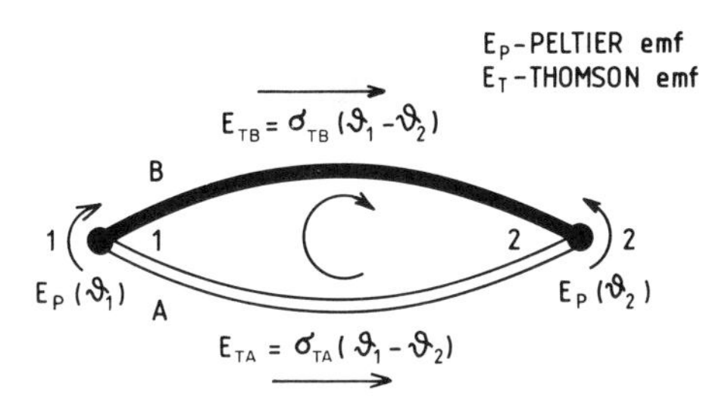

Figure 4.1
Closed thermoelectric circuit.

In the application of a circuit of two metals, A and B, for temperature measurement, it is rather difficult to use a function of two variables. For this reason it will be assumed that the temperature, ϑ_2, of one of the junctions, called the reference junction, will be held constant at the reference temperature, ϑ_r. Equation (4.5) may now be written as:

$$E_{AB}(\vartheta_1, \vartheta_2) = E_{AB}(\vartheta_1, \vartheta_r) = e_{AB}(\vartheta_1) - e_{AB}(\vartheta_r) = f_2(\vartheta_1) \tag{4.6}$$

Equation (4.6) indicates that the e.m.f. of the circuit of two metals A and B is a function of the measured temperature, ϑ_1, of the measuring junction, junction 1. It is of cardinal importance as it is the foundation of modern thermoelectric thermometry.

All of the previous considerations have been for the case of homogeneous materials. Inhomogeneities of either a physical or chemical nature cause additional parasitic e.m.fs. These should not be confused with normally occurring Thomson e.m.fs, even though they are a function of the temperature distribution along the conductors.

4.1.2 Law of the third metal

The practical application of a circuit with the two metals A and B, for the purposes of temperature measurement, also requires the introduction of an e.m.f. measuring device. Insertion of such a voltage measuring instrument necessarily involves the connection of a third metal C as the circuit of Figure 4.2 exposes. The internal leads of the instrument and its interconnecting leads with the circuit to be measured constitute this third metal. This assumes that the additional circuit is of the same continuous metal. Analysing Figure 4.2 shows that the sum of the circuital e.m.fs is:

$$E = e_{AB}(\vartheta_1) + e_{BC}(\vartheta_r) + e_{CA}(\vartheta_r) \tag{4.7}$$

When it is assumed that $\vartheta_1 = \vartheta_r$, then $E = 0$ and eventually:

$$e_{BC}(\vartheta_r) + e_{CA}(\vartheta_r) = -e_{AB}(\vartheta_r) \tag{4.7a}$$

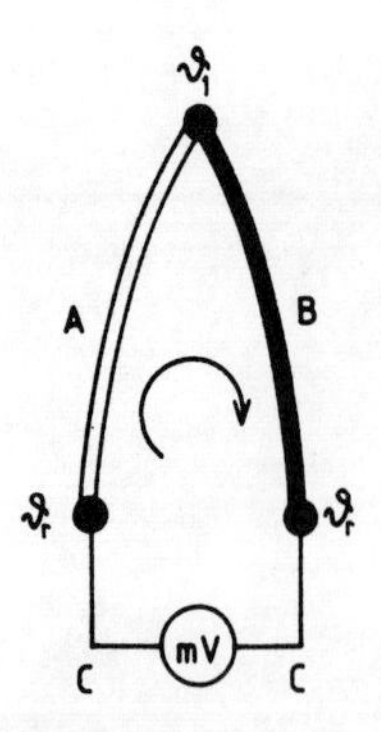

Figure 4.2
Thermoelectric circuit of metals A and B, third metal C inserted at the junction.

Substituting Equation (4.7a) into Equation (4.7) shows that:

$$E = e_{AB}(\vartheta_1) - e_{AB}(\vartheta_r) \tag{4.8}$$

The law of the third metal is based upon the result indicated in Equation (4.8). This law may be stated as:

Introducing a third metal, C, into a circuit, of the two metals A and B, does not alter the resulting e.m.f. in the circuit, provided that both ends of the third metal C are at the same temperature.

The third metal, C, may be introduced into a circuit at any point. For example, if metal B is cut and the indicating instrument, which is regarded as the third metal C, is connected at that point the resulting e.m.f. in the circuit shown in Figure 4.3 would be:

$$E = e_{AB}(\vartheta_1) + e_{BC}(\vartheta_2) + e_{CB}(\vartheta_2) + e_{BA}(\vartheta_r) \tag{4.9}$$

As $e_{BC}(\vartheta_2) = -e_{CB}(\vartheta_2)$ this equation may be rewritten in the form:

$$E = e_{AB}(\vartheta_1) - e_{AB}(\vartheta_r) \tag{4.10}$$

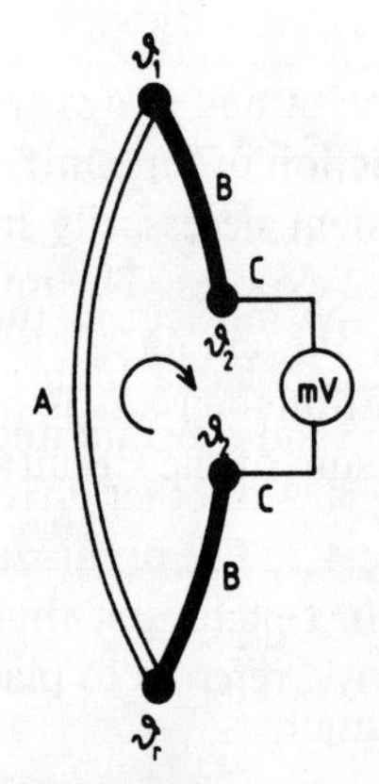

Figure 4.3
Thermoelectric circuit of metals A and B, third metal C inserted across the ends of metal B, cut in two.

Equation (4.10) is seen to be identical to Equation (4.6) for a circuit consisting only of the two metals A and B.

Figure 4.4 illustrates the situation when the junctions of metals B and C are at the respectively different temperatures of ϑ_2 and ϑ_3. In that case the resulting e.m.f. is:

$$E' = e_{AB}(\vartheta_1) + e_{BC}(\vartheta_2) + e_{CB}(\vartheta_3) + e_{BA}(\vartheta_r) \tag{4.11}$$

A comparison of Equations (4.10) and (4.11) indicates that the difference between the e.m.f. in the two cases under consideration is given by:

$$\Delta E = E' - E = e_{BC}(\vartheta_2) - e_{BC}(\vartheta_3) \tag{4.12}$$

Consequently the temperatures at both junctions of the metals B and C should be equal to ensure that the circuit gives the correct value of e.m.f.

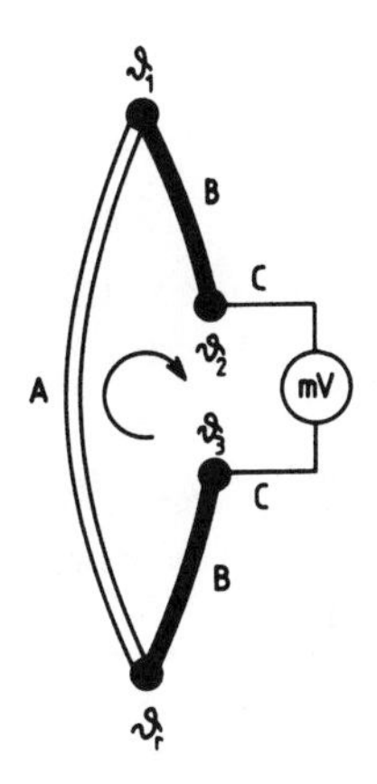

Figure 4.4
Thermoelectric circuit of metals A and B, third metal C inserted at any point, its ends being at unequal temperatures ($\vartheta_2 \neq \vartheta_3$).

4.1.3 Law of consecutive metals

Different kinds of metals are widely employed in thermoelectric thermometry. Platinum is universally taken as the reference metal as it has a high resistance to atmospheric influences, its physical properties are stable and constant and its melting temperature is high. To allow a comparison of the properties of other different metals, their e.m.fs against that of platinum are given in Table 4.1. The organization of this table is by increasing magnitude of e.m.f. values against platinum, thus forming a so-called *thermoelectric series*. Different metals and alloys, referred to platinum, have the e.m.fs given in their application ranges in Figure 4.5.

If the e.m.fs of different materials against platinum are known, it is easy to determine their e.m.fs relative to a combination of any two forming a pair.

Consider three thermoelectric circuits of the metals A, B and C shown in Figure 4.6. The equations for each circuit may be written as:

$$E_{BA}(\vartheta_1, \vartheta_r) = e_{BA}(\vartheta_1) - e_{BA}(\vartheta_r) \tag{4.13a}$$

$$E_{CA}(\vartheta_1, \vartheta_r) = e_{CA}(\vartheta_1) - e_{CA}(\vartheta_r) \tag{4.13b}$$

$$E_{BC}(\vartheta_1, \vartheta_r) = e_{BC}(\vartheta_1) - e_{BC}(\vartheta_r) \tag{4.13c}$$

Table 4.1
Value of e.m.f. for different metals referred to platinum at 100 °C and reference at 0 °C (Roeser and Wensel, 1941; Lieneweg, 1975).

Metal	e.m.f. (mV)	Metal	e.m.f. (mV)
Constantan	−3.51	Iridium	+0.65
Nickel	−1.48	Rhodium	+0.70
Cobalt	−1.33	Silver	+0.74
Alumel (95%Ni+Al, Si, Mn)	−1.29	Copper	+0.76
Palladium	−0.57	Zinc	+0.76
Platinum	0	Gold	+0.78
Aluminium	+0.42	Tungsten	+1.12
Lead	+0.44	Molybdenum	+1.45
94%Pt, 6%Rh	+0.614	Iron	+1.98
90%Pt, 10%Rh	+0.645	Chromel (90%Ni, 10%Cr)	+2.81
70%Pt, 30Rh	+0.647		

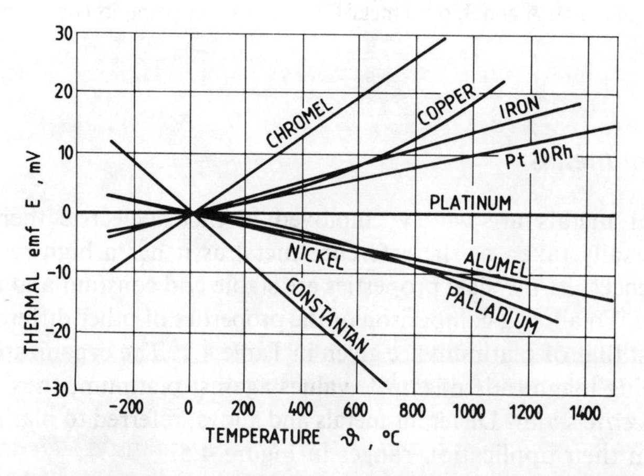

Figure 4.5
Thermal e.m.fs of different metals against platinum, displayed versus temperature.

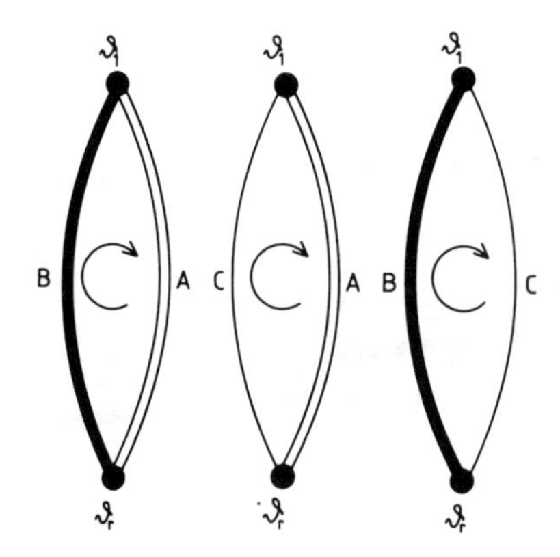

Figure 4.6
Law of consecutive metals.

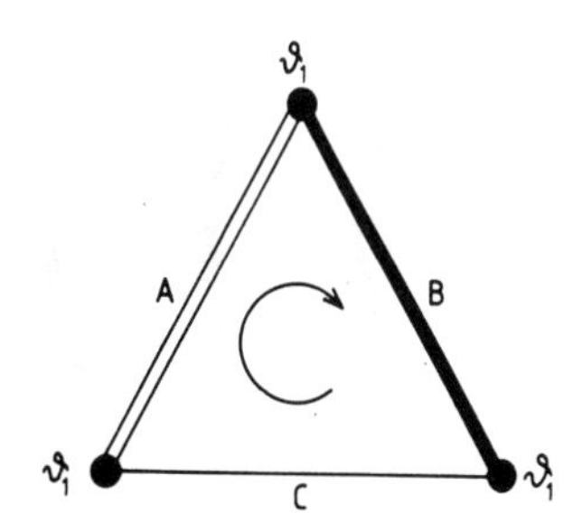

Figure 4.7
Circuit of three metals.

Subtracting on both sides of Equation (4.13a) and Equation (4.13b) yields:

$$E_{BA}(\vartheta_1, \vartheta_r) - E_{CA}(\vartheta_1, \vartheta_r) = e_{BA}(\vartheta_1) - e_{BA}(\vartheta_r) - e_{CA}(\vartheta_1) + e_{CA}(\vartheta_r) \qquad (4.14)$$

Consider the thermoelectric circuit given in Figure 4.7. If the three junctions of the metals A, B and C are at the same temperature ϑ_1 then

$$e_{AB}(\vartheta_1) + e_{BC}(\vartheta_1) + e_{CA}(\vartheta_1) = 0$$

or

$$e_{BC}(\vartheta_1) = e_{BA}(\vartheta_1) - e_{CA}(\vartheta_1) \qquad (4.15)$$

When the temperature ϑ_1 is replaced by the reference temperature ϑ_r Equation (4.15) may be used to obtain:

$$-e_{BC}(\vartheta_r) = e_{CA}(\vartheta_r) - e_{BA}(\vartheta_r) \qquad (4.16)$$

Substituting Equations (4.15) and (4.16) into Equation (4.14) gives:

$$E_{BA}(\vartheta_1, \vartheta_r) - E_{CA}(\vartheta_1, \vartheta_r) = e_{BC}(\vartheta_1) - e_{BC}(\vartheta_r)$$

or finally:

$$E_{BC}(\vartheta_1, \vartheta_r) = E_{BA}(\vartheta_1, \vartheta_r) - E_{CA}(\vartheta_1, \vartheta_r) \qquad (4.17)$$

This equation demonstrates another important thermoelectric law known as the law of consecutive metals, which may be stated as:

The e.m.f. of a circuit of metals B and C at given temperature difference of both junctions equals the difference of e.m.fs of the circuits which metals B and C would form with metal A, at the same temperature difference.

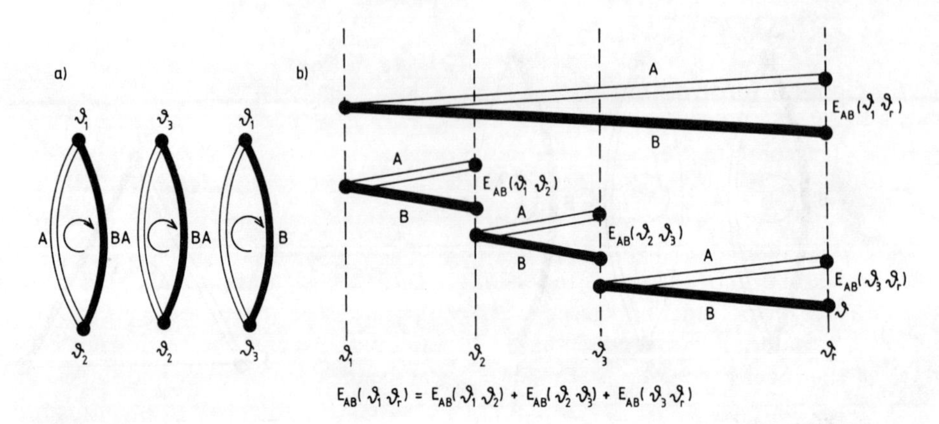

Figure 4.8
Law of consecutive temperatures; (a) thermoelectric circuits, (b) determination of resulting e.m.f. by linear superposition.

4.1.4 *Law of consecutive temperatures*

Now consider three thermoelectric circuits of metals A and B having junction temperatures as shown in Figure 4.8(a). The e.m.fs of the particular circuits are:

$$E_{AB}(\vartheta_1, \vartheta_2) = e_{AB}(\vartheta_1) - e_{AB}(\vartheta_2) \tag{4.18a}$$

$$E_{AB}(\vartheta_3, \vartheta_2) = e_{AB}(\vartheta_3) - e_{AB}(\vartheta_2) \tag{4.18b}$$

$$E_{AB}(\vartheta_1, \vartheta_3) = e_{AB}(\vartheta_1) - e_{AB}(\vartheta_3) \tag{4.18c}$$

When Equation (4.18b) is subtracted on both sides of Equation (4.18a) it is seen that:

$$E_{AB}(\vartheta_1, \vartheta_2) - E_{AB}(\vartheta_3, \vartheta_2) = e_{AB}(\vartheta_1) - e_{AB}(\vartheta_3)$$

Using this in Equation (4.18c) yields:

$$E_{AB}(\vartheta_1, \vartheta_3) = E_{AB}(\vartheta_1, \vartheta_2) - E_{AB}(\vartheta_3, \vartheta_2) \tag{4.19}$$

This relation is known as the law of consecutive temperatures which states:

The e.m.f. of a circuit whose measuring junction is at a temperature ϑ_1 and whose reference junction is at a temperature ϑ_3 equals the difference in the e.m.fs of this circuit when its reference junction is at a temperature of ϑ_2 and of a circuit whose measuring junction temperature is ϑ_3 and its reference junction temperature is ϑ_2.

This law of consecutive temperatures allows the application of the principle of superposition in the determination of the overall e.m.f. as shown in Figure 4.8(b).

4.2 THERMOCOUPLES

4.2.1 General information

The combination of two dissimilar conductors, which may be metals, alloys or non-metals, connected at one end is known as a *thermocouple*. Their point of connection is called the *measuring junction* and their free ends are referred to as the *reference junction*. This junction is also misleadingly referred to as the cold junction which could imply a restriction of the use of the thermocouple to situations when the temperature of the measuring junction exceeds that of the reference junction (Figure 4.9).

Materials, which are placed as far apart as possible from each other in the thermoelectric series of Table 4.1, form the most suitable thermocouples. In this way as high an e.m.f. as possible is ensured for a given temperature difference. Thermocouple materials should be characterized by:

- high melting temperature,
- high permissible working temperature,
- high resistance to oxidation and atmospheric influences,
- properties which are stable with time,
- properties which are repeatable in manufacture,
- low resistivity,
- low thermal coefficient of resistance,
- continuous and possibly linear dependence of e.m.f. with temperature,
- low thermal conductivity where possible.

In practice the commonly used thermocouple materials make a compromise between the different demands quoted above.

The e.m.f. versus temperature values of the more commonly used thermocouples are shown in Figure 4.10. *In this figure, and throughout this book, the convention of quoting the positive conductor first in the specific name of the thermocouple is used.* The majority of the thermocouples which will be described below are given in the standard IEC 584, Part 1 of the International Electro-technical Commission. Moreover they also conform to the national standards of many other countries such as France, Germany, Italy, Japan, Poland, UK, USA and USSR.

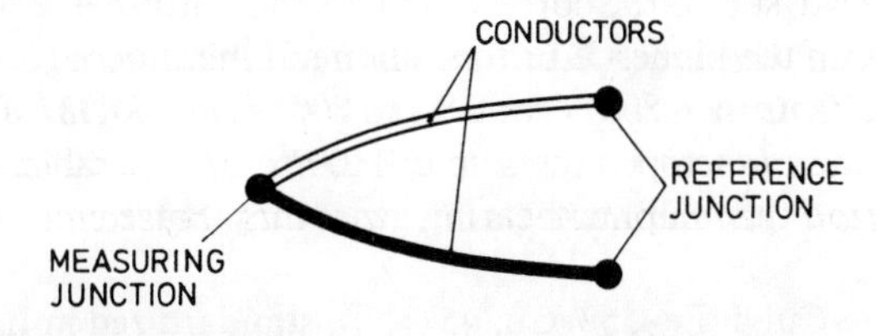

Figure 4.9
Thermocouple.

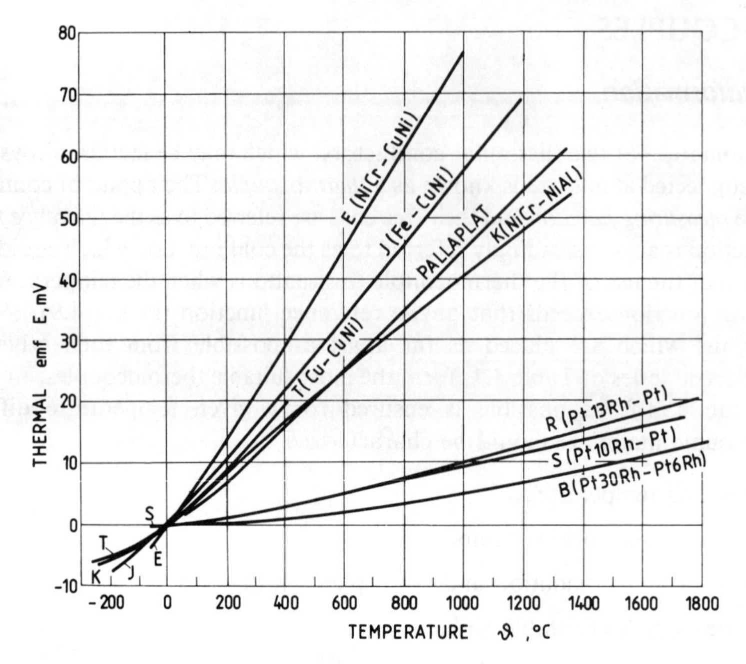

Figure 4.10
Thermal e.m.f. E of commonly used thermocouples versus temperature, ϑ.

4.2.2 Properties of commonly used thermocouples

Copper/constantan, code T or Cu–CuNi (Cu–55%Cu, 45%Ni). This thermocouple is made to IEC 584-1 standard. It is rated for continuous use within the temperature range from −250 °C to +400 °C and for short-term readings up to 500 °C. Similar in specification is the U-type. The U-type conforms to DIN 43710 standard. Although both the T and U types of thermocouples, which are commonly used in laboratory measurements, resist oxidizing and reducing atmospheres at temperatures up to 350 °C quite well, they cannot stand either oxidizing atmospheres at high temperatures or prolonged neutron flux radiation. They become brittle in hydrogen atmospheres above 370 °C. Their calibration is dependent upon the purity of the copper wire.

Nickel–chromium/constantan, code E or NiCr–CuNi (90%Ni, 10%Cr–55%Cu, 45%Ni), specified in IEC 584-1 standard, corresponds to the Chromel/Constantan structure popular in USA. With the highest e.m.f. of all metal thermocouples, it has a range of application from −270 °C to +800 °C or 0 °C to 800 °C in BS 4937. For short-term readings in oxidizing conditions temperatures up to 1100 °C are tolerable. Zysk and Robertson (1972) have reported that sulphur-bearing gases may cause embrittlement.

Iron/constantan, code J or Fe-CuNi (Fe–55%Cu, 45%Ni), standardized in IEC 584-1, corresponds to type L in DIN 43710. It is a popular low priced unit which has a high e.m.f. and good resistance to oxidizing and reducing atmospheres. It may be used up to 1000 °C but rapidly becomes covered with scale over 600 °C. In this case bigger wire diameters are necessary. Small additions of Al, Si and Ti to the CuNi wire raise its corrosion resistance without affecting its e.m.f. versus temperature dependence (Omega Engineering Inc., 1982). It is necessary to protect it from moisture, oxygen and sulphur containing gases. The temperature range is −200 °C to +850 °C and up to 1200 °C for short-term readings. In the GOST standard of the USSR the CuNi wire is replaced by Kopel (56%Cu, 44%Ni) having nearly identical characteristics.

Nickel-chromium/nickel-aluminium, code K or NiCr–NiAl (90%Ni, 10%Cr–95%Ni, balance Al, Si, Mn), also known as chromel/alumel in some countries, is defined in IEC 584-1. The K-thermocouple is the most commonly used base-metal type with a specified operating temperature range of −200 °C to 1100 °C extended to 1300 °C for short duration readings. As characterized in Figure 4.10, it has a nearly linear dependence of e.m.f. upon temperature. This thermocouple is resistant to oxidizing atmospheres but is affected by reducing and sulphur containing gases at higher temperatures. The negative, Ni wire, containing Al, Si and Mn, is less corrosion resistant than the positive wire and therefore some modified types such as Thermo-Kanthal or Nicrosil-Nisil, code N, reported in Burley *et al.* (1975) and Burley and Jones (1975), are also used. From the work of Brookes *et al.* (1985) it is clear that these latter types ensure uniform corrosion of both wires and have a more stable e.m.f. versus temperature dependence. To attain this some additions of Si are used. Although the type-K thermocouple used in Germany has a slightly different composition (85%Ni, 12%Cr–95%Ni, 3%Mn, 2%Al, 1%Si), its calibration is identical. Ageing of the thermocouple results in an increase in the value of e.m.f., especially in air. A further difficulty, particularly with units which are not well stabilized, is the existence of hysteresis at temperatures above 500 °C.

Platinum-rhodium/platinum, code S or Pt10Rh–Pt (90%Pt, 10%Rh–Pt), as defined in IEC 584-1, is the most popular of all the rare-metal thermocouples. Although it is specified for normal continuous use in the range −50 °C to 1300 °C and extended to 1600 °C for short-term readings, its importance was emphasized as it was the standard interpolation instrument in IPTS-68 within the temperature range from 630.74 °C to 1064.43 °C. As its structure gives a high resistance to corrosion, quite thin wires may be used, resulting in low cost and low thermal inertia of the thermocouple. In spite of its good resistance to oxidizing gases up to 1200 °C, Kinzie (1973) has pointed out that it is easily affected by Si and Fe and cannot tolerate any kind of reducing atmospheres. The presence of SiO_2 in the insulation and sheath materials, leads to the formation of metallic Si at higher temperatures. Any resulting diffusion of Si into the thermocouple causes a change in its calibration and embrittlement of its wires. Hence it is advisable to use Al_2O_3 as the sheath material instead. Even traces of S or C may accelerate the above-mentioned processes. Before installation, cleanliness is important. This may be accomplished by washing both wires with alcohol and avoiding manual contact of any kind.

Although the type-S thermocouple is quite stable, the Rh may diffuse from the Pt10Rh wire into the Pt wire. It is also possible for Rh sublimation to occur from the Pt10Rh wire. Both of these effects influence the calibration of the thermocouple by introducing errors which may be as high as 2 °C after exposure in oxidizing atmospheres for about 100 hours to temperatures in the region of 1200 °C.

Prior to 1922 in the UK, two different classes of Pt10Rh–Pt thermocouples, whose e.m.fs differed by about 10%, were available. When it was discovered that one of them contained a large amount of impurities, manufacturers began to supply a new type of Pt13Rh–Pt, which gave roughly the same e.m.f. versus temperature relationship as the older Pt10Rh–Pt unit.

Platinum–rhodium/platinum, type R or Pt13Rh–Pt (87%Pt, 13%Rh–Pt) as specified in IEC 584-1, has similar properties to the type-S unit described above.

Platinum–rhodium/platinum–rhodium, type B or Pt30Rh–Pt6Rh (70%Pt, 30%Rh–94%Pt, 6%Rh), was formerly known as PtRh18 alluding to its application limit of 1800 °C. It has the following advantages over the type-S unit:

- a higher upper temperature range of 1800 °C,
- a higher resistance to chemicals,
- metallic impurities affect its calibration to a lesser extent,
- the diffusion of Rh from the positive to the negative wire proceeds more slowly,
- in the temperature range, up to about 100 °C, the e.m.f. is negligibly small so that no reference temperature stabilization is needed.

All the above thermocouples are quoted in IEC 584 as well as in the standards of many other countries. Some other kinds of thermocouples, which are not quoted in any standard, are also available and will now be described.

A good example is the *Pallaplat* type, which is manufactured by Heraeus GmbH. It is composed of 95%Pt, 5%Rh–52%Au, 46%Pd, 2%Pt, giving an e.m.f. at 1300 °C of 60.29 mV which is around four times higher than that of the type S. It is quite resistant to oxidation and also has a long lifetime.

Other kinds of thermocouples such as *Nickel–cobalt/alumel* and *Nickel–iron/nickel* display negligibly small e.m.f. values in the temperature range from 0 °C to 200 °C. Consequently no stabilization or compensation of changing reference junction temperature is demanded. They can be used up to about 1000 °C with a sensitivity above 200 °C which is similar to the type-K thermocouple.

Comprehensive listings of calibration data for the standardized thermocouples appear at the end of the book in Table II to Table XI. Tolerances upon their output values are given in Table XII. Table XIII presents power series expansions and polynomials used for simulating the output signals of standardized thermocouples in computer applications. The resultant errors, while using these, are less than the least significant digit in the corresponding calibration tables.

Properties of metals and alloys for thermocouples are given in Table 4.2.

Characteristic data of standardized thermocouple wires is given in Table XV. These data are for the most commonly used wire diameters and also quote the maximum permissible working temperatures in pure air.

As only Fe and Ni are ferromagnetic, this property may be used to assist with the identification of thermocouples whose identifying data has been lost. Another way to identify the polarity of the wires in such an event is based upon the difference in their hardness. For the most common types this is given below.

Thermocouple	Positive wire	Negative wire
Pt10Rh–Pt	Hard	Soft
NiCr–NiAl	Hard	Soft
Fe–CuNi	Hard	Soft

4.2.3 Measuring junctions

There is a large number of sources of information about the measuring junction of a thermocouple. Typical contributions appear in Baker *et al.* (1953), Considine *et al.* (1957), Kortvelyessy (1987) and Zanstra (1975). Such a junction, which is placed at the location whose temperature is to be measured, is the joining point of the two dissimilar conductors comprising it. Good electrical contact must be ensured between the materials at the time of fabrication. A low electrical resistance and high mechanical strength are also essential. Junctions may be formed by brazing, soft and hard soldering, welding, such as by spot, flame and arc, and also by twisting and rolling as illustrated in Figure 4.11.

In the case of brazed and soldered junctions the presence of a third metal between the two thermocouple conductors, does not influence the resulting e.m.f. as long as the whole junction has a uniform temperature. This is based on the law of the third metal given in § 4.1.2.

Electric-arc welding, without the use of welding flux, is preferable for rare metal thermocouples. The a.c. electric arc should be formed between two graphite electrodes. Conductors, which should be free from any obvious mechanical defects such as indentations, kinks and non-uniformity of diameter, must also be clean. If cleanliness is achieved, as a result of previous degreasing and other precautions, poisoning of the resulting junction will be prevented. Before welding, both conductors should be twisted together to form about two turns. When the two wires are introduced into the arc for fabrication of the junction, carbonizing and thus poisoning could occur. This will be avoided if it is ensured that there is no direct physical contact between the thermocouple materials and the electrodes. A correctly formed junction should have the shape of a spherical bead with a smooth surface. Overheating of the junction may cause surface porosity.

A non-reducing, oxy-acetylene flame is the preferred welding environment for the type-J, Fe–CuNi, thermocouple. Both thermocouple wires should be cleaned by rubbing them with fine emery paper followed by twisting them together to form two to three turns. Before welding, both ends have to be heated to a red colour in a neutral flame, succeeded by immersion in borax. That conductor with the highest melting

Table 4.2 Physical properties of thermocouple metals and alloys.

Property	Unit	Thermocouple conductor					
		Cu	CuNi	Fe	NiAl	NiCr	Pt
Chemical composition			55%Cu, 45%Ni		95%Ni, 2%Al 2%Mn, 1%Si	85%Ni, 12%Cr	
Density	g/cm^3	8.9 [1]	8.85 [1]	7.85 [1]	8.7 [1]	8.55 [1]	21.4 [1]
Resistivity	10^{-6} Ωm	0.017 [4]	0.49–0.51 [4]	0.11–0.13 [4]	0.20–0.25 [4]	0.70–0.80 [4]	0.107 [4]
Thermal coefficient of resistance	$\times 10^{-3}$/°C	4.3 [4]	0.05 [4]	9* [4]	1.8* [4]	0.25* [4]	3.1 [4]
Mean value in range	°C	20–600	20–600	20–600	20–1000	20–1000	20–1600
Thermal conductivity	W/m °C	389 at 20 °C 356 at 500 °C	41.8 0–300 °C	75.3 at 20 °C 33.5 at 800 °C	58.6 20–700 °C	12.5 0–300 °C	79.4 at 20 °C
Specific heat	J/kg °C	481 at 20 °C 435 at 500 °C	418.7 0–300 °C	461 at 20 °C 712 at 800 °C	544 20–400 °C	418.7 0–300 °C	134.0 at 0 °C
Coefficient of linear thermal expansion	$\times 10^{-6}$/°C	1.8	16.8	14.6	16	15.7	9.3
Mean value in range	°C	20–600	20–600	20–600	20–600	20–600	20–800
Melting temperature	°C	1083 [3]	1270 [3]	1536 [3]	1400 [2]	1430 [2]	1769 [2]
Magnetic properties		None	None	Yes	Yes	None	None

continued

Table 4.2 *continued.*

Property	Unit	Thermocouple conductor							
		Pt6Rh	Pt10Rh	Pt13Rh	Pt30Rh	Kanthal P	Kanthal N	Chromel	Alumel
Chemical composition		94%Pt, 6%Rh	90%Pt 10%Rh	87%Pt 13%Rh	70%Pt 30%Rh	90%Ni 10%Cr	98–97%Ni 2–3%Si	90%Ni, 10%Cr	95%Ni 2%Mn, 2%Al
Density	g/cm^3	20.5 [3]	20 [1]	19.55 [3]	17.6 [2]	8.68 [5]	8.69 [5]	8.73	8.60
Resitivity	10^{-6} Ωm	0.185 [4]	0.20 [4]	0.189 [3]	0.21 [4]	0.69 [5]	0.22 [5]	0.72 [3]	0.29 [3]
Thermal coefficient of resistance	$\times 10^{-3}$/°C	1.8 [4]	1.4 [4]	1.6	1.3 [4]	0.5 [5]	0.3 [5]	3.2	1.88
Mean value in range	°C	0–100	20–1600	0–100	0–100	20–100	20–100	0–100	0–100
Thermal conductivity	W/m °C		30.2 at 20 °C			23.8 at 100 °C	33.3 at 100 °C	21.9 at 100 °C	33.8 at 100 °C
Specific heat	J/kg °C		146.5 at 0 °C			128 at 20 °C	139 at 20 °C	124.5 at 20 °C	144.8 at 20 °C
Coefficient of linear thermal expansion	$\times 10^{-6}$/°C		9.0			14	13.5	13.1	12
Mean value in range	°C		20–800			20–100	20–100	20–100	20–100
Melting temperature	°C	1810 [2]	1830 [2]	1840 [3]	1910 [2]	1430 [2]	1410 [2]	1430 [3]	1400 [3]
Magnetic properties		None	None	None	None	None	Yes	None	Yes

*Approximate values.
Sources: [1]DIN 43712; [2]Burns and Hurst (1972); [3]Omega Engineering catalogue; [4]Westhoff (1965); [5]Kanthal AB catalogue.

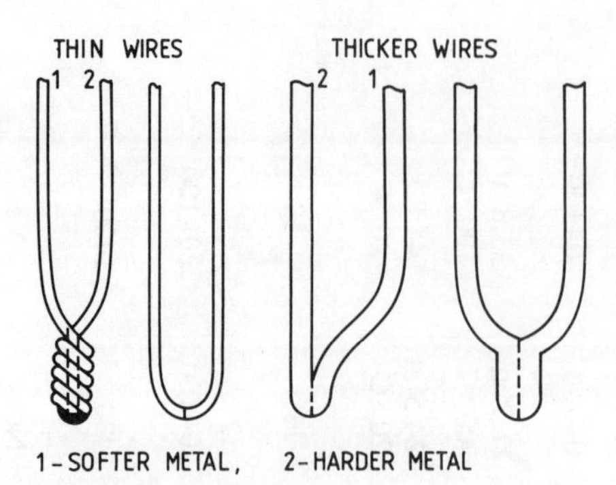

Figure 4.11
Thermocouple junctions.

point should be heated first during welding. After welding, residual flux has to be removed by washing. It is also possible to manufacture this type of thermocouple using either electric-arc welding or hard soldering. In the case of electric-arc welding the thermocouple acts as one electrode whilst the other is graphite. The hard-soldering fabrication method is only suitable for thermocouples to be applied in measuring temperatures not exceeding 600 °C.

Type-K, NiCr–NiAl, thermocouples may be manufactured using the same process as for the Fe–CuNi types described above. Spot or contact welding may also be used.

Soft or hard soldering is suitable for type-T, Cu–CuNi thermocouples, depending upon maximum temperature. Units for use up to 600 °C may be hard soldered whereas those for up to 200 °C may be soft soldered. Arc-welding with no flux may also be used. Typical hard and soft solders and their corresponding properties are given in Table XVI.

In some instances the need arises to lengthen the thermocouple conductors. This is particularly the case in rare-metal units where even very short lengths must be used to keep the unit cost reasonable. To avoid the generation of parasitic e.m.fs, it is recommended that these pieces be inserted at those parts of the structure furthest from large temperature gradients. This implies as far away from the measuring junction as possible.

After welding, all thermocouples should be stress-relieved to stabilize their properties and neutralize any inhomogeneities. These may arise from the welding formation of junctions, in the further steps of the fabrication process or as a result of transportation. There are many different opinions concerning the best manner of stress-relieving. Nevertheless, it is commonly accepted that the thermocouple should be maintained at its highest permissible working temperature for some hours.

4.3 THERMOCOUPLE SENSORS

4.3.1 Construction

A thermocouple sensor is a structure consisting of a thermocouple, which has associated wires, their insulation and terminal head, mounted in a suitable protective tube, sometimes called a well or sheath. A thermocouple sensor of this type is almost always referred to simply as a thermocouple.

The most common industrial thermocouple assemblies are general purpose (straight) thermocouples, for application at atmospheric pressure. As shown in Figure 4.12, its leads are insulated with ceramic tubes or beads. Adequate protection from chemical and mechanical influences is provided by a tubular, metallic or ceramic sheath, which is fastened directly at the terminal head comprising a terminal block and connecting screws. The terminal head is also equipped with a tight sealing gland having a rubber grommet, through which the compensating or connecting cable enters the head. If the sheath is ceramic, it is also normal to use an additional head adapter. This adapter is fastened by a flange to say the wall of a furnace. Double thermocouple assemblies in a single sheath are also available. For example, they find application where simultaneous temperature measurement and temperature control are needed.

Standardized wire diameters for thermocouples (see Table XV) may vary between 0.2 mm up to 4 mm for base metal components and from 0.1 mm to 0.5 mm for rare metal units. When choosing a particular diameter it is necessary to take account of the mechanical strength, temperature tolerance, electrical resistance and thermal inertia of the wire. Thermocouple wires should be protected from bending and indentation. They should also be kept clean to prevent the generation of parasitic e.m.fs.

Ceramic wire insulators insulate the conductors from each other and from the metal sheath. They are produced as insulating beads or as insulating tubes with one, two, four or more bores. An ability to withstand the highest working temperature of the thermocouple without losing their insulating properties or poisoning the thermocouple, is an essential requirement of these insulators. More information about different insulating materials can be found in §4.3.2.

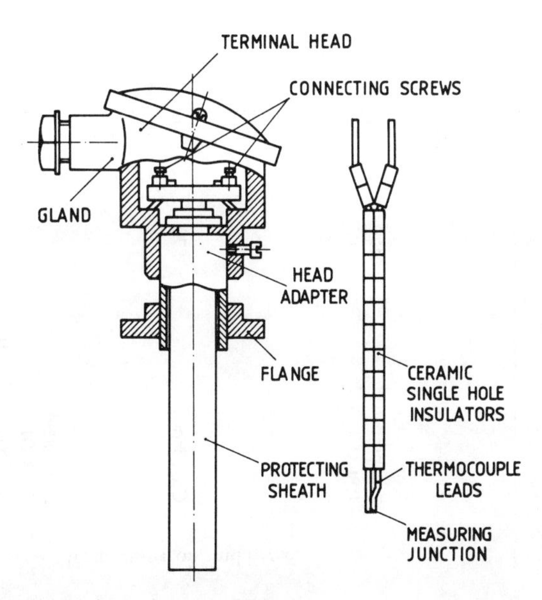

Figure 4.12
Industrial thermocouple assembly.

Table 4.3
Properties, applications and operating characteristics of materials for protective sheaths and thermowells for temperature sensors.

No	Material: General group	Material: Specific type	Maximum operating temperature in air (°C)	Resistance to: Oxidizing atmospheres	Resistance to: Reducing atmospheres	Resistance to: Sulphur-bearing atmospheres	Resistance to: Gases with N_2 low in O_2	Resistance to: Carburization	Applications and operating conditions, Temperatures in °C: Gases and steams	Applications and operating conditions, Temperatures in °C: Liquids, molten metals and salts
1	2	3	4	5	6	7	8	9	10	11
1	Non-ferrous metals	Copper	150	f	g	p			Steam, light duty environments	
2	Non-ferrous metals	Brass	*	p	f	p				
3	Non-ferrous metals	70/30 Copper-nickel to BS 2871: Pt 1–3, 1971/72	400						Corrosive environments especially under wet conditions	
4	Iron, cast iron, mild steel	Iron (technically pure)	550	p	p	p	f			Saltpetre baths up to 550 Cyanide baths up to 950 Magnesium, aluminium containing magnesium up to 700 Zinc up to 600
5	Iron, cast iron, mild steel	Cast iron	870	p	f	p			Reducing atmospheres up to 870	Bearing metal up to 600 Lead up to 600 Aluminium up to 700 Concentrated sulphuric and caustic solutions
6	Iron, cast iron, mild steel	Mild steel BS 3601 BS 3602 ST 410 AP15L Grade B ASTM106 Grade B A53 Grade B DIN 1629 ST35	550	p	p	p	f		Drying kilns Low temperature ovens	Oil baths Saltpetre baths up to 550 Cyanide baths up to 950 Bearing metal up to 600 Lead up to 700 Zinc up to 480

7	Stainless steel	316 Stainless steel BS 970: Pt 4, 1970 BS 3605 316 AISI T316	800	g	f	f			Steam up to 540 Drying kilns Low temperature ovens	Water up to 370 Oil baths Chemical solutions (nitric, phosphoric and acetic acid)
8		310 Stainless steel BS 970: Pt 4, 1970 BS 3605 310 AISI T310	1100	g	f	f			Sulphurous atmospheres	
9	Nickel chromium iron alloy	BS 3074, 1974 Grade NA14 ASTM B167 ASTM SB167 DIN NiCrl5Fe {Trade name Inconel 600}	1100	g	f	p	g	g	Carburizing up to 950 Nitriding up to 1000 Normalizing up to 1000 Case hardening up to 1000 General heat treatment furnaces and ovens	Annealing baths up to 600 Cyanide baths up to 1000 Lead up to 950
10	Iron nickel chromium alloy	BS 3074, 1974 Grade NAl5 ASTM B163 B407 ASME SB 1635SB407 DIN X10NiCrAlTi 3220 {Trade name Incoloy 800}	1100	g	f	g	g	g	Carburizing up to 950 Nitriding up to 1000 Normalizing up to 1000 Case hardening up to 1000 Sulphur-bearing atmospheres	Annealing salt baths up to 600 Cyanide baths up to 1000 Lead up to 950
11	Chromium iron alloy	ASTM TP446 ASTM TP446 AISI 446 DIN X18CrNi28	1150	vg	g	vg	g	f	Oil firing Annealing ovens up to 950 Billet heating ovens up to 950 Kiln flues up to 950 Hardening furnaces up to 1100 Sulphur-bearing atmosphere	Salt baths (neutral) up to 750 Salt baths (high speed) up to 1250

*, Composition dependent; vg, very good; g, good; f, fair; p, poor

continued over the page

Table 4.3 (*continued*)

1	2	3	4	Structure	Resistance to thermal shock	Mechanical strength	10	11
12	Ceramics and others	Impervious aluminous porcelain	1400	Impervious	Excellent	High	Pottery kilns Hardening furnaces	Highly resistant to fluxes and slag attack
13		Impervious recrystallized alumina	1800	Impervious	Fair	High	Reducing and carbonaceous atmospheres Alkaline and other fluxes Highly resis-tant to chemicals	Glass tanks up to 1450 Glass feeder up to 1300 Aluminium up to 700 Zinc up to 600 Copper up to 1250
14		Impervious mullite	1600	Impervious	Good	Very high	Sulphurous and carbona-ceous atmospheres. Kiln furnaces	Highly resistant to flux attack Aluminium up to 700 Copper brazing up to 1150
15		Silicon carbide SiC	1450	Porous	Excellent	High	Reducing atmospheres Billet heating up to 1350 Forge furnaces up to 1300 Highly resistant to flames	Aluminium Copper Zinc Slag
16		Silicon nitride SiN		Impervious	High	High but not shock resistant		Aliuminium Zinc
17		Ceramic-metals (cermets)	1400–2200	Little porous or impervious	High or very high	High		Various metals
18		Graphite	2300					Aluminium up to 700

Sources: Catalogues of TC Ltd, Leeds & Northrup Int., Omega Engineering Inc., Zysk and Robertson (1972), British Standards, DIN Standards.

4.3.2 Sheath materials and their properties

Protecting sheaths are used to protect the thermocouple against chemical or mechanical influences. The following properties have to be considered in choosing the sheath material:

- resistance to high temperatures,
- mechanical strength,
- corrosion resistance,
- chemical resistance,
- non-contamination of the given thermocouple,
- resistance to thermal shocks,
- imperviousness to gases,
- specific density, thermal conductivity and specific heat.

As the sheath material influences the constancy of calibration, the length of service and the dynamic properties of the thermocouple, it is important to make the right choice of material. Typical sheath materials for temperature sensors, which are presented in Table 4.3 may be mainly metallic or ceramic.

Metal sheaths provide good protection for all base metal thermocouples used at temperatures up to 1100 °C. As rare metal thermocouples are liable to be contaminated by metal vapours, an additional inner ceramic sheath is necessary. The application ranges of metal sheaths can be considerably increased by applying an additional protecting external layer. For example the Kawecki Chemical Co. (USA) produces tantalum sheaths protected by a plasma deposited layer of zirconium diboride. These sheaths can stand in molten aluminium at temperatures around 760 °C up to 14 000 hours. Metal protection tubes used for temperature measurement in flowing media, have to stand not only the static pressure but also dynamic forces, causing vibrations. Conically shaped wells are therefore preferred to cylindrical ones, especially at higher flow velocities. Protecting metallic sheaths intended for operation in air at atmospheric pressure, can be manufactured either as seam welded or seamless tubes with welded bottom. For temperature measurement in molten metals or salts, as well as for use under high pressures, a type of well, which is machined from solid bars, will be necessary. When long sheaths are required, savings in the quantity of expensive sheath materials used may be accomplished by using cheaper materials in those regions of the long sheath outside the high temperature zone.

Ceramic sheaths, used for temperatures over 1100 °C, do not possess such good resistance to mechanical and thermal influences as metal sheaths. Usually, the ceramic material is impervious to gases. As they are susceptible to thermal shocks, they require some protection. This may be adequately provided by a second sheath with good resistance to thermal shock.

Porous ceramics can only be used as primary sheaths, which enclose a secondary impervious tube. An example of a regularly applied combination is an impervious mullite sheath protected on the outside by a silicon carbide sheath. Mullite is suitable for use with rare metal thermocouples at high temperatures. In the case of impervious aluminous porcelain the use of a double sheath is also recommended. Recrystallized

Table 4.4
Industry standard thermocouple sensors.

Profile classification		Materials for sheath types	
Profile outlines	Profile features	Principal sheath structures	Specific materials (Table 4.3)
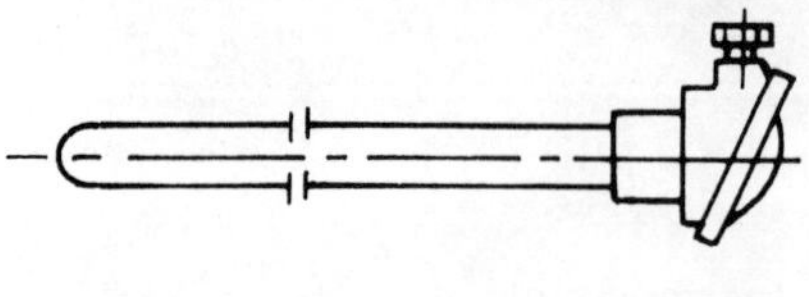	Straight sheathed	Metal primary no secondary	6, 7, 9, 10, 11
	Straight sheathed	Metal primary Ceramic secondary	9, 10, 11 14
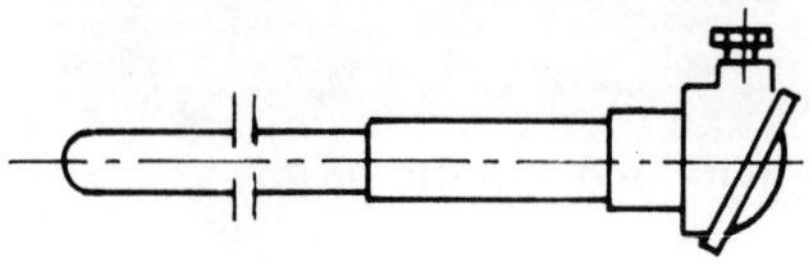	Straight sheathed	Ceramic primary no secondary	12, 13, 14, 15
	Straight sheathed	Ceramic primary Ceramic secondary	12, 13, 14, 15 12, 13, 14
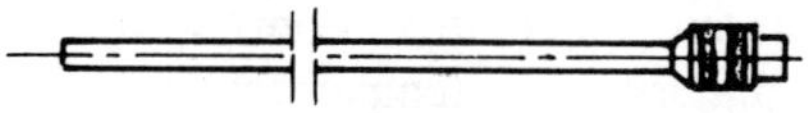	Flexible	Metal primary Mineral insulated	3, 6, 8, 9, 10, 11
	Uniform	Fe-Sheath acts as plus conductor	4
	Uniform	Metal sheathed	6, 9, 11
	Angle sheathed		
	Joined	Metal sheathed	4, 5, 6, 7, 9, 10, 11
	Joined	Ceramic or graphite sheath	12, 13, 14, 15, 16, 17, 18
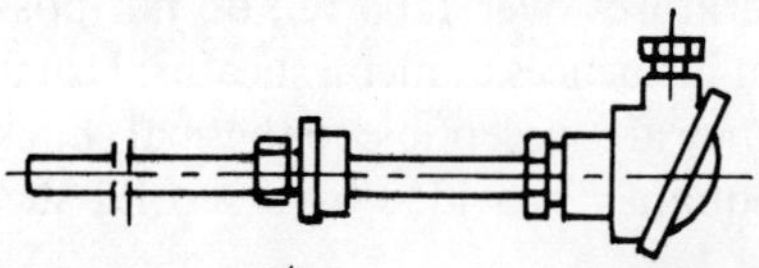	Thermowells	Straight drilled or welded thermowell, with thermocouple insert	2, 6, 7, 9, 10
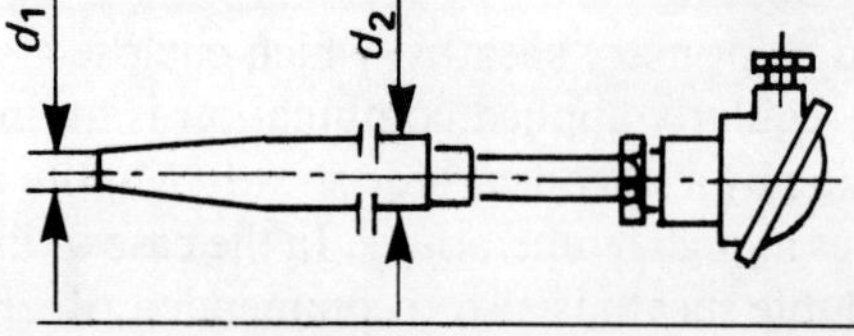 d_1 d_2	Thermowells	Conical drilled thermowell, with thermocouple insert	2, 6, 7, 9, 10

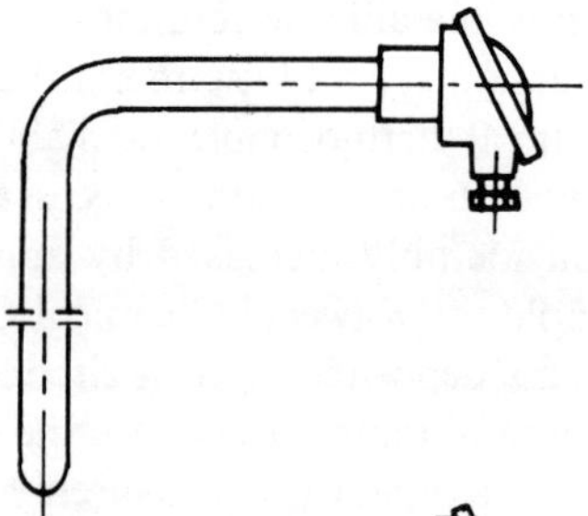

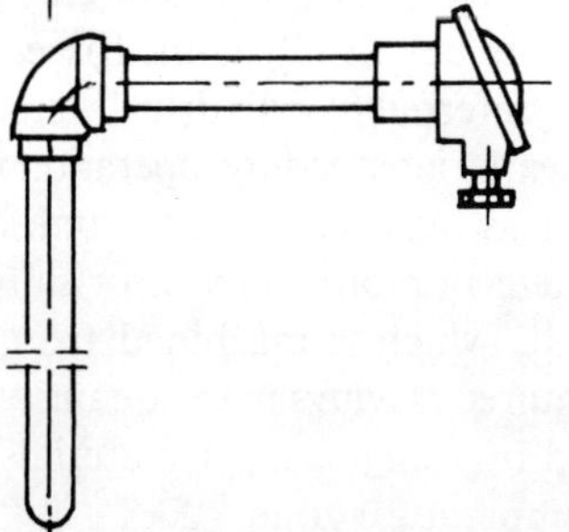

Dimension (mm)			Thermocouples		Applications
Immersion length	Total length	Diameter	Descriptions given in Section 4.2.2	Operating pressure	Industries, installations, processes
200–2000	200–2000	10–25	K, J, T, E. N	Normal	Heat treatment furnaces and ovens Open containers and ducts
300–1500	300–1500	20–30	R, S, B		
300–1500	300–1500	10–30	K, R, S, B, N		
300–1500	300–1500	20–50	R, S, B		
Any length 10 mm–200 m		0.25–6	K, J, T, E, N R, S, B, W-WRe		Chemical installations Nuclear power stations
300–1000	300–1500	15–25	J		Salt, oil, and other baths Electrode ovens Metal melting ovens
150–1000	300–3000	15–25	K, J		
300–3000	600–4000	15–30	K, J		
300–1000	600–2000	15–50	K, R, S, B		
100–400	200–600	6–20	K, J, T, E	Medium	Tanks and pipe-lines
60–300	200–500	$d_2 = 15–25$ $d_1 = (0.4–0.7)d_2$	K, J, T, E	High	

alumina, which is impervious, can be used as a single sheath, and also for insulating wires. As mutual reaction may occur between the sheath material and the insulation at high temperature, they must be correctly chosen. In principle, it is thus advisable to make the wire insulation and the protection sheath from the same material.

Cermets or ceramic-metals are regarded as the best insulating materials since they possess a combination of the best properties of both ceramics and metals. They are characterized by high mechanical strength, high resistance to thermal shock and high operating temperatures. As they are mostly impervious they do not need any protective sheath. For example, the external sheath of molybdenum and zirconium from Metalwerk Plansee (Austria) with an Al_2O_3 tube inside have been successfully used in molten steel. Protective sheaths, based upon cermets with a constitution of chromium and Al_2O_3 from Union Carbide Corporation (USA) are impervious, hard, abrasion resistant and have a thermal conductivity equal to that of stainless steel. They are recommended for measuring the temperature of molten copper up to 1150 °C, and liquid aluminium and sulphur containing gases up to 1095 °C. They also find application in chemical reactors up to 1375 °C. The proprietary cermet, called Metamic, from Morgan Refractories Ltd (USA), is composed of molybdenum and aluminium oxide. When used in the temperature measurement of liquid steel up to 1900 °C, a second internal sheath of Al_2O_3 must be added to protect the thermocouple from poisoning. Cermets of chromium, Al_2O_3 and SiO_2, which can be used up to 1450 °C, necessarily require an additional inner sheath when used with rare metal thermocouples.

4.3.3 *Review of thermocouple assembly structures*

A wide range of constructional forms, which are typical for industrial temperature sensors, is shown in Table 4.4. The table includes the dimensions, sheath material and applications for each of the thermocouple structures given.

The most popular industrial, *straight thermocouples* have already been described in §4.3.1 with representative illustration in Figure 4.12. Figure 4.13(a) demonstrates that they can be mounted on the wall of the furnace either by flanges or special mounting fittings. Sometimes, in furnaces with a protecting gas atmosphere, they must be mounted using gas-tight methods (Figure 4.13(b)).

Right-angle (corner) thermocouples are used for measuring the temperature of molten metals or in salt baths. Their shape ensures that their terminal head is not placed directly over the high temperature medium. In this manner it is protected against both high temperature radiation and any aggressive vapours. Their flanges, or other installation fittings, are mounted, so that they may be suitably moved, either on the vertical part to be immersed or on the horizontal supporting part. Protecting sheaths for this type of unit are mostly constructed using two different materials. The horizontal supporting part is made of constructional steel, while the immersible vertical part is made of a material capable of withstanding immersion in the molten medium. Uniform, one-material sheaths are often built of pure iron, which constitutes the positive conductor of an iron–constantan (Fe–CuNi) thermocouple. The constantan lead is then welded into the bottom of the iron sheath giving the sensor a very low

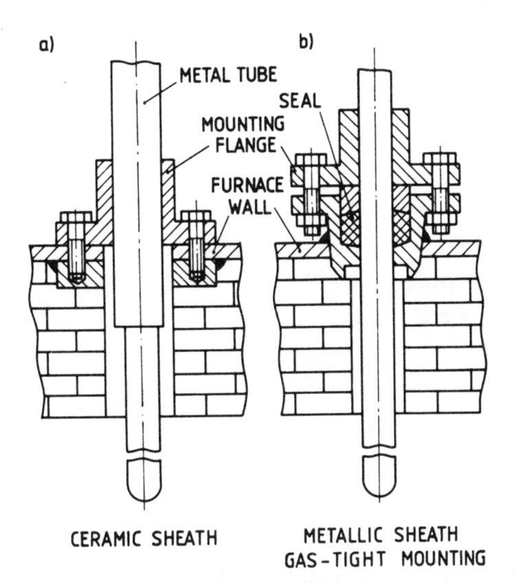

Figure 4.13
Thermocouple mounting in a furnace.

thermal inertia. Although uniform thermocouple sensors are more expensive, they possess the advantage of a longer working lifetime.

High-pressure thermocouple sensors, shown in Figure 4.14, are composed of a so-called thermocouple insert placed in a thermocouple well. Thermocouple inserts, like those manufactured by TC Ltd for example, normally contain a short mineral insulated (MI) metal sheathed thermocouple attached to a standard terminal block. Inserts, which are now popular easily exchanged replacement sub-assemblies, are pressed into the thermocouple well and fitted with standard terminal heads. As they are spring loaded, they have improved dynamic behaviour with good resistance to

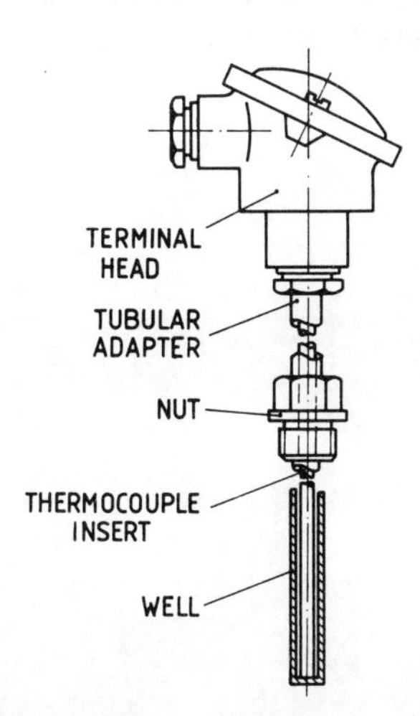

Figure 4.14
High-pressure thermocouple.

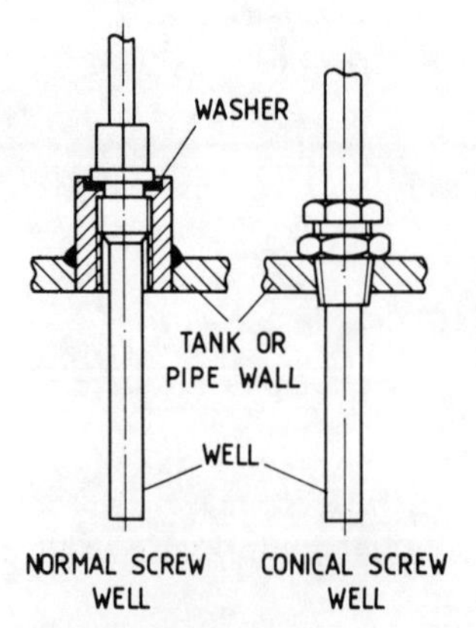

Figure 4.15
Screw-in thermocouples.

vibration. The permissible temperature, pressure and flow velocity in different media, all of which depend upon the constructional details, dimensions and well material, may be determined using special diagrams such as those in DIN 43763.

Cylindrical screw-in thermocouples are normally mounted in tank or pipe-line walls using welded fixing attachments similar to the representation of Figure 4.15(a). The washer secures the tightness of the connections which may be either straight or conical screw connections (Figure 4.15(b)), made directly in the tank wall.

Conical well thermocouples, which are machined from solid bars in a variety of materials (Figure 4.16), may be used for high-pressure measurements. They can be welded to or screwed into the walls.

Special type thermocouples are manufactured for different applications. They find use in measuring the temperature of media such as gases or of solid surfaces in for example nuclear power plants as well as in locations such as foundries, laboratories and so on. One example of special thermocouples is the miniature, short thermocouple

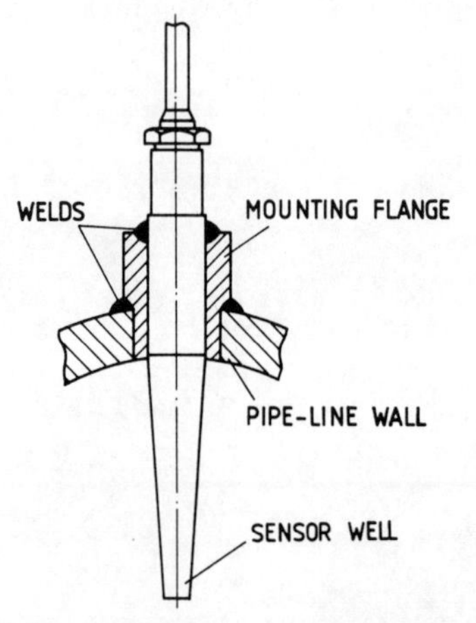

Figure 4.16
Thermocouple in a conical well machined from solid bar.

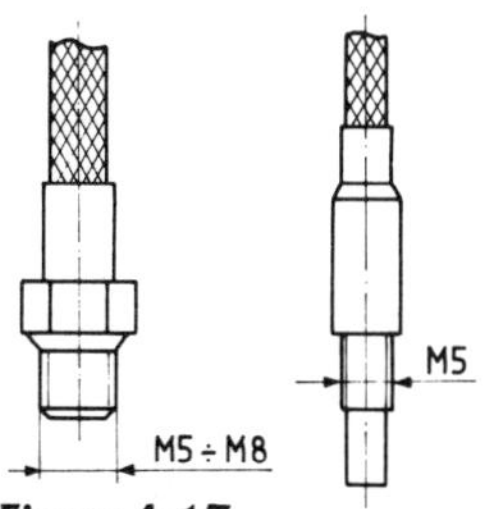

Figure 4.17
Short, miniature thermocouples for screwing in.

shown in Figure 4.17. Their screw mounting does not compromise applications which may require small penetration depth.

Mineral insulated (MI) thermocouples, which have a metal sheath, were originally designed in 1957 for temperature measurement in gas-cooled nuclear reactors. As they have important properties, they are now widely applied in almost all other locations where their properties may be of benefit. A considerable advantage is their thin, robust flexible structure which permits easy bending and twisting. Their high resistance to corrosion and vibration, which is specified in DIN 43721 and considered by Barzanty and Hans (1979), Bliss (1972) and Thomson and Fenton (1975), combined with a short response time, has made them indispensable in both research and industry. Bare unsheathed thermocouples are being increasingly replaced by them.

Figure 4.18 illustrates different forms of mineral insulated thermocouples. A continuous protective metallic sheath contains its wires, which are placed in a densely compacted mineral insulating powder. Variations may contain single, double, as shown respectively in Figures 4.18(a) and 4.18(b), or sometimes triple MI thermocouples. The sheaths may have diameters of between 0.2 mm and 8 mm while they may be in lengths as long as hundreds of metres.

Concentric MI thermocouples, which are shown in Figure 4.18(c) and Figure 4.19(e), have the terminals on both sheath ends. They have the following advantages when compared with the normal types. As their insulation layer is thicker they possess a

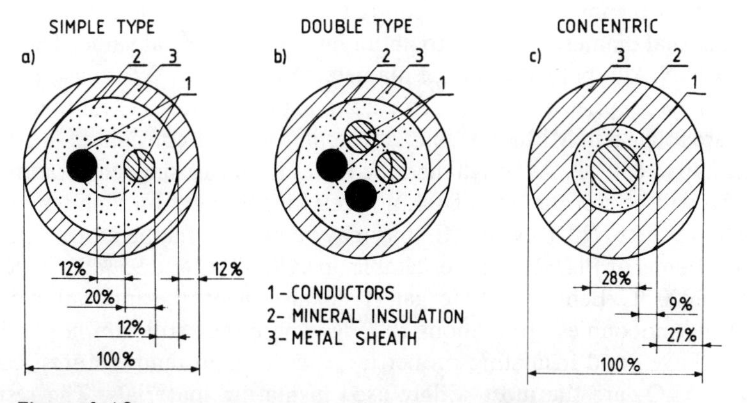

Figure 4.18
Mineral insulated thermocouples, enlarged cross-section (Thomson & Fenton, 1975; Weichert and others, 1976). Typical dimension ratios in percent.

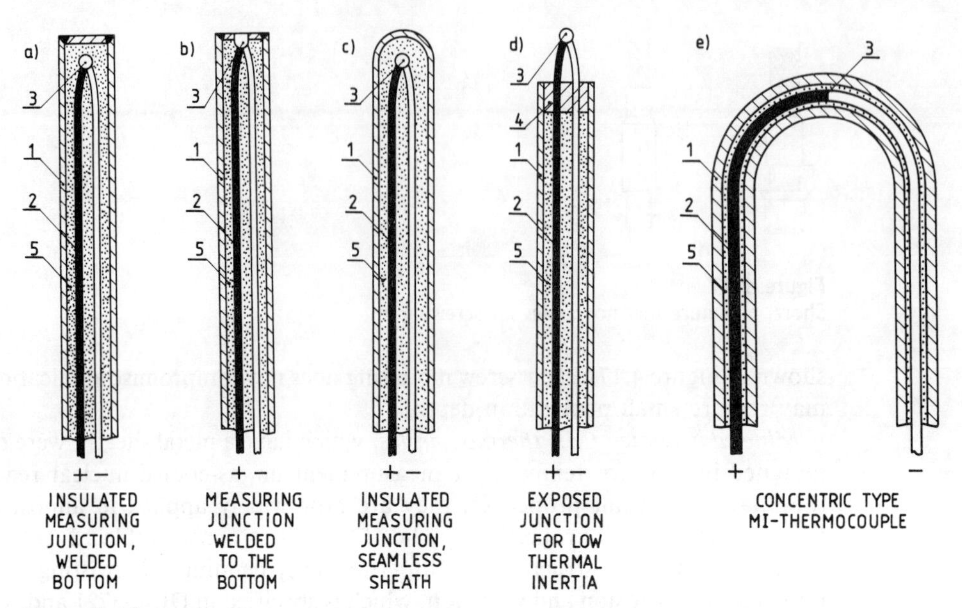

Figure 4.19
Types of MI thermocouples. 1: metal sheath; 2: conductors; 3: measuring junction; 4: refractory cement; 5: mineral insulation.

higher electrical resistance between the sheath and the conductors. Longer working life and a higher resistance to corrosion is ensured by the thicker walls of the sheath. This is also responsible for increased thermal conductivity and electrical resistance, due to a higher pressure on the insulation layer. Smaller physical size is accompanied by two important benefits. Firstly, there is greater ease in hermetically sealing the ends. In addition, the low resulting thermal inertia gives short response time. A final benefit of this form of MI thermocouple, pointed out by Thomson and Fenton (1975), is the possibility which it affords for the provision of two measuring junctions for the measurement of temperature difference.

If very short response times are required, the end part of the thermocouple may have its external diameter reduced to about half of the normal value. Thermocouples of the NiCr–NiAl type, with outside diameters in excess of 0.5 mm, are suitable for this class as the proprietary Thermocoax brand manufactured by Philips demonstrates.

There are four main MI thermocouple types in use. Iron-constantan (Fe–CuNi) may be applied up to 850 °C. Although thermocouples for measuring temperatures in excess of 1000 °C will be considered in §4.4, it is relevant to mention high temperature MI thermocouples here. NiCr–NiAl, MI units may be used for temperatures up to 1200 °C, whereas MI units of Pt10Rh–Pt are suitable up to 1500 °C and 95%W,5%Re–74%W, 26%Re to 2000 °C. Bentley and Morgan (1986) also report the application of Nicrosil–Nisil MI thermocouples up to about 1000 °C where they still exhibit good stability.

As they have good insulating properties, even at high temperatures, compressed MgO and Al_2O_3 are the most widely used insulating materials. The resistance of MgO between the thermocouple and the metal protective sheath (Figure 4.20) may be over $10^{12}\Omega/\mathrm{m}$ at 20 °C with the ability to withstand some 250 V. Grain sizes for

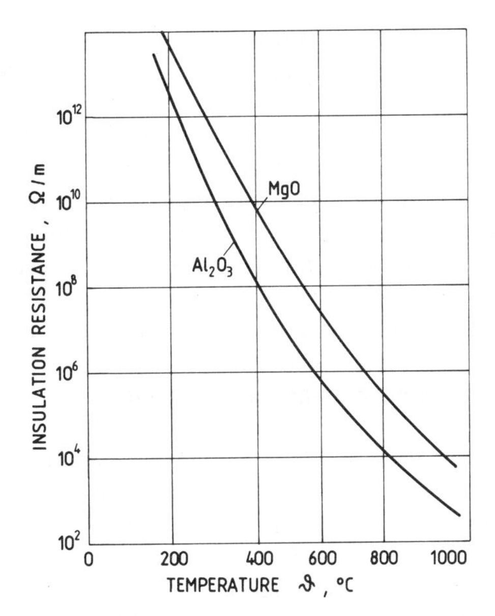

Figure 4.20
Insulation resistance between conductors and sheath versus temperature for thermocoax MI thermocouples. (Courtesy of Philips Ind. Automation, Pye Unicam Ltd, UK.)

insulation need to be in the order of 1 μm. As the hardness of Al_2O_3 makes working to this grain size difficult it is not so popular. In exceptional circumstances, when the temperatures to be measured are up to 2500 °C, BeO and ThO_2 are used as the insulating medium.

Moisture causes a reduction of the resistance of insulating materials as well as exerting an accelerating influence on wire corrosion. Thus, as Kinzie (1973) states, the insulation of all MI thermocouples should be protected from it. Even short-term exposure, for as little time as some minutes, to humid atmospheres may considerably shorten the life-time of MI thermocouples. It is also impracticable to attempt to dry the insulation after such exposure. For this reason MI thermocouples have both terminations hermetically sealed to prevent the ingression of undesirable moisture.

Metal sheaths for use in MI thermocouples may be made of Cr, Ni and Ti stabilized stainless steel for temperatures up to 800 °C. Temperatures up to about 1100 °C require inconel sheaths whilst Nicrosil may be used for Nicrosil–Nisil thermocouples up to 1250 °C. At higher temperatures Pt, Pt10Rh, Ni, Ta, Rh and Mo are applied. Additional protection of metal sheaths may be arranged when thin outer layers of Cr or Al_2O_3 are deposited using the plasma method. Alternatively increased thermocouple lifetime is achieved by using two-layer metal sheaths.

The minimum bending radius is usually 10 to 12 times the sheath diameter. Some producers, such as TC Ltd, permit the bending radius of MI thermocouples to be as small as four times sheath diameter, if carefully handled.

In the process of manufacture the diameter of the thermocouple sheath is reduced by drawing. Simultaneously, this also proportionately reduces the diameter of the conductors. To ensure that the conductors maintain their concentric position within the sheath, two, four or six bore ceramic insulating tubes are used. During the drawing process, the insulating tubes are crushed and compressed. A commensurate increase in their thermal conductivity takes place. Any reduction in diameter should not be greater than about 20 to 1. Exceeding this ratio causes wide non-uniformity of conductor cross-section and also a 'spreading' or distortion of the measuring junction. Concentric MI thermocouple units are much more easily drawn and have a longer life.

MI thermocouples may be applied for innumerable purposes. Examples occur in nuclear reactors, in industries for the manufacture and production of automobiles, aircraft, glass and chemicals, in medicine, in research laboratories of all kinds, and so on. Typical proportions of the components of MI thermocouples are given in Figure 4.18. The link between the thermoelectric wires and extension cables is accomplished with a range of different plug-and-socket connectors as shown in Figure 4.21.

Needle thermocouples are made, for example, of enamel insulated thin Cu–CuNi thermocouple soldered on the end of a stainless steel tube. The thermocouple is connected to extension wires inside the head as shown in Figure 4.22. This type of unit may also be made from NiCr tube serving as a positive conductor and an insulated CuNi wire soldered into the tube. They are used predominantly for the temperature measurement of plastic, liquid and granular substances, in medicine and research. The needles are 40 to 60 mm long and have diameters of 1 to 2 mm. They are being replaced to an increasing extent by MI units.

Laboratory thermocouples, which are widely applied in the temperature measurement of laboratory furnaces, are mostly designed to the requirements of specific customers. The demands for accuracy which are placed upon them for the measurement of both steady-state and transient thermal processes, places a limit upon their physical size. Although their application is somewhat specialized there is no need for either robustness or longevity. Figure 4.23 shows that base metal laboratory thermocouples are frequently applied without a protective sheath. Contrastingly, the rare metal laboratory units of Figure 4.24 require such a protective sheath as insurance against any influence which exposure to chemicals might cause.

Up to temperatures around 1000 °C the sheaths are made from quartz which allows the transmission of infra-red radiation to the measuring junction. In this way their dynamic properties are much improved over corresponding industrial grade units, which employ metallic protective tubes. At still higher temperatures, beyond 1000 °C,

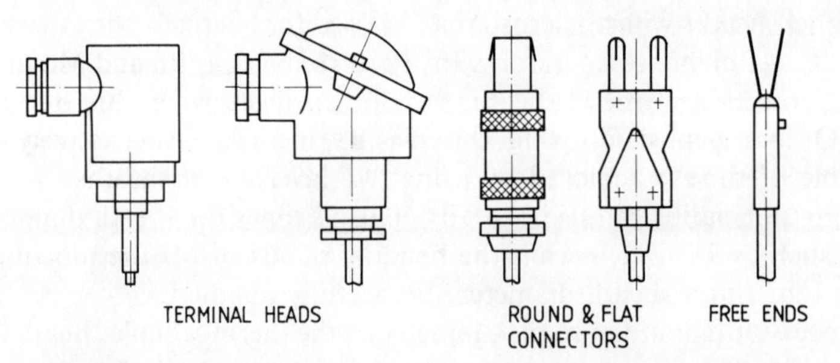

Figure 4.21
Terminals of MI thermocouples.

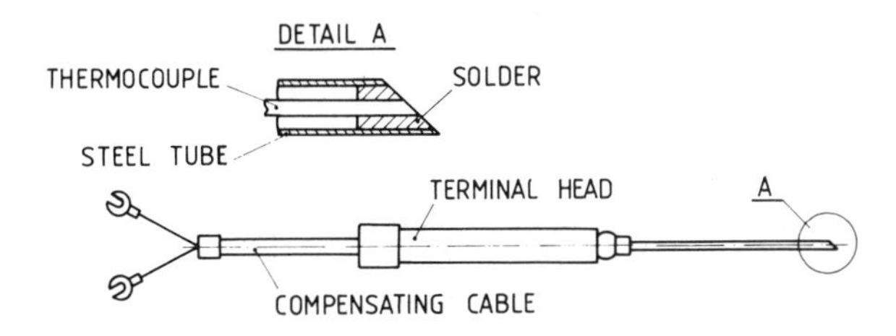

Figure 4.22
Needle thermocouple.

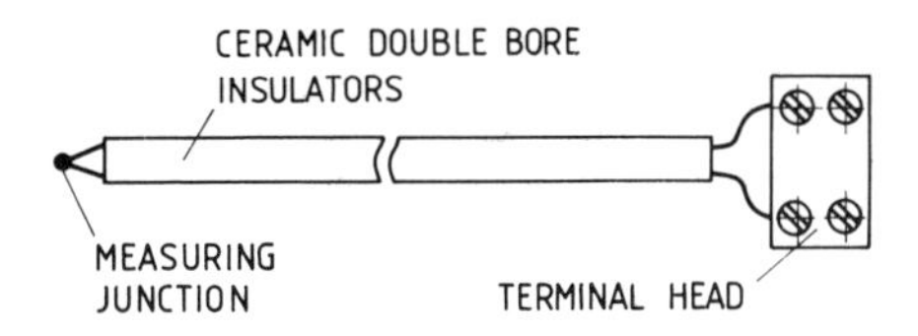

Figure 4.23
Unprotected laboratory thermocouple.

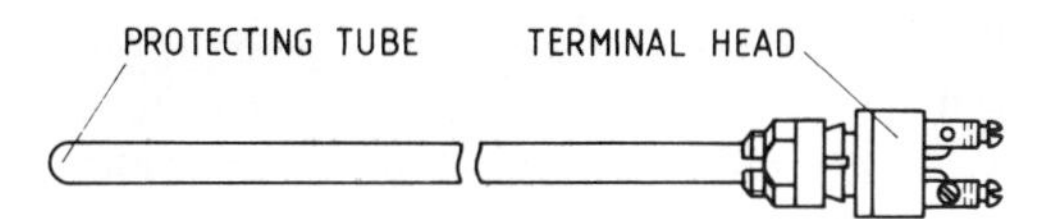

Figure 4.24
Rare-metal laboratory thermocouple.

ceramics, of the gas-impervious types detailed in Table 4.3, are used. Sheaths for this type of unit may be about 7 to 8 mm in diameter and 150 to 500 mm long.

Another type of laboratory thermocouple has an elastic constitution with insulation of PVC, polyethylene, nylon, teflon, silicon caoutchouc and glass braid. Special thermocouples for the measurement of solid surface temperature may be fastened to the solid body by screws, solder or adhesive. In spite of the higher cost, MI units are very convenient for laboratory use.

4.4 HIGH TEMPERATURE THERMOCOUPLES

At high temperatures above 1000 °C, special non-standardized metallic or non-metallic thermocouples are used. Those, which have not been mentioned already in §4.2, will now be described. The e.m.f.-temperature characteristics of the most popular high temperature thermocouples are graphically illustrated in Figure 4.25 and tabulated in Table XIV.

W–Mo and *W–MoW* thermocouples are characterized by high working temperatures and low e.m.f. values. Although W–Mo types cannot be used in oxidizing atmospheres, they are suitable for use in reducing environments up to about 2400 °C

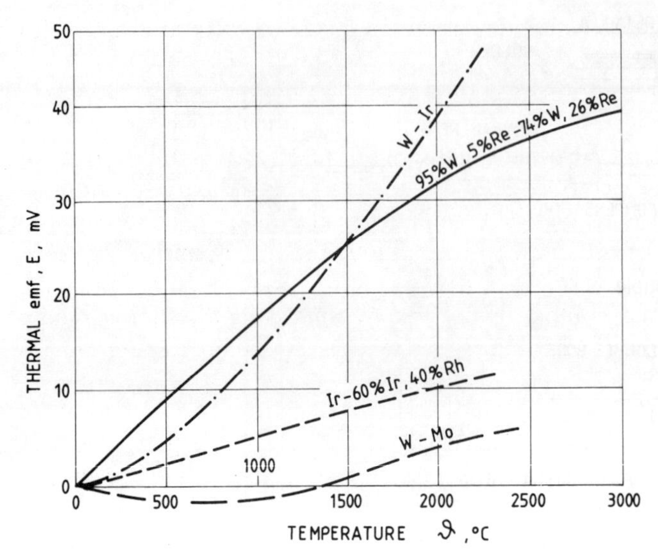

Figure 4.25
Thermal e.m.f. E of high temperature thermocouples versus temperature ϑ.

(Caldwell, 1962). As a reaction between Mo and Zr, Mg or Al_2O_3 above 1600 °C gives severe e.m.f. variations, the inner and outer insulation has to be chosen carefully. The improved resistance of W–MoW is accompanied by lower mechanical strength, which may lead to easy breaking. In addition, instability of e.m.f. values, with polarity reversal in the temperature range 1400–2600 °C, also limits their application (Caldwell, 1962; Kinzie, 1973; Lachman and McGurty, 1972).

W–WRe and *WRe–WRe* have recently become more widely employed due to the better mechanical properties of Re-alloys than W and Mo. Different alloy compositions are used such as W–74%W, 26%Re and 95%W, 5%Re–74%W, 26%Re. Although both have similar e.m.f.-temperature dependence, the latter is more frequently employed as it is more linear at lower temperatures and more ductile. Replacing the W-electrode with one of composition 95%W, 5%Re provides the opportunity to stabilize the resulting thermocouple by a short duration heating process, which causes no embrittlement of the positive conductor. Burns and Hurst (1972) have stated that this conductor may be made still more ductile if its alloy is enriched with traces of other elements which inhibit grain growth. Units of this type, that may be used up to temperatures of 2500 °C, resist poisoning in both neutral and hydrogen-containing atmospheres as well as in vacuum. However, Anderson and Bliss (1972) have pointed out that they cannot tolerate the presence of oxygen, H_2O, CO or CO_2. Further details of the properties of W and Re alloys are given by Vertogradsky and Chekhovskoy (1972).

As insulating materials such as MgO and Al_2O_3 may only be employed up to 1500 °C their use in thermocouple construction is restricted to those parts of the wire structure whose temperature never exceeds this value. Although BeO can be applied up to 2000 °C it is poisonous to staff (Anderson and Bliss, 1972). Details of external protective sheaths, which are normally made of WRe and IrRh alloys, appear in Anderson and Bliss (1972).

Typical applications of high temperature thermocouples made of W and WRe include the temperature measurement of the thermal insulation of spaceships, nuclear reactor cores and the exhaust gases of gas turbines. Thermocouples from W and Re alloys, which are considered to give the best types at high temperature units, are also manufactured as MI types as exemplified by the Philips Thermocoax proprietary form. These MI thermocouples, which, although rather expensive, are not very flexible, have wire insulation of MgO, BeO and HfO_2. A reduction in cost, and compensation for the lack of flexibility, is possible by connecting the shorter rigid part of about 50 cm to a more flexible stainless steel sheathed compensation cable, which can be used up to 870 °C.

Ir–IrRh thermocouples, which can have different proportions of Rh in their negative electrode, are utilized for temperatures up to 2000 °C. They are not only expensive but have a short lifetime, poor mechanical properties and unstable e.m.f. versus temperature characteristic. The most popular combination is the Ir–60%Ir, 40%Rh thermocouple. Its high resistance to oxidation is accompanied by the ability to withstand direct contact with the exhaust gases of gas engines for about 16 to 24 hours at temperatures of 2000 °C without any protective sheath. In normal use it is usual to provide them with the protection of a sheath of Al_2O_3 or BeO.

Graphite–W as a thermocouple, which was first described by Watson and Abrams in 1925, is further detailed in Zysk and Robertson (1972). In vacuum, it can withstand temperatures up to 1800 °C, whereas it can withstand reducing environments at temperatures up to 2500 °C. Under the influence of moisture its calibration undergoes some variations, which can be restored to its original value by annealing. As shown in Figure 4.26(a) they are constructed as a tungsten wire in a graphite tube, which serves as both the positive electrode and the sheath (Caldwell, 1962; Franks, 1962; Kinzie, 1973). They are applied in the measurement of the temperature of molten metals, and also in carborizing atmospheres. A significant feature is polarity reversal at a temperature of about 1000 °C.

Graphite-SiC units are shown in Figure 4.26(b). A SiC tube contains a graphite rod inside. Both have a screw-in connection or may be secured by a cement of SiC. Oxidation is prevented by a thin Al_2O_3 layer sprayed on to the surface of the graphite tube. This thermocouple, which can be used at temperatures as high as 1800 °C in reducing atmospheres, has a much higher e.m.f. (typically 500 mV at 1700 °C) than any other thermocouple.

Graphite–graphite with 0.1% *to* 0.2%*Be*, gives a thermocouple which can be used for spot readings up to 2600 °C or continuously up to 2000 °C in reducing or neutral atmospheres without any protective sheath. At 2000 °C its e.m.f. is about 50 mV, which is unfortunately different from batch to batch. Constructed in the same screwed rod and tube form as the previous type, its resistance is about 0.5 Ω (Zysk and Robertson, 1972), while its e.m.f. is a function of both its exact chemical composition

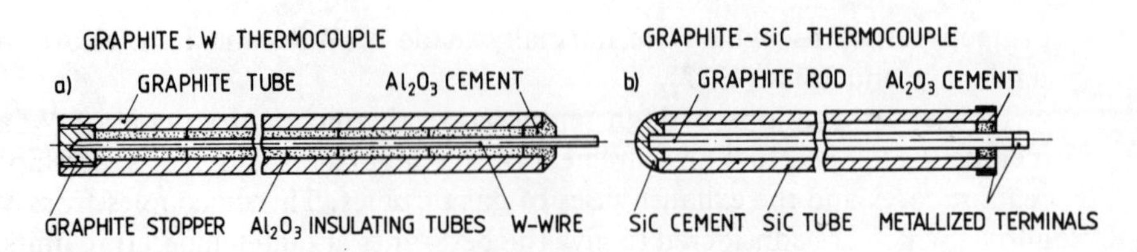

Figure 4.26
Non-metallic high temperature thermocouples.

and the technology used in its manufacture. A great diversity of relevant data appears in the literature (Caldwell, 1962; Burns and Hurst, 1975).

Other high temperature thermocouples, not described above, may be constructed from a wide range of other elements such as W, Mo, Rh, Re, Pt, Ir, Pd, C, SiC, Au, Nb, Te, Ta and Zr. More comprehensive descriptions and data can be found in the works by Anderson and Bliss (1972), Burns and Hurst (1972), Caldwell (1962), Franks (1962), Kinzie (1973), Lachman and McGurthy (1972), Vertogradsky and Chekhovskoy (1972) and Zysk and Robertson (1972). A list of different high temperature thermocouples may be found in the texts by Kinzie (1973) and Kortvelyessy (1987).

4.5 COMPENSATING CABLES

Compensating cables are used as an extension of the thermocouple from the terminal head, which is subject to big variations of temperature. This elongation ensures that the new reference junction is located in a position of constant temperature. The circuit, which is formed in this manner, conforms to the principle of linear superposition, as shown in §4.1.4. As the temperature of the thermocouple terminal head is never very high, e.m.f. versus temperature matching of the compensating cable to the thermocouple is only necessary in the temperature range 0–200 °C.

When the compensating leads are made from the same material as the thermocouple wires, they are called extension leads. Under these conditions there is no e.m.f. generated in the thermocouple junctions formed by connecting the extension leads to the thermocouple wires. However, because of the cost, which may be involved in rare metal units, it is more normal for the leads to be of a material different from that of the thermocouple. Another reason may be the high resistance of the extension leads. As the leads are mostly of different materials, it is essential, as a rule, to ensure that both connection points are at the same temperature, ϑ_2, shown in Figure 4.27. This guarantees that the resulting thermal e.m.fs generated at the connection points will cancel, as they will be of the same magnitude but opposite polarity.

Examination of Figure 4.27 shows that the e.m.f, E, across the terminals of the compensating leads is given by:

$$E = e_{AB}(\vartheta_1) + e_{BB'}(\vartheta_2) + e_{B'A'}(\vartheta_r) + e_{A'A}(\vartheta_2) \tag{4.20}$$

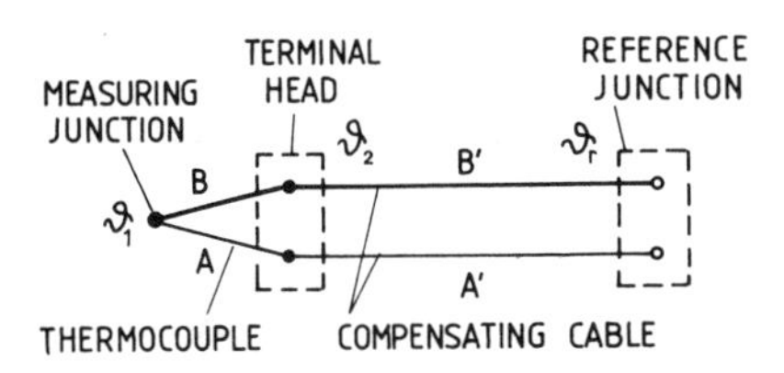

Figure 4.27
Thermocouple with compensating cable.

When no e.m.fs are generated at the connection points, corresponding to $e_{BB'}(\vartheta_2)=0$ and $e_{A'A}(\vartheta_2)=0$, or alternatively if they are equal but of opposite polarity, corresponding to $e_{BB'}(\vartheta_2)=-e_{A'A}(\vartheta_2)$, the resulting e.m.f. will be:

$$E=e_{AB}(\vartheta_1)-e_{A'B'}(\vartheta_r) \tag{4.21}$$

If it is assumed that the characteristics of the thermocouple AB and of the compensating leads A′B′ are identical, at the temperature ϑ_r it is apparent that $e_{A'B'}(\vartheta_r)=e_{AB}(\vartheta_r)$ so that finally:

$$E=e_{AB}(\vartheta_1)-e_{AB}(\vartheta_r) \tag{4.21a}$$

Equation (4.21a) shows that the resulting e.m.f. equals that value of e.m.f. exhibited by a thermocouple having its reference junction at the temperature ϑ_r in a similar manner as exposed by Equation (4.6). Hence it is apparent that the temperature, ϑ_r, may be called the reference temperature in this case also.

When applying compensating cables, attention must be paid to ensure that they are connected to the thermocouple leads with the correct polarity. Reversed polarity will lead to far larger errors than would occur in the case when no compensating cables are used. Moreover, Weichert *et al.* (1976) have shown that false matching between the compensating cables and the thermocouple may also cause large errors. For example, consider a NiCr–NiAl thermocouple matched with Pt10Rh–Pt compensating cable of correct polarity. When measuring a temperature of 1000 °C this combination will give an error of temperature value of about 85 °C when the terminal head is at a temperature of only 100 °C.

For Fe–CuNi and Cu–CuNi thermocouples, the extension cables are made of identical materials as the thermocouples themselves. In the case of Pt10Rh–Pt units the conductor pair compensating cables are made of Cu, Ni alloys with a typical constitution of either Cu–99.4%Cu, 0.6%Ni or 87.4%Cu, 12.3%Ni–80%Cu, 20%Ni. For NiCr–NiAl thermocouples it is possible to use either identical materials to the thermocouple or alternatively Cu–CuNi and Fe–CuNi wires may be used (Kinzie, 1973). Although a special compensating cable exists for the Pt30Rh–Pt6Rh thermocouple it is more usual to employ the cheaper Cu wires for industrial measurements. The errors which arise from this inadequate matching are given in Table 4.5.

Many countries publish standards, which specify the outputs of compensating cables conforming to IEC 584-1, such as BS 4937, ANSI/MC 96, NFC42-321, DIN 43710, JISC 1602 and PN/M-53854.

Table 4.5
Errors $\Delta\vartheta$(°C) resulting from using copper wires instead of special compensating cables for Pt30Rh–Pt6Rh, Type B thermocouples. The assumed nominal reference temperature, $\vartheta_{rn} = 20$ °C.

Measured temperature, ϑ_t (°C)	Reference temperature, ϑ_r (°C) 0	10	20	30	50	100	200	300
600					<0.5	6.0	30.2	71.8
800					<0.5	4.7	23.5	55.9
1000					<0.5	4.0	19.9	47.4
1200		$\Delta\vartheta < 0.02$ °C			<0.5	3.5	17.4	41.4
1400					<0.5	3.2	16.0	38.1
1600					<0.5	3.1	15.4	36.8
1800					<0.5	3.2	15.8	37.8

$$\Delta\vartheta = \vartheta_t - \vartheta_i$$

National colour codings for the insulation of compensating and extension cables are given in Table 4.6. It should be noted that the codings for the positive lead are not universal, different colours being adopted in different countries.

In an industrial environment the surrounding temperature may be as high as 500 °C as in the case, say, of measuring the temperature of molten steel using dip probes. Consequently, for type-S and R thermocouples, it is recommended by Bugden *et al.* (1975) and Kinzie (1973), that three lead cables are used. The positive lead is a special Cr/Ni/Fe alloy conductor while the negative one is a stranded conductor composed of two different numbers of wires of two different Cr/Ni/Fe alloys. If the thermocouple is type-S this form of compensation keeps the compensation errors well below ±2 °C in the whole range of reference temperature variation between 0 °C and 500 °C. One of the advantages of this method of compensation is the high resistance to oxidizing atmospheres possessed by three lead cables. This idea, which is as yet not very popular in practice, could be used to solve many non-typical problems.

Design of compensating cables uses the techniques associated with the technology of pairs of solid or stranded conductors with the total cross-sectional area of about 0.2 to 2.5 mm^2. Solid conductors are normally used for permanent installations while stranded cables are eminently suitable for portable instruments. Compensating cables have a resistance whose value may be calculated in the manner illustrated in Table 4.7. Other design factors and typical applications, which are given in Table 4.8, are based upon the information appearing in the catalogue of TC Ltd. Special purpose compensating cables are required in installations involving computers, locations where heavy duty armoured cable with high mechanical strength is necessary, in environments where electromagnetic or electrostatic interference are likely, or if there is a possibility of ingression by gas, steam, moisture or water. Multi-pair cables, with up to 50 pairs, collectively or individually screened and armoured are available. All forms of compensating cables are the subject of both national and international standards.

Thermocouple connectors, which are available for most standardized thermocouples, are used for reliable quick connections between thermocouples and compensating or extension cables. They replace the terminal head in many applications

Table 4.6
Colour coding for the insulation of thermocouple extension and compensating cables.

Thermocouple			Great Britain BS 1843		France NFC 42-323		FRG DIN 43714		Japan JIS C1610-1981		Poland PN/M-53859		USSR GOST		USA ANSI/MC 96.1	
Type			Insulation													
(Code)	Conductor	Polarity	Overall	Conductor	Overall	Conductor	Overall	Conductor	Overall	Conductor	Overall	Conductor	Overall	Conductor	Overall	Conductor
Cu–Cu Ni (T)	Cu	+	Navy blue	White	Navy blue	Yellow	Brown	Red	Brown	Red	Brown	Red	Brown	Red	Navy blue	Navy blue
	Cu Ni	–		Navy blue		Navy blue		Brown		White		Blue		Brown		Red
Fe–Cu Ni (J)	Fe	+	Black	Yellow	Black	Yellow	Dark blue	Red	Yellow	Red	Navy blue	Red	White	White	Black	White
	Cu Ni	–		Navy blue		Black		Navy blue		White		Blue		Brown		Red
NiCr–NiAl (K)	NiCr	+	Red	Brown	Yellow	Yellow	Green	Red	Navy blue	Red	Yellow	Red	Violet-black	Violet	Yellow	Yellow
	Ni	–		Navy blue		Violet		Green		White		Blue		Black		Red
NiCr–CuNi (E)	NiCr	+	Brown	Brown					Violet	Red					Violet	Purple-red
	CuNi	–		Navy blue						White						Red
NiCrSi–NiSi (N)	NiCrSi	+														
	NiSi	–														
Pt10Rh–Pt (S)	Pt10Rh	+	Green	White	Green	Yellow	White	Red	Black	Red	White	Red	Green	Red	Green	Black
	Pt	–		Navy blue		Green		White		White		Blue		Green		Red
Pt13Rh–Pt (R)*	Pt13Rh	+	Green	White					Black	Red					Green	Black
	Pt	–		Navy blue						White						Red
Pt30Rh–Pt6Rh (B)	Pt30Rh	+					Grey	Red	Grey	Red	Violet	Red			Grey	Grey
	Pt6Rh	–						Grey		White		Blue				Red

*Mostly for S and R thermocouples the same compensating cables are used.

Table 4.7
Resistance R^* in $\Omega\,m^{-1}$ at 20 °C of twin-run thermocouple extension and compensating cables $R = k/A\ \Omega\,m^{-1}$, k = constant, A = cross-sectional area of the conductor in mm^2.

Code	*k*-value
T	0.51
J	0.60
K	1.00
E	1.21
N	1.37
S	0.32
R	0.33
B	0.39

Source: TC Ltd catalogue

especially those involving multi-sensor applications. The pins and sockets of the plugs are made of alloys which are matched to the specific type of thermocouple so that interconnection errors are prevented. Pins may be round or flat with corresponding sockets which are standard colour coded against thermocouple type. For instance TC Ltd (UK) and Omega Engineering Inc. (USA) apply yellow for type-K, blue for type-T, black for type-J, violet for type-E, dark red for type-N and green for type-R and type-S. Green is also used for copper connectors while white is used for resistance thermometers. Different polarity pins have different sizes and are additionally marked for polarity. For temperatures around 210 °C connector envelopes are made from glass reinforced nylon. Higher temperatures, up to 650 °C, require connector envelopes made from ceramic material (Omega Engineering Inc., USA). Connector panels for up to 40 circuits, provided with matching colour codings, are also produced. When circumstances dictate, it may be more convenient to use terminal strips with plugs made from thermocouple grade alloys and marked with the code and polarity of the compatible units.

4.6 REFERENCE TEMPERATURE

4.6.1 Calculation of correction

Thermocouple characteristics, which are given by tables or by diagrams, correspond to a reference temperature of 0 °C. If the actual reference temperature, ϑ_r, is different from the nominal value, ϑ_{rn}, then errors will arise. It is easy to calculate any corresponding correction which is required by the readings. If the reference temperature is too high, $\vartheta_{r1} > \vartheta_{rn}$, as shown in Figure 4.28(a), then the measured value of the e.m.f., E', is too low. A correction, ΔE_1, should be added so that the corrected e.m.f. is:

$$E = E' + \Delta E_1 \tag{4.22}$$

The correction, ΔE_1, may be read from the thermocouple characteristics for the temperature difference $(\vartheta_{r1} - \vartheta_{rn})$. Figure 4.28(a) demonstrates how the true

Table 4.8
Thermocouple extension and compensating cables.

Operating temperature (°C)	Conductor insulation	Overall insulation: Optional overbraiding	Resistance to: Moisture	Resistance to: Abrasion	Additional notes
−30 to +100	PVC	PVC			
		Screened with Al tape in contact with Cu wire + PVC*	Very good	Good	Rejects electromagnetic electrostatic interference
		Screened with Al tape in contact with Cu wire + PVC + steel wire armour + PVC*	Very good	Good	Rejects electromagnetic + electrostatic interference + armouring
−273 to +250	Teflon	Teflon	Very good	Very good	
		Screened with Al tape in contact with Cu wire + teflon*	Very good	Very good	Rejects electromagnetic + electrostatic interference
⩽480	Glass fibre silicon varnished	Glass fibre, silicon varnished	Very good	Very good	Gas, steam and watertight Rejects electromagnetic + electrostatic interference
		Glass fibre, silicon varnished stainless steel wire armour	Fair	Fair	Rejects electromagnetic + electrostatic interference
⩽800	High temperature glass fibre silicon varnished	As above but HT glass fibre	Fair	Very good	
⩽1400	Ceramic, fibre, impregnated with ceramic binder		Fair	Very good	

*For computer installations. Also available with copper outer braiding.

temperature ϑ_t may be derived from the same characteristic. Similarly, when the reference temperature is too low, corresponding to $\vartheta_{r2} < \vartheta_{rn}$, the corrected e.m.f. will be:

$$E = E' - \Delta E_2 \tag{4.23}$$

Again the correction, ΔE_2, may be derived as shown in Figure 4.28(b). If the connected measuring instrument is calibrated in °C at the nominal reference temperature, ϑ_{rn}, then for any other reference temperature say, ϑ_{rl}, the true measured temperature, ϑ_t, may be found as follows. The e.m.f., E', corresponding to the indicated temperature, ϑ_i, as well as the error in the e.m.f., ΔE, for the difference in reference temperature $(\vartheta_{rn} - \vartheta_{rl})$, may both be found from the thermocouple characteristics as in Figure 4.28. Combining these two values gives the true value of the e.m.f. as $E = E' \pm \Delta E$, from which the corresponding true value of temperature, ϑ_t, may be found. In a similar manner the corrections may be found from the thermocouple tables.

Numerical example

A measuring instrument has been calibrated for a type-K thermocouple at the nominal reference temperature, $\vartheta_{rn} = 0$ °C. At the reference temperature, $\vartheta_r = 30$ °C, the indicated temperature was, $\vartheta_i = 830$ °C. Using tables, find the value of e.m.f. corresponding to ϑ_r. Hence deduce the corrected e.m.f. and find the corrected true temperature from tables.

Solution: From thermocouple tables, at a temperature $\vartheta_i = 830$ °C,

$$E' = 34.50 \text{ mV}$$

Also from tables corresponding to $\vartheta_r = 30$ °C,

$$\Delta E = 1.20 \text{ mV}$$

The corrected value of e.m.f. is thus:

$$E = E' + \Delta E = 34.50 + 1.20 = 35.70 \text{ mV}$$

From thermocouple tables for $E = 35.70$ mV true temperature, $\vartheta_t = 860$ °C.

Simplified formulae also exist for the calculation of the true e.m.f. values. These formulae, which are valid in the temperature range $0 < \vartheta_r < 50$ °C, have the universal form:

$$E = E' + k_1\vartheta_r + k_2\vartheta_r^2 \tag{4.24}$$

where E is the e.m.f. at $\vartheta_{rn} = 0$ °C, E' is the e.m.f. at $\vartheta_r \neq 0$ °C and ϑ_r is the actual value of the reference temperature in °C. The coefficients k_1 and k_2 have values which depend upon the specific thermocouple. They have the values given below.

Thermocouple	k_1(mV/°C)	k_2(mV/(°C)^2)
Pt10Rh–Pt	0.0054	0.012×10^{-3}
Pt30Rh–Pt6Rh	0	0
NiCr–NiAl	0.0404	0
Fe–CuNi	0.0532	0
Cu–CuNi	0.0406	0

To illustrate the application of this information the same numerical example as above may be considered. In this case the e.m.f., which may be calculated using Equation (4.24), has the value $E = 34.54 + 0.0404 \times 30 = 34.54 + 1.21 = 35.75$ mV. This is practically the same result.

Let the measuring instrument, calibrated in °C at the nominal reference temperature, ϑ_{rn}, indicate the temperature ϑ_i. If the pointer zero position lies on the nominal reference value, ϑ_{rn}, then the true temperature ϑ_t at $\vartheta_r \neq \vartheta_{rn}$, may be calculated as:

$$\vartheta_t = \vartheta_i + C(\vartheta_r - \vartheta_{rn}) \tag{4.25}$$

where the coefficient, C, is given in Figure 4.29 for different thermocouples.

Calculation of the corrections is only used in the laboratory measuring environment or in the case of short time industrial measurements.

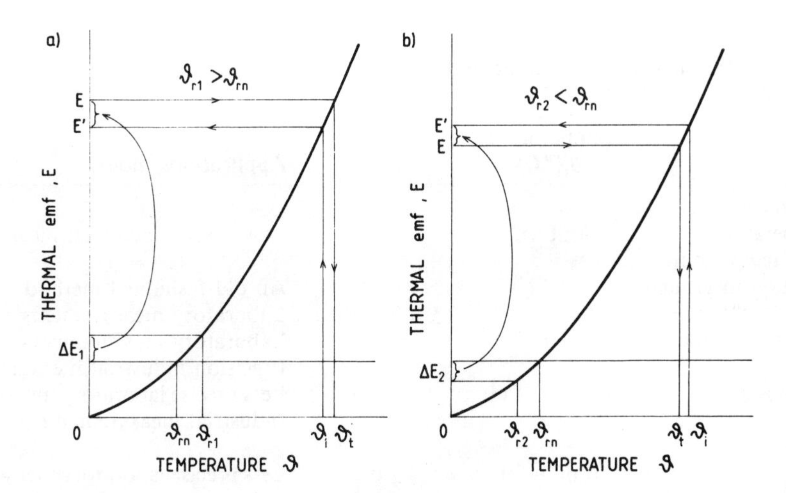

Figure 4.28
Graphic method to determine the correction ΔE to the thermocouple e.m.f., E, when the reference temperature ϑ_r differs from its nominal ϑ_{rn} value.

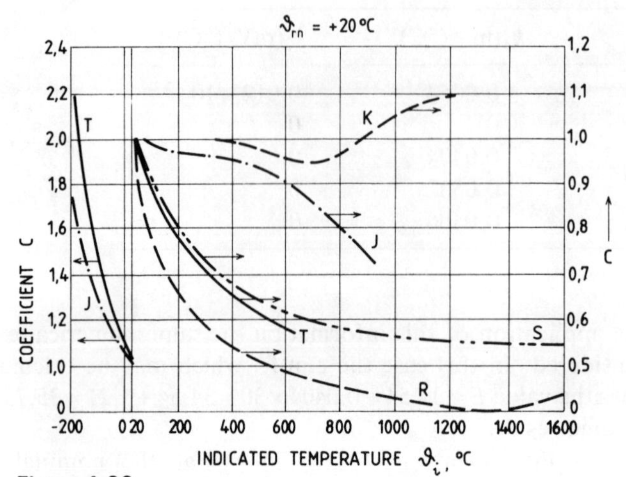

Figure 4.29
Coefficient C for correction calculation (Equation 4.25) versus indicated temperature ϑ_i for different thermocouple types.

4.6.2 Stabilization and correction techniques

Instead of calculating any correction, which may be necessary, it is more convenient to apply automatic reference junction stabilization or correction. A survey of the properties and applications of the devices for reference junction solutions, which is given in Table 4.9, shows that there are five compensation and three correction techniques. Although all of these will be described, it should be emphasized that the pointer of the indicating instrument for zero e.m.f. should be positioned at the assumed reference temperature.

Table 4.9
Reference junction stabilization and correction methods.

Method	Reference temperature ϑ_r(°C)	Error (°C)	Applications, notes
Stabilizing methods			
Reference temperature equals ambient temperature	Ambient	±5	Less precise industrial measurements
Reference junction in ground	+12	±2	An old-fashioned method
Ice point	0	±0.01–±0.1	Laboratory measurements
Peltier-element thermostat	0	±0.01	Laboratory measurements Precision industrial measurements
Electrical thermostat	+50 or +60	±0.5	Less precise laboratory measurements Industrial measurements
Correcting devices			
Bridge circuit	0 or +20	±0.5	Less precise laboratory measurements
Electronic	0	±0.8	Industrial measurements based upon standardized current signals
Built into measuring instruments	Depends upon particular design		Compensating cables connected directly to measuring instruments

The five reference temperature stabilization techniques, which may be utilized, will now be described.

1. *Reference temperature equals ambient temperature.* As the temperature of a room is fairly constant, at around 20 °C, with time over a year, it is popular industrial practice to locate the reference junction therein. Usually the relevant room temperature fluctuations do not exceed ±5 °C.
2. *Reference junction in the ground.* At a depth of about 2 m the temperature of the ground has a seasonally independent value of about ±12 °C. This old-fashioned method is rarely used.
3. *Ice-point (0 °C).* In this instance the reference junctions are placed in an oil-filled test-tube, immersed in a slush of melting ice contained in a Dewar vessel. Unfortunately, as it requires regular filling with ice at least once a day, the method is generally applied mainly in laboratories. It ensures a precision of from ±0.01 to ±0.1 °C. A detailed description of the preparation and handling of an ice-point is given by Sutton (1975).
4. *Peltier-element thermostat* (0 °C). A Peltier-element thermostat, described by Sutton (1975), is a fully automatic ice-point. Temperature control is based on variations in the volume of water and ice mixture. In the thermostat shown in Figure 4.30 the semiconductor cooling Peltier-elements refrigerate the water until a thin ice layer is formed. The water and ice volume increases and the membrane activates the switch, turning off the power to the cooling elements. Either on/off or continuous temperature control, with a precision of ±0.1 °C, is used. In the latter case, a solid state switch, such as a thyristor final control element, may be applied instead of a mechanically operated switch. Equalization of the temperature is assisted if the reference junctions of one or many thermocouples are placed in a copper block.
5. *Electrical thermostat at 50 or 60 °C.* This type of thermostat, which is popular in industry, requires an electrical heating element and a temperature controller, to maintain a constant reference temperature. As the ambient temperature in an industrial environment can reach even +40 °C, a reference temperature of +50 °C or +60 °C, with associated precision of control of about ±0.5 °C, is chosen.

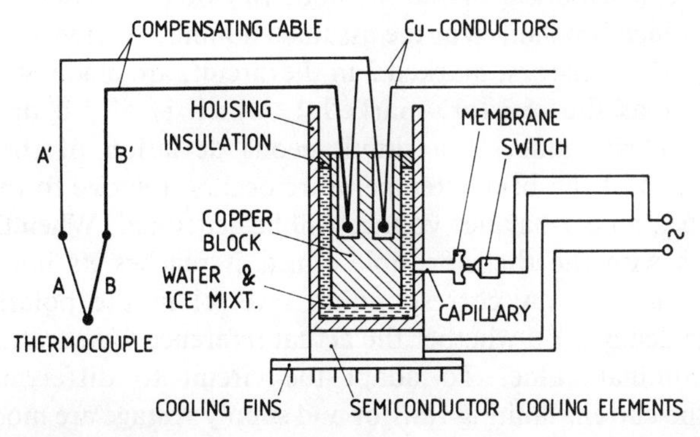

Figure 4.30
Peltier element thermostat (0 °C).

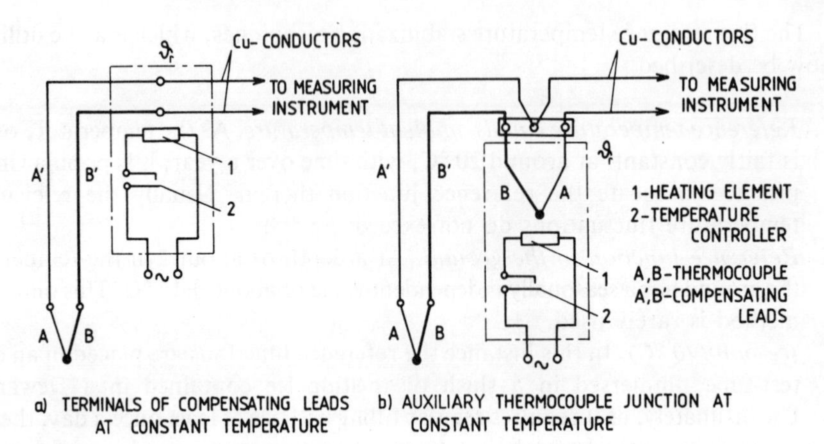

Figure 4.31
Reference junction thermostat (+50 °C).

As Figure 4.31(a) shows, the two terminals of the compensating leads are placed in the stable temperature chamber for which ϑ_r = constant. Alternatively, an auxiliary thermocouple inside the chamber may be employed as indicated in the scheme of Figure 4.31(b). In the method shown in Figure 4.31(b), all the terminals of the copper conductors should be maintained at the same but optionally chosen temperature. Both circuits produce equivalent results.

All the devices described so far can be built for one or many thermocouples.

It is also possible to eliminate the effect of fluctuating reference temperature on the readings from thermocouple thermometers by the use of one of three main kinds of correcting devices. These may be incorporated into thermocouple terminal heads, into the measuring instruments or may form a separate unit.

1. *Correcting bridge circuit*. This circuit is connected in series with the thermocouple leads. Correction is accomplished by generating an additional voltage which is a function of the actual value of the reference temperature. The circuit of Figure 4.32 is a representation of the method. In essence it consists of a d.c. resistance bridge which is balanced at the assumed nominal reference temperature (mostly +20 °C). The resistors, marked 2 in the circuit, are made of constantan or manganin whereas the resistor, marked 1, which is of Cu or Ni, has a temperature dependent value. If a simultaneous deviation of the reference temperature, ϑ_r, and of the bridge temperature occurs, relative to the assumed reference value, ϑ_{rn}, an off-balance voltage will be generated. When this voltage is added in series with the thermocouple e.m.f. it reaches its nominal value corresponding to $\vartheta_r = \vartheta_{rn}$. A corresponding reversal in the polarity of this generated voltage depends on whether the actual reference temperature is below or above the nominal value. To adapt the circuit to different types of thermocouples, the current limiting resistor and supply voltage are modified. The temperature sensitive resistor must be so placed that it always adopts the temperature of the terminals of the compensating leads. In most cases, at 20 °C,

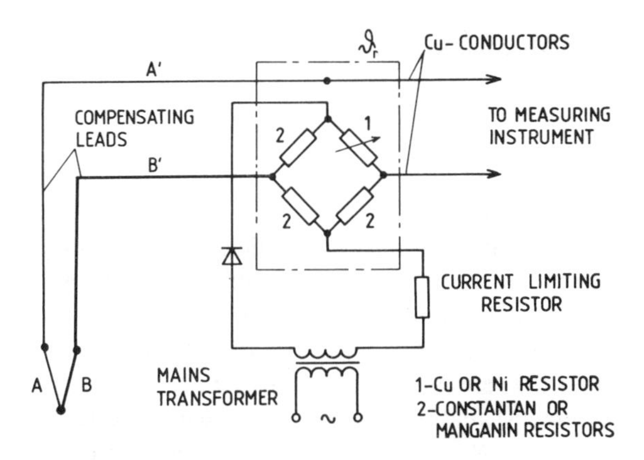

Figure 4.32
Correcting bridge circuit.

all of the bridge resistors have a resistance of 1 Ω. A precision of about ±0.5 °C in the correction of the readings is achievable using this circuit. More details may be found in Hunsinger (1966). There are also miniaturized bridge circuits ('Electronic Ice Point' by Omega Engineering Inc.). They include a self-contained battery, which is built into the thermocouple connectors, giving over 2500 hours continuous operation. These units correspond to the same colour codes as the connectors.

2. *Electronic correcting devices.* As expected, electronic technology has an attractive solution to the problem of correction for all these thermocouple circuits. An operational amplifier is used to amplify the e.m.f. of the thermocouple into direct signals with a span of ± 10 V. This is eminently suitable, as it allows compatible operation with modern computer processing techniques as well as contemporary signal communication and transmission circuits. The actual reference temperature is measured by an additional temperature sensor, which may be, for example, a transistor. After suitable conditioning of the transistor derived temperature signal a correcting signal is fed to the operational amplifier. These circuits can provide a precision of about ±0.8 °C within the reference temperature range of around 5–45 °C. A temperature of 0 °C is assumed as a nominal reference.
3. *Correcting devices built into the measuring instruments.* In this frequently used type the terminals of the compensating leads must then be connected directly to the instrument terminals. A more detailed description of these devices is given in §§4.7.3, 4.8.3 and 7.1.

4.6.3 Thermocouples needing no stabilization or correction

Some thermocouples, as for example Pt30Rh–Pt6Rh or NiCo–Alumel, have negligibly small e.m.f. values in the temperature range from 0 to around 100 °C. The variations of the reference temperature in this range exert no influence upon the readings, so that no compensating leads and no stabilizing or correcting devices are necessary (see Table 4.5).

4.7 DEFLECTION TYPE MEASURING CIRCUITS

4.7.1 *Principle of operation*

The direct deflection method embodies the oldest way of measuring electrothermal signals generated in thermocouples. By virtue of its simplicity and low cost this method is likely to continue to be used in many industrial applications when no great precision of measurement is required. A thermocouple measuring circuit has the six main constituents shown in Figure 4.33. The thermocouple leads AB, of resistance R_T, are connected to the rest of the circuit by compensating leads A′B′, of resistance R_c. A reference junction, at the reference temperature, ϑ_r, is subsequently connected to the measuring instrument, M, calibrated in °C and with internal resistance R_M using copper leads, of resistance R_{Cu}. An adjustable resistor R_a provides the facility to trim the loop resistance, R_l. The voltage, V, existing across the terminals of the indicating millivoltmeter instrument, M, in Figure 4.33 has the value:

$$V = E\frac{R_M}{R_T + R_c + R_{Cu} + R_a + R_M} = E\frac{R_M}{R_l + R_M} \tag{4.26}$$

where E is the electromotive force (more precisely thermoelectric force) at the measured temperature ϑ_t and reference temperature ϑ_r, and R_l is the loop resistance of the circuit excluding R_M.

The conditions under which the measuring instrument, calibrated in °C, indicates the true measured temperature require some emphasis. Firstly, the correct thermocouple type must be applied as indicated on the instrument scale. A constant loop resistance, R_l, which is equal to the nominal value, R_{ln}, also indicated on the instrument scale, is the second requirement. Finally, the reference temperature, ϑ_r, must equal the nominal value, ϑ_{rn}, for which the instrument has been calibrated.

4.7.2 *Loop resistance*

Loop resistance, which is the sum of the resistances of each element in the loop external to the measuring instrument, is given by

$$R_l = R_T + R_c + R_{Cu} + R_a \tag{4.27}$$

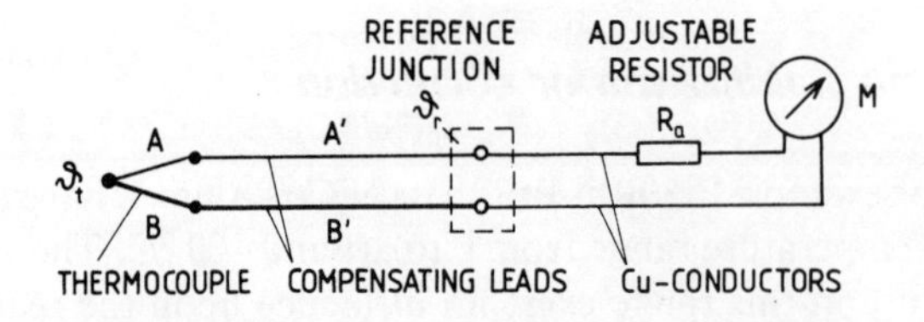

Figure 4.33
Deflection type measuring circuit.

As stated above, it must always be equal to its nominal value R_{ln}. In addition, if the circuit also comprises a correcting bridge circuit, its resistance must be considered and included in Equation (4.27) as well. The nominal standardized value of the loop resistance is, in most cases, 20 Ω. As the thermocouple resistance, R_T, is a function of the measured temperature and the resistances, R_c and R_{Cu}, are functions of ambient temperature, a precise satisfaction of the condition $R_l = R_{ln}$ is never possible.

Calculation of the particular resistances is managed in the following straightforward manner. Firstly, consider the thermocouple resistance R_T. A thermocouple, with the leads A and B, requires four pieces of data. Thus, consider that the leads A and B have the respective resistances R_{A20} and R_{B20} at a temperature of 20 °C and also the respective thermal resistance coefficients α_A and α_B. If the average temperature, in excess of the ambient temperature, is Θ_m then R_T is given by:

$$R_T = R_{A20}(1 + \alpha_A \Theta_m) + R_{B20}(1 + \alpha_B \Theta_m) \tag{4.28}$$

The values for R_{A20} and R_{B20} may be calculated from the data in Table XV while, Θ_m, the average thermocouple temperature, is approximately half of the measured temperature (over the ambient value).

Numerical example

Calculate the resistance of a NiCr–NiAl thermocouple using wires with a diameter 1.5 mm and length of 600 mm. The measured temperature is 1000 °C and the ambient temperature, ϑ_a, is 20 °C.

Solution: Using the data supplied in Table XV in Equation (4.28) yields:

$$R_T = 0.6 \times 0.410(1 + 0.25 \times 10^{-3} \times 980/2) + 0.6 \times 0.169(1 + 1.8 \times 10^{-3} \times 980/2)$$

i.e.

$$R_T = 0.47\ \Omega$$

For the approximate estimation of thermocouple resistances the data from Table 4.10 may also be used. The resistance is then calculated in terms of the resistance,

Table 4.10
Approximate estimation of thermocouple resistance.

Thermoelement		Average thermocouple temperature, ϑ_m(°C)	Resistance increase, δ_R(%/°C)
Cu–CuNi	Type T	400	0.02
Fe–CuNi	Type J	600	0.24
NiCr–NiAl	Type K	800	0.07
Pt10Rh–Pt	Type S	1300	0.20
Pt30Rh–Pt6Rh	Type B	1300	0.17

R_{20}, and its thermal sensitivity, δ_R, normally quoted in %/°C, and Θ_m, which is as defined for Equation (4.28). For these data the relation for R_T is then given by:

$$R_T = R_{20}(1 + 10^{-2}\delta_R\Theta_m) \tag{4.29}$$

Numerical example

Calculate the resistance of a NiCr–NiAl thermocouple at $\vartheta = 1000$ °C, when its resistance at 20 °C, $R_{20} = 0.35\ \Omega$. All of the other data is the same as in the previous example.

Solution: From Equation (4.29) and Table 4.10, it follows that:

$$R_T = 0.35 \times (1 + 10^{-2} \times 0.07 \times 980/2) = 0.47\ \Omega$$

Calculation of the other resistances of the circuit of Equation (4.27) is also possible. Using Table 4.7, the resistance of the compensating leads, R_c, and the resistance of copper leads, R_{Cu}, may be calculated for the assumed ambient temperature ϑ_a. Finally, the value of the resistance of the adjustable trimming resistor, R_a, must satisfy $R_a = R_{ln} - (R_T + R_c + R_{Cu})$. This resistor is built as a manganin coil of which a number of turns should be unwound until the required value for R_a is reached. In the case $(R_T + R_c + R_{Cu}) > R_{ln}$, either another instrument, with a higher R_{ln} should be used, or the cross-sections of the thermocouple, the compensating and the copper leads increased.

The measurement of the loop resistance, R_l, should be performed at the measuring and ambient temperatures and in the operating conditions under which it is intended that the scheme will perform its function. To prevent the thermocouple e.m.f. from influencing the results, while using a d.c. bridge, the measurements should be made twice with reversed polarities or, if possible, using a low frequency a.c. bridge. A median value of the two respective measurements for R_l' and R_l'' presents the final result for the loop resistance as:

$$R_l \approx 2\frac{R_l' R_l''}{(R_l' + R_l'')} \tag{4.30}$$

In the case when the measured values of R_l' and R_l'' do not differ from one another by more than 10%, the following simplified approximate relation may be used.

$$R_l \approx \frac{(R_l' + R_l'')}{2} \tag{4.31}$$

In the actual operating conditions the specific value of R_l may differ from R_{ln} by a small amount ΔR_l. The resulting error in the measured temperature, $\Delta\vartheta$, expressed in terms of R_l and the indicated and reference temperature, ϑ_i and ϑ_r respectively, is then given by:

$$\Delta\vartheta \approx \frac{\Delta R_l}{R_M + R_l}(\vartheta_i - \vartheta_r) \tag{4.32}$$

It is quite clearly seen from Equation (4.32) that increasing the internal resistance, R_M, of the measuring instrument, will decrease the errors.

4.7.3 Measuring instruments

In deflection type circuits, moving-coil millivoltmeters, calibrated in °C (or in °F) are used. Their internal resistance should be as high as possible, preferably between 80 Ω and 600 Ω. They should be vibration proof with a light and consequently low inertia moving element. Industrial types, which belong to class 1 or 1.5, may have a measuring range as small as 5 to 10 mV. As they are calibrated at 0, 20 or 50 °C nominal reference temperatures, they should be properly chosen for use in these actual operating conditions. Unless the instrument has a built-in correcting device, the pointer position for zero e.m.f. should agree with the actual reference temperature. Directly coupled amplifiers are being increasingly employed to boost electrothermal signals before they are measured by deflection type instruments. Further consideration of this question will be given in Chapter 8.

Built-in correcting devices

In instruments with built-in correcting devices, the ends of the compensating leads should be connected directly to the instruments' terminals in the manner of Figure 4.34. Built-in correcting devices take two main forms. A commonly occurring technique is the application of a bimetallic temperature sensitive element. In the first approach using this technique the element is so installed that it adjusts the readings for the zero e.m.f. condition as a function of ambient temperature. When it is used to adjust the position of a magnetic shunt as a function of ambient temperature in the second version, the driving torque of the instrument is consequently modulated as a function of ambient temperature. A final possibility is in the use of correcting bridge circuits, of the type described in §4.6.2. They may be incorporated inside the instrument housing.

4.7.4 Multi-point circuits

For temperature measurement at several points with one measuring instrument, multi-point switches are used. As the insulation failure of one circuit may influence the

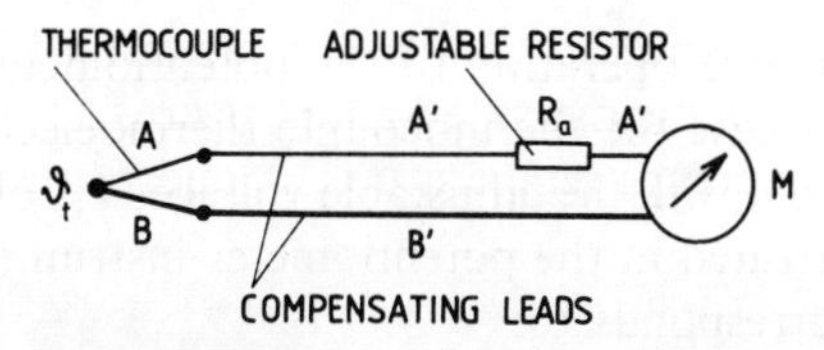

Figure 4.34
Deflection type measuring circuit, indicating instrument with built-in correcting device.

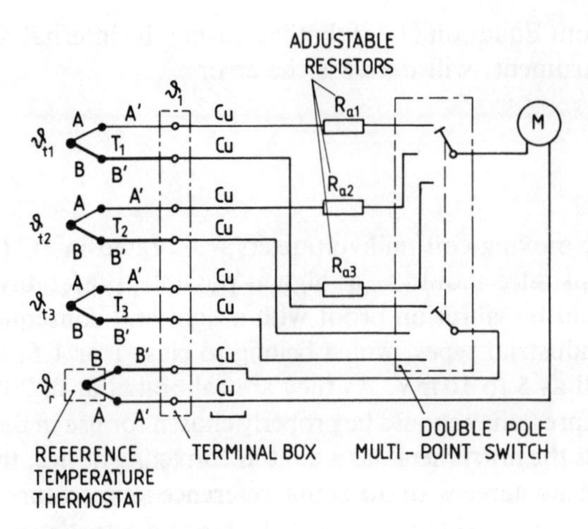

Figure 4.35
Multi-point deflection type measuring circuit.

others it is advisable to use double pole switches. In addition, parasitic thermoelectric forces, which could adversely affect the measurements, should be eliminated. This is possible using specially engineered switches in which parasitic e.m.fs of this nature are not generated. Either manually operated or automatic switches may be used.

Frequently, in multi-point circuits similar to that represented in Figure 4.35, one auxiliary thermocouple is used for several measuring circuits. This reduces the overall cost by removing the need for several compensating cables. The terminals of all of the thermocouples may then be gathered or herded together in one terminal box to simplify their maintenance at one common temperature. A thermostat, of a similar type to one of those described in §4.6.2, is used to provide a constant temperature environment for the measuring junction of the auxiliary thermocouple. This auxiliary thermocouple is acting as the reference for all of the other thermocouples.

Using the law of the third metal, which was proved in §4.1.2, it is easily shown that the e.m.f. of each circuit has a correct value of:

$$E = e_{AB}(\vartheta_t) - e_{AB}(\vartheta_r) \tag{4.33}$$

4.8 POTENTIOMETERS

4.8.1 Principle of operation

As shown in Figure 4.36, the principle of operation of the potentiometer method is the comparison of voltages. In this case the thermocouple thermoelectric force, E_x, which is to be measured, is compared with the adjustable voltage, V_c, of opposite polarity. This variable voltage is generated in the potentiometer instrument. From Figure 4.36 the balance condition corresponds to:

$$E_x - V_c = 0 \tag{4.34}$$

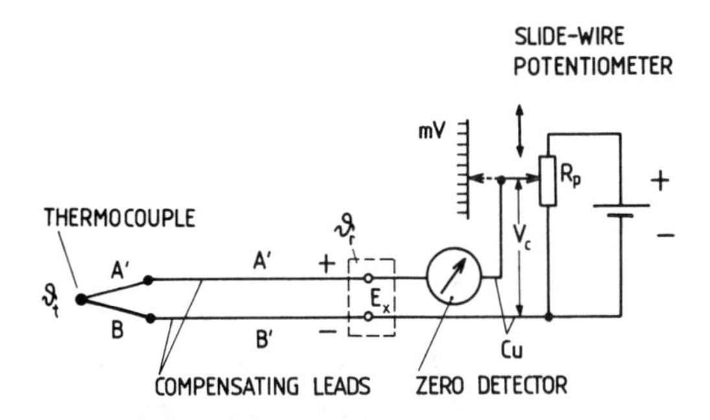

Figure 4.36
Thermoelectric thermometer, potentiometer method.

This balance condition occurs when a zero current is indicated by the zero current detecting galvanometer. The value of the e.m.f. is indicated by the slider position of the resistor, R_p. As potentiometer type indications are independent of the circuit resistance, they attain an accuracy which is higher than that of deflection type instruments. A limitation in loop resistance, which may be high, is only imposed by the necessary sensitivity. The two different types of potentiometers are the variable resistance potentiometers (Poggendorf type) and the variable current potentiometers (Lindeck–Rothe type).

4.8.2. *Variable resistance potentiometers*

In manually operated potentiometers, represented in Figure 4.37, the compensating voltage, V_c, is adjusted by setting the slide wire of the potentiometer, R_p, in such a way that zero is indicated by the zero current detector. The value of e.m.f. which is to be measured, is read from the slider position on a mV scale. In the early coarse

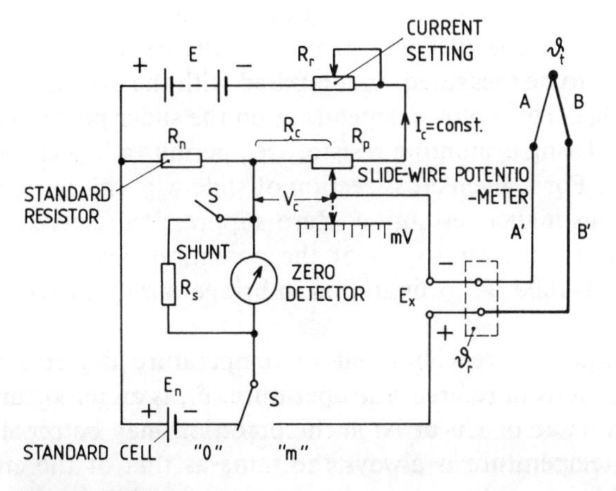

Figure 4.37
Hand-operated variable resistance potentiometer.

balancing stage the difference between E_x and V_c may be rather large. To protect the galvanometric detector from damage at this stage, it is shunted by a resistor, R_s. When the resulting coarse zero indication is reached, the switch, S, is opened, the system sensitivity increases and the circuit is balanced again. The balance condition is then given by:

$$E_x - I_c R_c = 0 \tag{4.35}$$

During the measurement the current, I_c, should have a constant, precisely prescribed value which should be checked before the measurement is made. The value of I_c is set by the resistor, R_r, with switch 'P' in the position marked 'O'. In this position the voltage drop, $I_c R_n$, is compared with the voltage, E_n, of the cadmium standard cell until the following balance condition is reached:

$$E_n - I_c R_n = 0 \tag{4.36}$$

In modern versions of this potentiometer a stabilized voltage source, which is fed from either mains or battery, is used. The error limit of the potentiometer is from 0.1 to 0.3% and its measurement range, divided into many subranges, is mostly zero to 50 mV. Manually operated potentiometers, which are usually portable, find common use in laboratories and for more precise short-term or industrial control measurements.

4.8.3 Automatic potentiometers

For continuous operation the process of voltage balancing should be automatic. Electronic servo-balanced potentiometers, which are built mainly as potentiometric recorders, are also commonly used in the measurement of other different physical quantities that might be transduced to electrical signals.

Most often they are based upon the principle of the resistance bridge potentiometer shown in Figure 4.38, which is different from the circuit in Figure 4.37. The thermoelectric force, E_x, to be measured, is compared with the off-balance voltage, V_{ab}, of a resistance bridge. This voltage depends upon the slider position of a slide-wire potentiometer, R_p. Using a shunting resistor, R_s, permits a higher value of the slide-wire resistance, R_p. For a given cross-section of slide-wire this implies a higher number of turns and thus higher resolution. Zero suppression, catered for by the resistor, R_1, reduces the temperature span of the potentiometer to, for example, 800–1000 °C instead of 0–1000 °C. A limitation in bridge supply current is ensured by the resistor R_2.

When such potentiometers are calibrated in temperature degrees, automatic compensation of fluctuations in reference temperature, ϑ_r, is an important facility. If the resistance, R_3, is made of Cu or Ni such correction may be obtained when it is arranged that its temperature is always the same as that of the ends of the compensating leads. This is mostly the potentiometer terminal box. Any change in the value of R_3 will exert an influence upon the division of the current between the

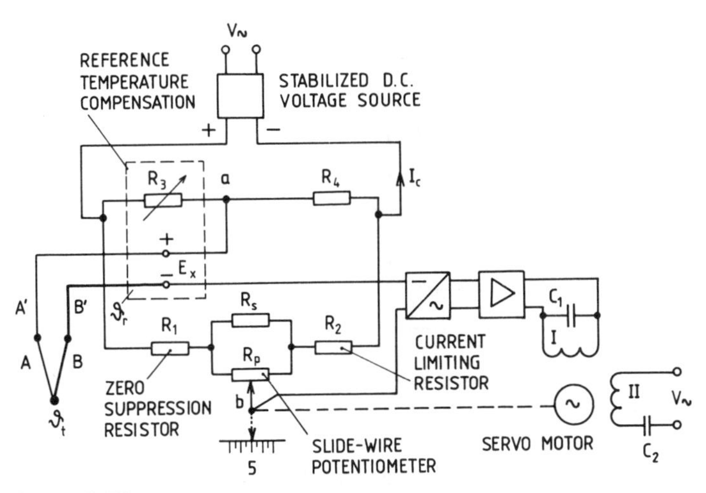

Figure 4.38
Automatic resistance bridge potentiometer.

bridge paths. To prevent any such influence, the resistor, R_2, is made of the same material as R_3. The bridge is fed from a zener diode stabilized voltage source, whose accuracy should be higher than 0.1%. Using a d.c.–a.c. converter, the off-balance voltage of the bridge may be converted into an alternating signal whose amplitude is proportional to this off-balance voltage and whose phase reverses by 180° with changing signal polarity. Amplification of this alternating signal in a transistor amplifier is followed by feeding it to winding I of a servo-motor. The other winding II is fed directly from the mains voltage through a phase-shifting capacitor, C_2. A proportional relationship between the rotational speed of the motor and the bridge off-balance voltage is assured by this approach. In addition, the direction of rotation depends upon the polarity of the bridge signal. The motor stops when the driven slider of the slide-wire potentiometer, R_p, attains a null-balance condition. Simultaneously, the motor drives the recording pen and the indicating pointer. The capacitor, C_1, forms a resonant circuit with the winding, I, whose resonant frequency corresponds to the fundamental frequency of the signal at the output of the converter. Conventional velocity feedback is fed to the amplifier to improve the dynamic performance of the system. Automatic potentiometers built mainly as temperature recorders, are described in Chapter 8.

4.8.4 *Variable current potentiometers*

In variable current potentiometers, as shown in Figure 4.39, E_x again represents the thermoelectric force to be measured. It is opposed by an adjustable voltage, V_c, which is generated as a voltage drop across a constant resistor, R_c. This voltage drop is adjusted by changing the compensating current I_c using a slide-wire resistor, R_r, satisfying the condition:

$$E_x - I_c R_c = 0$$

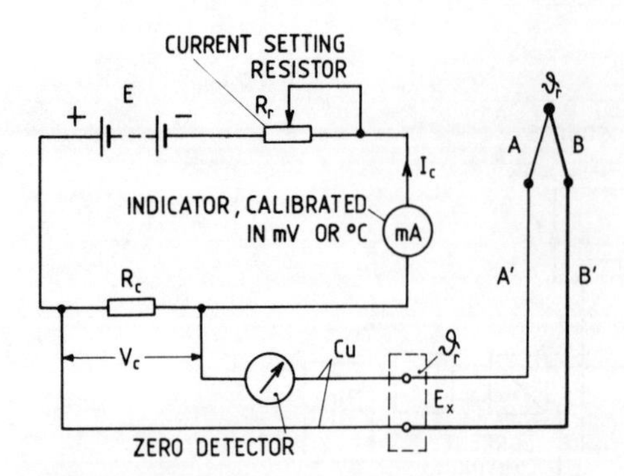

Figure 4.39
Variable current potentiometer.

The value of the current, I_c, is a measure of E_x to be determined. A zero-detector is used to indicate the null balance condition. The current indicating milliammeter, mA, can be calibrated in mV or °C. As milliammeters have a rather restricted accuracy, variable current potentiometers are not as precise as the former type, although they are simpler in construction and cheaper. In a laboratory it is easy to assemble such a potentiometer from typical instruments. Their accuracy is about 0.2 to 0.5% of the measuring range. They are also built as self-balancing potentiometers based upon an operational amplifier with feedback.

4.8.5 High precision laboratory potentiometers

For high precision in the measurement of e.m.f., only variable resistance potentiometers are used. When they are employed for the measurement of electrothermal e.m.fs their construction is beset by difficulties which are mostly caused by any form of parasitic thermoelectric forces generated at circuit contacts or by voltage drops on any of the selector switches employed. For this reason, it is essential to avoid any decade selector switches in the galvanometric circuit.

One of the most popular high precision potentiometers was designed in 1906 by Diesselhorst and Hausrath (Hall and Barber, 1964; Hunsinger, 1966; White, 1941). The simplified diagram of such a potentiometer, given in Figure 4.40, possesses only three, instead of five, resistor decades. A constant and known current, I, which is fed through a two-pole selector switch, S_1, to the double resistance decade R_1, R_1', is divided into I_1 and I_2. These currents correspondingly flow through R_2, R_2', R_4 and R_3, R_3', R_5. The circuit resistances of both paths are chosen in such a way, that $I_1/I_2 = 1/10$. In all of the positions of the switches, S_1, S_2, and S_3, the resistances of both paths to the currents I_1 and I_2 are constant. The voltage, V_{23}, opposed to the e.m.f., E_x, to be measured, is the difference between the two voltage drops caused by the currents I_1 and I_2 respectively. As there are no switch contacts between the terminals numbered 2 and 3, the voltage V_{23} is free from any parasitic thermoelectric forces. The current I is set using an additional compensating circuit comprising a standard cell which is not shown in Figure 4.40. Although there are

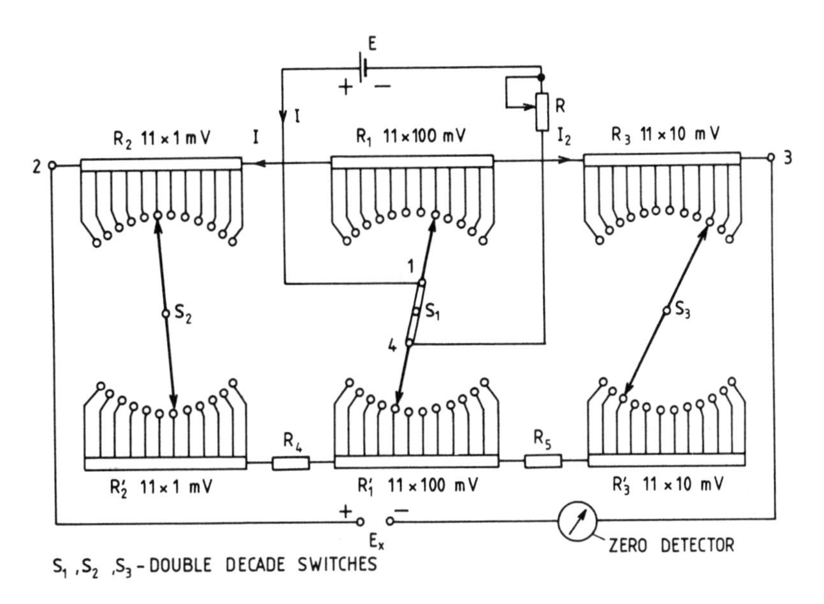

Figure 4.40
Diesselhorst and Hausrath potentiometer; simplified diagram.

no parasitic e.m.fs in the potentiometer circuit itself, their generation in the outside measuring loop must be prevented. With this in mind the terminal block of the potentiometer should be made in copper as well as all of the leads, which should be made of copper from the same batch. Under the influence of the heat emitted by the scale projection lamp, some e.m.fs may even be generated in the zero detector. To avoid the influence of these e.m.fs all the measurements should be made twice, simultaneously reversing the polarities of both E and E_x. The errors of the potentiometers described above are below the tenths of microvolts when they are constructed with five resistance decades.

Cascade potentiometers, of the type represented in Figure 4.41 (Hunsinger, 1966), can give similar accuracies. The decade resistors R_1 and R_3 are used as voltage dividers whilst the double-pole switch, S_1, links the resistor, R_2, with pairs of every other tap of the resistor, R_1. An additional resistor, which is not shown in Figure 4.41, is connected to obtain the necessary voltage drop across R_2. The compensating voltage, V_{23}, across the terminals 2 and 3 may be set by the switches S_1, S_2, S_3 respectively in steps of 0.1, 0.001 and 0.000 01 V. Positioning of the range switch, P_r, to the position $\times 1$ corresponds to the measuring range 1.9 V and to the position $\times 0.1$ to the measuring range 0.19 V. The contact resistance of the switches, S_2 and S_3, though they are in the loop of the galvanometer, do not contribute any errors, as the measurements are made under zero current conditions. As the currents in the resistor, R_2, are extremely small the contact resistances of the switch, S_1, have no real influence on the measurement results. Some small thermoelectric forces occur in this type of potentiometer. Because the whole circuit is composed of different metals, all of which may be subjected to some temperature differences, such parasitic effects are unavoidable. The current, I, is set using a standard cell circuit which is not shown in Figure 4.41.

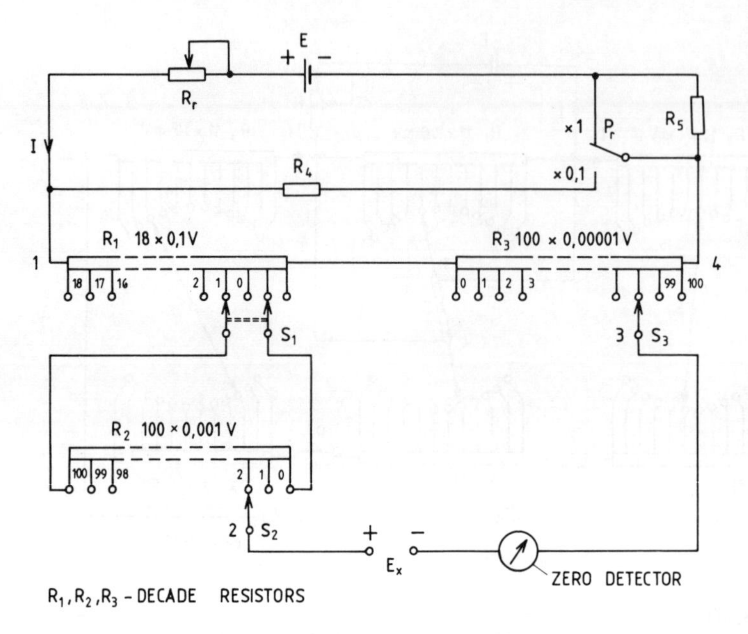

Figure 4.41
Cascade potentiometer; simplified diagram.

Characteristic data of a cascade potentiometer can be for example:

Measuring range, V	Voltage settings	Minimum voltage steps
1.9	1.8×0.1, 100×0.001, $100\times0.000\,01$	0.001% or 10 μV
0.19	as above $\times$ 0.1	0.002% or 1 μV
0.019	as above $\times$ 0.01	0.002% or 0.1 μV

Similar precision is possessed by double potentiometers. In this type the compensating voltage may be set in 0.1 μV steps throughout the complete measuring range up to 2 V. As shown in Figure 4.42, the circuit comprises two cascaded potentiometers, whose compensating voltages are consequently added.

The switches S_1 and S_2 are used for setting the values of the compensating voltages of the potentiometer, I, in steps from 0.1 to 0.001 V. Similarly the switches S_3 and S_4 enable the attainment of voltage steps of the potentiometer, II, as low as 0.000 01 to 0.000 0001 V. The current, I_1, should be set using a standard cell, while the current, I_2, is set by R_{r2} in the process of reaching the condition $I_2R_{s2}=I_1R_{s1}$. During the setting of I_2 the connections indicated by the dashed lines should be made. It should also be noted that the switch S has to be open for the setting of both I_1 and I_2. As none of the current flows through the movable contacts in this potentiometer, there is no influence of any contact resistance upon the measurement. In the case of very precise

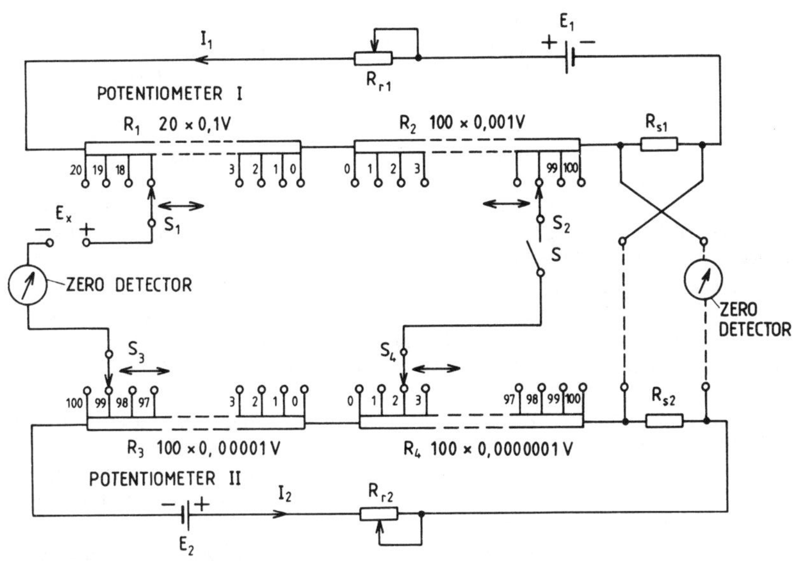

Figure 4.42
Double potentiometer; simplified diagram.

measurements the potentiometer should remain in that temperature at which it has been calibrated. This temperature will generally be +20 °C.

REFERENCES

Anderson, T. M. and Bliss, P. (1972) Tungsten-rhenium thermocouples. Summary Report appearing in *Temperature: Its Measurement and Control in Science and Industry*, Vol. 4, Part 3, Instrument Society of America, Pittsburgh, pp. 1735–1746.

Baker, H. D., Ryder, E. A. and Baker, N. H. (1953) *Temperature Measurement in Engineering*, vol. 1, John Wiley and Sons, New York.

Barzanty, J. and Hans, R. (1979) Fehlerfrueherkennung an Mineralstoff isolierten Mantel-thermoelementen und Mantelmessleitern, *Techn. Messen.*, **46**(6), 239–248.

Bentley, R. E. and Morgan, T. L. (1986) Ni-based thermocouples in the mineral insulated metal-sheathed format: thermoelectric instabilities to 1100 °C. *J. Phys. E: Sc. Instr.*, **19**, 262–268.

Bliss, P. (1972) Fabrication of high-reliability sheathed thermocouples, *Temperature: Its Measurement and Control in Science and Industry*, Vol. 4, Part 3, Instrument Society of America, Pittsburgh, pp. 1797–1803.

Brookes, C., Chandler, T. R. D. and Chu, B. (1985) Nicrosil-nisil: a new high stability thermocouple for the industrial user. *Measurement and Control*, **18**(9), 245–248.

Bugden, W. G., Tomlinson, J. A. and Selman, G. L. (1975) Improved compensating lead system for platinum-base thermocouples. *Temperature Measurement: Conference Series No. 26.* Institute of Physics, London, 1975, pp. 181–187.

Burley, N. A., Burns, G. W. and Powell, R. L. (1975) Nicrosil and Nisil, their development and standardisation. *Temperature Measurement: Conference Series No. 26.* Institute of Physics, London, 1975, pp. 162–171.

Burley, N. A. and Jones, T. P. (1975) Practical performance of Nicrosil-Nisil thermocouples. *Temperature Measurement: Conference Series No. 26.* Institute of Physics, London, 1975, pp. 172–180.

Burley, N. A. (1987) A novel advanced integrally-sheathed Type N thermocouple of ultra-high thermoelectric stability. *Thermal and Temperature Measurement in Science and Industry*, TEMPMEKO 87, 3rd Int. Conference, Sheffield, The Institute of Measurement and Control, London, pp. 115–125.

Burns, G. W. and Hurst, W. S. (1972) Studies of performance of W-Re type thermocouples. *Temperature: Its Measurement and Control in Science and Industry*, vol. 4, Part 3, Instrument Society of America, Pittsburgh, pp. 1751–1766.

Burns, G. W. and Hurst, W. S. (1975) Thermocouple thermometry. *Temperature Measurement: Conference Series No. 26*, Institute of Physics, London, 1975, pp. 144–161.

Caldwell, F. R. (1962) Thermocouple materials. *Temperature: Its Measurement and Control in Science and Industry*, Vol. 3, Part 2, Reinhold Publ. Co., New York, pp. 81–134.

Considine, D. *et al.* (1957) *Process Instruments and Controls Handbook*. McGraw-Hill, New York.

Franks, E. (1962) High-temperature thermocouples using non-metallic members. *Temperature: Its Measurement and Control in Science and Industry*, Vol. 3, Part 2, Reinhold Publ. Co., New York, pp. 189–194.

Hall, J. A. and Barber, C. R. (1964) Calibration of temperature measuring instruments. *Ed 3 Notes on Applied Science No. 12*, 1964, National Physical Laboratory, London.

Howard, J. L. (1972) Error accumulation in thermocouple thermometry, *Temperature: Its Measurement and Control in Science and Industry*, Vol. 4, Part 3, Instrument Society of America, Pittsburgh, pp. 2017–2029.

Hunsinger, W. (1966) *Temperaturmessung: Handbuch der Physik*. Springer-Verlag, Berlin.

Kinzie, P. A. (1973) *Thermocouple Temperature Measurement*. John Wiley and Sons, New York.

Kortvelyessy, L. (1987) *Thermoelement Praxis*, 2nd edition. Vulkan-Verlag, Essen.

Lachman, I. C. and McGurty, I. A. (1972) The use of refractory metals for ultra-high temperature thermocouples, *Temperature: Its Measurement and Control in Science and Industry*, Vol. 4, Part 2, Reinhold Publ. Co., New York, pp. 177–187.

Lieneweg, H. (1975) *Handbuch, Technische Temperaturmessung*, F. Viemeg, Braunschweig.

Omega Engineering Inc., USA (1982) *Temperature Measurement Handbook*.

Parsegian, V. L. and Fairchild, C. O. (1941) Performance characteristics of recording potentiometers. *Temperature: Its Measurement and Control in Science and Industry*. Reinhold Publ. Co., New York, pp. 639–645.

Pesko, R. N., Ash, R. L., Cupshalk, S. G. and Germain, E. F. (1972) Theory and performance of plated thermocouples. *Temperature: Its Measurement and Control in Science and Industry*, Vol. 4, Part 3, Instrument Society of America, Pittsburgh, pp. 2009–2016.

Reed, R. P. and Ripperger, E. A. (1972) Deposited thin film thermocouples of microscopic dimensions. *Temperature: Its Measurement and Control in Science and Industry*, Vol. 4, Part 3, Instrument Society of America, Pittsburgh, pp. 2209–2222.

Rienecker, W. (1983) Messung mit Thermoelement ohne 0 °C-Referenz. *Elektronik*, **32**(1), 29–33.

Roeser, W. F. (1941) Thermoelectric thermometry, *Temperature: Its Measurement and Control in Science and Industry*. Reinhold Publ. Co., New York, pp. 180–205.

Roeser, W. F. and Wensel, H. T. (1941) Appendix *Temperature: Its Measurement and Control in Science and Industry*. Reinhold Publ. Co., New York, pp. 1308–1310.

Schaller, A. (1972) Temperaturmessungen mit elektrischen Berührungsthermometern. *Messen und Prüfen*, **8**(3), 159–163.

Sutton, G. R. (1975) Thermocouple referencing. *Temperature Measurement: Conference Series No. 26*. Institute of Physics, London, 1975, pp. 188–194.

Thomson, A. and Fenton, A. W. (1975) High-integrity, small-diameter mineral-insulated thermocouples. *Temperature Measurement: Conference Series No. 26*. Institute of Physics, London, 1975, pp. 195–202.

Toenshoff, A. A. and Zysk, E. D. (1972) Material preparation and fabrication techniques for the production of high-reliability thermocouple devices. *Temperature: Its Measurement and Control in Science and Industry*, Vol. 4, Part 3, Instrument Society of America, Pittsburgh, pp. 1971–1796.

Vertogradsky, V. A. and Chekhovskoy, V. Ya. (1972) Thermal properties of Tungsten-Rhenium alloys used in high-temperature thermocouples. *Temperature: Its Measurement and Control in Science and Industry*, Vol. 4, Part 3, Instrument Society of America, Pittsburgh, pp. 1747–1749.

Weichert, L. *et al.* (1976) *Temperaturmessung in der Technik-Grundlagen und Praxis*. Lexika-Verlag, Grafenau.

Westhoff, G. (1965) *Grundlagen und Praxis der Temperaturmessung und Temperaturregelung*. Vulkan-Verlag, Essen.

White, W. P. (1941) Potentiometers for thermoelectric measurements, *Temperature: Its Measurement and Control in Science and Industry*, Reinhold Publ. Co., New York, pp. 265–278.

Zanstra, P. E. (1975) 'Welding uniform sized thermocouple junctions from thin wires. *J. Phys. E: Sc. Instr.*, **8**, 262–263.

Zysk, E. D. (1962) Platinum metal thermocouples. *Temperature: Its Measurement and Control in Science and Industry*. Vol. 3, Part 2, Reinhold Publ. Co., New York, pp. 135–136.

Zysk, E. D. and Robertson, A. R. (1972) Newer thermocouple materials. *Temperature: Its Measurement and Control in Science and Industry*, Vol. 4, Part 3, Instrument Society of America, Pittsburgh, pp. 1697–1734.

5 Resistance Thermometers

5.1 PHYSICAL PRINCIPLES

In 1871 C. W. Siemens delivered a lecture to the Royal Society in which he presented the possibility of temperature measurement by measuring the corresponding resistance variations of a metal conductor. The sensor presented by him, which was proposed to be of platinum wire, was subsequently investigated during 1874 by a committee. Two of the members of this committee, among others, were W. Thomson (Lord Kelvin) and James Clerk Maxwell. In these investigations, it was discovered that the resistor proposed by Siemens changed its resistance after heating up to a high temperature followed by a cooling. It was concluded that the resistor could not be used for temperature measurement because of this behaviour.

H. L. Callendar, who was not discouraged by these negative results, continued to research the problem. Eventually, in 1887, he published the paper entitled 'On practical measurement of temperature', which may be regarded as the beginning of resistance thermometry.

The principle of a resistance thermometer is based upon the dependence of the resistance of metal conductors upon temperature. As the temperature of a metal increases, the amplitudes of the thermodynamic vibrations of its atonomic nuclei increase. Simultaneously, the probability of collisions between its free electrons and bound ions undergo corresponding increases. These interruptions of the motion of the free electrons, due to crystalline collisions, cause the resistance of the metal to increase. The resistance increase may be described with the aid of a temperature coefficient of resistance or resistance temperature coefficient, which is expressed most commonly as an average value, α, in a given temperature range. In terms of the resistances, R_{100} and R_0, at the respective temperatures 100 °C and 0 °C, the defining equation for α has the form

$$\alpha = \frac{1}{R_0} \frac{R_{100} - R_0}{100} \tag{5.1}$$

5.2 RESISTANCE THERMOMETER DETECTORS (RTD)

5.2.1 *General information*

Resistance thermometer detectors, also called resistance thermometer elements in BS1041, or further simply RTDs, consist of a resistance conductor wound or deposited on an insulating support or former. The nominal resistance, R_0, of an RTD is its resistance at the reference temperature of 0 °C. In industrial applications, although the most commonly used RTDs have a nominal resistance of 100 Ω, elements of 50 Ω, 500 Ω and 1000 Ω are also manufactured. For calibration purposes, Pt-RTDs, with R_0 equal to 10 Ω and 25 Ω, are also employed.

Metals used for RTD should exhibit the following properties:

- high resistance temperature coefficient, to ensure high sensitivity,
- high resistivity to enable the construction of physically small RTD,
- possibly high melting temperature,
- stable physical properties,
- high corrosion resistance,
- easy reproducibility of the metal with identical properties,
- continuous and smooth dependence of resistance versus temperature without any hysteresis,
- sufficient ductility and mechanical strength.

In practice only pure metals are used, since they guarantee perfect reproducibility and easy standardization. As platinum fulfils all of these requirements in the best way, it is the most commonly used metal. The temperature dependence of the resistance of the three materials (platinum, nickel and copper) is shown in Figure 5.1. Although nickel and copper are also used, such use is restricted to lower temperatures. Tables XV to XVIII present the resistance versus temperature relationship and the permissible tolerances of Pt-RTD and Ni-RTD resistance elements as standardized by IEC and DIN. The same relationship for non-standardized Cu sensors is given in Table XIX.

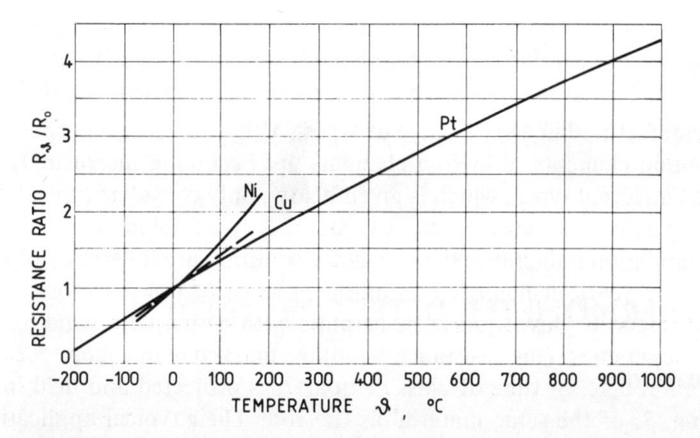

Figure 5.1
Resistance R_ϑ at ϑ °C related to R_0 at 0 °C versus temperature for Pt, Ni and Cu

5.2.2 Properties of different metals

Platinum, which is stable, forgeable and corrosion resistant, can be used at temperatures up to 1000 °C in neutral atmospheres. However, sublimation may cause resistance variations at higher temperatures. It is used in a high purity form for the manufacture of resistance thermometer elements. The ratio, $R_{100} : R_0$ of its resistance at 100 °C to that at 0 °C provides a means of judging its purity, as shown in Table 5.1. In the temperature range from 0 to 600 °C the resistance versus temperature relationship, which is also given by Equation (1.15), has the form:

$$R_\vartheta = R_0(1 + A\vartheta + B\vartheta^2)$$

Nickel, having the highest resistance temperature coefficient among all of the metals used for RTDs, is also relatively corrosion and oxidation resistant. It is normally applied in the range up to about 180 °C. An inflection at about 350 °C, changes its resistance versus temperature dependence.

Copper is rarely applied because of its poor oxidation resistance. It is sometimes used about ambient temperatures and also in refrigeration engineering. One of its popular applications is the temperature measurement of the windings of transformers and electrical machines, where the windings themselves constitute the resistance element.

Table 5.1
Metals used for resistance thermometers.

Metal	Operating temperature (°C)		Resistivity ($\mu\Omega$ m)	$\frac{R_{100}}{R_0}$
	Normal	Special		
Platinum	−220 to +850	−260 to +1000	0.10 to 0.11	1.385
Nickel	−80 to +250	−50 to +350	0.09 to 0.11	1.617
Copper	−100 to +100	−200 to +150	0.017 to 0.018	1.426

5.2.3 Construction

RTDs are manufactured in a wide range of types. Although general purpose detectors have wire-wound elements, thin film elements are becoming increasingly popular. The review of different types, which is given below, only considers typical detectors.

Wire-wound detectors, represented in Figure 5.2, form the majority of all RTDs. To ensure a minimum inductance the resistance windings are invariably of the bifilar form. For many years, it has been common practice to use detectors made from cylindrical formers of glass, quartz or ceramic with a fine wire winding of about 0.01–0.1 mm diameter. This resistance winding, marked 1 in Figure 5.2(a), which is formed on a rod, 2, or tube of glass or quartz, is protected and held in place by a thin coating, 3, of the same material as the rod. Their typical application range is from −200 °C to +500 °C for Pt-in-glass elements, up to +600 °C for quartz glass and up to +1000 °C for ceramic types (Evans, 1972). Figure 5.2(b) presents

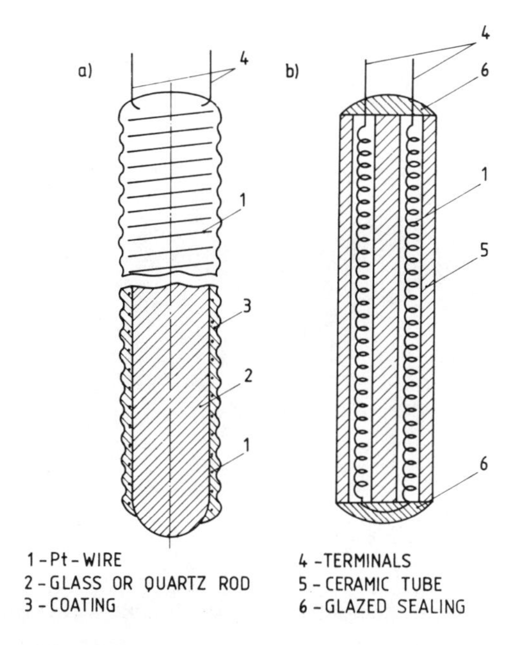

Figure 5.2
Pt-wound resistance thermometer detector.

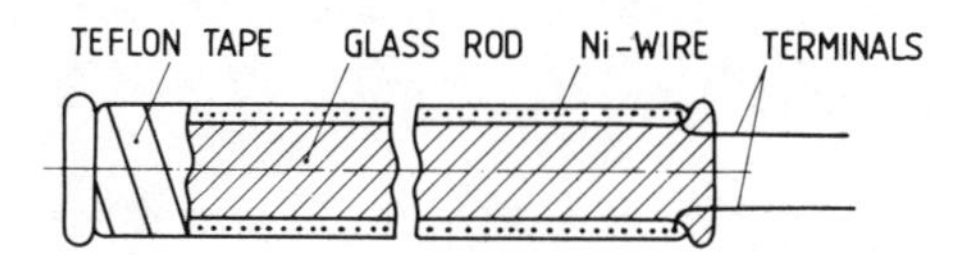

Figure 5.3
Ni-wound resistance thermometer detector.

another construction of an RTD in which small diameter, helical-shaped, resistance windings are placed in the bores of a ceramic tube, 5. Johnston (1975) has pointed out that this structure is especially vibration resistant. Multiple detectors, having two or three independent windings, are used for simultaneous temperature measurement and control, as well as in some bridge circuits to increase the system sensitivity. In the temperature range from -60 to $+150$ °C, Ni-wire wound detectors are used. In this type, shown in Figure 5.3, the resistance wire is insulated and protected by teflon tape or glass fibre.

In thin-film RTDs, which have been considered by Clayton (1988), the platinum material is deposited upon a suitable ceramic substrate. Some Pt-100 Ω detectors, which may be used up to about 600 °C for gas and surface temperature measurement, are as small as $10 \times 3 \times 1$ mm. This small size gives a time constant $N_T = 11$ s (see Chapter 9) in an air flow with a velocity of 1 m/s (Heraeus GmbH, Germany). They may be either glued or soldered to the body surface (§10.3.2.). The smallest thin-film RTDs, which can be used from -50 to $+200$ °C, have the dimensions $2 \times 2.3 \times 1$ mm (Omega Inc., USA). Miniature resistance detectors are shown in

(a)

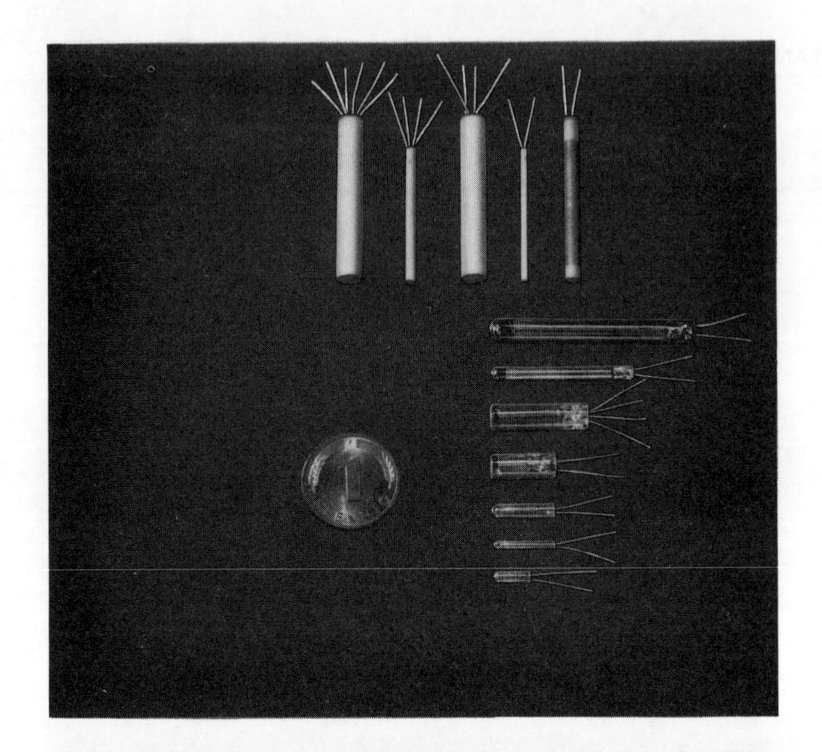

(b)

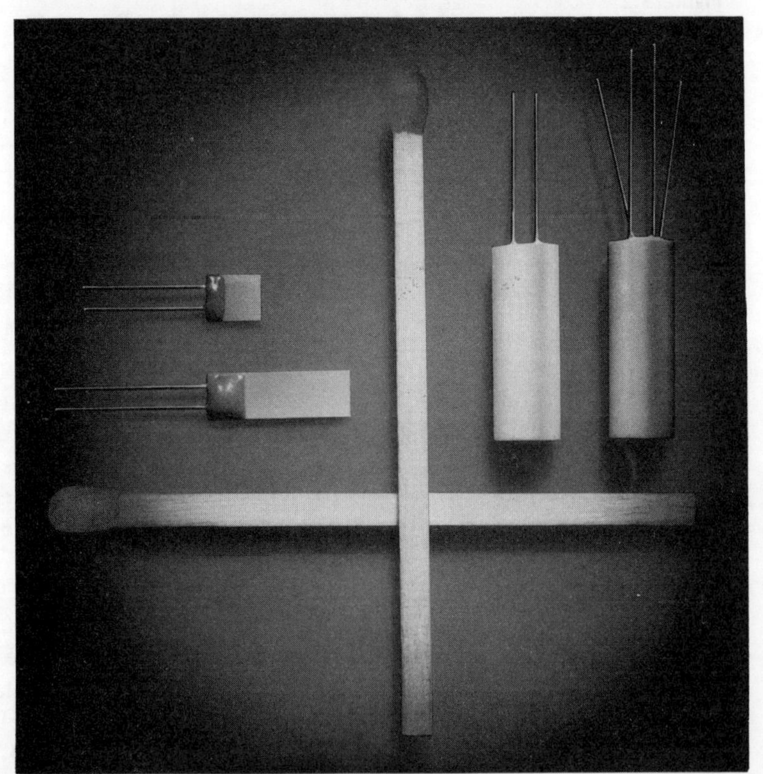

Figure 5.4
Miniature resistance thermometer detectors. (a) Pt-wire wound in ceramic and quartz, (b) Pt-thin film in ceramic (Courtesy of Heraeus GmbH, Germany).

Figure 5.4. New manufacturing technology has led to the production of 1000 Ω ceramic thin film RTDs, whose resistance is adjusted by a computer controlled laser (Hycal Engineering, USA). The high detector resistance markedly reduces the influence of connecting lead resistances upon the readings of the thermometer. They are used in the temperature range from −200 up to +540 °C.

A great diversity of special purpose elements also exist. Table 5.2 presents the properties of typical resistance detectors.

Measuring current

To measure the actual resistance of an RTD an electric current is passed through the detector. As this current produces a heating effect, which may cause additional measuring errors, it must be limited to a permissible value. The maximum permissible current is a function of the heat exchange from the detector, which depends on the physical shape of its surface, the material of its protective sheath and on the surrounding medium. The permissible measuring current, $I_{T,max}$, may be expressed in terms of the maximum allowable self-heating temperature error, $\Delta\vartheta_{max}$ K, the resistance of the detector, R_ϑ, at ϑ °C and the heat dissipation constant, C in W/K, by the equation

$$I_{T,max} = \sqrt{\frac{\Delta\vartheta_{max} C}{R_\vartheta}} \tag{5.2}$$

The heat dissipation constant, C, is defined as that heating power which causes, in the thermal steady-state, the self-heating of a resistance element by 1 K. It is also possible to use the constant S, expressed in K/mW, instead of C. In that case S gives the self-heating error related to the resistance detector power in the thermal steady-state. The constant C in still water is about forty times bigger than in still air. Some values for C, in still and flowing air, are given in Table 5.2. Measuring currents of about 1–2 mA do not usually cause any significant self-heating errors.

Numerical example

A Pt-100 Ω resistance detector without any protective tube is located in still air at 60 °C. Calculate the permissible measuring current, $I_{T,max}$, to ensure that the self-heating error is less that 0.5 K. Assume that the heat dissipation constant, C, has a value of 4×10^{-3} W/K.

From Equation (5.2):

$$I_{T,max} = \sqrt{\frac{0.5 \times 4 \times 10^{-3}}{123.24}} = 4 \text{ mA}$$

5.3 RESISTANCE THERMOMETER SENSORS

Sometimes resistance detectors may be immersed directly in the medium whose temperature is to be measured. In general, however, they require protection against

Table 5.2
Typical resistance thermometer detectors.

Construction		Detector	Dimensions (mm)	Temperature range (°C)	Heat dissipation constant A (mW/°C)		Dynamic properties			
							in flowing water $v = 0.2$ m/s		in flowing air $v = 1$ m/s	
Inner insulation	Outer insulation				in still air	in flowing air $v = 1$ m/s	$t_{0.5}$ (s)	$t_{0.9}$ (s)	$t_{0.5}$ (s)	$t_{0.9}$ (s)
Cylindrical[1]										
Ceramic	Ceramic	Pt–100 Ω wire	Ø 1 to 2.8 l 25 to 50	−220 to +750 short-term to +850	1.5 to 9	12 to 50	0.2 to 0.6	0.6 to 1.9	2.5 to 12	9 to 40
Cylindrical[1]										
Ceramic	Ceramic	2×Pt–100 Ω wire	Ø 1.7 to 2.8 l 25 to 50	−220 to +750 short-term to +850	3 to 8	12 to 20	0.4	1.6	4.5 to 10	14 to 37
Cylindrical[1]										
Glass	Glass	Pt–100 Ω wire	Ø 2.5 to 5 l 12 to 60	−220 to +600	4	13	0.8	2.5	10	30
Cylindrical[1]										
Glass	Glass	2×Pt–100 Ω wire	Ø 3.6 to 5 l 20 to 60	−220 to +600	4	13	0.7	2.4	10	30
Cylindrical[1,2]										
Glass	Ceramic glaze	Pt–100 Ω wire	Ø 1.8 to 3.8 l 20 to 30	−220 to +500	2	10	0.4	2.5	8	20
Cylindrical[1,2]										
Glass	Silicon	Pt–100 Ω wire	Ø 1 to 1.8 l 12 to 23	−60 to +350	1.2	5	0.2	0.6	2	7

Flat[1]										
Ceramic	Glass	Pt–100 Ω thin film	10.2 × 3.2 × 1	−200 to +500	1.9	75 (in well stirred water)	0.35	—	7.8	—
Cylindrical[2]										
Ceramic	Ceramic	Pt–100 Ω thick film	Ø 3.1 *l* 28	−70 to +500	—	50 (in well stirred water)	0.08 at $v = 1$ m/s	0.5 at $v = 1$ m/s	—	—
Cylindrical[2]										
Ceramic	Ceramic	Pt–100 to 2000 Ω thin film	Ø 3 *l* 7 ÷ 12	−50 to +600	—	—	—	—	—	—
Flat[2]										
Ceramic	Ceramic	Pt–100 to 1000 Ω thin film	2 × 23 × 1 2 × 10 × 1	−50 to +200	—	—	—	—	—	—
Cylindrical[1]										
Ceramic	Plastics	Ni–100 Ω wire	Ø 5.2 *l* 30	−60 to +150	4	16.6	—	—	—	—

1. Heraeus, GmbH, Germany.
2. Omega Engineering Inc., USA.

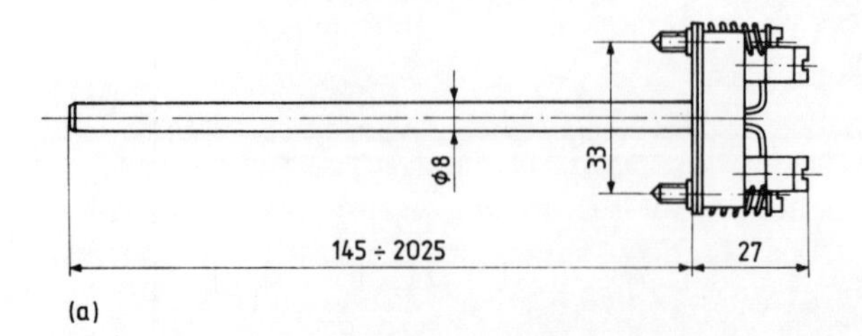

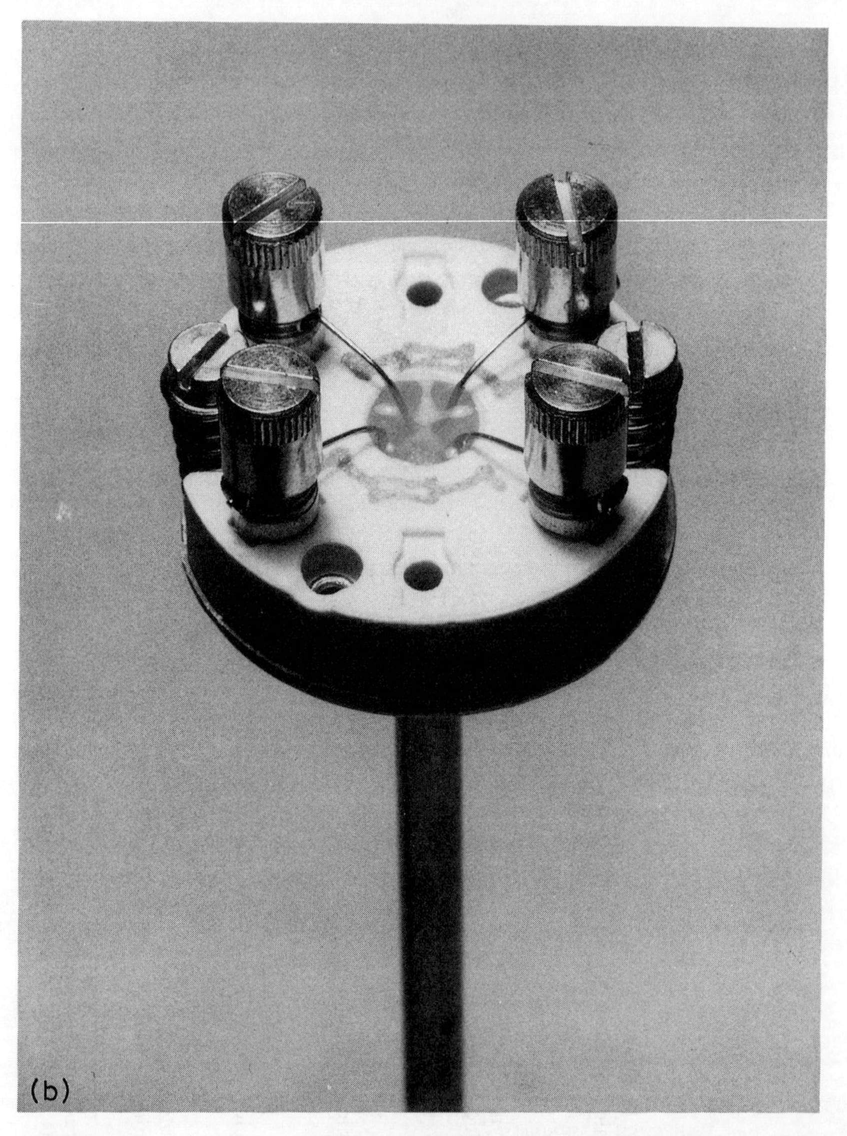

Figure 5.5
Resistance thermometer insert (a) general view (b) spring mounted terminal plate (Courtesy of Heraeus GmbH, Germany).

moisture, chemical influences and mechanical damage. A resistance detector with its protection sheath forms a resistance thermometer sensor or assembly. The combination, which is illustrated in Figure 5.5, consisting of the detector with its connecting leads, ceramic insulation, terminal block and a thin steel sheath, forms a resistance thermometer insert. Compact ceramic insulation, which protects the resistance detector from vibrations, simultaneously ensures good thermal conductivity. Fitted with standard terminal heads, these inserts constitute easily exchangeable resistance thermometer sensor assemblies when combined with a protective sheath. Variations of sheath and thermowell structures for resistance thermometer assemblies have the same general properties as those for thermocouples given in Table 4.4.

Vibrations and shocks, caused by working in a variety of different machine environments or by the flow of liquid media around them, are the principal causes of damage to resistive sensors. Therefore, resistance thermometer inserts should be placed inside a protective sheath in a shock-proof but elastic manner. Such an assembly should also ensure good heat conduction between the resistance element and the environment within which the measurement is to be made. Protection against high-frequency vibration or high acceleration may be achieved by pressing the insert against the sheath so that no relative displacement occurs between them (Figure 5.6).

Mineral insulated (MI) resistance thermometer sensors, similar to MI thermocouples, have the Pt-resistance detector and its connecting leads embedded in a compacted powder of MgO in a steel sheath as shown in Figure 5.7. Their outer diameters, which are 3 to 8 mm, allow a normal bending radius of 12 times the sheath diameter, while their total length may be 200 m (TC Ltd, UK). The end part of the sheath, of around 100 mm, should neither be bent nor worked in any way. An application range of − 100 to + 500 °C means that they are widely used in the chemical industry and also in power engineering. Normal forms are either 100 Ω or double 2 × 100 Ω sensors.

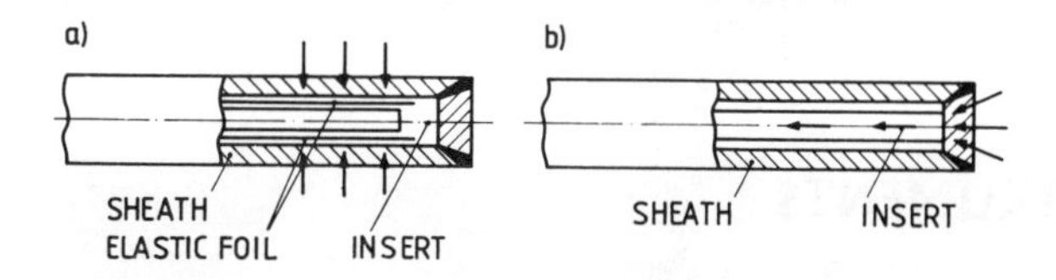

Figure 5.6
Resistance thermometer detector. Heat transfer from sheath to detector. (a) from cylindrical surface, (b) from bottom.

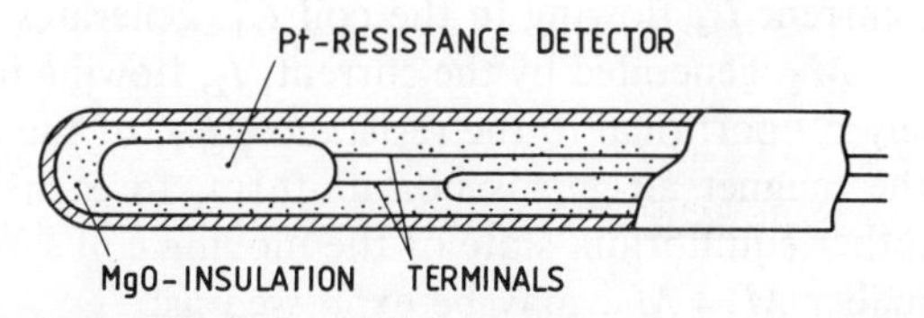

Figure 5.7
Mineral insulated resistance sensor.

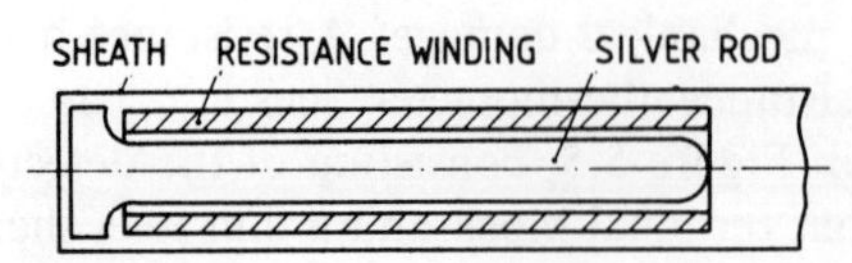

Figure 5.8
Resistance thermometer detector with a silver rod for improved dynamics.

Dynamic behaviour of resistance temperature detectors is very important. The following constructional forms can improve this performance:

- A silver rod of high thermal conductivity may be inserted into resistance detectors with an axial bore as shown in Figure 5.8.
- The detector may be soldered into the sheath if its surface is metallized (Heraeus GmbH, Germany).
- A reduction in thermal capacity is allowed if the resistance detector is hollow (Johnston, 1975).

For gas temperature measurements, resistance detectors, which have ventilated or ribbed covers, ensure a maximum heat transfer intensity.

There is a diversity of special purpose resistance sensors for temperature measurement. A variety of application areas such as in extruding processes, nuclear power stations, silos, water tanks, pipe-linings and so on make some especially stringent demands upon the detector assemblies. In a laboratory environment, it is normal practice to employ detectors which are either protected by placing them in a quartz tube or left completely unprotected.

Resistance sensors are produced in 2-, 3- or 4-wire configurations. This is also the case when there are two or three detectors in the same assembly. The purpose of the multiple wires and multiple detectors will be explained as each different measuring circuit is considered.

5.4 QUOTIENT INSTRUMENTS

For many years quotient or cross-coil measuring instruments were the popular resistance thermometer indicators. Their operating principle is shown in Figure 5.9. Coils C_1 and C_2, which are rigidly bound to each other, rotate freely, with no restoring torque, in the magnetic field of a permanent magnet N–S, whose air-gap is non-uniform. A constant current I_1, flowing in the coil C_1, generates a torque, M_1, which opposes the torque, M_2, generated by the current, I_2, flowing in the coil, C_2. The current I_2, is inversely proportional to the resistance, R_T, of the resistance thermometer detector. As the magnet air-gap is non-uniform, to each measured temperature corresponds another equilibrium state of the moving coil system. This state, which satisfies the equality $M_1 = M_2$, may be expressed as:

$$B_1(\alpha)I_1n_1 = B_2(\alpha+\beta)I_2n_2 \tag{5.3}$$

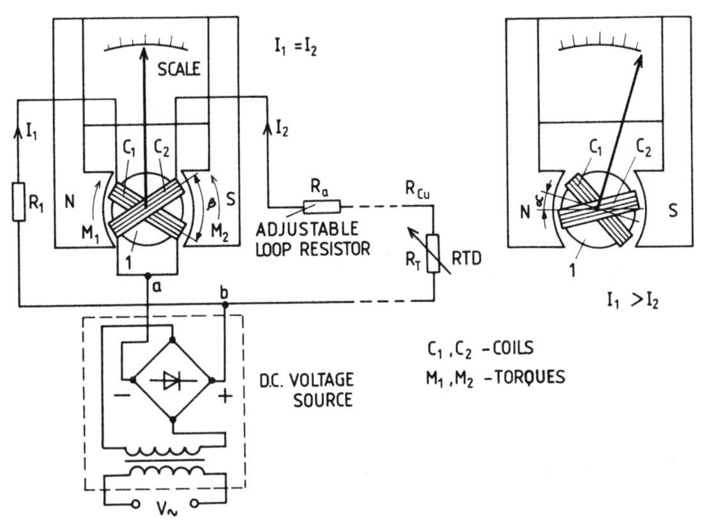

Figure 5.9
Quotient instrument, two-wire system.

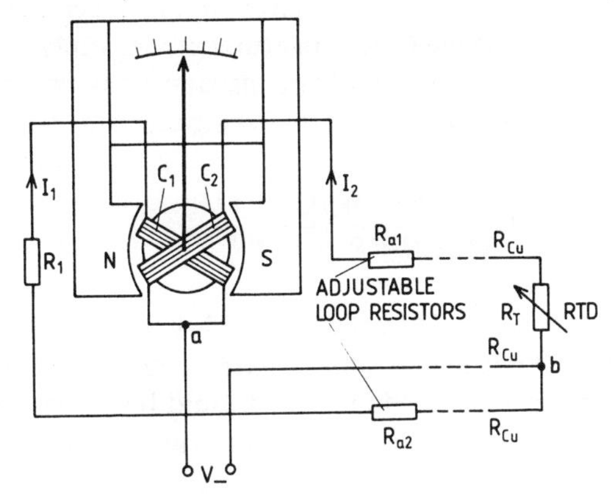

Figure 5.10
Quotient instrument, three-wire system.

In this equation, I_1 and I_2 are the coil currents, n_1 and n_2 are the numbers of turns in each coil, $B_1(\alpha)$ is the magnetic flux density in the air-gap of the magnet for each position of the coil, C_1, and $B_2(\alpha+\beta)$ is the magnetic flux density in the air-gap of the magnet for each position of the coil, C_2. The pointer deflection angle is denoted by α.

It is easily shown that the pointer deflection angle, α, and thus the scale reading, is proportional to the value of detector resistance, R_T, and is thus a function of the measured temperature. The scale division is non-linear. A first advantage of the instrument is its large deflection torque, which makes it insensitive to shock and vibration. Secondly, as the coils are fed from the same supply voltage, the readings are independent of variations in the supply voltage. Consequently a stabilized supply voltage is not required.

The resistance of the copper connecting leads influences the thermometer readings. This resistance, R_{Cu}, should not exceed the value of the nominal loop resistance, R_{ln}, given on the instrument scale. To match the actual loop resistance to its nominal value, R_{ln}, additional adjustable resistance, R_a, is used. A standard reference resistor, whose resistance corresponds to the RTD resistance at the given known temperature, is also supplied. Matching of the reading of the indicating instrument with the temperature given at the standard resistor should be achieved by adjusting the value of the resistance, R_a, while the RTD is temporarily replaced by the standard resistor.

Influences on the instrument readings, due to any variations in the resistance of the copper connecting leads, may be eliminated using a three-wire system shown in Figure 5.10. It can be seen from this diagram that the loop resistances of both coils vary in the same way.

5.5 DEFLECTION TYPE BRIDGE CIRCUITS

In the Wheatstone bridge of Figure 5.11 the resistance thermometer detector, R_T, is switched as one of four bridge resistors. If the bridge is energized from a stabilized voltage source, Z, its off-balance voltage, V_{ab}, is a measure of the temperature. The bridge output signal is measured by the measuring instrument, M, which is calibrated in temperature units. If the internal resistance of the instrument is far greater than the bridge equivalent resistance, the voltage V_{ab} is given by:

$$V_{ab} = V\frac{R_1 R_T - R_2 R_3}{(R_1 + R_2)(R_3 + R_T)} \tag{5.4}$$

As a three-wire circuit is being used, it has been assumed that the resistance of the copper connecting leads and the equalizing resistances have no influence on the bridge readings. Point b, where the third wire is connected, should be placed as near the sensor's RTD as possible. This occurs well inside the protective sheath of the sensor.

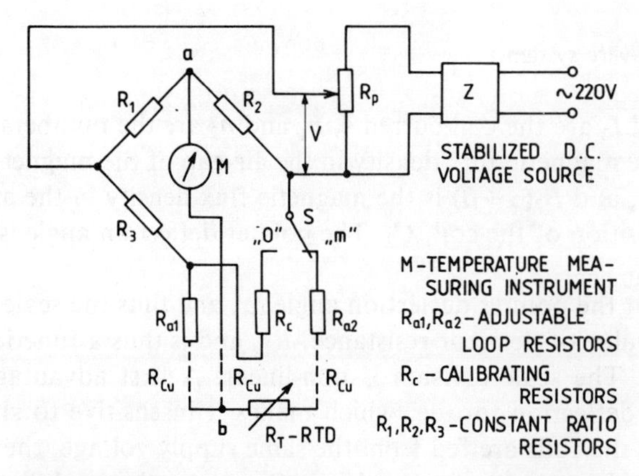

Figure 5.11
Deflection type bridge circuit.

The readings depend upon supply voltage, V. Hence, means are provided for manual voltage adjustment by setting of the resistor, R_p. For this purpose the switch S should be set in the position '0', which connects the calibrating resistor, R_c, in the bridge circuit. The voltage, V, is properly adjusted when the readings of instrument, M, are brought to a marked scale position. Periodic readjustment should be performed to the recommendation of the producer. It is possible to dispense with the voltage adjustment, if the stability of the supply voltage is guaranteed. The error limit of the indicator scale, which is nearly linear, is about $\pm 1\%$ to $\pm 1.5\%$ of full scale. As the bridge output signal is often preamplified by a solid state operational amplifier, robust instruments may be used as indicators. Under these conditions, it is then easy to achieve both extremely narrow temperature measuring ranges, even about 1 °C, and any accompanying zero suppression which may be required. The error limits are then about ± 0.05 to $\pm 0.1\%$ of the full scale deflection. As in all similar cases a standardized transmitter system, of the type described in Chapter 8, can be used. Digital indicating meters are also used instead of deflection type instruments.

In the design of a bridge circuit, Klempfner (1976) has emphasized that care should be taken to limit the measuring current in the measuring detector so as not to risk any self-heating errors. Bridge circuits with linear scale divisions are also possible (Diamond, 1970; Kraus, 1975; Bolk, 1979; Carius, 1981).

5.6 QUOTIENT-BRIDGE CIRCUITS

Indications of a deflection type bridge circuit depend upon the supply voltage as shown in Figure 5.11. Although quotient instruments have voltage independent indications, they do not allow narrow operating temperature ranges. However, they satisfy a reasonable compromise between the two circuits mentioned above. The

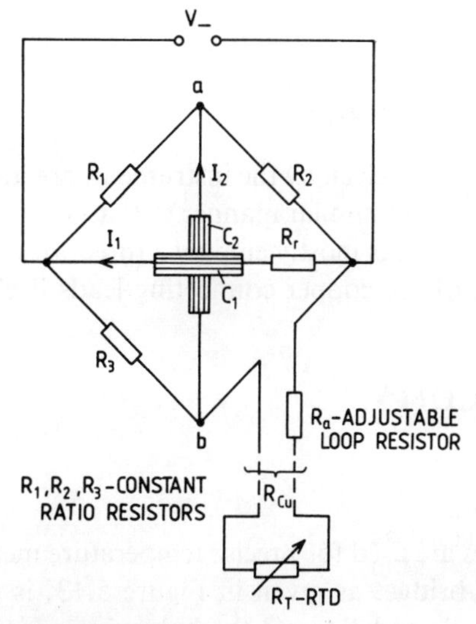

Figure 5.12
Quotient bridge circuit.

moving coil instruments, M, from Figure 5.11, is replaced by a quotient instrument as in Figure 5.12. In this case, the restoring torque, M_1, of the coil, C_1, is proportional to the bridge supply voltage, V, whereas the driving torque, M_2, is proportional to the bridge off-balance voltage, V_{ab}. Each torque opposes the other. Both the torque, M_1, and the system measuring range, can be set simultaneously by changing the resistance, R_r. The currents I_1 and I_2 in the respective coils, C_1 of resistance R_{C_1}, and, C_2 of resistance R_{C_2}, are given by:

$$I_1 = \frac{V}{(R_{C_1} + R_r)} \tag{5.5}$$

$$I_2 = \frac{V_{ab}}{R_{C_2}} \tag{5.6}$$

Combining Equations (5.4) and (5.6) yields:

$$I_2 = \frac{V}{R_{C_2}} \frac{(R_1 R_T - R_2 R_3)}{(R_1 + R_2)(R_3 + R_T)} \tag{5.7}$$

The pointer deflection angle, α, of the quotient instrument is a function of the current ratio. As the physical form of the instrument may be accounted for by a constructional constant, K, this gives:

$$\alpha = K f_1 \left(\frac{I_1}{I_2} \right) \tag{5.8}$$

Thus from Equations (5.5) to (5.8) it follows that:

$$\alpha = K f_1 \left\{ \frac{R_{C_2}(R_1 + R_2)(R_3 + R_T)}{(R_{C_1} + R_r)(R_1 R_T - R_2 R_3)} \right\} \tag{5.9}$$

Equation (5.9) proves that the indications, given by the instrument, are independent of supply voltage as well as its variations. In a similar manner to that of the deflection type bridge circuit, quotient type bridge circuits mostly employ a three-wire connection so that the influence of the resistance of the copper connecting leads is eliminated.

5.7 BALANCED TYPE BRIDGE CIRCUITS

5.7.1 Manually balanced bridges

Manually balanced Wheatstone bridges are used for precise temperature measurement in laboratory practice. The two-wire bridge, as given in Figure 5.13, is composed of two constant value ratio resistors, R_1 and R_3, and the balancing resistor, R_2. A resistance temperature detector, R_T, is connected to the bridge by copper connecting

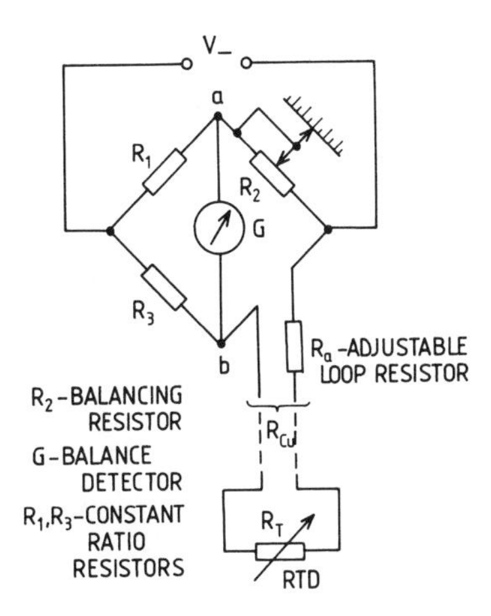

Figure 5.13
Manually-balanced bridge circuit.

leads with a resistance, R_{Cu}, while an additional adjustable resistor, R_a, is inserted in the sensor circuit. In this manner, the sum of the resistances, R_a and that of the connecting leads, equals the nominal loop resistance, R_{ln}. By adjusting the slide-wire resistor, R_2, the balance detector, G, is brought to a null-balance condition. Subsequently, at this null, the measured temperature is read on the associated scale from the slider position. A galvanometer or an electronic amplifier can be used as the detector of the balance condition, which is described by the equation:

$$R_1[R_T+(R_{Cu}+R_a)]=R_2R_3 \tag{5.10}$$

which yields:

$$R_T=\frac{R_2R_3}{R_1}-(R_{Cu}+R_a) \tag{5.11}$$

The indicated value is correct only in the case when $R_{Cu}+R_a=R_{ln}$. As changes in ambient temperature may cause variations in R_{Cu}, reading errors may arise. For maximum sensitivity of the circuit, the galvanometer internal resistance should have the value given by Hunsinger (1966) as:

$$R_G=\frac{(R_1+R_3)(R_2+R_T+R_a+R_{Cu})}{R_1+R_2+R_3+R_T+R_a+R_{Cu}} \tag{5.12}$$

Supply voltage variations of up to about $\pm 20\%$ may be tolerated as they exert only a slight influence upon the sensitivity of the circuit. A two-wire system, as shown in Figure 5.13, can only be used in the case when the resistance of the copper connecting leads, R_{Cu}, is nearly constant. Otherwise a three-wire balanced

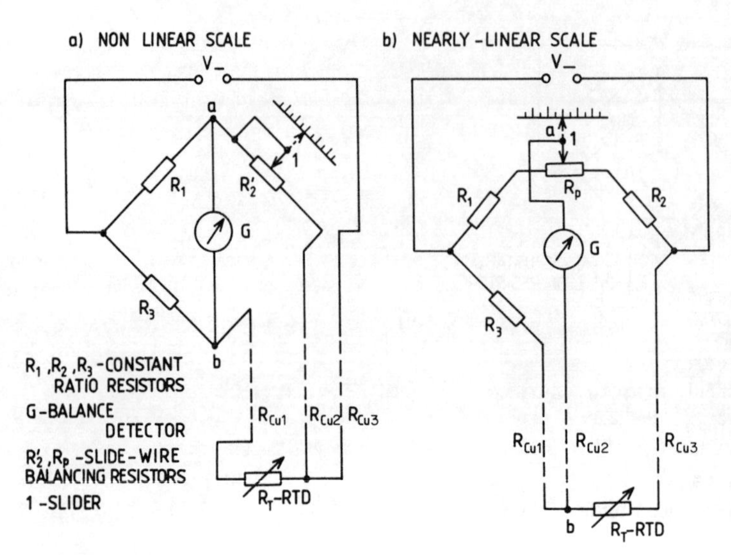

Figure 5.14
Three-wire manually-balanced bridge circuit.

bridge, as given in Figure 5.14(a), has to be applied. The balance condition for such a bridge is:

$$R_1(R_T + R_{Cu1}) = R_3(R_2' + R_{Cu2}) \tag{5.13}$$

Usually:

$$R_{Cu1} = R_{Cu2} = R_{Cu} \tag{5.14}$$

where R_{Cu} is the resistance of any one of the three wires. Consequently, the detector resistance R_T is then given by:

$$R_T = \frac{R_3 R_2'}{R_1} + R_{Cu}\frac{R_3 - R_1}{R_1} \tag{5.15}$$

In symmetrical three-wire bridges, having $R_1 = R_3$, Equation (5.15) becomes:

$$R_T = R_2' \tag{5.16}$$

Equation (5.16) shows that the measured value of R_T is not influenced by the resistances of the connecting leads.

A modification of the arrangement, which has been described above, is shown in Figure 5.14(b). In this circuit it is clear that the contact resistance at the slider of the resistor, R_p, is introduced into the balance detector circuit. Hence, this parasitic contact effect exerts absolutely no influence upon the accuracy with which the bridge may be balanced. Simultaneously, a more linear scale is achieved.

Manually balanced bridges are mostly used in laboratory measurements where the error limit is about ±0.1% of the full scale. Four-wire bridges are also used. In this configuration, an additional blind loop is added alongside the normal two-wire connection to the sensor RTD as described in BS 1041: Part 3.

5.7.2 *Automatically balanced bridges*

In both industry and laboratories the use of automatically or self-balanced bridges is becoming more frequent. Most manufacturers produce them as a special purpose modification of the automatic resistance bridge potentiometers previously described in §4.8.3. The modification occurs only in the input measuring circuit as Figure 5.15 indicates.

The sensor, with resistance, R_T, is connected to the bridge by a three-wire system as shown in Figure 5.15(a). Balancing of this bridge is accomplished by a symmetrically arranged slide-wire resistor, R_p, which is shunted by a resistor, R_s. This arrangement permits the use of a slidewire resistor, R_p, with a higher resistance value. As such an increase in resistance is realized with a higher number of turns, the resolution is commensurately increased. Setting of the measurement range is made using the resistor, R_s. A chopper amplified bridge off-balance voltage, V_{ab}, drives the servo-motor in precisely the same way as has already been described in §4.8.3. The scale is nearly linear. The adjustable resistors, R_a, which are only sometimes necessary, depend upon the values of R_1, R_2 and R_3. Instructions for the adjustment of the values, R_a, are provided by the equipment producer.

In four-wire systems, as shown in Figure 5.15(b), no adjustable resistors are needed. The sensing RTD, R_T, is connected to the bridge by a two-wire line having the resistance R_{Cu1}. An identical blind loop, lying alongside the sensor wire pair, is

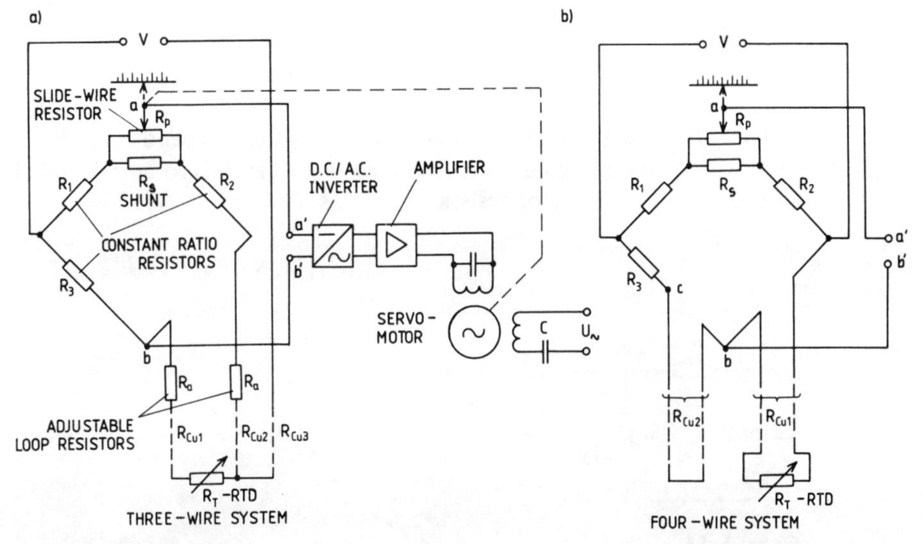

Figure 5.15
Automatic balanced bridge circuit.

connected to the terminals cb. The blind loop has an identical resistance to the sensor wire pair so that $R_{Cu2} = R_{Cu1}$. By this means perfect compensation of ambient temperature variations is achieved.

Automatically balanced bridges are usually supplied from a d.c. source. In this way, any indication errors from the induced parasitic e.m.fs at the frequency of the mains supply voltage, are avoided. A more detailed description of similarly constructed automatic potentiometers is given in §4.8.3.

The smallest measured temperature ranges are about 5 °C and the permissible resistance of the copper connecting leads is about 10 Ω. Close similarity in the construction of automatic potentiometers and bridges, allows the provision of separate terminals for thermocouples and RTDs in the same multi-point recorder, so increasing their flexibility of application.

5.8 VOLTAGE DIVIDER CIRCUITS

Simple resistance voltage divider circuits, of the type shown in Figure 5.16, are also used in industry. The circuit, supplied from a stabilized voltage source, employs the voltage drop, V_T, across the detector resistance, R_T, as the output voltage. This is given by:

$$V_T = V\frac{R_T}{R_s + R_T} \tag{5.17}$$

Maximizing the output voltage, V_T, is possible by correctly choosing the value of R_s. Suppose the temperature measurement range is from ϑ_1 to ϑ_2. If the corresponding values of R_T vary from R_{T1} to R_{T2}, then, according to Klempfner (1976), the maximum valuc of V_T is obtained when the series resistance R_s has the value given by:

$$R_s = \sqrt{R_{T1}R_{T2}} \tag{5.18}$$

The maximum supply voltage, V_{max}, which depends upon the maximum permissible measuring current, $I_{T,max}$, of the RTD, has to be calculated on the basis of the minimum detector resistance R_{T1} as:

$$V_{max} = I_{T,max}(R_{T1} + \sqrt{R_{T1}R_{T2}}) \tag{5.19}$$

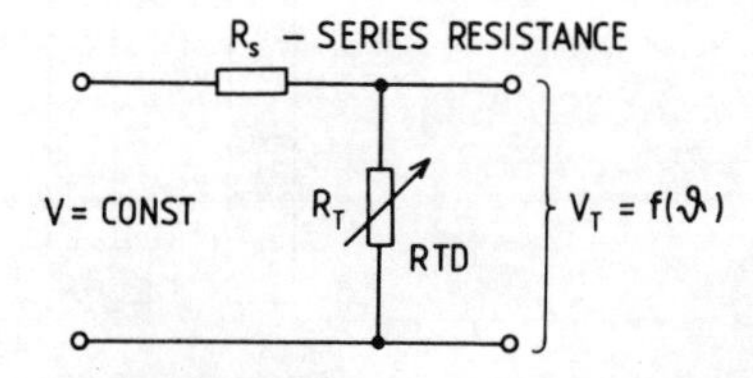

Figure 5.16
Voltage divider circuit.

When these values are satisfied the output voltage of the circuit then varies between the limits:

$$V_{T,min} = V_{max}\left(\frac{R_{T1}}{R_{T1} + \sqrt{R_{T1}R_{T2}}}\right) \tag{5.20}$$

$$V_{T,max} = V_{max}\left(\frac{R_{T2}}{R_{T2} + \sqrt{R_{T1}R_{T2}}}\right) \tag{5.21}$$

Usually, the output signal of the circuit is amplified in an operational amplifier. At the same time the amplifier also permits the setting of the signal zero to be set to correspond with $R_{T,min}$.

Numerical example

Calculate the range of the variations in the output signal of a voltage divider circuit to be used for temperature measurement in the temperature range from $\vartheta_1 = 0$ °C to $\vartheta_2 = 500$ °C. The permissible measuring current $I_{T,max} = 0.003$ A. For the Pt-100 Ω RTD used at $\vartheta_1 = 0$ °C, $R_{T1} = 100\ \Omega$, and at $\vartheta_2 = 500$ °C, $R_{T2} = 280.94\ \Omega$.

From Equation (5.19) it follows that:

$$V_{max} = 0.003 \times (100 + \sqrt{100 \times 280.94}) = 0.8\ \text{V}.$$

Finally from Equations (5.20) and (5.21) one obtains:

$$V_{T,min} = 0.29\ \text{V and } V_{T,max} = 0.50\ \text{V}.$$

5.9 HIGH PRECISION LABORATORY CIRCUITS

Four terminal platinum elements of $R_0 = 10\ \Omega$ or $25\ \Omega$, are usually applied as resistance thermometer elements for high precision instruments (see also §16.2.2). For the most commonly used 25 Ω elements, the resistance variation, corresponding to a temperature difference of 1 °C, is about 0.1 Ω. The smallest achievable error limits of resistance measurements are estimated to be about some 10^{-4}% to 10^{-5}% of the measured values. Voltage comparison methods, the modified Wheatstone bridge and the double Kent bridge are the three main methods for the measurement of resistance.

A voltage comparison method, as in Figure 5.17, permits the total elimination of any influence due to the resistance of the copper connecting leads. This arises in the configuration of Figure 5.17 because the voltage drops across the resistance detector, R_T, and the standard resistor, R_s, are measured under conditions of zero current. Both resistors carry the same current, I_1. The potentiometer circuit for voltage drop

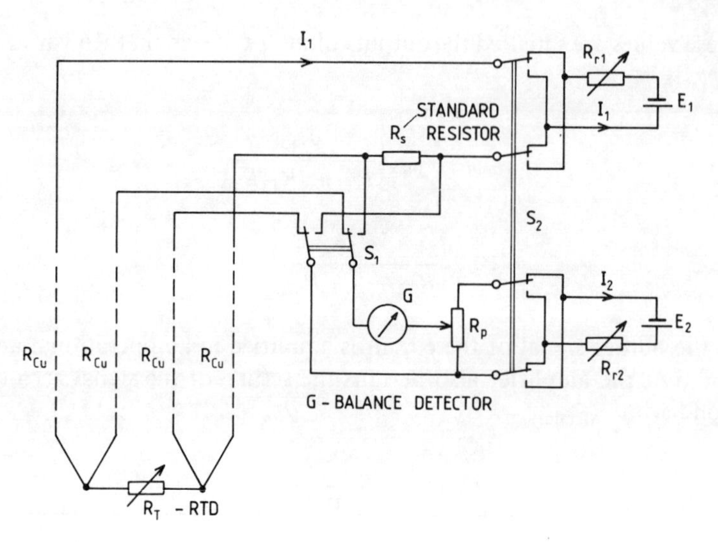

Figure 5.17
Voltage comparison method.

measurements, comprising the potentiometric resistor, R_p, balance detector, G, and the circuit for the auxiliary current, I_2, has been presented in a very simplified way. The voltage drops, V_T and V_s, across the detector resistance, R_T, and R_s are given respectively by:

$$V_T = I_1 R_T \tag{5.22}$$

$$V_s = I_1 R_s \tag{5.23}$$

The two equations yield the measured value of R_T as:

$$R_T = R_s \frac{V_T}{V_s} \tag{5.24}$$

To eliminate any parasitic thermoelectric e.m.fs, the voltage drops, V_T and V_s, should be measured twice by reversing the direction of the current flow, I_1, using the switch, S_2. Thus, for the measurement of one value of R_T, four consecutive measurements have to be made. If the temperature to be measured varies in time, some additional errors may occur.

This method is very convenient if the resistance, R_T, to be measured varies over a wide range. Such variations are possible. For example, when a thermometer, having a resistance of 25 Ω at 0 °C, is used at very low temperatures, its resistance may be even less than 0.1 Ω. By using a standard resistor of suitable resistance, and choosing the measuring current properly, the desired precision can easily be obtained. A nearly constant percentage of precision of measurement may be achieved over a large range of values of the resistance of R_T. By contrast, bridge methods only give constant absolute error, while their percentage precision increases for smaller measured resistances. Another advantage of the voltage comparison method is the elimination of the resistances of the copper connecting leads. As these lead resistances may be of the same order of magnitude as small measured values of R_T, such a benefit is significantly important. Usually, the same low voltage potentiometers are used as for the measurement of thermocouple e.m.fs. Mueller (1941) has established that the error limit may be kept well under 0.001% of the measured value.

A modified Wheatstone bridge, also known as the Mueller bridge (Mueller, 1941; Baker *et al.*, 1961; Hunsinger, 1966), is shown in Figure 5.18 in simplified form. The two ratio resistors, R_1 and R_3, have to be made equal, before the measurement, by adjusting the slide-wire resistor r, so as to obtain $R_1+r_1=R_3+(r-r_1)$. Four copper leads, having resistances R_{Cu1}, R_{Cu2}, R_{Cu3} and R_{Cu4} connect the resistance thermometer detector, R_T, to the switch, S. The resistance detector, R_T, is also connected in series with the balancing resistor, R_4, by which the bridge balance is achieved. Balancing of the bridge is performed twice, in the two positions of the switch, S, giving the two balance conditions as:

$$R_T+R_{Cu3}-R_{Cu2}=R_2-R_4' \tag{5.25}$$

and

$$R_T+R_{Cu2}-R_{Cu3}=R_2-R_4'' \tag{5.26}$$

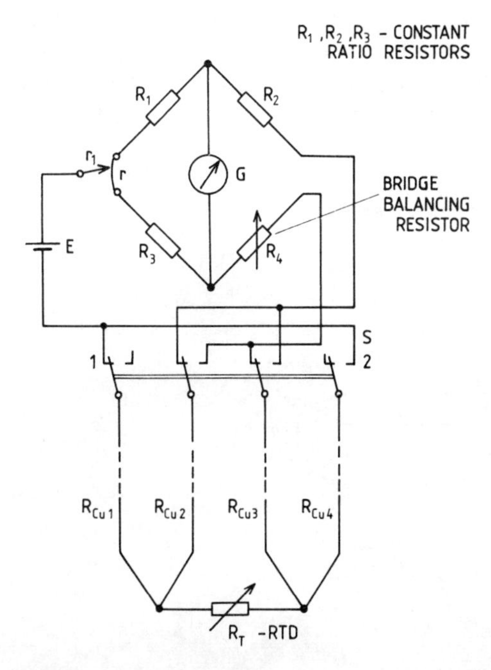

Figure 5.18
Mueller bridge.

From Equations (5.25) and (5.26), it follows that the resistance of the detector, R_T, may be expressed in terms of the values of the balancing resistor, R_4' and R_4'', in the respective positions 1 and 2 of the switch, S, as

$$R_T = \frac{(R_2 - R_4') + (R_2 - R_4'')}{2} \tag{5.27}$$

As the balancing resistor, R_4, and the sensor element, R_T, are in the same bridge arm, the external loop resistance of the galvanometer is constant. After balancing the bridge by suitable setting of the six decade resistor, R_4, this property makes interpolation of the galvanometer readings possible. These are directly proportional to the difference between the actual value of R_T and the value found from the bridge balance condition.

To reach the highest possible measurement precision, the temperature of the ratio resistors should be stabilized to within ± 1 °C. The resistors themselves should be made from manganine in a wire-wound form which is as tension free as possible. In practice the highest possible precision of the method is about 0.0001%. Commonly, precisions in the order of 0.005% to 0.05% are achieved in laboratory applications.

The Thomson double bridge has been adapted for temperature measurement by F. E. Smith. This bridge, whose schematic is given in Figure 5.19, is known as the Smith Bridge Type III (Hall, 1956; Stimson, 1956; Hall and Barber, 1964). Once again the resistance detector, R_T, is connected to the bridge circuit by four connecting leads with respective resistances R_{Cu1}, R_{Cu2}, R_{Cu3} and R_{Cu4}. The ratio resistors have the values $R_1 = 1000\ \Omega$ and $R_2' = 10\ \Omega$. Also the resistor R_2'' and part of the resistor R_4 have constant values. For a given value of R_2'', the adjustable part of R_4 is always equal to the adjustable resistance, R_3. In this manner the bridge can be balanced for any

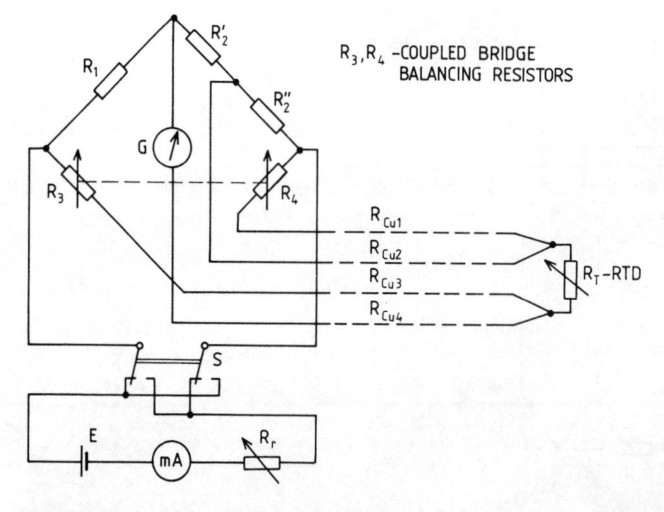

Figure 5.19
Smith Bridge, type III.

value of R_3. The adjustable part of R_4, is mechanically coupled with R_3 and both are simultaneously changed. The bridge is balanced when:

$$R_T = R_3 \frac{R_2'}{R_1} + (R_{Cu3} - R_{Cu2}) \frac{R_2'}{R_1} \tag{5.28}$$

and when at the same time

$$R_4 = R_2'' \frac{R_2' + R_3}{R_1 - R_2'} \tag{5.29}$$

The influence of the resistance of the connecting leads on the results of the measurement is accounted for by the term $(R_{Cu3} - R_{Cu2})(R_2'/R_1)$ in Equation (5.28). Even in the case where this term is neglected the error would be less than about 0.0002 °C assuming $R_1 = 1000\,\Omega$ and $R_2' = 10\,\Omega$. The elegance of this circuit, as compared with those formerly mentioned, is that only a single reading is required for an R_T measurement.

In measurements of the highest precision it is always necessary to eliminate the resistance of the connecting leads. In the Smith Bridge Type III, this can be achieved by the preliminary adjustment of the resistances of the connecting leads, so that when their sequence is changed from 4, 3, 2, 1 to 1, 2, 3, 4 the bridge balance is maintained. Very precise measurement of R_T, independent of the resistance of the connecting leads, can also be achieved by making measurements twice. This is allowed by a reversal of the switch, S, which changes the supply voltage polarity, and taking the median value of the two measurements as the final result.

In the most precise bridges, the resistors, R_3 and R_4, have seven decades. Inside the bridge housing the temperature is stabilized with a tolerance of ± 0.01 °C and its uniformity is better than 0.04 °C. An overall precision in the measurement of R_T of about $10^{-5}\%$ is possible. For laboratory applications, smaller, more compact variations of the Thomson double bridge are built with a precision approaching 0.1 to 0.01%.

Up-to-date versions are totally automatic a.c. bridges, with integral microcomputer and digital display. An example is that manufactured by Automatic Systems Laboratories Ltd (UK). This bridge, which is based upon a four-wire sensor, has an accuracy amounting to $+0.01$ °C when using a Pt-100 Ω detector. The bridge may be incorporated into a computer system.

REFERENCES

Baker, H. D., Ryder, B. A. and Baker, N. H. (1961) *Temperature Measurement in Engineering*, Vol. 2, John Wiley and Sons, New York.

Bolk, W. T. (1979) Linearisierung der Kennlinie von Pt-Widerstandsthermometern. *Techn. Messen.*, **46**(10), 375–376.

Carius, W. (1981) Vielstellen-Temperaturmesschaltung mit Linearisierung für Pt-100-Widerstandsthermometer. *ATM*, (4) 127–130.

Clayton, W. A. (1988) Thin-film platinum for appliance temperature control. *IEEE Trans. Ind. Appl.* **24**(2), 332–336.

Diamond, J. M. (1970) Linearisation of resistance thermometers and other transducers. *Rev. Sc. Instr.*, **41**(1), 53–56.

Evans, J. P. (1972) High temperature platinum resistance thermometry, *Temperature: Its Measurement and Control in Science and Industry*, Vol. 4, Part 2, Instrument Society of America, Pittsburgh, pp. 899–906.

Hall, J. A. (1956) The international temperature scale, *Temperature: Its Measurement and Control in Science and Industry*, Vol. 2, Reinhold Publ. Co., New York, pp. 115–140.

Hall, J. A. and Barber, C. R. (1964) Calibration of temperature measuring instruments. *Ed 3 Notes on Applied Science No. 12*, National Physical Laboratory, London.

Hunsinger, W. (1966) *Temperaturmessung: Handbuch der Physik*. Springer-Verlag, Berlin.

Johnston, J. S. (1975) Resistance thermometry. *Temperature Measurement: Conference Series No. 26*, Institute of Physics, London, 1975, pp. 84–90.

Klempfner, F. (1976) *Widerstandsthermometerschaltungen*, Hartmann and Braun AG, Bericht 02/3522, Frankfurt.

Kraus, E. (1975) Eine neue Methode zur Linearisierung eines Platinum-Widerstandsthermometers. *ATM*, (11), 187–190.

Mueller, E. F. (1941) Precision resistance thermometry. *Temperature: Its Measurement and Control in Science and Industry*, Reinhold Publ. Co., New York, pp. 162–179.

Stimson, H. F. (1956) Precision resistance thermometry and fixed points. *Temperature: Its Measurement and Control in Science and Industry*, Vol. 2, Reinhold Publ. Co., New York, pp. 141–168.

6 Semiconductor Thermometers

6.1 INTRODUCTION

Semiconductor thermometers (Sachse, 1975) are made from materials which are neither conductors nor insulators. Research of the thermal properties of semiconductors was first reported by William Faraday in 1834. Their industrial production was started at the Bell Telephone Company and, simultaneously, at Osram in 1930. It is apparent from the work of many authors such as Sze (1969) and van der Ziel (1968), among others, that these materials may have an intrinsic, or pure form, a compound form or a doped form. Compound and doped semiconductors are often called extrinsic semiconductors.

Thermometers of this type, which may use bulk material temperature dependencies or junction effect carrier density relations, may be classified by the number of electrodes and number of junctions possessed per sensor. This ordering is based upon that used by Sze (1969) in the classification of semiconductor devices. There are two main groups of semiconductor thermometers, whose operation is based upon different manifestations of the same physical principles. Bulk effect two-electrode sensors, which belong to the resistive group, possess no semiconductor junctions. They are thermistors or silicon-RTDs, also called Silistors by Hyde (1971). Junction device temperature sensors are either diodes with one junction and two terminals, transistors, with two junctions and three terminals, or integrated circuit sensors with multiple junctions and numbers of terminals.

Semiconductors exhibit strong temperature dependent behaviour. From fundamental physical considerations it can be shown that extrinsic semiconductors possess three main regions of temperature dependence. In doped materials at temperatures below about 150 K, and particularly within the cryogenic range, there are practically no minority carriers as most material impurities are 'frozen out'. The other two regions correspond to what may be called 'normal' (200 K to 500 K) and intrinsic (above 600 K) ranges (van der Ziel, 1968; Sze, 1969). As these effects can be very tightly controlled and predicted for doped materials their use in temperature measurement is inevitable.

In the temperature range between about 200 K and 500 K, where 'normal' semiconductor behaviour occurs, carrier mobility has a sensitivity to both doping and temperature which is well described by an empirically derived analytical expression (Arora *et al.*, 1982). Bulk effect semiconductor temperature sensors arise from this temperature dependence of mobility as well as the temperature dependent density of carriers in the bulk homogeneous regions of a material. Junction and monolithic temperature sensors depend upon the relations between carriers across junctions for their temperature dependent behaviour. At temperatures above about 600 K extrinsic materials behave in a similar manner to intrinsic materials.

6.2 THERMISTORS

6.2.1 General information

Thermistors are non-linear (Stanley, 1973) temperature dependent (Droms, 1962; Hyde, 1971) resistors with a high resistance temperature coefficient. In practice, only thermistors with a negative temperature coefficient (NTC type) are used for temperature measurement. Thermistors having positive temperature coefficient (PTC type) are only used for the binary detection of a given temperature value.

The production of thermistors, which is very complicated, uses ceramic manufacturing technology, consisting of high pressure forming and sintering at temperatures up to 1000 °C. Although the process for the manufacture of both types is similar, they are made from different materials (Roess, 1984). PTC types have a fundamental composition based upon barium titanate. Mixtures of different powdered oxides of Mn, Fe, Ni, Cu, Ti, Zn and Co are used to make NTC thermistors. Their properties depend upon their heat treatment temperature and atmosphere, as well as on the manner in which they are subsequently annealed. After the thermistor has been metal coated and trimmed to adjust its resistance, its connecting leads are then attached before encapsulation. At 20 °C the resistance of a thermistor may be in the range of some kΩ to about 40 MΩ.

It is now well established that when extrinsic semiconductor materials are in the normal temperature range they have an equilibrium product of hole carrier density, p, and electron carrier density, n, which is approximately equal to the square of the carrier density, n_i, in intrinsic material (van der Ziel, 1968; Sze, 1969). In n-type material the density of electrons is therefore:

$$n = (N_c N_d)^{1/2}\, e^{-(V_g/2kT)} \tag{6.1}$$

where N_c is the density of conduction band states, N_d is the density of donour impurities (per cm^3), V_g is the material energy gap, T is the absolute temperature (K), and k is the Boltzmann constant.

Again considering an n-type semiconductor the conductivity, σ, may be expressed as:

$$\sigma = nq\mu \tag{6.2}$$

where n is the density of electrons, q is the charge on an electron, and μ is the carrier mobility.

The expression for the mobility as a function of doping and temperature, as given by Arora *et al.* (1982), includes carrier scattering effects due to many sources. Although it may be applied for thermistors, it is easier to see the principles of temperature modulated conductivity if only lattice scattering is considered. Under these conditions, to a good approximation, the mobility may be written as:

$$\mu = k_1 T^d \tag{6.3}$$

where k_1 is a material dependent constant and d is a small-valued negative number.

Becker *et al.* (1946) combined Equations (6.2) and (6.3) to obtain the resistivity of n-type material as:

$$\rho = k_2 T^{-c}\, e^{(B'/T)} \tag{6.4}$$

where k_2 is a material dependent constant and c is a small-valued number, which may be positive or negative.

Hyde (1971) shows how the best fit to this basic relationship for NTC thermistors gives the commonly used approximation to the resistance versus temperature characteristic of a thermistor in the form:

$$R_T = R_\infty\, e^{(B/T)} \tag{6.5}$$

where T is the thermistor temperature in K, R_T is the thermistor resistance at temperature T, R_∞ is the limit value of R_T as $T \rightarrow \infty$, and B is the constant ($\neq B'$ of Equation (6.4)), depending on the thermistor material, in K, which gives the best straight line fit to a logarithmic plot of Equation (6.4) (Hyde, 1971).

Although attempts have been made to provide a better approximation (Bosson *et al.*, 1950), the approximate form given in Equation (6.5) will be used exclusively in this book. As the value, R_∞, is impossible to determine, Equation (6.5) can be expressed in terms of its resistance, R_{Tr}, at some reference temperature, T_r, usually 293 K, in the more readily usable form:

$$R_T = R_{T_r}\, e^{B[1/T-(1/T_r)]} \tag{6.6}$$

The other quantities in Equation (6.6) are the same as in Equation (6.5). Define the thermistor's resistance temperature coefficient as:

$$\alpha_T = \frac{1}{R_T}\frac{dR_T}{dT} \tag{6.7}$$

Differentiating Equation (6.6) and inserting the result together with the value of R_T from Equation (6.6) into Equation (6.7) leads to:

$$\alpha_T = -\frac{B}{T^2} \tag{6.7a}$$

From Equation (6.7a) it is evident that the absolute value of α_T, and the sensitivity of the thermistor both decrease with increasing measured temperature. The coefficient,

α_T, is usually expressed in %/K. Using Equation (6.7), and its symbols notation, it is possible to represent Equation (6.6) in another frequently used form given by

$$R_T = R_{T_r}\, e^{[\alpha_{T_r} \Delta T (T_r / T)]} \tag{6.8}$$

where α_{T_r} is the resistance coefficient at T_r, and ΔT is temperature difference (i.e. $\Delta T = T - T_r$).

The main parameters of thermistors are controlled by their composition. For normal applications in the temperature range −50 °C to 200 °C, all types contain Mn and Ni. If the percentage of these components is varied by adding Co and Cu, the specific resistivity can be varied between 10 Ω cm and 10^5 Ω cm with a corresponding increase in the B coefficient from 2580 K to 4600 K. At the reference temperature of 293 K, the value of α_T usually lies between −2 %/K and −6 %/K. As these normal NTC materials have phase transitions above 500 °C, they cannot be used to manufacture devices for use in this range. However, rare earths may be used up to temperatures around 1500 °C. A summary of the properties of NTC materials is given in Table 6.1. Carlson (1984) has reported a stable high temperature thermistor.

Figure 6.1 presents the ratio, R_T/R_{T_r}, as a function of temperature with the coefficient, α_{T_r}, as parameter at a reference temperature taken as $T_r = 293$ K (or 20 °C). For comparative purposes, the characteristic of a Pt-100 Ω RTD is also shown.

The voltage–current characteristic of a thermistor is defined as the voltage drop across the thermistor expressed as a function of the current flowing in it, with the ambient temperature of a given surrounding medium as a parameter. Typical voltage–current characteristics, for a thermistor in still air at the ambient temperature, ϑ_{a1}, are shown in Figure 6.2. The characteristic of the same thermistor in still water at the temperatures ϑ_{a1}, ϑ_{a2}, ϑ_{a3} are also shown in this figure. Initially, the thermistor voltage drop is directly proportional to its current. With increasing current, the resulting self-heating of the thermistor is accompanied by a commensurate decrease of its resistance, so causing the voltage versus current characteristic to decrease. On the $V = f(I)$ curves, for each current value, the corresponding temperature increases, $\Delta\vartheta_1, \Delta\vartheta_2, \ldots, \Delta\vartheta_3$, are also indicated. These values may be used for the estimation

Table 6.1
Properties of materials used for NTC thermistors (Roess, 1984). (Reproduced by permission of Butterworth-Heinemann Ltd, UK.)

Temperature range (°C)	Composition (main oxides)	Resistivity ρ_{25} (Ω cm)	Constant B (Equation 6.5)	Temperature coefficient $-\alpha$ (%/K)
−50 to 250	Mn56 Co8 Ni16 Cu20	10^1	2580	3.0
	Mn65 Co9 Ni19 Cu7	10^2	3000	3.5
	Mn70 Co10 Ni20	10^3	3600	4.5
	Mn85 Ni10	10^4	4250	5.0
	Mn94 Ni16	10^5	4600	5.4
>300	Sm70 Tb30	$2.5\ 10^3$*	6500*	0.6*

*At 600 °C.

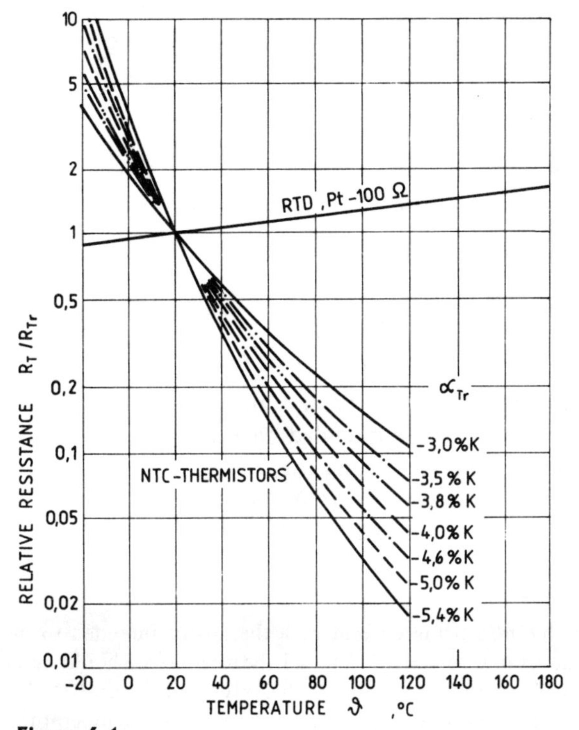

Figure 6.1
Resistance R_T of a temperature sensor at temperature T to R_{T_r} at 293 K (20 °C) versus temperature, ϑ.

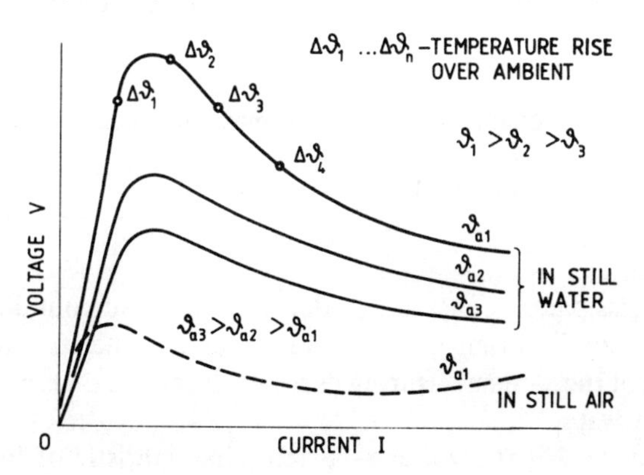

Figure 6.2
Voltage–current characteristics of a thermistor for different ambient temperatures ϑ_a and media.

of self-heating errors. With increasing ambient temperature, which is also the measured temperature, the resistance of the thermistor decreases so that its characteristics are shifted downwards.

Thermistors possess a heat dissipation constant, C, given in W/K, similarly defined as for the RTD in §5.2.3 and used in Equation (5.2). This constant permits a similar determination of the permissible measuring current, $I_{T,max}$, of a thermistor of resistance, R_T, for a given assumed self-heating error, $\Delta\vartheta_{max}$, as:

$$I_{T,max} = \sqrt{\frac{\Delta\vartheta_{max} C}{R_T}} \tag{6.9}$$

Conversely, the self-heating error, $\Delta\vartheta$, at the measuring current, I_T, can be evaluated as

$$\Delta\vartheta = \frac{I_T^2 R_T}{C} \tag{6.10}$$

The permissible measuring current, $I_{T,max}$, must always be calculated at the minimum possible value of R_T in the intended measuring range. This value occurs at the upper temperature of the measuring range.

Numerical example

Calculate the permissible measuring current of a thermistor intended to measure air temperature in a range from 0 to 100 °C. The self-heating error should be kept below 0.5 °C. In air the heat dissipation constant, C, has a value of 0.8×10^{-3} W/K, while the thermistor resistance at 20 °C is $R_T = 8.5$ kΩ. Also, at this temperature of 20 °C (293 K) the resistance coefficient, α_T, has a value of −4%/K or −0.04 1/K.

Solution: From Equation (6.8) at a temperature $T = 373$ K or 100 °C:

$$R_T = R_{T_r}\, e^{[\alpha_{T_r} \Delta T (T_r/T)]} = 8364\ \Omega$$

Inserting this value of R_T into Equation (6.9) yields the maximum measuring current as:

$$I_{T,max} = 2.18 \times 10^{-4}\ \text{A} \approx 0.22\ \text{mA}$$

The value of the heat dissipation constant, C, depends on the surrounding medium of the thermistor. For example, in air C has a value which is smaller than its value in water. Consequently, at the same measuring current, the errors due to self-heating are larger in air than in water.

Only the initial, linear part of the voltage–current characteristic of Figure 6.3 is used for temperature measurement. The static value of the resistance, R_T, of a thermistor at the given temperature, ϑ_{a1}, can be calculated, from the values of current and voltage in Figure 6.3, as:

$$R_{T1} = V_1/I_1 \tag{6.11}$$

A comparison of the advantages and disadvantages of NTC thermistors and of metallic resistance detectors is given in Table 6.2. The list of advantages provides

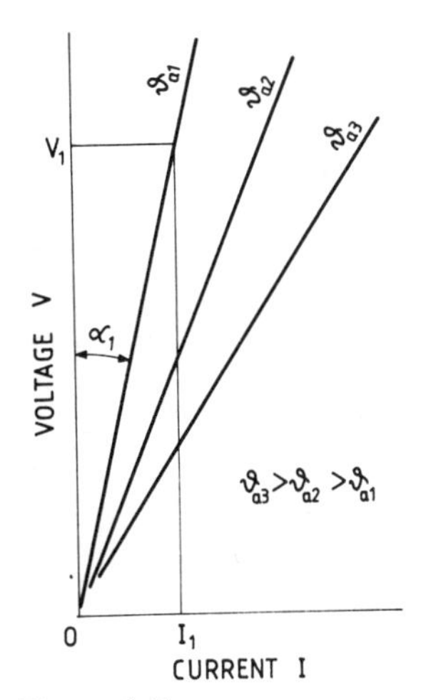

Figure 6.3
Initial linear parts of voltage–current characteristics of a thermistor, used in temperature measurements.

Table 6.2
Advantages and disadvantages of NTC thermistors as compared with metallic resistance detectors.

Advantages	**Disadvantages**
Small detector dimensions	
Higher detector resistance, (less affected by the resistance of the connecting leads which can be much longer as a consequence)	Non-linear resistance versus temperature characteristic
Higher temperature sensitivity	Non-standardized characteristics
Narrower measuring temperature range	Lower measuring temperature range
Measurement of small temperature differences	Susceptibility to permanent decalibration at higher temperatures
Low thermal inertia of the sensor	

a rational basis for the choice between using a thermistor and a resistance detector.

Thermistors of the PTC type, which may be used as binary temperature sensors are also produced in thin film technology (Morris and Filshie, 1982; Nagai *et al.*, 1982). They are used to protect semiconductor devices and electrical machinery. At preset temperatures, ϑ_B, such as for example, 35, 55, 75, 95 °C, the resistance of these PTC thermistors may increase from about 100 Ω to about 100 kΩ with increasing temperature. The properties of materials for use as PTC thermistors are given in Table 6.3.

6.2.2 Thermistor design

The most popular thermistor designs, which have been used for over forty years, are in the shape of beads and disks. More recently chip thermistors have been used.

Beads are made by allowing evenly spaced minute droppings of oxide slurry to fall upon two parallel stringed platinum alloy wires. Owing to the high surface tension of the slurry, the drops maintain their ellipsoidal shape. After drying, the drops are

Table 6.3
Composition and properties of PTC thermistors (Roess, 1984). (Reproduced by permission of Butterworth-Heinemann Ltd, UK.)

Composition (main components)	Nominal temperature (typical ϑ_B values) (°C)	Voltage range (V)
Ba-Titanate	120	6–40
Ba-Ca-Titanate	115	6–100
Ba-Sr-Titanate	−40 to 100	6–40
Ba-10%-Pb-Titanate	160	12–220
Ba-20%-Pb-Titanate	200	12–220
Ba-60%-Pb-Titanate	350	110–220

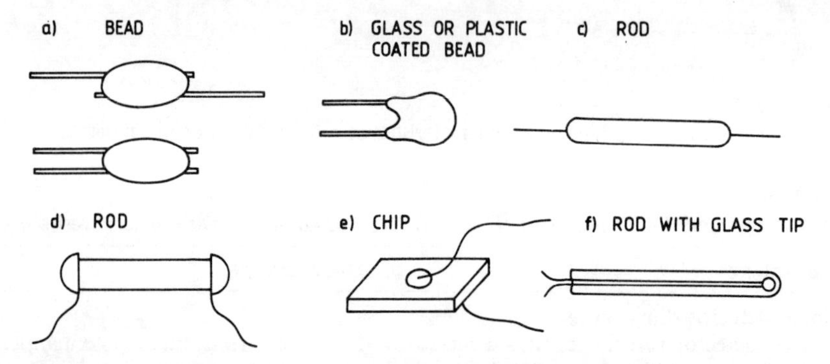

Figure 6.4
Typical thermistors.

sintered at temperatures between 1100 °C and 1400 °C. During the sintering process they shrink, so adhering to the wires with a well formed good electrical contact. Subsequently, they are cut, as shown in Figure 6.4(a), before being hermetically sealed with a glass or teflon layer which protects them from oxidation and environmental influences. The wires have a diameter of about 0.0125 to 0.125 mm while the beads vary in diameter from about 0.1 to 2 mm (Spoff, 1972; Weichert *et al.*, 1976).

Disk thermistors are produced by pressing oxide powders under several tons of pressure in a round die (Sierracin/Western Thermistor, USA). After sintering they are covered by a silver layer to permit soldering of the terminal wire. The thermistors, shown in Figure 6.4(e), which are wholly protected by an epoxy layer, have diameters from 1 to 10 mm and thicknesses ranging from 0.1 to 2 mm. Square plate thermistors, also called chip thermistors, have dimensions of 0.5×0.5 mm to 3×3 mm and thicknesses of 0.025 to 0.05 mm.

Different shapes of thermistors, whose typical properties are listed in Table 6.4, are represented in Figure 6.4. Although thermistors are normally applied in the temperature range from −100 to +300 °C, some types for application at high temperatures and at low temperatures are also available. The high temperature types may be used at temperatures up to 1200 °C while the low temperature components find application in the range from 5 to 200 K.

Table 6.4
Typical NTC thermistor thermometers.

Type (Figure 6.4)	Dimensions (mm)		Reference temperature (T_r) (K)	Resistance (R_{T_r}) (Ω)	Resistance temperature coefficient (α_{T_r}) (%/K)	Constant B (Equation (6.5)) (K)	Heat dissipation constant A (mW/K): In still air	Heat dissipation constant A (mW/K): In stirred oil	Time constant N_T: In still air (s)	Time constant N_T: In stirred oil (s)	Maximum operating temperature (°C)	Stability of R_{T_r} (%/year)
Bead	Φ; 0.06 to 1	—	293				~1	~8	0.4 to 25	1 to 2.5	250	0.1 to 0.5
Bead (glass-coated)	Φ; 0.1 to 2	—	293	40 Ω to 40 MΩ	−2 to −6	500 to 20 000	0.8	—			250 (350)	0.05 to 0.25
Rod	Φ; 0.5 to 5	l; 5 to 50	293						~20	—	250	
Disk	Φ; 1 to 10	t; 0.1 to 2	293	40 Ω to 1 MΩ			0.02 to 30		10 to 50	—	120	
Square plate (chip)	l×b; 0.5×0.5 up to 3×3	t; 0.025 to 0.05	293							—	120	0.5 to 3
Rod (with glass tip)	Φ; 1.5 to 3	l; 10 to 20	293	2 kΩ to 10 kΩ			~1	~1	3	—	250	

l, length; t, thickness; Φ; diameter.

Tolerances of the value of R_T for a given type of thermistor are usually around 5% to 20%, whereas tolerance for the constant, B, of Equations (6.5)–(6.6) is around 5%. These large tolerances are regarded as the main disadvantage in thermistor applications. Selected thermistors, divided into various groups of narrow tolerances, are available. This ensures total interchangeability, with temperature errors kept below +0.1 to 0.2 °C (Omega Engineering Inc., USA). Their prices, are of course, much higher.

Long time instability of thermistors, which is mainly attributed to their resistances, is caused by lattice structure changes due to oxidation and thermal tensions or by changes in the resistance of the metallized contact. This last cause seems to be the most important. The most stable types are glass-covered bead thermistors, whose resistance does not change more than 0.05 to 0.25% per year, as compared with 0.5 to 3% per year for disk and rod thermistors. These resistance changes are usually easily compensated for in the measuring circuits by periodic calibration checks.

Bare thermistors are rarely used, except occasionally in laboratory measurements. In most cases thermistors are used with a protective sheath.

Portable thermistor sensors, in the form of probes, with extendable coiled cables, are produced for almost all types of likely applications such as in the temperature measurement of air, liquids, surfaces of solids, meat, fruit and chemicals. More specialized areas of application are in biology and medicine. In the medical field, thermistor probes are disposed of after only one use to avoid the possibility of cross-contamination. This is not unreasonable as they are comparatively inexpensive. Their 90% rise time is about 1 to 3 s.

Stationary thermistor sensors are used in the temperature measurement of extruders, storage tanks and containers, in chemical apparatus and in grain silos as 3 to 6 sensor sets.

6.2.3 Correction and linearization of thermistor characteristics

There are two main methods of assuring the interchangeability of thermistor sensors. Production control methods allow the selection and division of thermistors into groups with a small scattering of the thermistor characteristics. Subsequently they may be separated into components with narrow temperature tolerances. This may be either over a range of temperatures or at a single temperature. Tolerances may be, for example, ±0.05 °C, ±0.1 °C, ±0.2 °C and ±1 °C which are marked on the component by a colour code (Sierracin/Western Thermistors, Oceanside, USA). Array configuration methods employ the ideas associated with other resistance array manufacturing techniques (Connolly, 1982; Costlow, 1983). Thus it is possible to correct and linearize the thermistor characteristics using a computer program to calculate the compensation values based upon the measured thermistor characteristics at three given temperatures. Such a procedure is carried out during production.

The non-linear resistance versus temperature characteristic is regarded as the main disadvantage of thermistors. This functional dependence, as given by Equation (6.5), results in decreased thermistor sensitivity at higher temperatures. McGhee (1989) has pointed out that linearization of temperature sensor and particulary thermistor

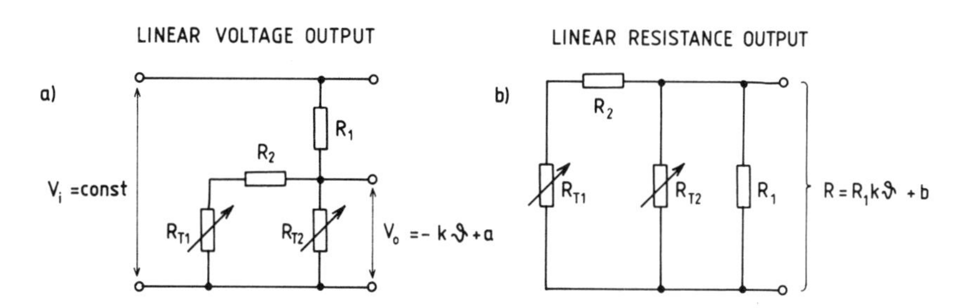

Figure 6.5
Linear output thermistor assemblies R_{T1}, R_{T2} thermistors, *a*,*b*-constants at $\vartheta = 0$.

characteristics is a special case of the more general sensor linearization problem. Linearization may use analogue linearizing circuits or it may be digital (McGhee, 1989). The digital approach uses a number of different circuits.

Thermistors whose linearization is based upon using analogue circuit techniques, are available. The linearization is mainly based upon the most convenient and classical method given by Beakley (1951) and Hyde (1971). For example, Omega Engineering Inc. (USA) produce 'linear output thermistor assemblies'. They consist of two or three thermistors packaged as a single sensor and of additional film resistors. They are produced either as linear voltage versus temperature as given in Figure 6.5(a), or linear resistance versus temperature, as in Figure 6.5(b). White (1984) also provides a review of the techniques used for the linearization of resistance thermometers. The linearity is extended over a certain temperature range in which the non-linearity errors do not exceed from ± 0.03 to ± 1.1 °C. An assembly may have a sensitivity as high as 30 mV/K, which is many times greater than that of a thermocouple. For multi-point temperature measurement, one resistor set can be used for many thermistor assemblies. In the circuit, given in Figure 6.5(a), both positive or negative slope output voltage signals are possible. Player (1986) describes an extension of this technique to give a wide range thermistor thermometer. In every 10 °C sub-range the compensating network of the thermistor is changed. Although it is not possible to arrange ideal compensation in one overall range, the technique exchanges a deviation of roughly cubic form for a deviation with a quadratic form over a smaller sub-range. As thermistor characteristics are exponentially deterministic, a logarithmic amplifier may be used for linearizing purposes (Patranabis *et al.*, 1988).

Digital compensation methods fall into various main groups. A general method applying one-, two- and three-point digital methods to a number of electrical output temperature sensors, including thermistors, is considered by Bolk (1985). The technique of using an analogue-to-digital converter described by Iglesias and Iglesias (1988) may be adapted to suit thermistors. Pulse generators whose frequency is related to the resistance of the thermistor are also applied. The principle of operation of the basic circuit shown in Figure 6.6(a), is based upon temperature to frequency conversion. The frequency of the square wave output signal is:

$$f = \frac{1}{2R'C\ln(1 + 2R_2/R_1)} \tag{6.12}$$

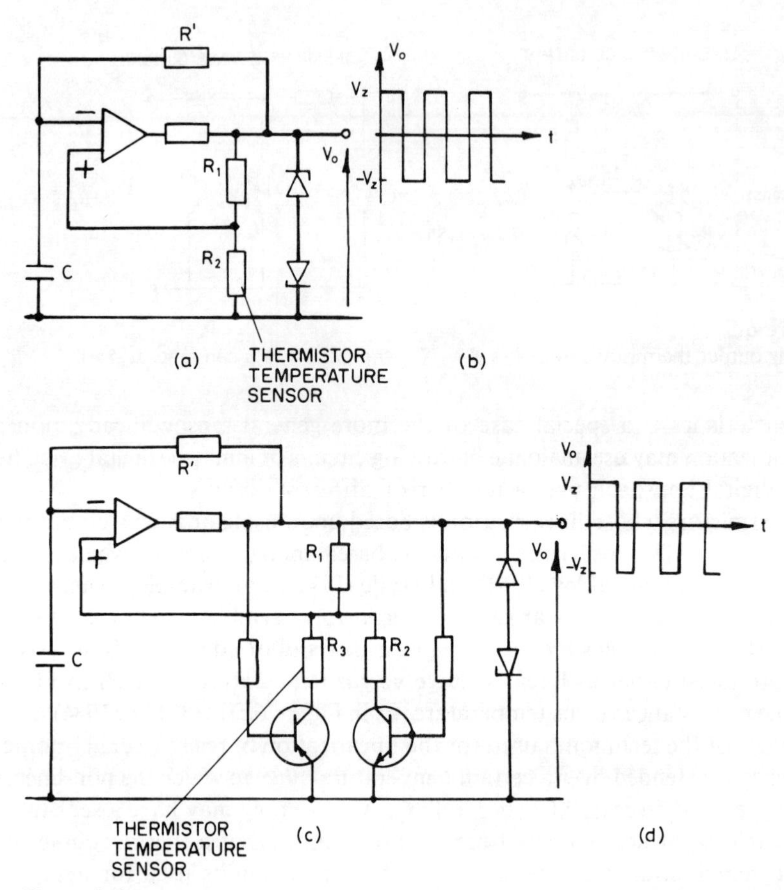

Figure 6.6
Linearized temperature to frequency converter for thermistor sensors. (a) Basic diagram, (b) modified linearizing circuit.

Since the resistance versus temperature characteristic of the thermistor has an exponential form, replacing R_2 by the thermistor resistance allows cancellation of the exponential behaviour by the logarithmic term in the expression. Although complete cancellation cannot be achieved with this simple circuit, good linearity over a limited temperature range is possible. Sengupta (1988) shows how the linearity may be extended by including additional switching transistors (Figure 6.6(c)). The transistors switch different resistors (R_3 is the thermistor) into the circuit to give different time constants for charging and discharging of the capacitor. In this manner the output voltage is saturated for longer at one supply rail voltage than it is at the other.

A final group of methods uses post-conversion techniques based upon a ROM look-up table/software routine (Brignell, 1985).

6.2.4 Measuring circuits of thermistors

The measuring circuits of thermistor thermometers are most commonly deflection type bridge circuits like that shown in Figure 6.7. Thermistor bridge design is

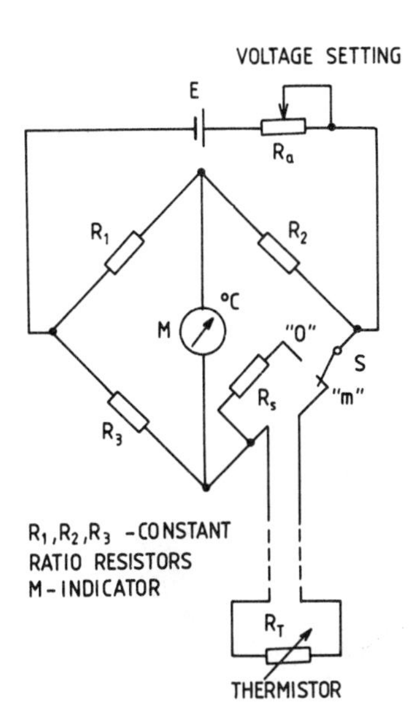

Figure 6.7
Thermistor thermometer in deflection type bridge circuit.

considered by Bentley (1984). The bridge energy source may be a battery cell or a rectified a.c. supply voltage. To ensure that the supplying voltage remains constant, a standardizing resistor, R_s, is provided. In the position 'O' of the switch, *S*, where R_s temporarily replaces the thermistor, R_T, the value of R_a is adjusted in such a way that the readings of the meter, *M*, are brought to a marked scale position. This is not necessary when a stabilized voltage source is used. Measuring temperature ranges of 30 to 50 °C, may easily be achieved. The whole measuring range is divided into several selectable sub-ranges. Most producers now supply thermistor thermometers based on a deflection type bridge circuit with an output amplifier and either analogue or digital meters such as that illustrated in Figure 6.8.

For lower precision of measurements, the simple series connected thermistor thermometers, shown in Figure 6.9, are also used. They comprise a current limiting resistor, R_1, and a microammeter, *M*, graduated in temperature degrees. A standardizing resistor, R_s, and switch, *S*, are also provided. The permissible measuring current of the thermistor should not be exceeded.

Thermistors, which are generally supplied with their indicating meters by the same manufacturer, have many applications. Their large signal makes them especially appropriate in almost all applications within their somewhat limited temperature range between about −50 °C to about 300 °C. Physical and biological fields frequently use thermistors such as in the food industry or in medicine as detailed by Spoff (1972). Other important areas of application are in air and liquid temperature measurement as well as in the temperature measurement of small electronic elements and machine parts.

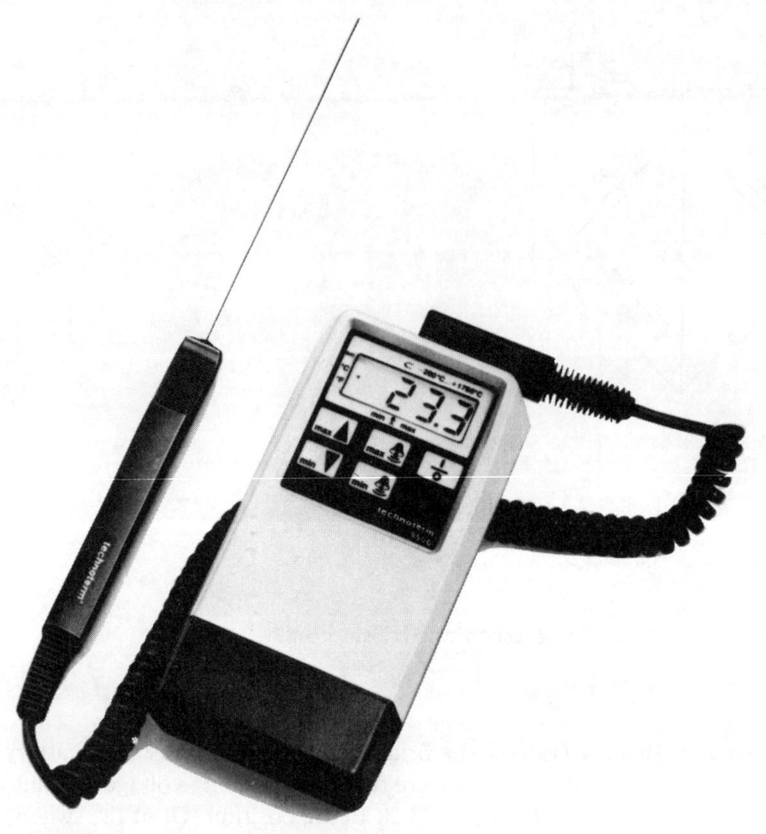

Figure 6.8
Microprocessor based thermistor thermometer with min/max storage and programmable alarms. (Courtesy of Testoterm, Germany.)

6.3 SILICON RESISTANCE THERMOMETER DETECTORS: Si-RTD

6.3.1 General information

Silicon resistance thermometer detectors or Si-RTDs, also called Silistors by Hyde (1971), are PTC silicon resistors. They should not be confused with the binary type of PTC thermistor described in §6.2 whose temperature behaviour is significantly different. Their manufacturing technology is based upon the familiar planar technology which has proved extremely successful in the manufacture of other semiconductors. There are four principal steps in their production (Philips Components Ltd, UK). A neutron transmutated doped (NTD) silicon wafer of ^{30}Si is irradiated with neutron radiation to produce ^{31}Si. This ^{31}Si then decays to produce the *n*-type dopant ^{31}P. Extremely low spreads in dopant density, which are required for the tight tolerance in sensor resistance, result from the use of this process. The growth of a glass layer, during the subsequent n^+ diffusion, and silicon nitride

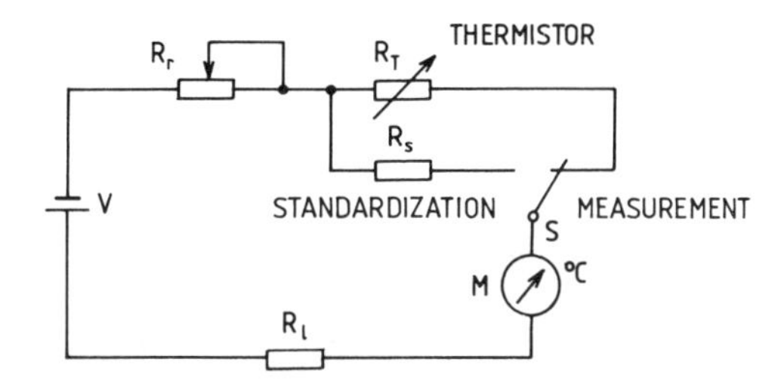

Figure 6.9
Series connected thermistor thermometer.

passivation ensure long-term stability. Metallization prevents contamination and migration. Finally, galvanic growth of silver mushroom contacts is completed. This ensures good pressure contact for the KTY83/KTY85 series. When these contacts are plated with a copper–tin–silver alloy they ensure reliable high temperature operation of the KTY84 high temperature detectors.

Doped Si, in the 'normal' region of semiconductor behaviour, has a resistivity given by:

$$\rho = \rho_{25}\left(\frac{T}{298}\right)^{2.3} \qquad (6.13)$$

Equation (6.13) gives a power series expansion for the resistance, which is similar to that of a Pt-RTD, in the form:

$$R_T = R_{Tr}[1 + A(T - T_r) + B(T - T_r)^2] \qquad (6.14)$$

The constants A and B depend upon sensor type as shown in Table 6.5 while R_{Tr} is the nominal resistance of the sensor at the reference temperature, T_r. Hence, it is a simple matter to store a calibration table in ROM for high precision microprocessor-based applications. A resistance–temperature coefficient can be defined for these devices in the same way as for the Pt-RTD. Table 6.5 gives details of a selection of Si-RTDs, which may be applied in simple or precision temperature measurement or in temperature compensation. The resistance versus temperature characteristics of Si-RTDs are shown in Figure 6.10.

6.3.2 Design of Si-RTDs

The principles of the KTY series of PTC silicon resistors, manufactured by Amperex and Philips Components, are represented in Figure 6.11. They consist of an n-type silicon cell metallized on one side with a small contact layer on the other. When used in the single cell structure they are of a more basic type. However, as they are polarity sensitive bilateral connection is not possible. The formal double sensor structure, illustrated in Figure 6.11, consists of two single sensors with opposite polarity in series. This ensures that the sensor has bilateral terminal properties.

Table 6.5
Important parameters for KTY silicon temperature sensors. (Reproduced by permission of Philips Components Ltd, UK.)

Series	Nominal resistance R_{Tr} (Ω)	T_r (K)	Measuring temperature range (°C)	Sensor constants (Equation (6.14)) A (%/K)	B (%/K² × 10⁻³)	Operating current I (mA)
KTY81-1	980 to 1050 (7 types)	298	−55 to 150	0.7874	1.874	1
KTY81-2	1960 to 2100 (7 types)	298	−55 to 150	0.7874	1.874	1
KTY83-1	950 to 1050 (7 types)	298	−55 to 175	0.7635	1.731	1
KTY84-1	950 to 1050 (4 types)	373	0 to 300	0.6116	1.025	2
KTY85-1	950 to 1050 (7 types)	298	−40 to 125	0.7635	1.731	1
KTY86-2	1990 to 2010 (1 type)	298	−40 to 150	0.7646	1.752	1

All of the sensors may be supplied with different types of encapsulation such as SOD-70. This has a recommended operating range of −55 to +150 °C. The single sensor basic structure may be supplied in a compact DO-34 encapsulation with a polarity marking. This form, which is the KTY83-1 type, has a nominal resistance of 1000 Ω. Although it is a single sensor it can operate at temperatures up to 175 °C. Extension even of this range to +300 °C, is allowed with the KTY84 series which is supplied in a DO-34 encapsulation again with a nominal resistance of 1000 Ω measured at 100 °C. The higher nominal resistance of the devices means that terminal and contact resistances exert a negligible influence upon their operation.

As sampled acceptance testing is made to military standard MIL STD 105D, the reliability and stability of Si-RTDs are comparatively good. A significant contribution is made to these properties by the planar manufacturing process itself which helps to ensure high quality and exceptional reliability. Reliability testing is performed under maximum rated operating conditions, as part of the product acceptance screening at the production level. Tests include constant operation to estimate stability, temperature cycling and storage at high and low temperatures. The typical value of drift, measured at 150 °C after 2000 hours of operation, lies between 0.13 and 0.15 K with the maximum in the range from 0.38 to 0.66 K. Estimated lifetimes in the range 155 000 to 250 000 hours indicate the reliability of Si-RTDs.

6.3.3 *Measuring circuits of Si-RTDs*

Whether Si-RTDs are applied for temperature measurement or for compensating against unwanted thermal effects, it is recommended that other components should have as low a temperature coefficient as possible. This means that fixed resistors should preferably be metal film types and potentiometers should be cermet types.

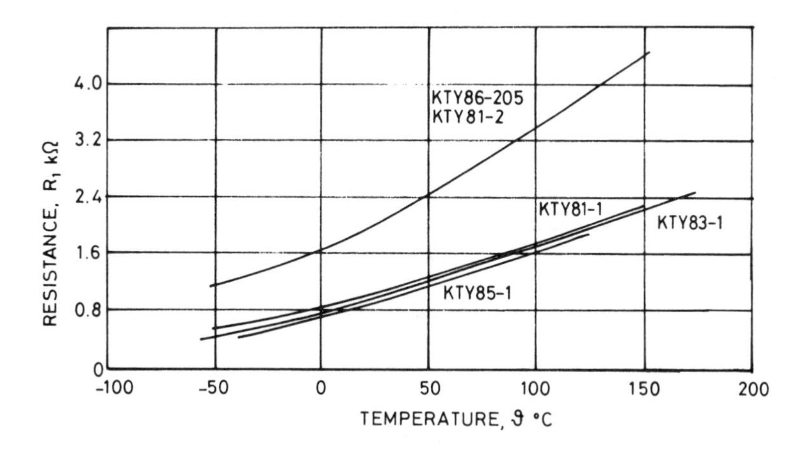

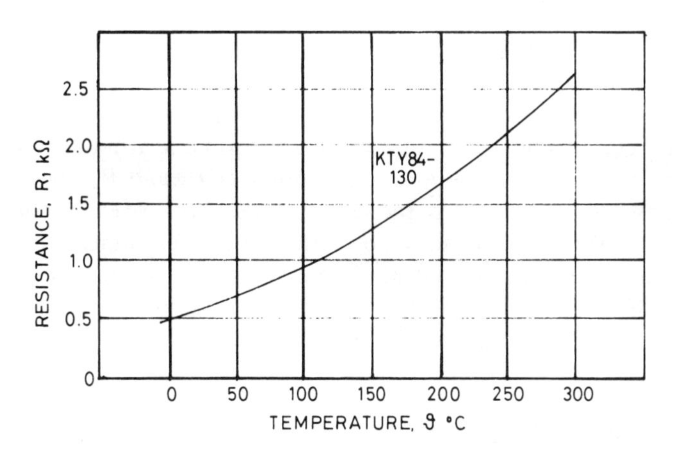

Figure 6.10
Resistance versus temperature characteristics of some Si-RTDs. (Courtesy of Philips Components Ltd.)

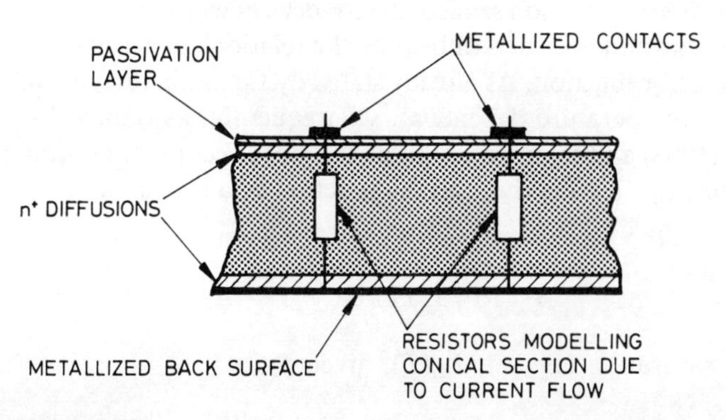

Figure 6.11
Structure of a double silicon RTD. (Courtesy of Philips Components Ltd.)

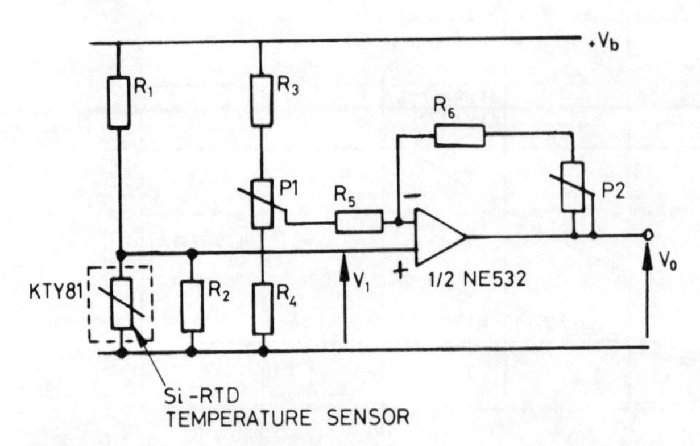

Figure 6.12
Simple temperature measuring circuit using a KTY81 sensor. (Courtesy of Philips Components Ltd.)

Figure 6.12 shows a circuit diagram for a simple temperature measurement application using a Si-RTD. This circuit may be used to measure the temperature of rooms, ovens, electric irons and domestic and industrial water heaters. Resistor R_1 and the parallel combination of R_2 and the sensor form one arm of a Wheatstone bridge. The other arm is formed by R_3, a potentiometer P_1 and R_4. The values of R_1 and R_2 are chosen so as to linearize the sensor characteristic over the temperature range of interest. Calibration is achieved by setting the potentiometer P_1 until V_o is 1 V when the sensor temperature is 0 °C. At a higher temperature, say 100 °C, P_2 is adjusted until the output voltage, V_o, is at the correct span level. Notice that adjustment of P_2 does not affect the zero of the scale.

6.4 DIODES AND TRANSISTORS

6.4.1 *General information*

Diodes and transistors are junction semiconductor devices whose current versus voltage characteristics are substantially determined by the relations between carriers on each side of a semiconductor junction. As already stated by Equation (6.1) in §6.2.1, carrier density is strongly temperature dependent. Consequently, as quoted by Sze (1969) and van der Ziel (1968) and taking account of the more accurate notation of Tsividis (1980), the current, I_d, flowing through the junction of a semiconductor diode may be written as:

$$I_d(T) = I_{so}(\mathrm{T})\ \mathrm{e}^{(qV_d/kT)} \qquad (6.15)$$

with the reverse saturation current, $I_{so}(T)$, given by:

$$I_{so}(T) = \frac{qAT^3\,\overline{D}(T)\mathrm{e}^{(-qV_g(\mathrm{T})/kT)}}{N_B} \qquad (6.16)$$

where A is the base-emitter area of the diode junction, k is the Boltzmann constant, q is the electron charge, T is the absolute temperature in K, $V_g(T)$ is the temperature dependent energy gap for a given material, N_B is the Gummel number (total number of impurities per unit area in the base of the diode), $\bar{D}(T)$ is the temperature dependent 'effective' minority carrier diffusion constant in the base, and V_d is the forward voltage across the device.

As diodes may be operated in either reverse bias or forward bias modes it is possible to measure temperature by measuring either reverse saturation current or forward voltage. Take logarithms on each side of Equation (6.16), then differentiate explicitly to obtain the temperature coefficient of the reverse saturation current as:

$$\frac{dI_{so}}{I_{so}}=\left(3+\frac{d\bar{D}}{\bar{D}}+\frac{qV_g(T)}{T}-\frac{q}{kT}\frac{dV_g}{dT}\right)\frac{dT}{T} \tag{6.17}$$

It is easy to calculate that the temperature coefficient of Equation (6.17) doubles for every 10 °C rise in temperature.

Under conditions of temperature independent constant forward current the reverse saturation current can be neglected. The temperature coefficient, $\partial V_d/\partial T$, of V_d may be found as follows. Differentiate Equation (6.15) with respect to temperature to find:

$$\frac{\partial I_d}{\partial T}=\left[\frac{\partial I_{so}}{\partial T}+I_{so}\left(-\frac{q}{kT^2}V_d+\frac{q}{kT}\frac{\partial V_d}{\partial T}\right)\right]\mathrm{e}^{(qV_d/kT)} \tag{6.18}$$

Since the forward current is independent of temperature, Equation (6.18) can be solved for $\partial V_d/\partial T$ to obtain:

$$\frac{\partial V_d}{\partial T}=\frac{k(\partial I_{so}/I_{so})}{q\,(\partial T/T)}-\frac{q}{kT}V_d\approx\begin{cases}-2\,\mathrm{mV/K} & \text{for Si}\\ -1.25\,\mathrm{mV/K} & \text{for Ge}\end{cases}\ (\text{at } T=300\,\mathrm{K}) \tag{6.19}$$

The effects of temperature in transistors, which are similar to those in diodes, are detailed by Sah (1961). His expression for the short-circuit collector current, I_{cs}, in many kinds of transistor has been given more detailed attention by Tsividis (1980) who wrote the equation in the form:

$$I_C(T)=I_s(T)\,\mathrm{e}^{(qV_{BE}/kT)} \tag{6.20}$$

where I_C is the collector current, $I_s(T)$ is the temperature dependent reverse saturation current of the base-emitter junction and V_{BE} is the base-emitter voltage.

Equation (6.20), which is valid for both diffusion and drift devices, has a similar notation as in Equation (6.15).

6.4.2 Diodes and their measuring circuits

Diode thermometers, fabricated from the compound semiconductor, GaAs, may be used in the temperature range from 2 K to 300 K. Cohen *et al.* (1963), who reported GaAs diode thermometers in detail, have noted that the lower limit of the

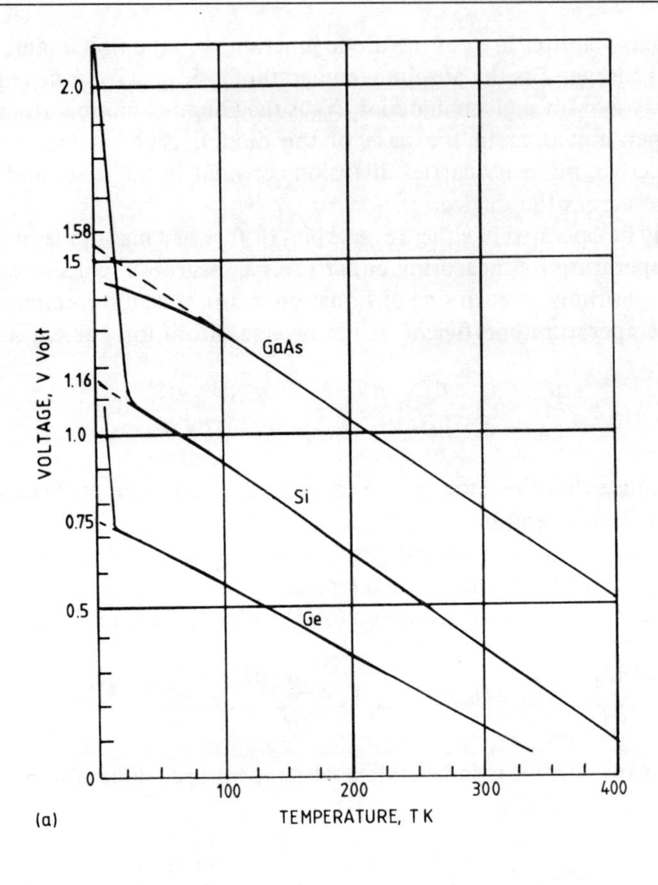
2.0
1.58
1.5
1.16
1.0
0.75
0.5
0
GaAs
Si
Ge
VOLTAGE, V Volt
100
200
300
400
TEMPERATURE, T K
(a)

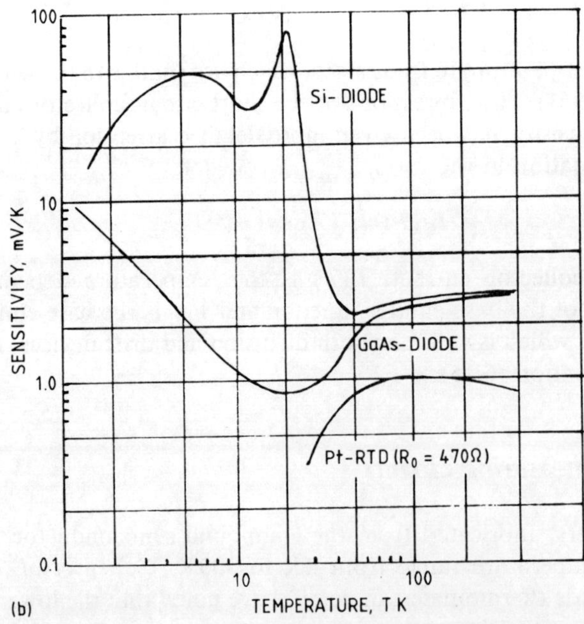
100
10
1.0
0.1
Si-DIODE
GaAs-DIODE
Pt-RTD (R0 = 470Ω)
SENSITIVITY, mV/K
10
100
TEMPERATURE, T K
(b)

linear temperature versus forward voltage characteristics of Equation (6.15), occurs at the 'freeze out' temperature (see §6.1). Below that temperature their sensitivity exhibits a minimum which is a serious disadvantage. For that reason Si diodes have been introduced which have a sensitivity approximately ten to fifty times greater than that of the GaAs device. A comparison between the forward voltages of GaAs and Si diodes as a function of temperature is given in Figure 6.13(a). Their temperature coefficient (sensitivity) of forward voltage, compared with that of a Pt-RTD with a current of 500 μA flowing in it, is shown in Figure 6.13(b) (Swartz and Swartz, 1974). The characteristics of GaAs diodes, which became commercially available in 1966 (Swartz and Gaines, 1972), may be modelled using empirically fitted equations (Pavese, 1974).

Temperature limit detectors and rate of change temperature detectors are required, for example, in a fire alarm system (Aleksic and Vasiljevic, 1986). Although thermistors, forward biased diodes and transistors are commonly used, their application is usually based upon the detection of voltage levels which therefore require voltage comparators. In some cases a limit detector which consumes little power, is required. It is possible to design temperature switches, to the basic circuit shown in Figure 6.14, using the temperature dependence of reverse saturation current of diodes (Aleksic and Vasiljevic, 1986). For temperatures below T_0, the reverse saturation current of the diode, I_s $(T < T_0)$, is less than the bus-voltage limited constant current I_o, so that the diode voltage V_d is approximately equal to V_{cc}. If the reverse current of the diode increases beyond I_o, owing to a change in temperature, the voltage V_d drops to a low value (almost 0 V) causing the CMOS inverter to change level. The threshold temperature is fixed by choosing a suitable value of fixed biasing current I_o.

Diode thermometers normally use the nearly linear forward voltage as a function of temperature for the output signal. As the voltage temperature coefficient for Si diodes is larger than for Ge types, Si diodes are more often used. Talpe *et al.* (1987) describe the use of common diodes for cryogenic thermometry. Typical measuring circuits may be of the simple form shown in Figure 6.15 with non-linearity errors of about ± 1 to $\pm 3\%$ of full scale deflection. Krause and Dodrill (1986) note that the reproducibility of diode thermometers, which can approach ± 50 mK, may be badly affected if precautions are not taken to ensure that the biasing source is sufficiently well regulated and that adequate screening and guarding are used. These sources of interference as well as those due to improper electrical grounding and unexpected ground loops can cause reproducibility errors of typically a few tenths of a kelvin to deteriorate to as much as 4 K in an extreme case. For accurate measurements over the range -50 to $+30$ °C (Griffiths *et al.*, 1974) sophisticated circuits should perform three main functions. A voltage offset circuit is usually required to maintain the anode of the sensing diode at a fixed voltage depending upon measuring range. As it is usual to operate the diode with a constant current forward bias, a constant current source is required. The final function, which is the display, presentation and reading of the temperature values, may be performed

Figure 6.13
(a) Voltage versus temperature characteristics of some semiconductor junctions after Rao and others (1983). (Courtesy of Butterworth-Heinemann Ltd, UK.) (b) Sensitivity versus temperature of some temperature sensors after Swartz and Swartz (1974). (Courtesy of Butterworth-Heinemann, Ltd, UK.)

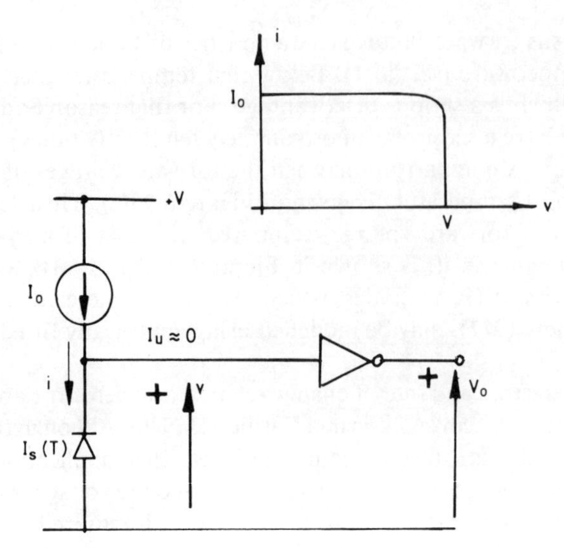

Figure 6.14
Schematic diagram of a semiconductor temperature switch.

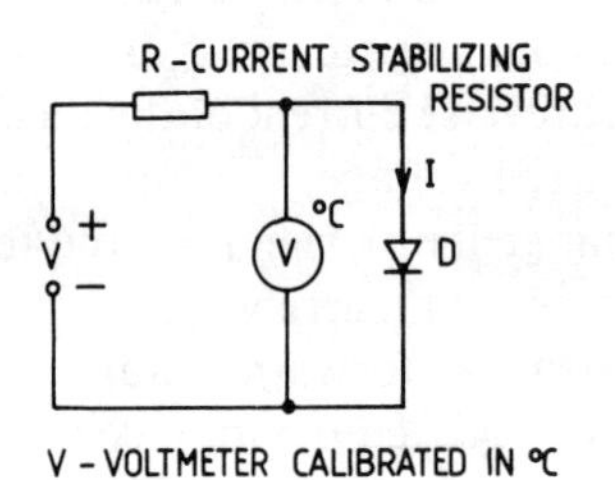

Figure 6.15
Diode as a temperature sensor-measuring circuit.

using specially designed sensing and display circuits based upon millivoltmeters (Griffiths *et al.*, 1974). Alternatively, Weichert *et al.* (1976) have stated that high input impedance analogue or digital voltmeters may be used to read out the temperature values.

6.4.3 Transistors and their measuring circuits

Transistor thermometers are normally fabricated using Si. The basic principles of bipolar junction transistors as absolute thermometers, which were given by Felimban and Sandiford (1974) for the 2N2222A device, may be more fully understood from an accurate description of their behaviour with temperature (Tsividis, 1980). From the relations in Equation (6.20) it can be seen that temperature may be obtained by measuring either the short circuit collector current or the base-emitter voltage of a transistor. Both methods have been used in practice.

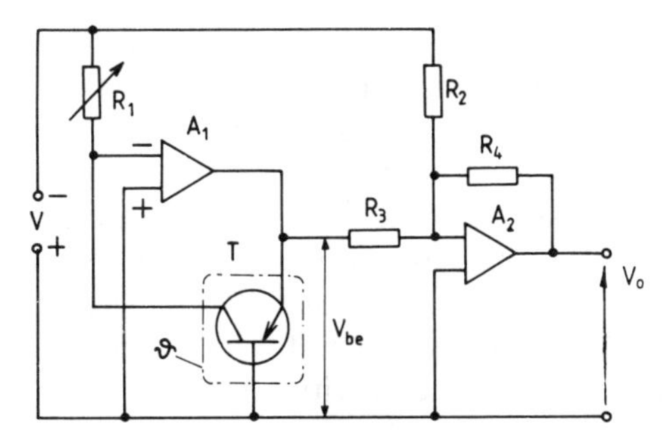

Figure 6.16
Transistor T as a temperature sensor.

When used to measure temperature, transistors are normally connected in the negative feedback path of an operational amplifier. In such a circuit, which has a logarithmic response over 9 decades as described by Gibbons and Horn (1964), the feedback converts the transistor into an approximately ideal transdiode from the input biasing point of view. Figure 6.16 presents a frequently used development of this measuring circuit described by Swartz and Gaines (1972). Integrated circuit operational amplifier, A_1, should be characterized by small variations in input offset current, whilst A_2 should have small variations in input offset voltage. The resistor, R_1, is used for setting the transistor collector current, with the setting of resistor, R_2, fixing the reference temperature. Matching of the circuit gain to the indicating meter, in most cases a digital voltmeter, is allowed by the resistor, R_4. Every time the sensor is changed, the resistor, R_1, has to be reset. The base-emitter voltage, V_{be}, is a nearly linear temperature function. A similar circuit, described by Ruehle (1975), uses the transistor as part of the bridge. The additional operational amplifier simultaneously ensures that the base and collector currents are held constant. The base-emitter voltage, V_{be}, is then a linear function of temperature in the range from 1 to 5 V, having non-linearity errors smaller than 0.05%. As all diode and transistor sensors exhibit a non-linear behaviour with temperature, their signals require linearization (Davies and Coates, 1977). Verster (1968, 1972) uses a technique of linearization based upon a circuit which switches the short circuit collector current of a 2N1893 device between 57.3 μA and 5.0 μA. Over the range -100 °C to $+100$ °C the temperature may be resolved to within 0.01 °C. Ohte and Yamagata (1977) describe how the application of feedback compensation can linearize the voltage versus temperature characteristic. The overall deviation from linearity is within ± 0.1 °C over the temperature range -50 to $+125$ °C, with long-term drift of some ± 0.2 °C in a 24 000 hour test. The thermometer time constants in stirred water were 0.2 s for the minisensor and 0.9 s when used with a needle sheath compared with 7 s with a stick probe.

An application for temperature measurement in a meteorological sounding balloon, using the 2N2222 *n-p-n* Si transistor, has been reported by Sridaran *et al.* (1987). The circuit, operating at a constant emitter-base voltage of 0.6 V, is based upon the temperature dependence of the collector current, which increases from 10^{-11} to

10^{-4} A for temperature variations from 170 to 310 K. An error limit of ±1 °C is reached. Feedback is also used to improve the linearity and stability of this thermometer.

Transistor pairs, packaged in a single sheath, are also used. Verster (1972) has reported a special electronic circuit which maintains both stabilized and constant collector currents. Their output voltage, which is a linear temperature function, can be easily adapted to a new sensor by exchanging one resistor in the measuring circuit. If the circuit is designed as an integrated unit then self-heating errors may be a problem. A digital voltmeter is the typical instrument for indicating the temperature values acquired by the circuit which has a sensitivity 10 mV/K.

6.5 INTEGRATED CIRCUIT TEMPERATURE SENSORS

When two transistors are operated at a constant ratio of unequal emitter currents the difference between their base-emitter voltages is proportional to absolute temperature often abbreviated to PTAT. Integrated circuit, IC, or monolithic temperature sensors, which are designed on the basis of this effect, provide either an output current or voltage referred to as IPTAT and VPTAT respectively. They were developed following the seminal work of Hilbiber (1964) and Widlar (1965) in the realization of monolithic voltage standards which are of particular significance in electronic equipment using A-to-D and/or D-to-A converters (see §8.4).

Taking Equation (6.20) into account in considering the circuit of Figure 6.17, follow the approach of Timko (1976). It can be shown that the voltage, V_T, across the resistor, R, in terms of the emitter current ratio, r, as the base-emitter voltage difference of T_2 and T_4, is

$$V_T = T\frac{k}{q}\ln r \qquad (6.21)$$

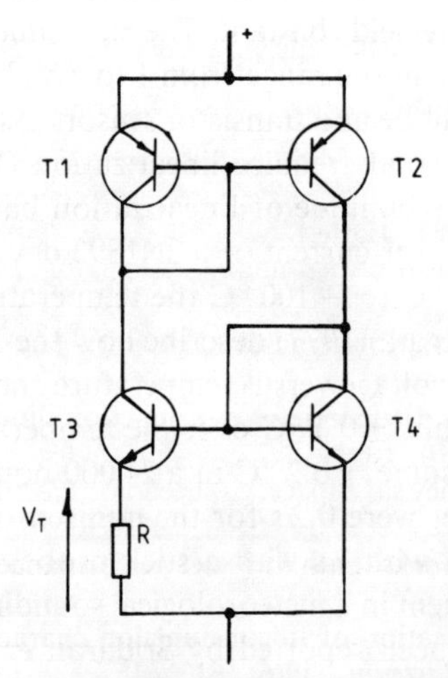

Figure 6.17
Basic PTAT circuit.

If this voltage were buffered and amplified it would provide a signal directly proportional to absolute temperature. This is the basis of operation of integrated circuit VPTAT temperature sensors as well as band-gap references. Current mirror-based VPTAT circuits like that shown in Figure 6.17 have formed the basis of a number of developments in IC temperature sensors (de Haan and Meijer, 1980; Meijer, 1980; Regan, 1984; R. S. Components Ltd, 1987; Meijer *et al.*, 1989; National Semiconductors Corp., 1989; PMI Ltd, undated). On the other hand, the current, I, drawn from each side of the circuit, which is equal to that flowing in the resistor, R, is equal to

$$I = T\frac{k}{qR}\ln r \tag{6.22}$$

Under the same assumptions as for the VPTAT output it is apparent that a IPTAT output is also possible. This is the basis of operation of the sensor developed by Timko (1976) and now available from Analog Devices Inc. (1989) and second sourced by R. S. Components Ltd (1983).

REFERENCES

Aleksic, Z. J. and Vasiljevic, D. M. (1986) Low-power temperature detector and rate of temperature change detector. *IEEE Trans. Instr. Meas.*, **IM-35**(4), 571–575.

Arora, N. D., Hauser, J. R. and Roulston, D. A. (1982) Electron and hole mobilities in silicon as a function of concentration and temperature. *IEEE Trans Elec. Dev.*, **ED-29**(2), 292–295.

Beakley, W. R. (1951) The design of thermistor thermometers with linear calibration. *J. Sci. Instrum.*, **28**(6), 176–179.

Becker, J. A., Green, C. B. and Pearson, G. L. (1946) Properties and uses of thermistors-thermally sensitive resistors. *Trans. AIEE*, **65**, 711.

Bentley, J. P. (1984) Temperature sensor characteristics and measurement system design. *J. Phys. E: Sci. Instrum.*, **17**, 430–439.

Bolk, W. T. (1985) A general linearising method for transducers. *J. Phys. E: Sci. Instrum.*, **18**, 61–64.

Bosson, G., Gutmann, F. and Simmons, L. M. (1950) A relationship between resistance and temperature of thermistors. *J. Appl. Phys.*, **21**, 1267–1268.

Brignell, J. E. (1985) Interfacing solid state sensors with digital systems. *J. Phys. E: Sci. Instrum.*, **18**, 559–565.

Carlson, R. O. (1984) Stable high temperature (500 °C) thermistors. *Rev. Sci. Instrum.*, **55**(12), 1999–2006.

Cohen, B. G., Snow, W. B. and Tretola, A. R. (1963) GaAs p–n junction diodes for wide range thermometry. *Rev. Sci. Instrum.*, **34**(10), 1091–1093.

Connolly, E. (1982) Focus on resistor networks: they save space, connections. *Electronic Design*, March 31, 125–136.

Costlow, T. (1983) Precision resistors track networks' path to better performance. *Electronic Design*, Feb. 17, 99–102.

Davies, C. E. and Coates, P. B. (1977) Linearisation of silicon junction characteristics for temperature measurement. *J. Phys. E: Sci. Instrum.*, **10**(6), 613–616.

de Haan, G. and Meijer, G. C. M. (1980) An accurate small-range IC temperature transducer. *IEEE J. Sol. St. Circ.*, **SC-19**(5), 1089–1091.

Droms, C. R. (1962) Thermistors for temperature measurement. *Temperature: Its Measurement and Control in Science and Industry*, Vol. 3, Part 2, Reinhold Publ. Co., New York. pp. 339–346.

Felimban, A. A. and Sandiford, D. J. (1974) Transistors as absolute thermometers. *J. Phys. E: Sci. Instrum.*, **7**, 341–342.

Gibbons, J. F. and Horn, H. S. (1964) A circuit with logarithmic transfer response over 9 decades. *IEEE Trans. Circ. Theory*, **CT-11**, 378–384.

Griffiths, B., Stow, C. D. and Syms, P. H. (1974) An accurate diode thermometer for use in thermal gradient chambers. *J. Phys. E: Sci. Instrum.*, **7**, 710–714.

Hilbiber, D. F. (1964) A new semiconductor voltage standard. *ISSCC Dig. Tech. Papers*, pp. 32–33.

Hyde, F. J. (1971) *Thermistors*. Iliffe Books, London.

Iglesias, E. G. and Iglesias, E. A. (1988) Linearization of transducer signals using an analog-to-digital converter. *IEEE. Trans. Instr. Meas.*, **IM-37**(1), 53–57.

Krause, J. K. and Dodrill, B. D. (1986) Measurement system induced errors in diode thermometry. *Rev. Sci. Instrum.*, **54**(4), 661–665.

McGhee, J. (1989) The impact of electronics and information technology upon measurement and control in thermal systems. Keynote lecture IMEKO TC-12 Symp. *Microprocessors in Thermal and Temperature Measurements*, Lodz, Poland.

Meijer, G. C. M. (1980) An IC temperature transducer with an intrinsic reference. *IEEE J. Sol. St. Cir.*, **SC-15**(3), 370–373.

Meijer, G. C. M., van Gelder, R., Nooder, V., van Drecht, J. and Kerkvliet, H. (1989) A three terminal integrated circuit temperature transducer with microcomputer interfacing. *Sensors and Actuators*, **18**, 195–206.

Morris, W. M. and Filshie, J. H. (1982) Thin film thermistor. *J. Phys. E: Sci. Instrum.*, **15**(5), 411–414.

Nagai, T., Yamamoto K. and Kobayashi, I. (1982) SiC-thin film thermistor. *J. Phys. E: Sci. Instrum.*, **15**(5), 520–524.

National Semiconductor Corp. (1989) Data Sheets for LM34 and LM35 Series of temperature transducers.

Ohte, A. and Yamagata, M. (1977) A precision silicon temperature thermometer. *IEEE Trans. Instr. Meas.*, **IM-26**(4), 335–341.

Patranabis, D., Ghosh, S. and Bakshi, C. (1988) Linearizing transducer characteristics. *IEEE Trans. Instr. Meas.*, **IM-37**(1), 66–69.

Pavese, F. (1974) An accurate equation for the V–T characteristic of a GaAs diode thermometer in the 4–300K range. *Cryogenics*, **14**(8), 425–428.

Player, M. A. (1986) A simple wide-range thermistor thermometer. *J. Phys. E: Sci. Instrum.*, **19**, 787–789.

PMI (undated) Application Note 18. Thermometer applications of the Ref-02.

Rao, M. G., Scurlock, R. G. and Wu, Y. Y. (1983) Miniature silicon diode thermometers for cryogenics. *Cryogenics*, **23**(12), 635–638.

Regan, T. (1984) Applying the LM35 temperature sensor. *Electronic Prod Des.*, November, 85–86.

Roess, E. (1984) Properties and uses of thermistors. *Electronic Prod Des.*, June, 179–185.

R. S. Components Ltd, London (July 1983) Data Sheet 3992 for Semiconductor temperature sensor stock number 308–809.

R. S. Components Ltd, London (Nov. 1987) Data Sheet 8307 for LM35Z and LM35DZ Temperature sensor ic.

Ruehle, R. A. (1975) Solid state temperature sensor outperforms previous transducers. *Electronics*, **48**(6), 127–130.

Sah, C. T. (1961) A new semiconductor tetrode—the surface-potential controlled transistor. *Proc. Inst. Radio Engrs*, **49**, 1623–1634.

Sachse, H. B. (1975) *Semiconducting Temperature Sensors and Their Applications*, John Wiley and Sons, New York.

Sengupta, R. N. (1988) A widely linear temperature to frequency converter using a thermistor in a pulse generator. *IEEE Trans. Instr. Meas.*, **IM-37**(1), 62–65.

Spoff, M. (1972) Thermistors for biomedical use. *Temperature: its Measurement and Control in Science and Industry*, Vol. 4, Part 3, Instrument Society of America, Pittsburgh, pp. 2109–2124.

Sridaran, S., Das, S. R. and Raghavarao, R. (1987) A novel temperature probe for middle atmospheric and meteorological applications. *J. Phys. E: Sci. Instrum.*, **20**(10), 1198–1201.

Stanley, K. W. (1973) Non-linear resistors. *Radio and Electronic Eng.*, **43**(10), 1-4.

Swartz, J. M. and Gaines, J. R. (1972) Wide range thermometry using gallium. *Temperature: Its Measurement and Control in Science and Industry*, Vol. 4, Part 2, Instrument Society of America, Pittsburgh, pp. 1117–1124.

Swartz, D. L. and Swartz, J. M. (1974) Diode and resistance cryogenic thermometry: a comparison. *Cryogenics*, **14**, 67–74.

Sze, S. M. (1969) *Physics of Semiconductor Devices*, John Wiley and Sons, New York.

Talpe, J., Stolovitzky, G. and Bekeris, V. (1987) Cryogenic thermometry and level detection with common diodes. *Cryogenics*, **27**, 693–695.

Timko, M. P. (1976) A two-terminal IC temperature transducer. *IEEE J. Sol. St. Circ.*, **SC-11(6)**, 784–788.

Tsividis, Y. (1980) Accurate analysis of temperature effects in I_c-V_{BE} characteristics with application to bandgap reference sources. *IEEE J. Sol. St. Cir.*, **SC-15**(6), 1076–1084.

Verster, T. C. (1968) *p–n* junction as an ultralinear calculable capacitor. *Electronics Letters*, **4**(9), 175–176.

Verster, T. C. (1972) The silicon transistor as a temperature sensor. *Temperature: Its Measurement and Control in Science and Industry*, Vol. 4, Part 2, Instrument Society of America, Pittsburgh, pp. 1125–1134.

Weichert, L. *et al.* (1976) *Temperaturmessung in der Technik-Grundlagen und Praxis*. Lexika-Verlag, Grafenau.

Widlar, R. J. (1965) Some circuit design techniques for linear integrated circuits. *IEEE Trans. Circ. Theory*, **CT-12**, 586–590.

White, D. R. (1984) The linearisation of resistance thermometers. *J. Phys. E: Sci. Instrum.*, **17**, 381–385.

Ziel, A. van der (1968) *Solid State Physical Electronics*. Prentice-Hall, Englewood Cliffs, NJ.

7 Optical Pyrometers

7.1 INTRODUCTION

The simplest and oldest non-contact way of estimating the temperature of a radiating body is by observing its colour. Table 7.1 summarizes the relationship between temperature and colour. Using this method experienced practitioners can estimate temperatures over about 700 °C, with a precision sufficient for simpler heat-treatment processes. Pyrometers, also known as infrared thermometers, are non-contact thermometers, which measure the temperature of a body based upon its emitted thermal radiation. No disturbance of the existing temperature field occurs in this non-contact method. In pyrometry the most important radiation wavelengths which are situated between from 0.4 to 20 μm belong to the visible and infrared (IR) radiation bands. A pyrometer is composed of four main parts. An optical system, concentrates the radiation on a radiation detector, which may be either the human eye or a thermal or photoelectric sensor. The detected radiation is conditioned by a signal converter before being displayed in a read-out instrument, which may have additional analogue or digital output as options.

Those pyrometers to be described in this chapter are classified mainly according to their spectral response.

1. *Total radiation pyrometers* use thermal radiation detectors, which are heated by the incident radiation.

2. *Partial or band radiation pyrometers*, operate over a restricted wavelength band in which the output signal is only generated by the impact of a stream of photons on a photoelectric detector. As no detector heating due to any incident radiation occurs, they are called photoelectric pyrometers. In the case of very narrow operating wavelength bands they become spectral or monochromatic pyrometers.

3. *Disappearing filament pyrometers* are based upon matching the luminance of the background and of the filament by adjusting the lamp current. The observer's eye is the detector. They are also within the group of spectral or monochromatic pyrometers.

4. *Ratio pyrometers*, also known as two-colour pyrometers, deduce the temperature from the ratio of the radiation intensity emitted by the object in two different spectral wavebands.

Table 7.1
Colours of radiating bodies.

Temperature (°C)	Colour	Temperature (°C)	Colour
550–580	Black/purple	830–880	Dark orange
580–650	Brown/purple	880–1050	Orange
650–750	Purple	1050–1150	Yellow/orange
750–780	Dark carmine	1150–1250	Yellow
780–800	Carmine	1250–1320	White/yellow
800–830	Orange/carmine		

7.2 RADIATION, DEFINITIONS AND LAWS

Before giving a detailed description of particular pyrometers some main definitions and laws, relevant to radiation heat transfer, will be given. A more detailed discussion of these problems may be found in the more specialized publications of Gröber *et al.* (1963), Hackforth (1960), Harrison (1960), Jakob (1957) and Rohsenow and Harnett (1973).

Thermal radiation is a part of electromagnetic radiation. Let us assume that a radiant heat flux, Φ, a heat quantity in a unit time, is incident on the surface of a solid. Of this heat flux, the portion, Φ_α is absorbed, whilst Φ_ρ is reflected and Φ_τ is transmitted. The following definitions are introduced:

$$\begin{aligned} \text{absorptivity}, \ \alpha &= \Phi_\alpha/\Phi \\ \text{reflectivity}, \ \rho &= \Phi_\rho/\Phi \\ \text{transmissivity}, \ \tau &= \Phi_\tau/\Phi \end{aligned} \tag{7.1}$$

For every solid, applying the principle of energy conservation, gives:

$$\alpha + \rho + \tau = 1 \tag{7.2}$$

In the case of transparent bodies, as represented in Figure 7.1, many internal reflections cause additional absorption. For example, Harrison (1960) notes that the total reflected heat flux, Φ_ρ, is composed of the primary heat flux $\Phi_{\rho 1}$, and a secondary one $\Phi_{\rho 2}$.

There are three specific cases.

1. $\alpha = 1$, $\rho = 0$, $\tau = 0$—the body is a black body which totally absorbs all incident radiation
2. $\alpha = 0$, $\rho = 1$, $\tau = 0$—the body is a white body totally reflecting incident radiation.
3. $\alpha = 0$, $\rho = 0$, $\tau = 1$—the body is a transparent body as all of the incident radiation is completely transmitted.

The concept of a black body is very important in pyrometry. Figure 7.2 presents some configurations with properties approaching those of a black body.

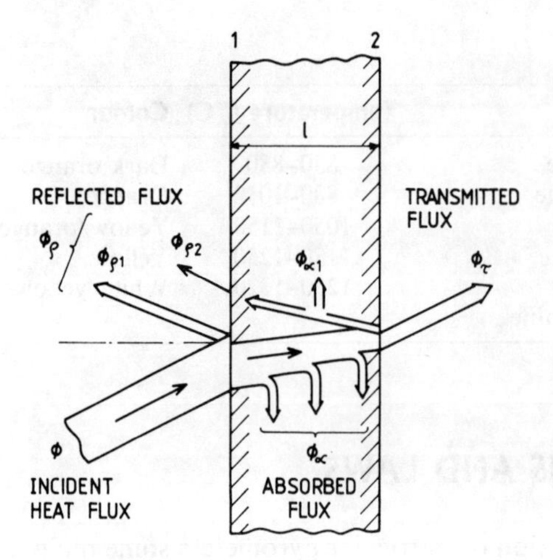

Figure 7.1
Decomposition of the heat flux Φ in a transparent body.

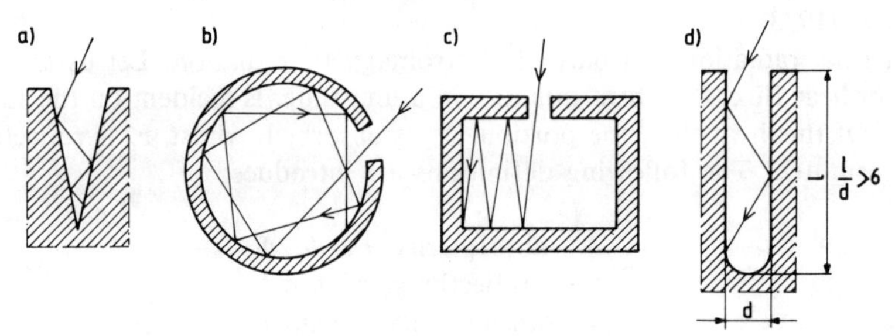

Figure 7.2
Models of black body.

Heinisch (1972) shows that in the cavities presented in Figure 7.2, total absorption of the incident radiation is reached by its multiple internal reflection.

Similarly to the α, ρ and τ factors, which are valid for the total radiaton, the spectral properties, α_λ, ρ_λ and τ_λ at the wavelength λ, may also be introduced:

$$\begin{aligned}\alpha_\lambda &= \Phi_{\lambda\alpha}/\Phi \\ \rho_\lambda &= \Phi_{\lambda\rho}/\Phi \\ \tau_\lambda &= \Phi_{\lambda\tau}/\Phi\end{aligned} \tag{7.3}$$

Equation (7.2) then becomes:

$$\alpha_\lambda + \rho_\lambda + \tau_\lambda = 1 \tag{7.4}$$

The values of α, ρ and τ depend upon the material, its surface state and temperature while α_λ, ρ_λ and τ_λ additionally depend upon the wavelength, λ.

The radiant intensity W or the radiant exitance is the heat flux per unit area expressed as the ratio of the heat flux $\mathrm{d}\Phi$ emitted from the infinitesimal element of the surface $\mathrm{d}A$, to the surface area $\mathrm{d}A$ itself:

$$W = \frac{\mathrm{d}\Phi}{\mathrm{d}A} \quad \mathrm{W/m^3} \tag{7.5}$$

In the same units as the radiant intensity, the heat flux density, q, of the incident radiation is given by:

$$q = \frac{\mathrm{d}\Phi}{\mathrm{d}A} \quad \mathrm{W/m^2} \tag{7.5a}$$

This also takes account of the conduction and convection heat flux in addition to the radiation heat flux.

The spectral radiant intensity, W_λ, is defined as:

$$W_\lambda = \frac{\mathrm{d}W}{\mathrm{d}\lambda} \quad \mathrm{W/m^2\,\mu m} \tag{7.6}$$

Planck's law gives the radiant flux distribution of a black body, as a function of the wavelength and of the body's temperature by the relation:

$$W_{o\lambda} \frac{c_1 \lambda^{-5}}{e^{c_2/\lambda T} - 1} \tag{7.7}$$

where $W_{o\lambda}$ is the spectral radiant intensity of a black body, $\mathrm{W/m^2\,\mu m}$ (the suffix 0 will be used in future to indicate a black body), λ is the wavelength, μm, T is the absolute temperature of the thermal radiator, K, c_1 is the first radiation constant, $c_1 = 3.7415 \times 10^{-16}\ \mathrm{W/m^2}$, and c_2 is the second radiation constant, $c_2 = 14\,388\ \mu\mathrm{m\,K}$.

For a given wavelength range, from λ_1 to λ_2, Equation (7.7) can be evaluated as:

$$W_{o,\lambda_1-\lambda_2} = \int_{\lambda_1}^{\lambda_2} \frac{c_1 \lambda^{-5}}{e^{c_2/\lambda T} - 1} \mathrm{d}\lambda \tag{7.8}$$

where $W_{o,\lambda_1-\lambda_2}$ is the band radiant intensity of a black body.

Hackforth (1960) has shown that if $\lambda T \ll c_2$, Planck's law of Equation (7.7) can be replaced using the same notation by a simpler Wien's law.

$$W_{o\lambda} = \frac{c_1 \lambda^{-5}}{e^{c_2/\lambda T}} \tag{7.9}$$

The spectral radiant intensity $W_{o\lambda}$ of a black body as a function of wavelength λ, at different temperatures, calculated from Planck's law, is shown in Figure 7.3.

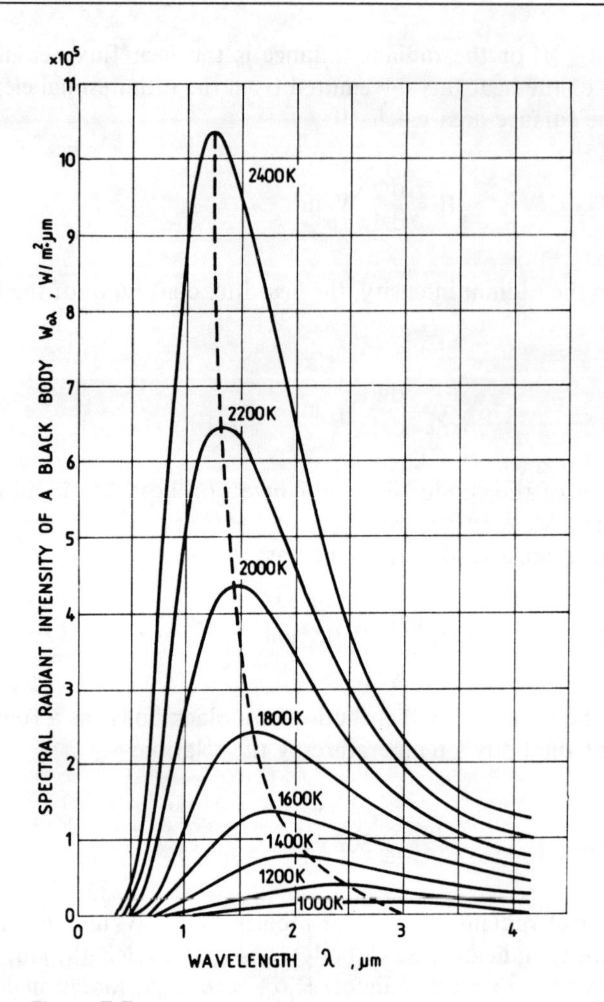

Figure 7.3
Spectral radiant intensity of a black body $W_{o\lambda}$ versus wavelength at different temperatures in accordance with Planck's law (Equation (7.7)).

At all temperatures of importance in radiation pyrometry, the errors, which result from replacing Planck's law by Wien's law, are negligibly small. The relative errors may be calculated from the relation:

$$\frac{\Delta W_{o\lambda}}{W_{o\lambda}} = \frac{W_{o\lambda,\mathrm{W}} - W_{o\lambda,\mathrm{Pl}}}{W_{o\lambda,\mathrm{W}}} = -\mathrm{e}^{-c_2/\lambda T} \tag{7.10}$$

where $W_{o\lambda,\mathrm{W}}$ is the spectral radiant intensity calculated from Wien's law and $W_{o\lambda,\mathrm{Pl}}$ is calculated as above from Planck's law.

The relative errors calculated from Equation (7.10) are presented in Table 7.2 as a function of the values of the product λT.

Table 7.2
Relative errors resulting from replacing Planck's law (Equation (7.7)) by Wien's law (Equation (7.9)) as a function of the values of λT.

λT (m K)	1.25×10^{-3}	1.5×10^{-3}	2×10^{-3}	3×10^{-3}
$\frac{\Delta W_{o\lambda}}{W_{o\lambda}}$ (%)	0.001	0.007	0.08	0.8

Figure 7.3 shows that with increasing temperature, the maxima of the spectral radiant intensity are displaced towards the shorter wavelengths. At the given temperature, T, where the maximum is reached, the wavelength λ_{max}, may be easily calculated from Wien's displacement law to obtain:

$$\lambda_{max} T = 2896\ \mu\text{m K} \tag{7.11}$$

For any given temperature, the area under the corresponding curve is a measure of the total power radiated at all wavelengths by a black body so that:

$$W_o = \int_{\lambda=o}^{\lambda=\infty} W_{o\lambda} d\lambda \tag{7.12}$$

The ratio of the spectral radiant intensity, W_λ, at the wavelength λ of a non-black body to the spectral radiant intensity of a black body, $W_{o\lambda}$, at the same temperature is called the spectral emissivity ϵ_λ.

$$\epsilon_\lambda = \frac{W_\lambda}{W_{o\lambda}} \tag{7.13}$$

If the spectral emissivity ϵ_λ of a given body is constant for each wavelength (i.e. ϵ_λ = constant) such a body is called a grey body. Similarly to Equation (7.13), if all wavelengths from 0 to ∞, are taken into consideration, the term total emissivity ϵ is used:

$$\epsilon = \frac{W}{W_o} \tag{7.14}$$

where W is the radiant intensity of any given body and W_o is the radiant intensity of a black body.

Following Kirchhoff's law, the spectral absorptivity α_λ of all opaque bodies equals their emissivity ϵ_λ, so that

$$\alpha_\lambda = \epsilon_\lambda \tag{7.15}$$

For a given wavelength band, from λ_1 to λ_2, Kirchhoff's law is expressed by:

$$\alpha_{\lambda_1-\lambda_2} = \epsilon_{\lambda_1-\lambda_2} \tag{7.15a}$$

where $\alpha_{\lambda_1-\lambda_2}$ is the band absorptivity and $\epsilon_{\lambda_1-\lambda_2}$ is the band emissivity.

When all wavelengths from $\lambda_1 \to 0$ to $\lambda_2 \to \infty$ are taken into consideration, the corresponding form for Equation (7.15a), which is also valid, then becomes:

$$\alpha = \epsilon \tag{7.15b}$$

where α is the total absorptivity, and ϵ is the total emissivity.

The Stefan–Boltzmann law which represents the dependence of the total radiant intensity W_o of a black body upon the temperature T, is expressed as:

$$W_o = \int_0^\infty W_{o\lambda}\, d\lambda = \sigma_o T^4 \tag{7.16}$$

where $W_{o\lambda}$ is the spectral radiant intensity of a black body as given by Forsythe (1941), and σ_o is the radiation constant of a black body with the value,

$$\sigma_o = 5.6697 \times 10^{-8}\ \text{W/m}^2\,\text{K}^4.$$

Equation (7.16) can be expressed in a more readily usable form as

$$W_o = C_o \left(\frac{T}{100}\right)^4 \tag{7.16a}$$

where C_o is the technical radiation constant of a black body, with the value

$$C_o = \sigma_o \times 10^8 = 5.6697\ \text{W/m}^2\,\text{K}^4$$

For grey bodies Equation (7.16a) becomes:

$$W = C_o \epsilon \left(\frac{T}{100}\right)^4 \tag{7.17}$$

where C_o is as before, and ϵ is the total emissivity.

In technical practice the majority of real bodies may be regarded as grey ones.

Knowledge of the total emissivity ϵ, and of the spectral emissivity ϵ_λ at $\lambda = 0.65\ \mu$m, for different materials, is necessary, to be able to calculate the corrections to be introduced when making pyrometric temperature measurements. The emissivity of different materials, which depends heavily upon the surface state, its homogeneity and temperature, may only be determined approximately. Worthing (1941) describes methods for the measurement of emissivity.

Comparison of the properties of different materials, independent of their surface state may be made using the specific total emissivity ϵ' and the specific spectral emissivity ϵ'_λ. The values of ϵ' and ϵ'_λ are determined for the direction normal to the surface for flat samples which should be polished and sufficiently thick. This last condition allows semi-transparent bodies to be regarded as totally opaque. The values of ϵ and ϵ_λ are also determined for the direction normal to the surface. Approximate values for the emissivity of different materials are given in Tables XXII and XXIII.

It must be stressed that uneven, rough and grooved surfaces may have much higher values of emissivity than are their specific emissivities.

Using the Maxwell theory of electromagnetism, Considine *et al.* (1957), following Drude, have proposed an approximate formula to calculate the specific spectral emissivity, ϵ'_λ, of metals as:

$$\epsilon'_\lambda \simeq K\sqrt{\frac{\rho}{\lambda}} \tag{7.18}$$

where $K = 0.365\ \Omega^{-1/2}$, ρ is the resistivity in Ω cm, and λ is the wavelength in cm.

Equation (7.18) which is valid for $\lambda > 2\ \mu$m, uses the original units of Drude. The emissivity of non-conductors which is a function of the material refractive index n_λ, is given in BS 1041, p. 5 by the formula:

$$\epsilon_\lambda = \frac{4n_\lambda}{(n_\lambda + 1)^2} \tag{7.19}$$

where n_λ which is the refractive index of the material, has a value in the range of 1.5 to 4 for most inorganic compounds and in the range 2.0 to 3.0 for metallic oxides. For most clean metals the emissivity is low, with a value of about 0.3 to 0.4, falling sometimes to 0.1 for aluminium. Spectral emissivities of metals become lower at lower temperatures where the wavelengths are longer. Non-metallic substances have emissivities of about 0.6 to 0.96, which do not vary greatly with temperature. It should be borne in mind, that the appearance of non-metals in visible light cannot be a basis for predicting their emissivities. Most non-metals, such as wood, brick, plastic and textiles at 20 °C have a value of total emissivity nearly equal to unity.

Radiant heat exchange between two bodies

Consider two parallel surfaces, having identical areas A and the respective temperatures and emissivities T_1, T_2, ϵ_1, ϵ_2, emitting towards each other with the intensities given by the Stefan–Boltzmann law in Equation (7.16a). The heat flux (power) Φ_{12}, exchanged between these surfaces, for $T_1 > T_2$, is given by:

$$\Phi_{12} = \frac{C_o A}{(1/\epsilon_1) + (1/\epsilon_2) - 1}\left[\left(\frac{T_1}{100}\right)^4 - \left(\frac{T_2}{100}\right)^4\right] \tag{7.20}$$

where C_o is the technical radiation constant, and A is the radiating area.

If one of the bodies of area A_1 is placed inside another one of area A_2 ($A_1 < A_2$), then Equation (7.20) becomes:

$$\Phi_{12} = \frac{C_o A_1}{(1/\epsilon_1) + (A_1/A_2)\,[(1/\epsilon_2) - 1]}\left[\left(\frac{T_1}{100}\right)^4 - \left(\frac{T_2}{100}\right)^4\right] \tag{7.20a}$$

In the very important practical case when $A_2 > 3A_1$, Equation (7.20) becomes:

$$\Phi_{12} = A_1 \epsilon_1 C_o \left[\left(\frac{T_1}{100} \right)^4 - \left(\frac{T_2}{100} \right)^4 \right] \tag{7.20b}$$

Lambert's directional law which describes the radiant intensity of a black body as a function of the radiation direction, is given by:

$$W_{o\varphi} \cong W_{o\perp} \cos \varphi \tag{7.21}$$

where $W_{o\varphi}$ is the radiant intensity of an element of area under the angle φ between the radiation direction and the direction normal to the surface, and $W_{o\perp}$ is the radiant intensity as before but in the direction normal to the surface.

Radiant intensity, $W_{o\perp}$, in the direction normal to the surface is π times smaller than the total radiant intensity.

$$W_{o\perp} = \frac{W_o}{\pi} \tag{7.22}$$

Equation (7.21) is only partially valid for non-black bodies. Large deviations from Lambert's law, which can be observed especially for polished metals when $\varphi > \pi/4$, are caused by the dependence of the emissivity upon the observation angle.

Some definitions, taken from lighting technique, are also used in optical pyrometry, in the case when the thermal radiation takes place in the visible wavelength range. Luminosity, I_φ, is the radiant flux propagated in an element of solid angle. Radiance, L, also called luminance, which is a density of luminosity of a surface in a given direction, is expressed as

$$L = \frac{dI_\varphi}{\cos \varphi \, dA} \tag{7.23}$$

where dA is the area of an element of the radiating surface and φ is the angle between the radiant flux direction and the direction normal to the surface.

Radiance is a deciding factor in the subjective impression of the body's brightness. Lambert's law of Equation (7.21) which is also valid for the luminosity, is:

$$I_\varphi = I_\perp \cos \varphi \tag{7.24}$$

where $I_\perp$ is the luminosity in the direction normal to the surface.

Combining Equations (7.24) and (7.23), yields:

$$L = \frac{dI_\perp \cos \varphi}{dA \cos \varphi} = \frac{dI_\perp}{dA} \tag{7.25}$$

From Equation (7.25) it follows that the radiance of a black body is independent of the viewing angle and is always the same as in the direction normal to the surface. For the majority of non-black bodies, the radiance is nearly constant for φ in the range from 0 to $\pi/4$.

7.3 TOTAL RADIATION PYROMETERS

7.3.1 General information

In total radiation pyrometers the temperature of a body is determined by the thermal radiation, which it emits over a large range of wavelengths. This radiation is concentrated onto a thermal radiation detector by a lens, mirror or light guide as shown in Figure 7.4. The thermal radiation detectors which are used are a thermocouple, or more often a group of series connected thermocouples called a thermopile, a metal or semiconductor bolometer, a pyroelectric detector or a bimetal. Heating of the thermal detector by the concentrated incident thermal radiation gives a detector output signal which is proportional to its temperature and thus at the same time to the value of the measured temperature.

A total radiation pyrometer using a lens, was first constructed by Féry (1902) who later (Féry, 1908) also used a concave mirror in 1904. Pyrometers may have an optical system with fixed or adjustable focal length. The former type is now more popular.

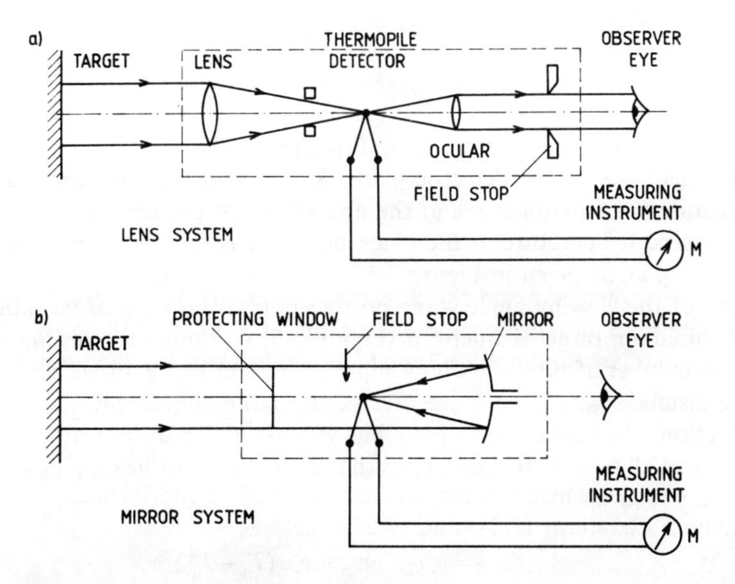

Figure 7.4
Basic diagrams of total radiation pyrometers.

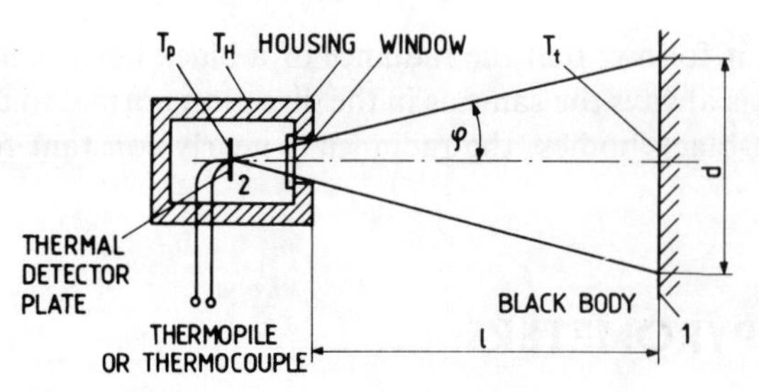

Figure 7.5
Total radiation pyrometer—simplified design.

7.3.2 Scale defining equation for black bodies

Consider a pyrometer shown in a simplified way in Figure 7.5. The thermal radiation, emitted by a black body 1, whose temperature, T_t, is to be measured, falls through a window onto a thermal detector plate 2. On its window side this plate is blackened to give as high an emissivity as possible while its other side should have as low an emissivity as possible. The incident thermal radiation heats-up the plate to a certain temperature T_p, which is measured by a thermocouple or a thermopile. A reference junction temperature for this thermopile is provided by the temperature, T_H, of the pyrometer housing. Although there is no concentrating optical system in the form of a lens or mirror in the vastly simplified Figure 7.5, neither the working principle nor the sensitivity of the pyrometer are altered. This is apparent as the existence of a concentrating optical system only reduces the necessary area of the radiating body.

On the surface of the detector plate, the heat flux density of the flux, emitted by the body and absorbed by the plate, is given by:

$$q_{1-2} = \sigma_o \epsilon_2 K_1 \sin^2 \varphi (T_t^4 - T_p^4) \tag{7.26}$$

where σ_o is the radiation constant from Equation (7.16), ϵ_2 is the total emissivity of the blackened side of the detector plate, K_1 is the coefficient depending on the construction of the pyrometer and the absorption of the optical system, T_t is the true, measured temperature of the black body, T_p is the plate temperature, and φ is the viewing angle given in Figure 7.5.

Instead of the viewing angle some producers give the ratio of working distance l to the minimum target diameter d (Figure 7.5) or simply the distance ratio. An increasing number of producers now supply diagrams of the target diameter versus working distance, which are more precise and more convenient.

In practice, when the detector plate has very small dimensions, its viewing angle, φ, is the same all over its surface. Thus the thermal flux, or heating power absorbed by the plate is given by:

$$\Phi_{1\to2} = \sigma_o \epsilon_2 K_1 A_p \sin^2 \varphi (T_t^4 - T_p^4) \tag{7.27}$$

where A_p is the one-side plate area and the other symbols are as in Equation (7.26).

As the area of the plate, A_p is much smaller than the inner area of the pyrometer housing, and neglecting the radiant heat exchange at the unblackened back side of the plate, the total heat flux transmitted from the plate to the pyrometer housing is expressed as:

$$\Phi_{2\to H}=\sigma_0\epsilon_2 A_p(T_p^4-T_H^4)+K_2(T_p-T_H) \tag{7.28}$$

where K_2 is the heat transfer coefficient by convection and conduction from the plate to the housing and the other symbols are as in Equation (7.27).

When the plate is in the thermal steady-state, the received radiant heat flux $\Phi_{1\to 2}$ equals the heat flux $\Phi_{2\to H}$ transferred to the pyrometer housing so that:

$$\sigma_0\epsilon_2 K_1 A_p \sin^2\varphi(T_t^4-T_p^4)=\sigma_0\epsilon_2 A_p(T_p^4-T_H^4)+K_2(T_p-T_H) \tag{7.29}$$

The output signal of the pyrometer, which is the thermal e.m.f., E, of the thermocouple or thermopile of Figure 7.5 is a linear function of the temperature difference between the plate temperature, T_p, and that of the housing, T_H, is thus given by:

$$E=K_e(T_p-T_H) \tag{7.30}$$

where K_e is the thermocouple gain, mV/K. The gain of a thermopile, composed of n thermocouples is, nK_e.

Calculation of the characteristic $T_p-T_H=f(T_t)$ of a pyrometer is based on the solution of Equation (7.29), whose complicated form as well as the temperature dependence of ϵ_2, K_1 and K_2, excludes the possibility for a practical analytical solution. In practice, the pyrometer characteristic, which is always determined experimentally, has the approximate form:

$$E\approx K(T_t^b-T_p^b) \tag{7.31}$$

in which the exponent, b, with a value between 3.5 and 4.5, and the constant, K, depend on the construction of the pyrometer. Equation (7.31) also concerns thermocouple and thermopile detectors. For resistance and semiconductor bolometers other formulae are used.

7.3.3 Influence of reference temperature

The readings of a total radiation pyrometer with thermocouple or thermopile detectors depend on the difference between the measuring junction or plate temperature T_p and the reference junction temperature, which equals the pyrometer housing temperature T_H. To make the readings independent of the housing temperature T_H, whose variations would affect the pyrometer readings, the thermoelectric radiation detector should be designed in such a way, that its heat losses to the housing are a linear function of the temperature difference T_p-T_H.

This can be explained by considering what happens if the housing temperature increases from T_H to T'_H. This increased housing temperature will cause a decrease in the e.m.f. of the detector owing to the lower value of the difference, $T_p - T_H$. A simultaneous decrease in the heat loss of the detector also results from the increase in housing temperature, which subsequently gives rise to an increase in T_p to T'_p. Properly designed pyrometers should meet the condition:

$$T_p - T_H = T'_p - T'_H \tag{7.32}$$

so that the pyrometer readings are independent of the housing temperature.

The described compensation method, sometimes causes an increase of the heat loss by the detector which results in a decrease of the detector sensitivity. Effective pyrometer design should be a compromise between the compensation ability and pyrometer sensitivity. Other compensation methods will be discussed later, when the various constructions of some total radiation pyrometers are described.

7.3.4 Influence of target distance

For pyrometer readings to be correct, the whole field of view should be filled by the target area, so that the whole detector plate is irradiated by the source radiation. This also means that the rotational cone base of Figure 7.5 is fully covered by the measured target surface. In this case, the total radiation energy received by the detector plate is the same for any target distance. No absorption of the radiant flux during its transit between the target and pyrometer has been considered so far.

7.3.5 Temperature measurement of non-black bodies

Total radiation pyrometers are calibrated under the assumption that the measuring target is a black body. From Equation (7.27), the radiant heat flux emitted by the target at the temperature T_t and absorbed by the detector plate is given by:

$$\Phi_{1\to 2} = \sigma_0 \epsilon_2 K_1 A_p \sin^2 \varphi (T_t^4 - T_p^4)$$

or introducing the coefficient K'_1, it will be:

$$\Phi_{1\to 2} = K'_1 (T_t^4 - T_p^4) \tag{7.33}$$

In practice, as usually $T_t \gg T_p$, Equation (7.33) becomes:

$$\Phi_{1\to 2} \approx K'_1 T_t^4 \tag{7.34}$$

where K'_1 is a constructional constant.

For example, for $T_t = 2000$ K and $T_p = 400$ K, $T_p^4 \simeq 0.0016\, T_t^4$.

For non-black bodies, having emissivity ϵ, the radiant flux absorbed by the detector plate will be:

$$\Phi'_{1\to2}=K'_1\epsilon T_t^4 \tag{7.35}$$

As a total radiation pyrometer is calibrated for black bodies, for use in measuring the temperature of non-black bodies, the indicated temperature value T_i, called the black temperature is lower than T_t. Since T_i is the temperature, at which the detector would get the same radiant flux from a black body, then:

$$\Phi'_{1\to2}=K'_1T_i^4 \tag{7.36}$$

Equating (7.35) to (7.36), shows that:

$$T_t=T_i\sqrt[4]{\frac{1}{\epsilon}} \tag{7.37}$$

Numerical example

When the temperature of a body of $\epsilon=0.6$ was measured by a total radiation pyrometer, the indicated temperature was $T_i=1200$ K. Calculate the true temperature of the body.

Solution: From Equation (7.37),

$$T_t=1200\sqrt[4]{\frac{1}{0.6}}\simeq 1370\text{ K}$$

Emissivity corrections, $\Delta\vartheta$, to be added to the readings of a radiation pyrometer which are calculated from

$$\Delta\vartheta=\vartheta_t-\vartheta_i=T_t-T_i=T_i\left(\sqrt[4]{\frac{1}{\epsilon}}-1\right) \tag{7.38}$$

are given in Figure 7.6

The corrections, calculated from Equation (7.38) or read from Figure 7.6, are not precise for the following reasons.

1. Real radiation pyrometers which are never truly total radiation pyrometers, behave more like band radiation pyrometers. For calculation of the correction the band emissivity $\epsilon_{\lambda_1-\lambda_2}$ should be considered instead of the total emissivity ϵ.
2. Values of emissivity ϵ which are temperature dependent, are given as approximated values in the technical literature. In addition they depend on the surface state and its degree of oxidation.

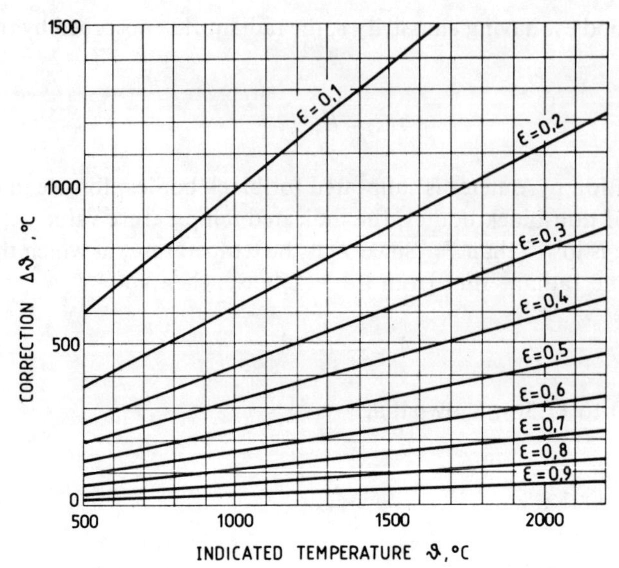

Figure 7.6
Corrections $\Delta\vartheta$ to readings of a total radiation pyrometer, measuring temperature of a body of total emissivity ϵ, as a function of indicated temperature value ϑ_i.

In practice the following procedures are advisable.

1. Measurement conditions should be arranged, so that the object, whose temperature is to be measured, is as near to a black body as possible. Using an auxiliary sighting tube which is immersed in the measured medium as shown in Figure 7.2(d), is one way of achieving these conditions.
2. In a continuous production process the apparent 'black' temperature of the object which is measured with a pyrometer, is compared with the same temperature measured by another method, giving the true temperature. This apparent temperature may serve as a reference for the repeatability of the process.
3. Readings obtained by another measuring instrument provide the basis for correcting the pyrometer readings if there is a possibility of setting the value of the emissivity ϵ. The pyrometer readings are then correct at a given temperature level, and as long as the object emissivity is constant.
4. An apparent increase of emissivity has been reported by Heimann and Mester (1975) using an additional reflecting plate of low emissivity placed above the target surface (Figure 7.7). The surface of the plate and of the target combine to give a model of a black body, which assures the correctness of the readings. As the reflecting surface is never a perfect mirror, even better results may be obtained by heating it to a temperature near the measured one. It is not always possible to use this arrangement.
5. Applying a polarizing filter whilst directing the pyrometer at a viewing angle about 45° to the measured surface, results in an increase in the apparent emissivity as compared with the direction normal to the surface. However, an optimum angle has to be found. In addition, as pointed out by Walter (1981), the necessity of a rigid filter and pyrometer mounting and radiation attenuation, decreasing the

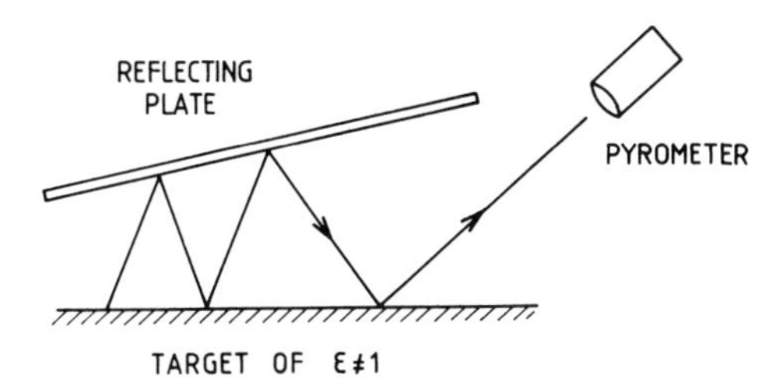

Figure 7.7
Method of apparent increase of the target emissivity.

pyrometer output signal, limit the application range of the method. The method can only be applied to metallic surfaces.

If an object of emissivity, ϵ, and temperature, T_t, is surrounded by walls of emissivity $\epsilon \simeq 1$ at temperature, T_w, such as occurs when measuring the temperature of a charge inside a furnace, the pyrometer readings depend not only on radiation from the object, but also on radiation from the walls reflected from the object. The true temperature, T_t, may then be calculated from:

$$T_t = T_i \sqrt[4]{\frac{1}{\epsilon}} \cdot \sqrt[4]{1-(1-\epsilon)\left(\frac{T_w}{T_i}\right)^4} \qquad (7.39)$$

where T_i is the indicated temperature.

If $T_w \ll T_i$, Equation (7.37) is to be applied (see §14.1.2).

7.3.6 Extension of measurement range

Extension of the measurement range towards higher temperatures is possible by weakening the radiant flux coming from the object. Grey filters are used for this purpose. The radiant flux absorbed by the detector plate is given by Equation (7.34) as:

$$\Phi_{1\to 2} = K_1' T_t^4$$

Let the corresponding pyrometer indication, T_i, remain the same while assuming that a grey filter with the transmission factor, τ_1, is used. Of course this is possible at another higher object temperature T_t', at which the radiant flux is:

$$\Phi_{1\to 2} = K_1' \tau_1 (T_t')^4 \qquad (7.40)$$

where τ_1 is the filter transmission factor and T_t' is the new object temperature.

By equating the formulae (7.34) and (7.40) it is apparent that

$$T_t^4 = \tau_1 (T_t')^4 \qquad (7.41)$$

As the indicated and true temperature for black bodies and for a pyrometer without filter are equal, it follows that:

$$T_i = T_t$$

so that eventually

$$T_i = T_t' \sqrt[4]{\tau_1} \quad (7.42)$$

where T_i is the reading of pyrometer with grey filter and T_t' is the measured temperature.

The grey filter, used for extension of the temperature range, may be pushed in and out so that the pyrometer has two temperature scales. One is the lower temperature range and the other used with the grey filter, is the higher temperature range. In some pyrometers exchangeable optics are used for changing the temperature range.

7.3.7 Optical systems

Lenses, mirrors and light guides which are used to concentrate the incoming radiant flux, reduce the pyrometer viewing angle and consequently the necessary object diameter. Lenses should be made of materials characterized by:

- high transmission factor over a wide wavelength range,
- high mechanical strength,
- possibly high working temperature,
- good resistance to atmospheric and chemical influences,
- good resistance to abrasion,
- good resistance to rapid temperature variations.

As it passes through the lens, incident thermal radiation is attenuated by absorption and reflection at both lens surfaces (Figure 7.1). The same effects occur at the sighting window. Normally it is enough to take only one internal reflection into account. Hackforth (1960) points out that coated lenses are used to reduce the surface reflection factor. He also notes that the overall lens transmission may be even doubled by correctly choosing the lens coating and its thickness. Materials such as SiO, ZnS, CeO_2, MgF_2 and so on, with thicknesses equal to one quarter of the wavelength of the incident radiation, are suitable. The application range of different optical materials depends upon their transmission factors as a function of the wavelength and on the thickness of the lens or window. In pyrometry, the upper cut-off wavelength of incident infrared radiation, caused by the lens material, is extremely important. Following Wien's displacement law given in Equation (7.11), this long wavelength transmission limit determines the lowest temperature which the pyrometer can measure. Figure 7.8 presents some of the transmission limits of the more popular materials used for lenses and sighting windows of radiation pyrometers. Different

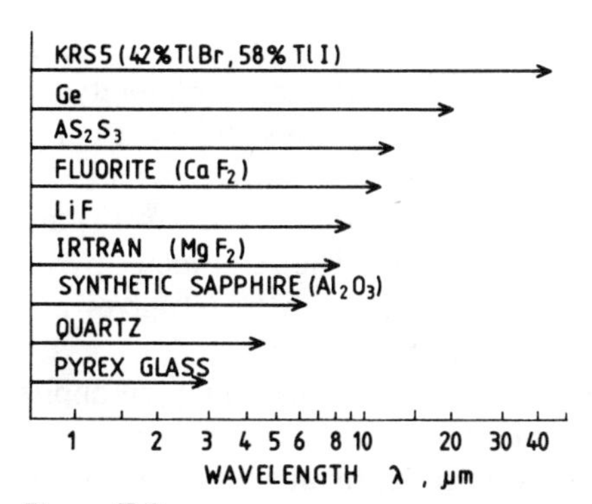

Figure 7.8
Transmission limits of some materials used for pyrometer lenses.

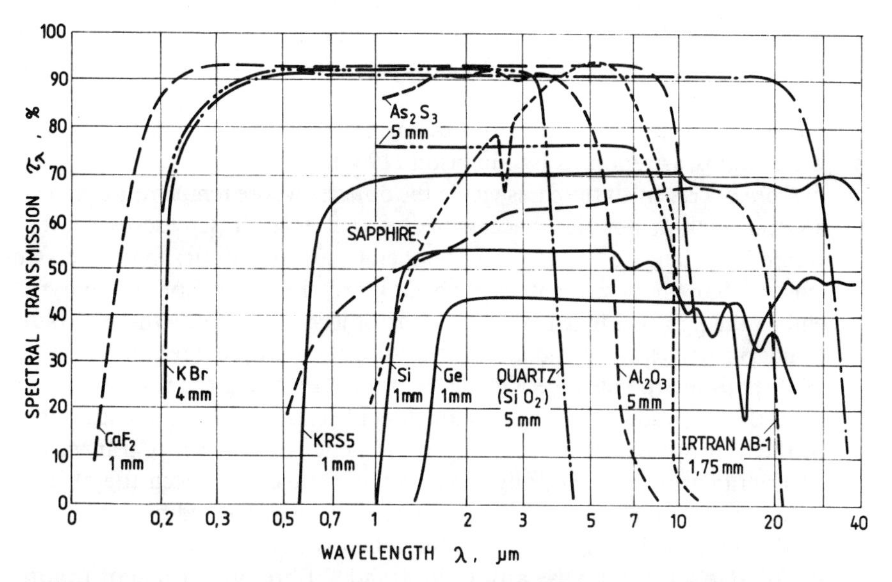

Figure 7.9
Spectral transmission τ_λ of plates made of materials used for pyrometer lenses.

plates of known thickness, made of materials used for pyrometer lenses have their relative spectral transmission, τ_λ, displayed as a function of the wavelength λ in Figure 7.9.

Commonly used lens materials are now described. Pyrex glass, transparent from 0.3 to 3 μm, is used when high mechanical and chemical resistance is necessary. Quartz (SiO_2), transparent from 0.2 to 4 μm, can withstand temperatures higher than those of glass, has high mechanical and chemical resistance and may also withstand rapid temperature variations. Synthetic sapphire (Al_2O_3), transparent from 0.2 to 5 μm, is hard and abrasion resistant. As it can be applied up to about 1000 °C it is also used for light guides. Unfortunately it is easily broken and cannot stand rapid temperature variations. Fluorite (calcium fluoride, CaF_2), transparent from 0.1 to

9.5 μm, can be used for measured temperatures as low as +50 °C. Its applications are limited by low mechanical strength, softness and poor workability. KRS-5 (42% TIBr, 58% TII), transparent from 0.5 to 36 μm, is now the most commonly used material for the lenses of low temperature pyrometers, starting from −50 °C, where its mechanical strength is adequate. Silicon, transparent from 8 to 14 μm, sometimes replaces KRS-5, for low temperature pyrometers. The Ardometer pyrometer from Siemens AG (Germany) uses this material. Hackforth (1960) gives more detailed information on lens and window materials.

At the lowest measured temperatures, where no lenses may be applied, mirrors are used. As described in §7.2, they are made of metals of good electrical conductivity, which are characterized by a high reflection factor at low temperatures and long wavelengths. Although mirrors absorb less infrared radiation than lenses, this advantage is partially cancelled by the need to use a protecting window. They are made mainly of polished gold, silver or aluminium of high reflectivity. Gold has good resistance to atmospheric and chemical influences, while the other metals have to be covered by a protective coating, transparent to infrared. This coating cannot be too thin, otherwise it will not act as a reflection reducing layer. Figure 7.10 gives the specific spectral reflectivities ρ_λ' of different metals as a function of radiation wavelengths as reported by Harrison (1960).

Fibre optics. In all the cases where the objects, whose temperature is to be measured, are too small or not easily accessible, as well as in those cases when the pyrometer would be endangered by excessive temperatures, fibre optics may successfully replace lenses. Figure 7.11 presents the working principle of a fibre-optic pyrometer. The end of the light guide is placed near the object which emits thermal radiation. This radiation arrives at the radiation detector after multiple internal reflections from the inner polished rod surface. Owing to absorption along the rod, imperfect reflection from the rod walls and reflection losses at the entrance and exit ends of the rod, some of the transmitted energy is lost. The efficiency of the energy transmission depends on the radiation entrance angle, φ, on the distances between the object and rod as well as between the rod and the detector. It is also affected by the length and design of the light guide. Light guides are made of artificial sapphire (Al_2O_3) or quartz (SiO_2) as a solid rod or as a flexible stranded fibre-optic cable of thin fibres, up to 2 m long. A light guide can be bent, provided that the angle of incidence at the side wall is always greater than the critical angle. Harrison (1960) gives a detailed theory of light guides. Fibre optics is rarely used in total radiation pyrometers.

7.3.8 Radiation detectors

Thermal radiation detectors for use in total radiation pyrometers should have the following properties:

- high sensitivity, defined as the ratio of output signal to the incident radiation power,
- time stable properties,
- high resistance to shocks and vibrations,

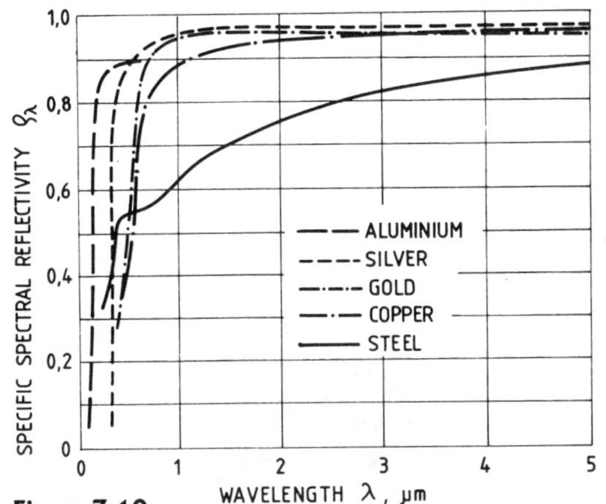

Figure 7.10
Specific spectral reflectivities ρ'_λ of metals used for pyrometer mirrors.

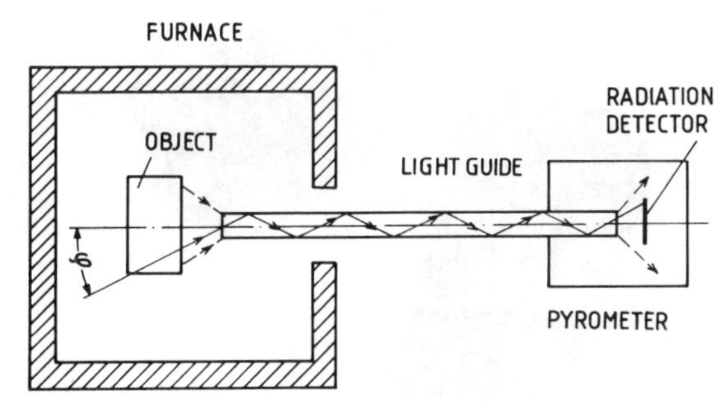

Figure 7.11
Working principle of fibre-optic pyrometer.

- low thermal inertia,
- output signal independent of the pyrometer position,
- high output signal-to-noise ratio,
- high emissivity,
- sensitivity independent of wavelength.

Thermopiles which are the most commonly applied thermoelectric detectors, possess all of these properties, together with an easily measurable or transformable output signal. These detectors are miniaturized elements in which the measuring junctions of a number of series connected thermocouples are exposed to the incident radiation from the object, whose temperature is to be measured. The reference junctions of the detector are kept at the pyrometer housing temperature.

As stated in §7.3.3, the resistance to any heat conduction between the measuring junctions and the pyrometer housing, is a compromise between the influence of the ambient temperature and the pyrometer sensitivity. Lieneweg (1975) asserts that a

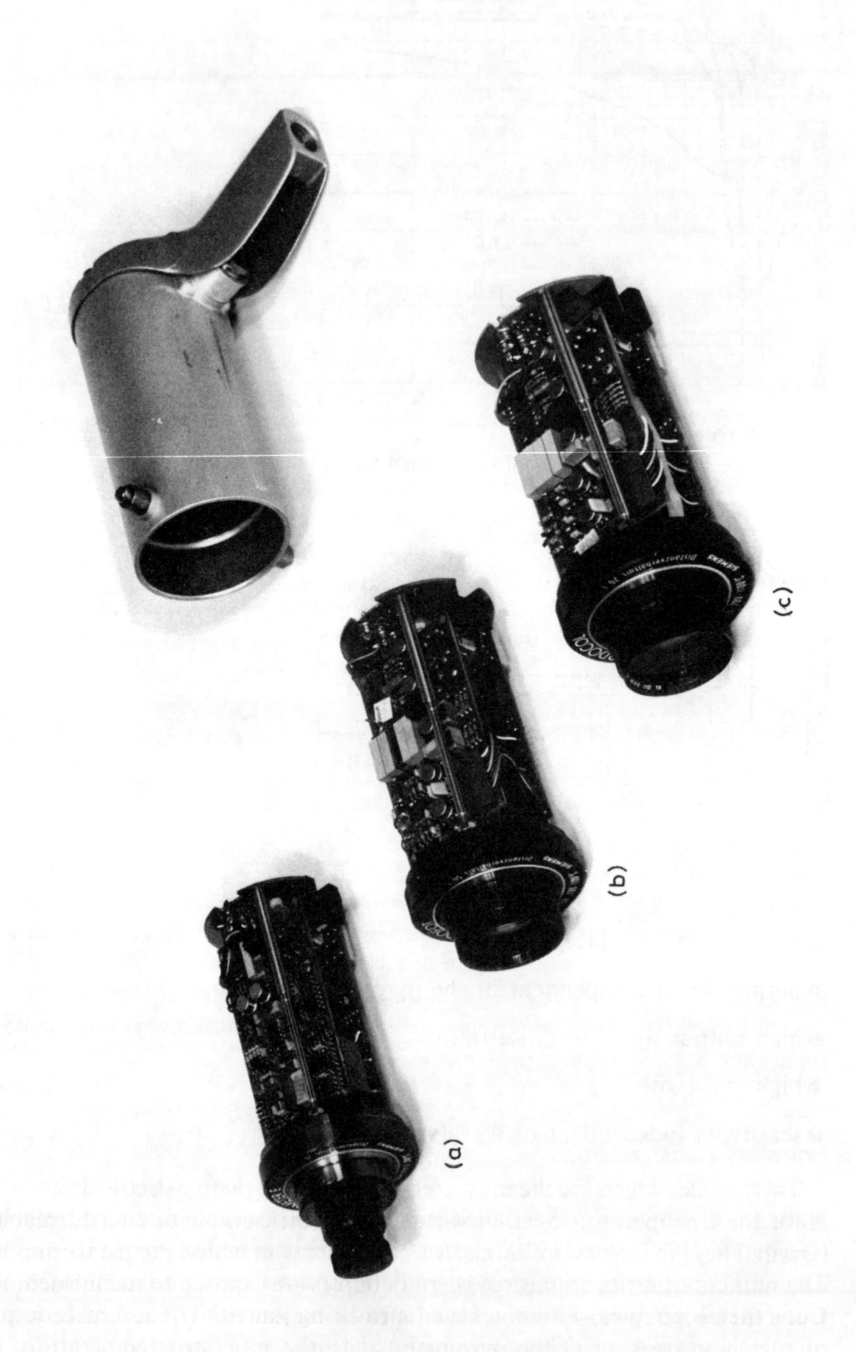

Figure 7.12
(a) Total radiation pyrometer Ardometer E. (b) Photoelectric pyrometer Ardofot. (c) Two-colour pyrometer Ardocol. (Courtesy of Siemens AG, Germany.)

good solution to the problem is the enclosure of the thermopile in an air evacuated glass-bulb. As well as increasing the detector sensitivity this eliminates convective heat exchange, making the pyrometer output signal totally independent of the pyrometer position.

Thermopile types which are used, are now described. Wire thermopiles made of thin thermocouple wires of diameter 0.1 to 0.15 mm with thin blackened plates, are used as radiation receivers. Ribbon thermopiles consist of thin thermocouple ribbons which are 0.025 mm thick and 0.5 mm wide. They are soldered or welded together with one surface blackened to form the radiation receiver. Thin film thermopiles are deposited on a non-metallic plate, which is the radiation receiver. These types of thermopile have extremely low thermal inertia. For example, the thin film thermopiles in the Rayotube Pyrometer by Leeds & Northrup, have a time constant of only 0.015 s. In the Ardometer E pyrometer of Siemens AG for measurement ranges up to 500 °C, the thermopile which consists of 15 thermocouples made of β-Tellur-bismuth, has a really high sensitivity of 600 μV/K. The low thermal conductivity of both metals, allows the use of thermocouple leads with larger cross-sections, thus preventing breakage due to brittleness. For the higher measurement ranges a thermopile, of 48 thermocouples made of gold-constantan, is compensated for ambient temperature variations by a Ni-resistor.

Thermistor and metal bolometers which are also applied, are constructed in thin film technology, with a resistance of 1 to 5 MΩ. In most cases they are used in a.c. bridge circuits to allow easy amplification of the output signals. Sometimes a d.c. bridge circuit, with the modulation of incident radiation is used. Baker *et al.* (1953) report that the time constants of bolometers are from 1 to 16 ms. Hoffman (1981) has pointed out that pyroelectric detectors are becoming increasingly popular. They are based on the phenomenon that the dipole moments of the charges in pyroelectric crystals, such as triglicine sulphate (TGS) change their orientation as a function of temperature. As the temperature varies, a temporary imbalance of charges appears so that an easily amplified a.c. voltage which is modulated by the incident radiation flux, is generated. Thermal radiation detectors have sensitivity proportional to the radiation intensity. As they have a lower sensitivity than photoelectric types they are better suited for applications at longer wavelengths and thus at lower temperatures. Their sensitivity is totally independent of the wavelength since they may be regarded as black or grey bodies.

7.3.9 Review of construction

A total radiation pyrometer called an *Ardometer* has been produced by Siemens AG (Germany) (1987) since 1920. Figure 7.12 shows an Ardometer in its present form. The lens concentrates the incident radiation on a thermopile, composed of 11 Ni–CuNi thermocouples, with blackened measuring junctions. The e.m.f. output signal of the thermopile, which is proportional to the measured temperature, is also a function of the temperature of the pyrometer housing. A Ni resistor, connected in parallel with the thermopile is used to compensate for this influence, in the housing temperature range from 0 to 180 °C or 0 to 100 °C for lower measuring

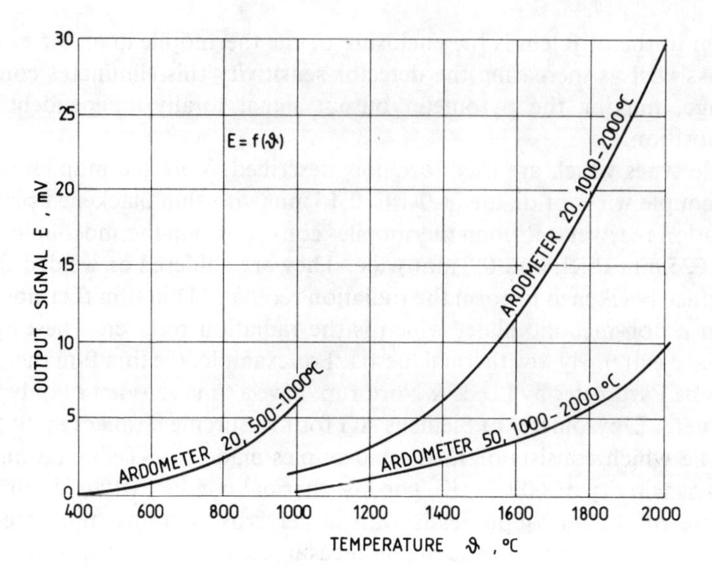

Figure 7.13
Output signal versus temperature of the Ardometer pyrometer. (Courtesy of Siemens AG, Germany.)

temperature ranges. The pyrometer which is directed at the target, observing it through the eyepiece, has measurement ranges of from 400 to 840 °C and from 1000 to 2000 °C.

Quartz lenses, transparent between 0.4 to 3.8 μm, or fluorite lenses transparent between 4 and 8 μm, are used for measurement up to 700 °C. Over 700 °C glass lenses, transparent between 0.4 and 2.5 μm, are used.

The distance ratio (§7.3.2) of the Ardometer 20 pyrometer is 20 and 50 for the Ardometer 50 instrument. In Figure 7.13 the e.m.f., E, of the pyrometer is displayed as a function of temperature. As can be seen, the value of E given by the pyrometer is greater for smaller distance ratio types. A special Ardometer 50N is intended for temperature measurement from very short distances of about 100 mm. In the whole possible distance range additional measurement errors resulting from distance variations are confined between -0.5% to $+0.5\%$ of full scale deflection (FSD). The 98% response time of the Ardometer is about 2.5 s while its half value time is around 0.5 s.

For black bodies and measured temperatures in the range between 1350 °C and 1750 °C, the overall measurement errors are $\pm 1\%$ FSD, whereas below 1350 °C and above 1750 °C these errors are $\pm 1.5\%$ FSD. Ardometer instruments, which can operate compatibly with indicators, recorders, transmitters and controllers, have non-linear scale division.

A new development of the pyrometer described above is the Ardometer E (Figure 7.12(a)) whose block diagram is given in Figure 7.14. By applying a new thermopile and new electronic circuitry, much lower temperatures can be measured with a generated thermal e.m.f. output signal which is a linear function of the measured temperature (Siemens AG, 1987). The lens concentrates the incident radiation on the thermopile

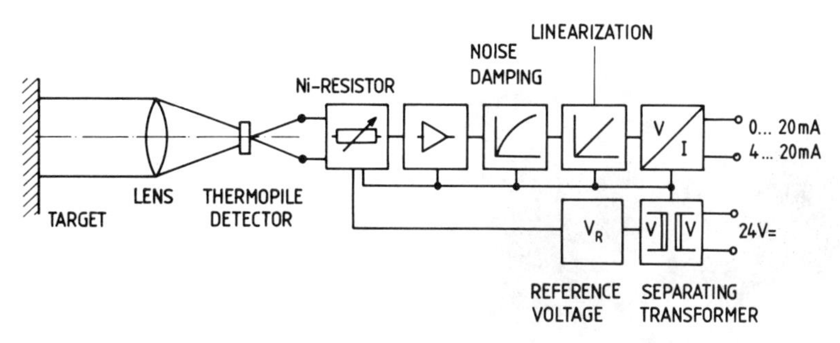

Figure 7.14
Block diagram of Ardometer E. (Courtesy of Siemens AG, Germany.)

whilst a Ni-resistor is used to compensate for variations of pyrometer housing temperature, as before. Amplification of the thermopile output signal gives an instrument signal span of 10 V. An adjustable resistor enables the necessary emissivity of the target to be set. A special circuit, which dynamically damps the noise signal, gives a noise-free linearized output signal as a linear 0–20 mA current signal using a V/I converter. The measured temperature ranges are from 0–60 °C to 800–1800 °C for which lenses of CaF, quartz, Si, and glass are used. For the temperature measurement of objects at a distance more than 2 m away in the presence of water vapour, an Ardometer E is used with a band filter, transparent in the atmospheric window from 8 to 14 μm. Different pyrometer mountings and assemblies are produced for stationary or portable service, which may also have special jackets provided for air or water cooling. Steel or ceramic sighting tubes may be used.

One of the products of Land Infrared Ltd (UK) is a portable total radiation pyrometer Cyclops 33 shown in Figure 7.15. This instrument has a thermal pyroelectric detector with a working range from -50 to $+1000$ °C. It has a spectral response within the 8 to 14 μm wavelength range, thus avoiding the atmospheric absorption bands. Its special features are:

- focusing reflex optics for accurate target definition at a distance from 750 mm to ∞,
- digital temperature display within the field of view,
- microprocessor based electronics,
- different measuring modes such as continuous, peak or valley picking, or mean value over a sampling period,
- BCD output for printer,
- accuracy $\pm 1\%$ of reading,
- target diameter say of 13 mm at a distance of 750 mm, corresponding to a distance ratio of about 58,
- response time 0.5–1 s,
- emissivity adjustable from 0.3 to 1.

Figure 7.15
Portable total radiation pyrometer Cyclops 33. (Courtesy of Land Infrared Ltd, UK.)

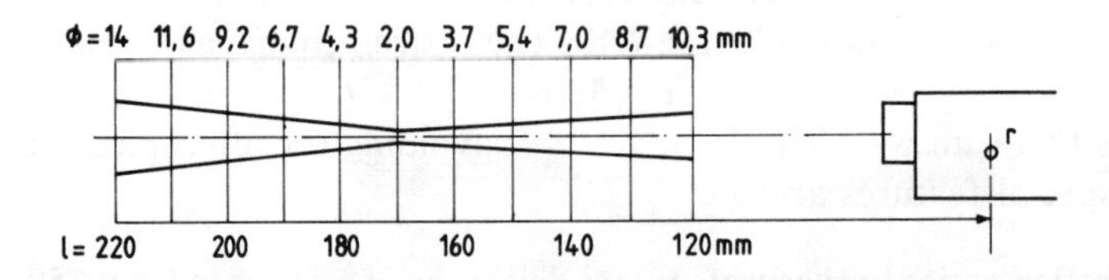

Figure 7.16
Target diameter ϕ versus target distance l of total radiation pyrometer Cyclops 300 S. (Courtesy of Land Infrared Ltd, UK.)

A modification of this pyrometer is the Cyclops 33 CF (Close Focus) instrument which is suitable for a small target diameter of typically 2 mm. When the distance between target and instrument is 170 mm, the distance ratio is thus about 85. Instead of a distance ratio, a diagram of the target diameter versus target distance of the same form, as shown in Figure 7.16 is often given. A fixed focus optical system is used. The temperature range is from 5 to +600 °C.

For temperature measurements at long distance, Land Infrared Ltd produces a Cyclops-Tele which uses a pyroelectric detector giving a temperature range of from 20 to +400 °C. Reflex focusable optics allow temperatures to be measured at

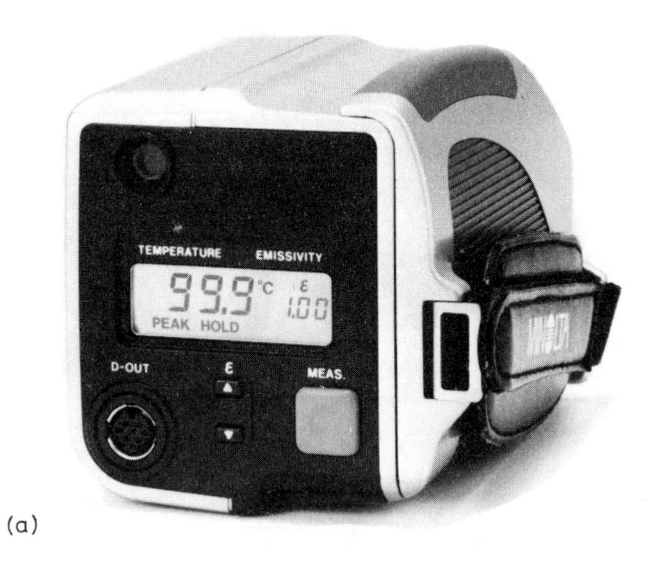

(a)

(b)

Figure 7.17
Mirror total radiation pyrometer Compac 3. (Courtesy of Land Infrared Ltd, UK.)

Figure 7.18
Mirror total radiation pyrometer PF-03. (Courtesy of Keller, GmbH, Germany.)

distances from 3 m to infinity. A nominal target area of $16 \times 52\ mm^2$ at a distance of 10 m, corresponds to a distance ratio of around 200. Typical applications of the Cyclops-Tele instrument are in temperature measurements to assist in the discovery of trouble spots in electrical distribution systems from ground to air, insulators and switch points. It may also be used in the detection of overheated bearings, forest fires, and icebergs, in the maintenance of nuclear reactors and chimneys, as well as for environmental measurements such as the temperatures of cloud and water surfaces.

An example of a rarely built mirror total radiation pyrometer is the Compac 3 of Land Ltd (undated) shown in Figure 7.17. This compact, microprocessor based, one hand operated pyrometer, has a pyrometer detector, giving a temperature range from -50 to $+500$ °C. In the continuous or peak measure modes, the readings are digitally displayed with a preset value for the emissivity. The minimum target diameter at a distance of 2 m is 35 mm with a circle in the viewfinder to indicate the necessary target diameter. Although it is designed as a portable instrument it can also be installed as a fixed pyrometer with a digital output.

Another type of mirror total radiation pyrometer is the PF-03 pyrometer made by Keller GmbH (Germany) shown in Figure 7.18. It is connected by a 5-m long line to a measuring amplifier. The working temperature ranges of the pyrometer are

Figure 7.19
Mirror total radiation pyrometer Rayotube. (Courtesy of Leeds & Northrup Co, USA.)

between 20–150 °C and 150–500 °C with a spectral response of from 0.3 to 10 μm. An analogue output signal of 0–20 mA is associated with a reading accuracy of $\pm 1.5\%$ of FSD at $\epsilon = 1$. Different mountings with air or water cooling are available.

Leeds & Northrup Instruments (USA) produces a Rayotube pyrometer (Figure 7.19) in which the radiation from the target is concentrated by a single-mirror optical system upon a blackened disk. Attached to the disk are the measuring junctions of a thin-film thermocouple. These Rayotube pyrometers are produced, prefocused to one of six focal lengths ranging from 4 in to 24 in to infinity. Rayotube is equipped with ambient temperature compensation for 0 to 100 °C without forced-air or water cooling. Its measuring ranges are from 200 to 1000 °F to 1250 to 2800 °C, with a maximum output e.m.f. of 24 mV. Rayotube can be supplied with a number of standard open-end sighting tubes or closed-end target tubes, 12 to 30 in long. The tubes are of Inconel, Fyrestan, Silicon Carbide, and Alumina, depending on the working atmosphere. Where hot gas or flame might reach the window and damage the pyrometer, either through an open-ended sighting tube or a broken ceramic target tube, a safety shutter, actuated by a bimetallic trip at 100 °C can provide protection.

Total radiation pyrometers to be precise, are band radiation pyrometers, similar to the photoelectric band pyrometers. Good examples are those radiation pyrometers for low-temperature measurement having a spectral response range of from 8 to 14 μm, so that any influence of reflected sunlight is eliminated.

7.4 PHOTOELECTRIC PYROMETERS

7.4.1 General information

Photoelectric pyrometers belong to the class of band radiation pyrometers which should be their proper name in strict agreement with the classification given in §7.1. However, as they are based on photoelectric detectors it has been decided to call them photoelectric pyrometers. This name is also compatible with British Standard 1041, Code for Temperature Measurements, Part 5.

The thermal inertia of those thermal radiation detectors described in §7.3 does not permit the measurement of rapidly changing temperatures. For example, whereas the smallest time constant of a thermal detector such as a bolometer is about 1 ms or of a thermopile is 15 ms, the smallest time constant of photoelectric detectors can be about 1 or 2 μs.

It has been pointed out by Larsen and Shenk (1941) that although H. E. Ives proposed the use of photoelements for temperature measurements as early as 1923, the first industrial photoelectric pyrometers did not appear on the market until 1932.

The radiation energy, emitted by an object, does not change the temperature of the detector. An electric output signal is brought about by the electron transitions resulting from the radiation. The range of the spectral response of photoelectric pyrometers depends on the spectral sensitivity of the photoelectric detector and on the spectral transmission characteristics of the optical system used in them. Figures 7.8 and 7.9 show the spectral transmission limits and the relative transmission abilities of plates of known thickness, made of different materials used for pyrometer lenses. Some commonly used photoelectric detectors have the spectral sensitivity shown in Figure 7.20.

The operating wavelength range of photoelectric pyrometers indicates that they can be either spectral or band pyrometers. An appropriate choice for the operating wavelength band of the pyrometer must take the following factors into account:

- the measured temperature range,
- the intended applications,
- the measuring conditions, including atmosphere absorption,
- the spectral emissivity of the target.

Figure 7.21 which shows the spectral transmission, τ_λ, of the atmosphere and also the absorption free atmospheric windows, is based upon information presented in

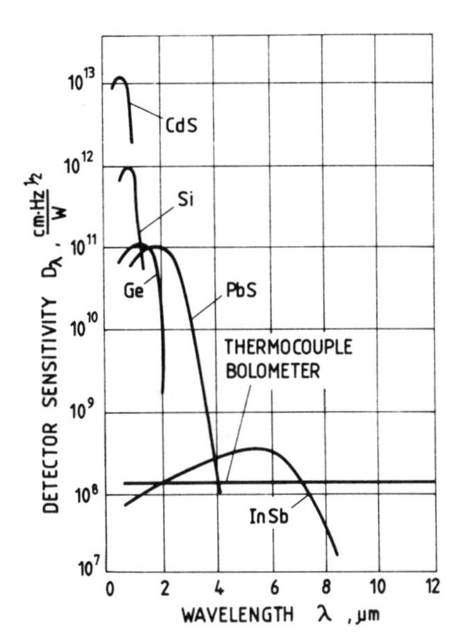

Figure 7.20
Spectral sensitivity, D_λ, of photoelectric radiation detectors.

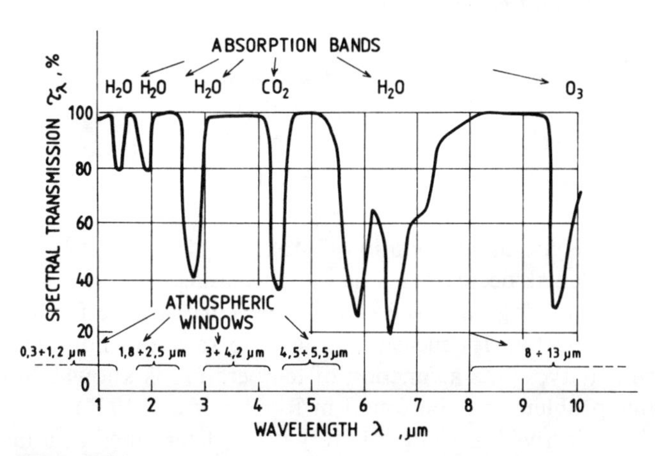

Figure 7.21
Spectral transmission, τ_λ, of 8 m atmosphere layer. Marked on the diagram are the atmospheric windows used in the design of photoelectric pyrometers and the absorption bands of some gases.

Lotzer (1976) and Warnke (1972). Many contemporary pyrometers operate in these wavelength ranges. The disturbing influences of sunlight can be avoided by choosing the 8 to 14 μm wavelength range.

A photoelectric pyrometer is characterized by a feature called its reference wavelength, at which the temperature is being measured. When the response band is not too wide, its weighted mean value or effective wavelength, λ_e, can be used. This wavelength is such that the calibration of the pyrometer over a certain range of temperatures is the same as that of a spectral (monochromatic) pyrometer responding to the radiation of that wavelength. The mean effective wavelength $\lambda_{e,(T_1-T_2)}$ for temperatures T_1 and T_2 is the wavelength for which

$$\frac{L_{\lambda_{T_1-T_2}}(T_1)}{L_{\lambda_{T_1-T_2}}(T_2)} = \frac{\int_0^\infty L_\lambda(T_1)\tau_\lambda S_\lambda \mathrm{d}\lambda}{\int_0^\infty L_\lambda(T_2)\tau_\lambda S_\lambda \mathrm{d}\lambda} \tag{7.43}$$

where $L_\lambda(T_1)$, $L_\lambda(T_2)$ are the spectral radiances of black-body radiation at wavelength λ and temperatures T_1 and T_2, $L_{\lambda_{T_1-T_2}}(T_1)$, $L_{\lambda_{T_1-T_2}}(T_2)$ are the same quantities for the mean effective wavelength $\lambda_{e,(T_1-T_2)}$, τ_λ is the relative, spectral transmissivity of the pyrometer optics, including the filter, and S_λ is the relative spectral detector sensitivity.

In the case when T_2 is approaching T_1, the limiting value of the effective wavelength, λ_e, which applies to temperature, T, is given by

$$\frac{1}{\lambda_e} = \lim_{T_2-T_1} \frac{1}{\lambda_{e,(T_1-T_2)}} \tag{7.44}$$

Basing on Equation (7.43) and Equation (7.44) the following equation can be derived (B.S. 1041, Pt. 5, Righini *et al.* 1972):

$$\frac{1}{\lambda_e} = \frac{\int_0^\infty (1/\lambda) L_\lambda(T)\tau_\lambda S_\lambda \mathrm{d}\lambda}{\int_0^\infty L_\lambda(T)\tau_\lambda S_\lambda \mathrm{d}\lambda} \tag{7.45}$$

where all the symbols are as in Equation (7.43).

The effective wavelength decreases a little with increasing temperature, this variation being larger if the measuring range is greater. Knowledge of the effective wavelength of a pyrometer is important for measuring the temperature of non-black bodies, if their spectral emissivity, ϵ_λ, as a function of temperature is known. More detailed discussion of this problem may be found in Righini *et al.* (1972).

In practice, the effective wavelength is very often determined as a median value of the wavelength range. The precision of this assumption is greater if the wavelength range is narrower.

7.4.2 Radiation detectors

Photoelectric pyrometers may use photoconductors, photodiodes, photovoltaic cells, or vacuum photocells as radiation detectors.

Photoconductors are built from glass plates with thin film coatings of thickness 1 μm, from the materials PbS, CdS, PbSe or PbTe. When the incident radiation has the same wavelength as the materials are able to absorb, the captured incident photons free photoelectrons, which are then able to form a conducted electric current. As

Table 7.3
Commonly used photoelectric radiation detectors (Warnke, 1972).

Detector	Wavelength, λ: Corresponding to maximum detector sensitivity (μm)	Wavelength, λ: Maximum operating wavelength (λ_{max}) (μm)	Resistance (Ω)	Time constant *n* (s)
Ge	1.2	1.8		1
Si	0.9	1.2	10^7	~1
PbS	2–2.4	2.3–3.1	10^6–10^7	150–500
InAs	3.4	3.7	20	2
InSb	6.0	7.0	4.5–9	<1

the resistance of a photoconductor which decreases with increasing radiation intensity also depends on its temperature, this phenomenon has to be considered in the construction of a pyrometer. If not irradiated the 'dark' resistance of a photoconductor is from 10^4 to 10^8 Ω. Considering that the photoconductor sensitivity depends on the radiation wavelength, the concept of the operating wavelength band can be introduced.

Since the sensitivity and spectral response of the photoconductors undergo some changes with ambient temperature and time, in most cases they are applied as a null detector. This may be achieved by comparing, for instance, two radiation intensities, falling alternatively upon its surface. In most cases the surface of a photoconductor has to be protected against atmospheric influences by covering it with a protective varnish layer of materials like polystyrene.

Photodiodes, in germanium or silicon are operated with a reverse bias voltage applied. Under the influence of the incident radiation their conductivity as well as their reverse saturation current is proportional to the intensity of the radiation within the spectral response band from 0.4 to 1.7 μm for Ge and 0.6 to 1.1 μm for Si. The high sensitivity of photodiodes permits the construction of pyrometers with high distance ratios. To compensate for the dark current, which occurs in the non-irradiated state, a second identical diode, protected from radiation, is used.

Photovoltaic cells which generate a voltage depending upon incident radiation, are constructed with a thin semiconductor film deposited on a metal plate. Under no-load conditions this generated voltage is a logarithmic function of the incident radiation intensity. They are simple and robust in construction. As photovoltaic cells generate strong output signals that can be utilized without any further amplification, there is no need to apply any external voltage. However, because their sensitivity is low in the infrared range, they can only be applied for higher temperatures. Materials used for photovoltaic detectors are selenium, silicon, indium antimonide (InSb) and indium arsenide (InAs).

Vacuum photocells operate on the principle that the incident infrared radiation causes the emission of electrons from a metallic photocathode which is placed in a vacuum glass bulb with an anode. At a given d.c. voltage between cathode and anode, the electric current is a function of the radiation intensity. Low pressure, neutral gas can also be used, in which case the photocell has a higher sensitivity and operates at a lower voltage. Although the vacuum cells have lower sensitivity they are more stable, more linear and are more rapid in response. Their spectral response band is below 2 μm.

Photomultipliers which are similar to photocells, are not used in industrial pyrometers because of their higher price and larger size. As photocells and photomultipliers are both influenced by the ambient temperatures, they must be kept at constant temperature unless a compensating circuit is used.

The properties of the most commonly used photoelectric radiation detectors, which are listed in Table 7.3, give a comprehensive view when combined with the comparison of the sensitivities of some chosen photoelectric detectors given in Figure 7.20. Following the approach of Warnke (1972) the information given in both Figure 7.20 and Table 7.3, is based upon an assumed figure of merit D, which allows an estimation of the sensitivity of photoelectric radiation detectors, using the relation:

$$D(\lambda, f_o) = \frac{(A\Delta f)^{1/2}}{\mathrm{NEP}_\lambda} \frac{\mathrm{cm\ Hz}^{1/2}}{\mathrm{W}} \tag{7.46}$$

where NEP is the spectral noise equivalent power, defined as that utilized radiation power, at wavelength λ which is equal to the noise power, A is the active area of the detector, f_o is the frequency of optical modulation, Δf is the utilized frequency range, and λ is the wavelength of radiation.

Index D is given for unit detector area and referred to unit frequency range as a function of the monochromatic radiation wavelength and at the applied frequency, f_o, of the optical modulation (up to about 1 kHz).

7.4.3 Scale defining equation for black bodies

The output signal of the photoelectric radiation detectors in photoelectric pyrometers is proportional to the number of photons per unit time, falling on the detector surface. Using Planck's law given in Equation (7.8), Warnke (1972) expresses this number by the equation:

$$N_{\lambda_1-\lambda_2} = K_1 \int_{\lambda_1}^{\lambda_2} \frac{c_1'\lambda^{-4}}{e^{c_2/\lambda T} - 1}\, d\lambda \frac{1}{\mathrm{s\ m}^2} \tag{7.47}$$

where K_1 is the constructional constant of the pyrometer, $c_1' = c_1/hc = 1.88 \times 10^{15}\ \mathrm{ms}^{-1}$, $c_1 = 3.7415 \times 10^{-16}\ \mathrm{Wm}^2$, $c_2 = 14388\ \mu\mathrm{m\ K}$, h is Planck's constant, $h = 6.6253 \times 10^{-34}$ Js, c is the velocity of light in vacuum, and λ is the wavelength in μm.

With increasing bandwidth (λ_1 to λ_2) the output signal of the radiation detector also increases. However, considering the wavelength dependence of the target emissivity, transmission of the atmosphere and the pyrometer optical system, as well as the sensitivity of the detector it is advisable to use wavelength bands which are as narrow as possible. In this way repeatable pyrometer readings may be obtained in an industrial environment.

In a narrow temperature range Equation (7.47) can be replaced by the simpler relation.

$$N = B_1 T^n \tag{7.48}$$

where B_1 and n are constructional constants.

In most photoelectric pyrometers the photoconductive radiation detectors are used, which are connected in series with the voltage source and the loop resistance of the measuring circuit. Assuming that the voltage and loop resistance are both constant, the output current, I_T, of the pyrometer, while measuring a black body at temperature, T, can be expressed as:

$$I_T \approx B_2 . T^n \tag{7.49}$$

where B_2 is a constructional constant, and n is a coefficient with a practical value between 5 and 12.

The above constants B_1, B_2 and n depend on the utilized wavelength band (λ_1 to λ_2). In the case of a narrower band, n increases and B_1 and B_2 decrease while in the extreme case, when the wavelength band $\lambda_2 - \lambda_1$ tends to zero, the pyrometer properties approach the properties of a spectral (monochromatic) pyrometer and $n \rightarrow 12$.

7.4.4 Temperature measurement of non-black bodies

To measure the temperature of non-black bodies, Reynolds (1961) has proved an analogous equation corresponding to Equation (7.37). For calculation of the true target temperature T_t, when using a photoelectric pyrometer, this relation is

$$T_t = T_i \sqrt[n]{\frac{1}{\epsilon_{\lambda_1 - \lambda_2}}} \tag{7.50}$$

where T_t is the true, measured temperature, T_i is the indicated temperature, $\epsilon_{\lambda_1 - \lambda_2}$ is the band emissivity, expressed as a mean value in the range from λ_1 to λ_2, and n is the coefficient given in Equation (7.49).

Calculation of the corrections is rather difficult, because the band emissivity is never precisely known. In practice, similarly as in the case of total radiation pyrometers, it is advisable either to create such measuring conditions that the target properties may approach those of a black body, or in the case of measurements in a continuous production process, measure the true temperature of the object by another method. In this case the pyrometer readings serve to ensure the repeatability of the process, as for instance in temperature measuring and limiting in induction or direct resistance heating, where total radiation pyrometers would be too inert.

By comparing the readings with those of another instrument and setting the corresponding emissivity value, a possibility of optical or electrical correcting of readings exists in some photoelectric pyrometers. From Equation (7.50) the relative emissivity errors can be calculated as:

$$\frac{T_i - T_t}{T_t} = \sqrt[n]{(\epsilon_{\lambda_1 - \lambda_2}) - 1} \tag{7.51}$$

Thus the relative error in the measurement of non-black-body temperatures decreases as the value of n increases. This conclusion holds true for all those pyrometers, whose scale is defined by Equation (7.49). The influence of both the exponent n and the

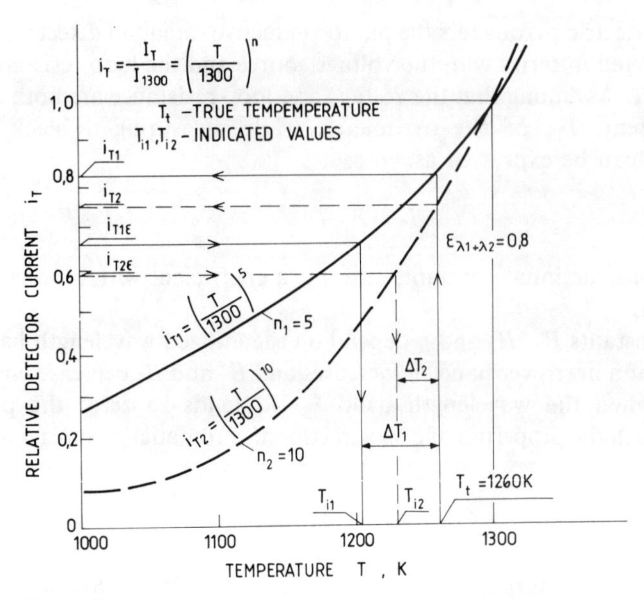

Figure 7.22
Influence of the exponent n (Equation (7.49)) on the emissivity errors, ΔT, in photoelectric pyrometers.

steepness of the scale characteristic on the emissivity errors is explained in Figure 7.22 where the relative values $i_T = f(T)$ of the photoelectric detector current, I_T, are displayed against the measured temperature T. This relative current, i_T, is the ratio of the detector current, I_T, at the temperature T, to the detector current at the arbitrarily taken reference temperature, say $T = 1300$ K.

For two different pyrometers, having respectively $n_1 = 5$ and $n_2 = 10$, at the measured true temperature $T_t = 1260$ K, the relative currents i_{T1} and i_{T2} have been determined. Then for the band emissivity $\epsilon_{\lambda_1 - \lambda_2}$, the new relative currents have been calculated as:

$$i_{T\epsilon} = i_T \cdot \epsilon_{\lambda_1 - \lambda_2} \tag{7.52}$$

In Equation (7.52), i_T is the relative current value of the pyrometer measuring the temperature of a black body ($\epsilon_{\lambda_1 - \lambda_2} = 1$) that is in normal calibration conditions.

The indicated temperatures T_{i1} and T_{i2} as well as the measurement errors ΔT_1 and ΔT_2 have been found in a graphical way, based on the calculated values of $i_{T1\epsilon}$ and $i_{T2\epsilon}$ as given in Figure 7.22. It is evident that the error ΔT_2 of the pyrometer with a higher value of n_2, is smaller than ΔT_1 of the other pyrometer with a lower value of $n_1 (n_1 < n_2,\ \Delta T_1 > \Delta T_2)$.

Reynolds (1961) observes that high n-values can be achieved by correctly choosing the spectral sensitivity of the detector and the spectral transmissivity of the pyrometer optics.

In most practical applications Leclerc (1976) has advised the use of pyrometers with as short effective wavelength λ_e as possible. He gives the following reasons:

Table 7.4
Measurement errors of a pyrometer calibrated for a black body as a function of effective wavelength and associated non-black body emissivity at $T_t = 1300$ K.

λ_e (μm)	ΔT (K) at $\epsilon_{\lambda_e} = 0.5$	$\epsilon_{\lambda_e} = 0.8$
0.8	−63	−21
2.3	−164	−58
5.2	−320	−124

1. A photoelectric pyrometer of a narrow wavelength range according to Worthing (1941) can be regarded as a spectral pyrometer of a given λ_e. In that case, from Equation (7.75) as the indicated temperature T_i is given by:

$$T_i = \frac{1}{(1/T_t) - (\lambda_e/c_2)\ln \epsilon_{\lambda_e}}$$

at the given measured temperature T_t and with the emissivity ϵ_{λ_e}, the indicated temperature T_i comes closer to the true measured value T_t. For example, at $T_t = 1300$ K, a pyrometer calibrated for a black body will have the measurement errors $\Delta T = T_i - T_t$ as a function of λ_e and ϵ_{λ_e} given in Table 7.4.
2. Following the Drude theory (Engel, 1974), most metals exhibit a higher emissivity at shorter wavelengths.
3. Pyrometers operating at shorter wavelengths are simpler, less expensive, do not need any special optical materials and mostly use radiation detectors with a greater output signal.

Kelsall (1963) describes a special method of automatically compensating the influence of emissivity. The radiation from the target of ϵ_1 together with the radiation from an auxiliary heating element of $\epsilon_2 = 1$, reflected from the target (Φ_1) and the direct radiation from the heating element (Φ_2) fall alternately on a radiation detector as shown in Figure 7.23. Adjustment of the temperature of the heating element is made in such a way, that the difference between the two detector signals equals zero. The temperature T_2 is then the same as the temperature to be measured T_1. From target to detector, the radiant heat flux is given by the equation:

$$\Phi_1 = k_1\epsilon_1 f(T_1) + k_1\epsilon_2\rho_1 f(T_2) \quad (7.53)$$

with the heat flux from the heating element given by:

$$\Phi_2 = k_2\epsilon_2 f(T_2) \quad (7.54)$$

When the two fluxes are equal, it follows:

$$k_2\epsilon_2 f(T_2) = k_1\epsilon_1 f(T_1) + k_1\epsilon_2\rho_1 f(T_2) \quad (7.55)$$

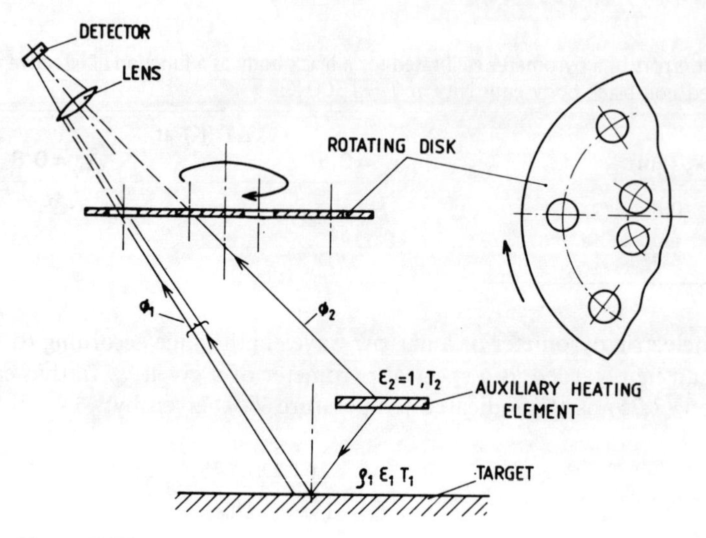

Figure 7.23
Automatic compensation of emissivity influence.

where ϵ_1, ϵ_2 are the emissivities of target and heating element, ρ_1 is the reflectivity of the target, k_1 is a coefficient, depending on the diameter of the apertures in the rotating disk 3, in the way of flux Φ_1, k_2 is as before for Φ_2.

Assuming $\epsilon_2 = 1$ and $k_1 = k_2$, it follows that

$$f(T_2) = \epsilon_1 f(T_1) + \rho_1 f(T_2) \tag{7.56}$$

$$f(T_2)(1 - \rho_1) = \epsilon_1 f(T_1) \tag{7.57}$$

Since the target transmissivity, τ, has a value of zero, Equation (7.2) becomes:

$$\alpha_1 + \rho_1 = 1 \tag{7.58}$$

and

$$\epsilon_1 = \alpha_1 = 1 - \rho_1 \tag{7.59}$$

so that finally:

$$f(T_2) = f(T_1) \tag{7.60}$$

giving

$$T_2 = T_1 \tag{7.61}$$

The temperature T_2, measured by a thermocouple, is equal to the measured temperature T_1.

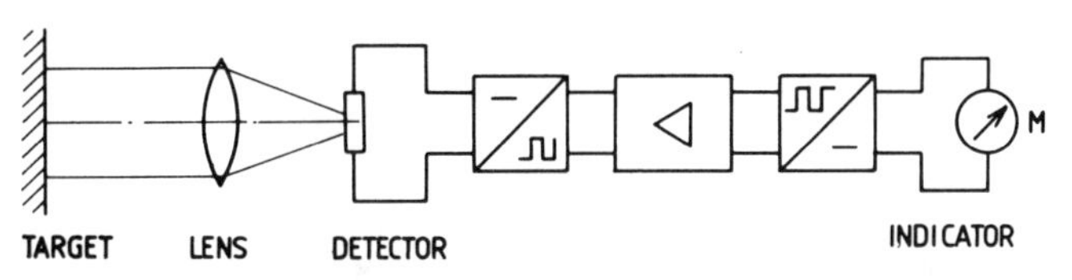

Figure 7.24
Block diagram of a photoelectric pyrometer with direct radiant flux.

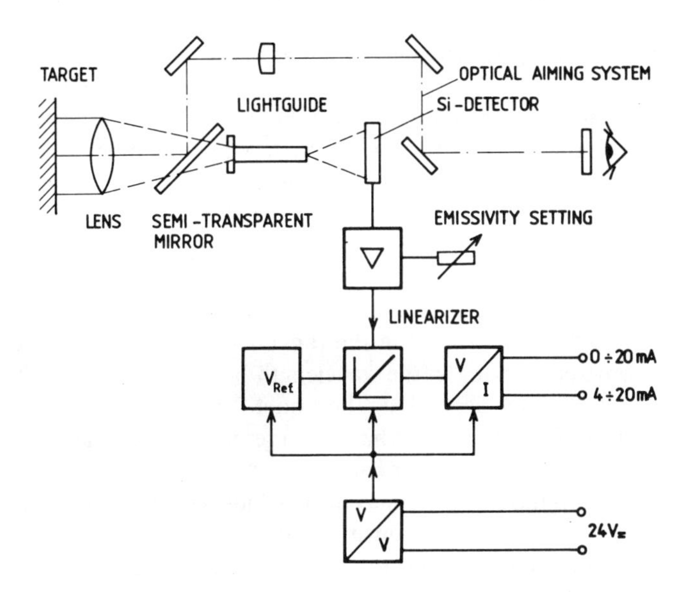

Figure 7.25
Basic diagram of a photoelectric pyrometer Ardofot by Siemens AG (Germany). (Courtesy of Siemens AG, Germany.)

In a similar automatic circuit by Toyota *et al.* (1972), the temperature T_2 of the heating element is automatically adjusted as a function of the detector output signal. Measuring the temperature of steel, brass and aluminium at 300 °C, the measuring errors were about −7 to +3 K. In the case of polished, stainless steel at 250 °C the errors were about ±15 K.

7.4.5 Review of construction

There are two main principles used in the design of photoelectric spectral and band pyrometers depending upon whether the pyrometer uses direct radiant flux, or modulated radiant flux.

In pyrometers with direct radiant flux the most commonly used detectors are photovoltaic cells. A block diagram of such a pyrometer which is shown in simplified form in Figure 7.24, may be given in more detail in the diagram of the Ardofot from Siemens AG (Germany) in Figure 7.25. Ardofot is used in the temperature range from

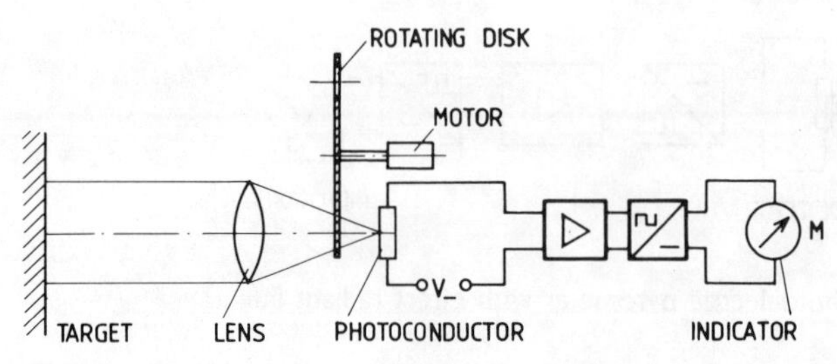

Figure 7.26
Basic diagram of a photoelectric pyrometer with modulated radiant flux.

400 to 2500 °C. The radiation from the target in Figure 7.25, is concentrated on the Si photovoltaic cell by the lens of the optical system which includes a lightguide. A semi-transparent mirror, which is also a part of the optical system, is used for aiming the pyrometer at the target. The amplified and linearized output signal of the detector is transformed in the converter into a current output of 4 to 20 mA or 0 to 20 mA. Calibration for a black-body target, as described in §7.3.5, is an important feature of this pyrometer. For measuring the temperature of a non-black body, the corresponding emissivity can be set on a special scale in the range from $\epsilon = 0.3$ to 1. Ardofot is a band radiation pyrometer working in the wavelength band from 0.6 to 1.2 μm, with a distance ratio of 10 to 75 which depends on the model. The pyrometer head is 185 mm long by 83 mm diameter. The pyrometer which has a response time of about 40 ms, can be connected to an indicator, recorder or controller. A separate unit, which contains the power supply and output transmitter may also include a peak picker or mean value indicator if required. The pyrometer housing is shown in Figure 7.12(b).

In pyrometers with modulated radiant flux other less stable photoelectric detectors such as a PbS-photoconductor or a InSb or InAs photovoltaic cell, can be used. The modulation of the incident radiant flux is obtained either by a rotating disk with apertures or by using a vibrating fork. In order to prevent any disturbances which may be synchronous with the mains frequency the modulation frequency has to be different from the mains frequency or its harmonics.

In the working principle of the pyrometer, shown in Figure 7.26, the pyrometer readings are made independent of the housing temperature, by alternately irradiating the radiation detector by the target and then by an auxiliary source of stabilized constant temperature. Photoelectric pyrometers from Uher AG (Germany), based on a Si photovoltaic cell and working in the range over 900 °C, have an extremely high distance ratio, up to 700 at the target distance of about 1 m. Special diagrams display the minimum target size as a function of its distance from the pyrometer.

For many special applications photoelectric fibre-optic pyrometers are built. They are used especially when

1. No cooling is possible in high ambient temperature, even up to about 250 °C.
2. The target is directly inaccessible.
3. The heated charge temperature in induction heating appliances has to be measured through the air gaps between the inductor turns.
4. Dust and smoke are present in the air.

Figure 7.27
Fibroptic pyrometer from Land Infrared Ltd. (Courtesy of Land Infrared Ltd, UK.)

A Fibroptic pyrometer from Land Infrared Ltd (UK) can be mentioned as an example. The small-size pyrometer head is connected to the Si or Ge cell in a separate amplifier unit by a flexible fibre-optic light guide in a protective metal monocoil of about 6 mm diameter. As shown in Figure 7.27, a small radiation concentrating lens of 12 mm diameter, encased in a cylindrical assembly of diameter 13.5 mm and 50 mm long, is mounted at the end of the fibre optics. Instead of the lens a glass rod probe clad in a ceramic sheath of diameter 6 mm may also be used. The length of the flexible fibre optics varies between 1 to 3.5 m, while the spectral response band is 0.7 to 1 μm for a Si detector and 0.9 to 1.8 μm for a Ge detector. Within the measurement ranges of the pyrometer, between 300 to 750 °C up to 1400 to 3350 °C, the repeatability of the readings is ±0.15% of the indicated value in K with a response time of 5 ms.

Similar parameters are possessed by the Fiberay Radiation Thermometer from Leeds & Northrup Instruments (USA), whose optical system at the end of a flexible fibre-optic guide can stand even +540 °C. The target size at a distance of 100 mm is of diameter 7.5 mm with a measurement range of 1000 to 1600 °C. Based on a Si photodiode, the spectral response of the pyrometer is 0.65 to 1.1 μm.

7.5 DISAPPEARING FILAMENT PYROMETERS

7.5.1 General information

The first disappearing filament pyrometer was built in 1901 by L. Holborn and F. Kurlbaum. Disappearing filament pyrometers, are spectral pyrometers. In them the brightness of a lamp filament is changed by adjusting the lamp current until the filament disappears against the background of the target, whose temperature is to be measured. In pyrometers of this kind, the eye of the observer is itself the detector. In another seldom applied type of pyrometer the brightness match is achieved by the attenuation of the target brightness, using a neutral grey filter.

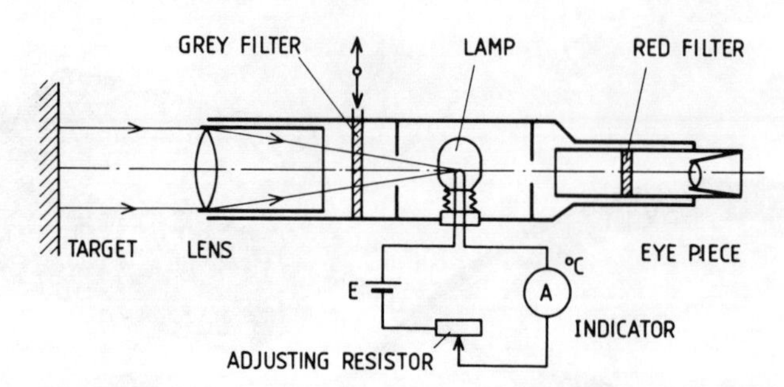

Figure 7.28
Principle of disappearing filament pyrometer.

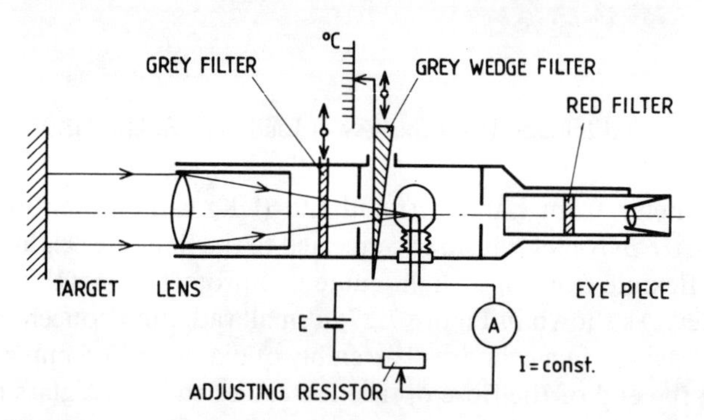

Figure 7.29
Principle of disappearing filament pyrometer with movable grey wedge filter.

In the first type shown in Figure 7.28, the observer sees the filament of the lamp against the target background, through the eyepiece and red filter. The lamp current is adjusted by the resistor until the filament picture disappears when the brightness (or radiance, L, of Equation 7.23) of filament and target are identical. The measured value of temperature is read from the ammeter, A, calibrated in temperature units. A comparison of the radiance, L, of the filament and the target occurs at one wavelength. A grey filter can be used to increase the measurement range. In the second type of disappearing filament pyrometer shown in Figure 7.29, a movable grey wedge filter is adjusted in such a way that the image of the lamp filament disappears against the background of the target. A pointer, coupled with the grey filter, indicates the measured temperature. An ammeter, A, is used to check that the lamp current, which can be adjusted by a resistor, remains constant. The advantages of this type are that the lamp which is run at a constant low value of current, is less likely to change its characteristics and additionally the replacement lamp must be adjusted to the same brightness at only one point.

Disappearing filament pyrometers, which are calibrated for black-body targets, have a lower limit of temperature range of about 700 °C, determined by the longwave visibility limit of the human eye.

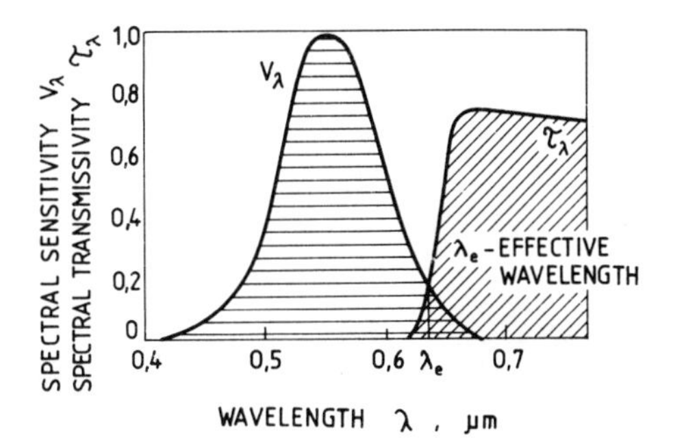

Figure 7.30
Spectral transmissivity τ_λ of red filter and spectral relative sensitivity V_λ of standard human eye versus wavelength λ.

7.5.2 Red filter

Applying a red filter has the following reasons and advantages.

The spectral transmissivity, τ_λ, of a red filter and the spectral relative sensitivity, V_λ, of a standard human eye are displayed in Figure 7.30 as a function of wavelength, λ. It may be clearly seen that there is a well defined small wavelength band, in which the brightness of the filament and that of the target can be compared. This wavelength band is narrow enough to assume that the comparison occurs at one wavelength, called the effective wavelength, λ_e, of a disappearing filament pyrometer. All of the pyrometer readings should be referred to this value of λ_e. Thus, as the comparison of the brightness of the lamp filament and that of the target takes place only at one colour, the subjective estimation of colour by different observers cannot influence the measurement results.

As shown in Figure 7.30 the effective wavelength λ_e, which can be found in a graphical way, has a value of 0.65 μm given by the abscissa of the centre of gravity of the common cross-hatched area under the curves $\tau_\lambda(\lambda)$ and $V_\lambda(\lambda)$. Because of the small width of the utilized wavelength band, the effective wavelength, λ_e, is nearly constant at all measured temperatures.

In practice when applying filters such as the Scholl RG2 and Jena 4512, λ_e, does not change by more than 0.003 μm in the measured temperature range of 1300 to 3600 K. Forsythe (1941) points out that the λ_e value which is also a function of the filter temperature, may be regarded as nearly constant. More detailed theoretical background and methods of experimental and calculative determination of λ_e may be found in Henning (1951) and Righini *et al.* (1972).

As there is a high percentage of the red colour in the emitted thermal radiation, in the temperature range below 800 °C, the application of a red filter is not advised. The resulting increase in observed radiation intensity consequently also increases the overall precision of the measurements.

The total emitted radiant intensity, W_0, following the Stefan–Boltzmann law given in Equation (7.16), and the spectral radiant intensity, $W_{0,\lambda=0.65}$, following Planck's law of Equation (7.7), which are displayed against the absolute temperature of a black

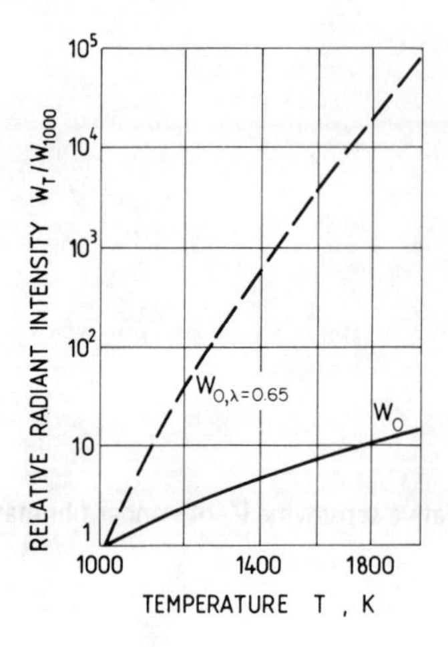

Figure 7.31
Spectral radiant intensity $W_{0,\lambda=0.65}$ and total radiant intensity W_0 versus black body temperature. Both values related to radiant intensities at 1000 K.

body in Figure 7.31, are both relative values of radiant intensity referred to those at 1000 K. From this figure it can be seen that the steepness of the curve $W_{o,\lambda=0.65}=f(T)$ is far greater than that of the curve $W_o=f(T)$.

Weichert *et al.* (1976) show that the spectral radiance $L_{o,\lambda}$ of a black body is directly proportional to the spectral radiant intensity so that:

$$L_{o\lambda}=CW_{o\lambda} \tag{7.62}$$

where C is a constant.

The same dependence is also valid for the total radiance L_0. From the above considerations and also from Figure 7.31, it follows that the spectral radiance difference corresponding to a unit temperature difference at $\lambda=0.65\ \mu$m is far greater than for the total radiance.

In summary, the reasons for the application of a red filter are:

- Comparison of the filament and target brightness takes place only at one wavelength or colour so eliminating the influence of any subjective colour estimation by different observers.
- The effective wavelength $\lambda_e=0.65\ \mu$m, which is still within the visible spectrum range, adjoins the infrared radiation band thus permitting the lowest possible temperatures to be measured.
- At $\lambda_e=0.65\ \mu$m the pyrometer sensitivity is higher than for the total radiation.
- It is relatively easy to produce good filters of $\lambda_e=0.65\ \mu$m which are stable in time.

7.5.3 Scale defining equation for black bodies

Disappearing filament pyrometers are calibrated for black bodies, whose spectral radiant intensity at the temperature T_t, follow Wien's law of Equation (7.9) in the manner:

$$W_{o\lambda}=c_1\lambda^{-5}\,e^{-c_2/\lambda T_t}$$

Conforming to Equation (7.62) the spectral radiance is:

$$L_{o\lambda}=Cc_1\lambda^{-5}\,e^{-c_2/\lambda T_t} \tag{7.63}$$

where C is a conversion factor.

If a black body is observed by a human eye, through a red filter of spectral transmissivity, τ_λ, the physiological feeling of brightness will be given by:

$$L'_{o\lambda}=Cc_1V_\lambda\tau_\lambda\lambda^{-5}\,e^{-c_2/\lambda T_t} \tag{7.64}$$

where T_t is the true temperature and V_λ is the relative spectral sensitivity of a standard human eye.

Assuming that the spectral emissivity of the filament is $\epsilon_{f\lambda}$, then, observing the filament through the same red filter, results in the feeling of brightness given by:

$$L''_{o\lambda}=Cc_1\epsilon_{f\lambda}V_\lambda\tau_\lambda\lambda^{-5}\,e^{-c_2/\lambda T_f} \tag{7.65}$$

where T_f is the filament temperature.

In Equations (7.64) and (7.65) the negligibly small lens and eyepiece attenuation of incident radiation does not need to be considered.

At the moment of reading the measured temperature value, the brightness of the filament and the target are equal. Combining Equations (7.64) and (7.65), yields:

$$L'_{o\lambda}=L''_{o\lambda} \tag{7.66}$$

For $\lambda=\lambda_e$, it follows that:

$$e^{-c_2/\lambda_e T_t}=\epsilon_{f\lambda_e}\,e^{-c_2/\lambda_e T_f} \tag{7.67}$$

or:

$$\frac{-c_2}{\lambda_e T_t}=\ln\epsilon_{f\lambda_e}-\frac{c_2}{\lambda_e T_f} \tag{7.68}$$

and finally:

$$\frac{1}{T_f}=\frac{1}{T_t}+\frac{\lambda_e}{c_2}\ln\epsilon_{f\lambda_e} \tag{7.69}$$

For any given pyrometric lamp, the filament temperature, T_f, is a function of the lamp current, I, so that:

$$T_f=f_1(I) \tag{7.70}$$

or more conveniently:

$$I = f_2(T_f) \qquad (7.71)$$

From Equations (7.69) and (7.71) it follows that:

$$I = f_3(T_t) \qquad (7.72)$$

This allows direct calibration of the ammeter of the pyrometer in temperature units. In BS 1041 the temperature found in that way is called the radiance or luminance temperature of the target at the wavelength λ_e.

The temperature scale of a disappearing filament pyrometer, which is not linear, has its scale division increasing at higher temperatures.

7.5.4 Temperature measurement of non-black bodies

When measuring the temperature of non-black bodies the pyrometer readings are too low. The radiance temperature of a target at a given wavelength is the temperature of a black body which exhibits the same spectral radiance, as the considered target. For a non-black body of spectral emissivity, $\epsilon_{\lambda e}$, at the temperature, T_t, and at the wavelength, λ_e, the spectral radiance is given by:

$$L_{\lambda e} = C\epsilon_{\lambda_e} c_1 \lambda_e^{-5} e^{-c_2/\lambda_e T_t} \qquad (7.73)$$

The pyrometer readings are then T_i, for which the radiance temperature of a black body is:

$$L_{\lambda_e} = C c_1 \lambda_e^{-5} e^{-c_2/\lambda_e T_i} \qquad (7.74)$$

Equating (7.73) and (7.74) yields:

$$\frac{1}{T_t} = \frac{1}{T_i} + \frac{\lambda_e}{c_2} \ln \epsilon_{\lambda_e} \qquad (7.75)$$

or

$$T_t = \frac{1}{(1/T_i) + (\lambda_e/c_2) \ln \epsilon_{\lambda_e}} \qquad (7.76)$$

The true temperature T_t of a non-black body of emissivity $\epsilon_{\lambda e}$ can easily be calculated from Equation (7.76), when the indicated temperature T_i is known. Substituting $c_2 = 1.4388 \times 10^{-2}$ m K from Equation (7.7) and $\lambda_e = 0.65\ \mu m = 0.65 \times 10^{-6}$ m, then (7.76) becomes

$$T_t = \frac{1}{(1/T_i) + (\log \epsilon_{\lambda_e}/9613)} \qquad (7.77)$$

Numerical example

Measuring the temperature of a body of $\epsilon_{\lambda_e} = 0.7$, a disappearing filament pyrometer indicated $T_i = 1300$ K (1027 °C). Calculate the true temperature of the body.

Solution: Inserting values into Equation (7.76) gives:

$$T_t = \frac{1}{(1/1300) + (\log 0.7/9613)} = 1327 \text{ K}$$

When measuring the temperature of non-black bodies the corrections to the readings can be read directly from the diagram of Figure 7.32. The necessary ϵ_λ values of different metals at $\lambda = 0.65\ \mu$m, are given in Table XXII.

As the ϵ_{λ_e} values are known but with an uncertainty of $\pm 10\%$ to $\pm 20\%$, the resulting errors of the corrections $\Delta\vartheta$ can be estimated from Figure 7.32. Moreover, as the steepness of the curve $W_{o,\lambda=0.65} = f(T)$ is far greater than that of $W_o = f(T)$ in Figure 7.31, errors in the temperature measurement of black bodies using disappearing filament pyrometers are smaller than those for total radiation pyrometers.

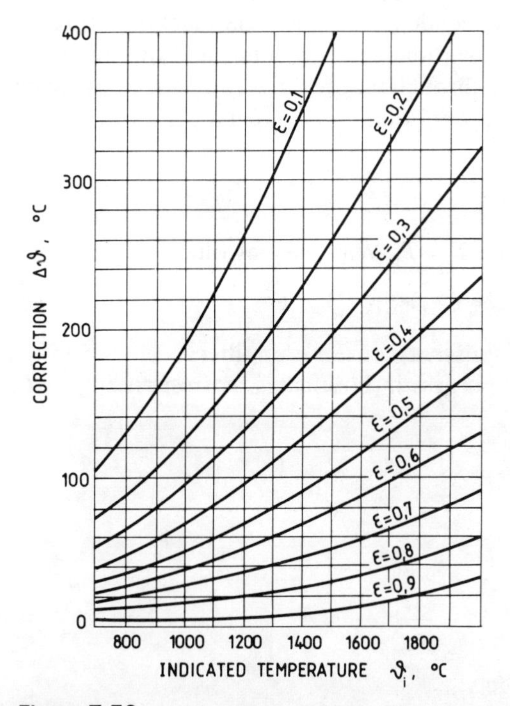

Figure 7.32
Corrections $\Delta\vartheta$ to readings of disappearing filament pyrometers of non-black bodies versus indicated temperature values ϑ_i. ϵ stands for $\epsilon_{\lambda=0.65}$.

In Equation (7.25) it has been proved that the radiance of black bodies does not depend on the viewing angle, φ. This is also true for non-black bodies, where only insignificant radiance changes are observed for $\varphi > \pi/4$. Taking into account the actual values of ϵ_{λ_e}, it follows that, the disappearing filament pyrometers can be directed at any viewing angle.

A method of measuring the temperature of metallic surfaces, using a polarizing filter also exists. At higher temperatures, metallic surfaces emit radiation which is polarized parallel to the surface at an angle $\pi/5$ to their normal, where the radiating surface approaches a black body. The pyrometer is then calibrated together with a polarizing filter which is introduced in front of the lens. In that way the measured radiance temperature nearly equals the true value of the surface temperature (Pepperhoff, 1960; Tingwaldt, 1960; Murray, 1972; Walter, 1981).

7.5.5 Extension of measurement range

The tungsten filament of a pyrometer lamp can only be used up to 1400 °C. At higher temperature tungsten sublimes, the filament resistance increases and a dark deposit is formed on the glass surface, gradually changing the lamp characteristic. To extend the pyrometer measurement range up to 2000 °C a grey filter which is placed between the pyrometer lamp and lens, reduces the target radiance without influencing that of the filament. Assuming that the spectral transmissivity of the grey filter is τ_λ, it is possible to determine the dependence between the black-body temperature T_t, equal to the indicated value T_i without the grey filter, and the true temperature T_t' of another black body observed through the grey filter. In both cases the physiological brightness feeling is the same. From Equation (7.64) at $\lambda = \lambda_e$, without the grey filter, the spectral radiance is:

$$L'_{o\lambda} = Cc_1 V_{\lambda_e} \tau_{\lambda_e} \lambda_e^{-5} \, e^{-c_2/\lambda_e T_i}$$

where the indicated temperature $T_i = T_t$. With the grey filter at $\lambda = \lambda_e$, this radiance is:

$$L'''_{o\lambda} = Cc_1 V_{\lambda_e} \tau_{\lambda_e} \tau'_{\lambda_e} \lambda_e^{-5} \, e^{-c_2\lambda_e/T_t'} \tag{7.78}$$

where τ'_{λ_e} is the spectral transmissivity of the grey filter.

For an equal feeling of brightness in both cases, corresponding to $L'_{o\lambda} \cong L'''_{o\lambda}$, it follows that:

$$\frac{1}{T_i} - \frac{1}{T_t'} = \frac{\lambda_e}{c_2} \ln \frac{1}{\tau'_{\lambda_e}} \tag{7.79}$$

Denoting:

$$\frac{\lambda_e}{c_2} \ln \frac{1}{\tau'_{\lambda_e}} = A \tag{7.80}$$

from (7.79) and (7.80) one gets:

$$T_i = \frac{1}{A + 1/T_t'} \tag{7.81}$$

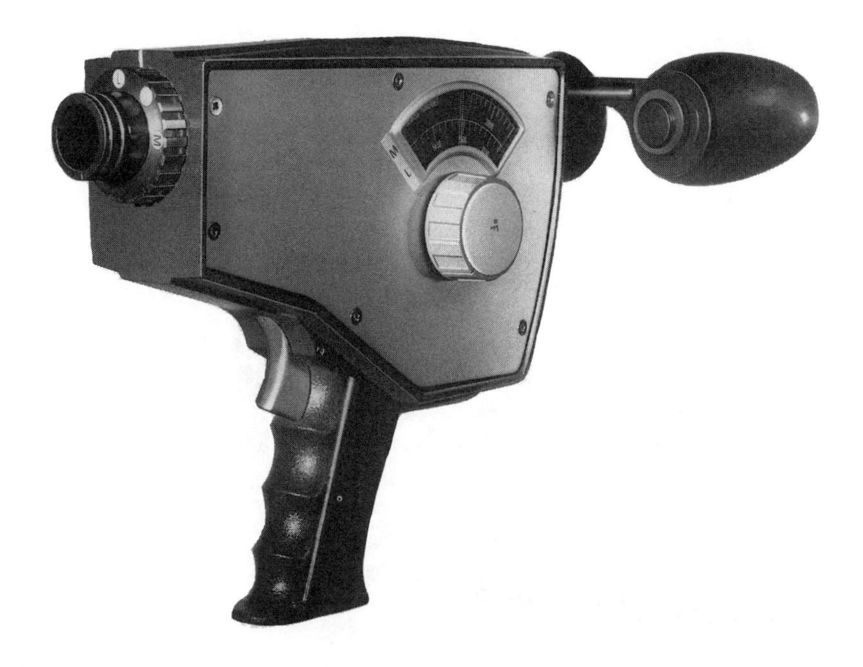

Figure 7.33
Disappearing filament pyrometer. (Courtesy of Leeds & Northrup, USA.)

The coefficient A describes the radiance reducing factor of the grey filter. Following Equation (7.81) it is possible to calibrate the disappearing filament pyrometer above the maximum filament temperature.

A seldom used method for the extension of the measurement range, employing rotating disks with apertures, is described by Griffith (1947).

7.5.6 Review of construction

An industrial, self-contained disappearing filament pyrometer, which is produced by Leeds & Northrup Instruments (USA) is shown in Figure 7.33. This pyrometer, which operates at $\lambda_e = 0.65\ \mu m$, has two or three measuring ranges of 775 to 1225 °C, 1075 to 1750 °C and 1500 to 2800 °C. Fahrenheit graduated scales are also offered. Repetitive readings taken under stable conditions usually agree within ± 5.5 °C. Powering of the instrument, which uses an electrical null-balance principle with the filament current set by a slidewire, is by rechargeable battery. With a standard lens at a distance of 200 mm the minimum target size is 0.75 mm. Leeds & Northrup also

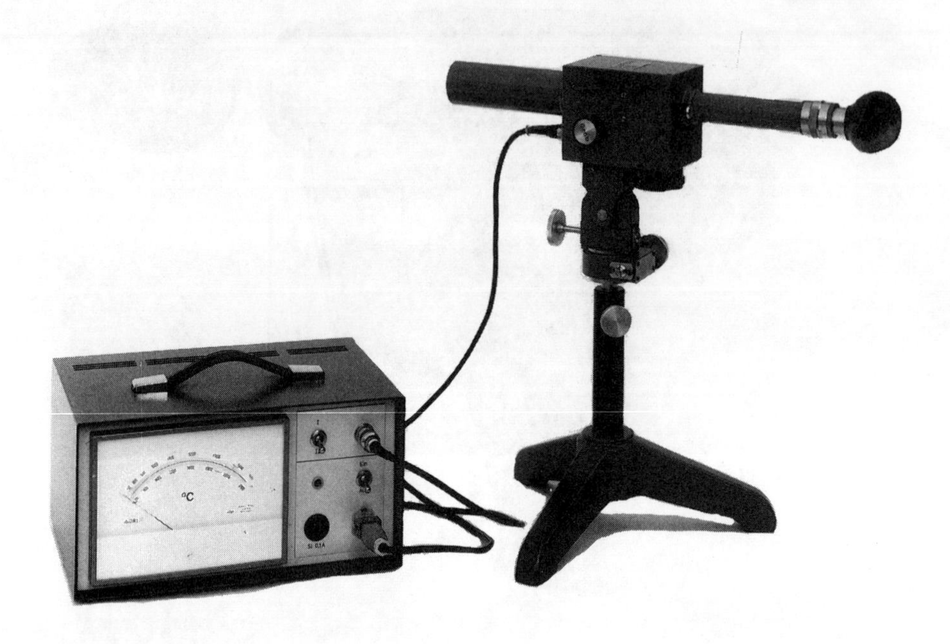

Figure 7.34
Disappearing filament pyrometer Micro-Pyrometer PB06. (Courtesy of Keller GmbH, Germany.)

offers an automatic microprocessor based type of self-balancing disappearing filament pyrometer incorporating an LED display in its field of view. It can be used as a thermometric standard for the calibration of other pyrometers in addition to normal applications in laboratories and industry.

The PB06 Micro-Pyrometer which is a special disappearing filament pyrometer, offered by Keller GmbH (Germany) is intended for very small target sizes of even 0.1 mm. This pyrometer shown in Figure 7.34 can be delivered for target distances of 2 to 10 m and with seven supplementary magnifying lenses for distances from 42 up to 1200 mm. It has a digital display for the three switchable measurement ranges of 650 to 1450 °C, 1450 to 2050 °C and 2000 to 5000 °C with an optional 0 to 20 mA output. The effective wavelength is 0.65 μm.

The same firm also produces a disappearing filament pyrometer shown in Figure 7.35 with a movable grey wedge filter. This Optix Type PB05 pyrometer is a small, constant current pyrometer with a diameter of 80 mm and a length of 225 mm. For each exchangeable pyrometer lamp its nominal, indicated value of current can be adjusted by a slidewire resistor, following the ammeter readings. The pyrometer is then ready for measuring the target temperature when it is observed through the eyepiece. Adjustment of the rotationally graded grey filter until a small spot in the view field disappears is followed by reading the temperature from the rotating scale. The pyrometer, which may have two to three different temperature ranges from 750 °C up to 2000 °C or 3000 °C, is suitable for the smallest

Figure 7.35
Disappearing filament pyrometer with movable grey wedge filter Optix Type PB05. (Courtesy of Keller GmbH, Germany.)

target area of diameter about 4 mm at a distance of 1 m. An error limit of ±1% of measured value for a black body occurs over its nearly linear scale. A special additional scale can be provided for temperature measurement of free radiating iron or steel melts, with calibrated emissivity of 0.4 at $\lambda_e = 0.65\ \mu m$.

In recent times, the application of disappearing filament pyrometers in industry has become less frequent. Instead they are being replaced by other pyrometers, still operating at $\lambda_e = 0.65\ \mu m$, such as the Ultimax supplied by Ircon Inc. shown in Figure 7.48.

7.6 TWO-COLOUR PYROMETERS

7.6.1 General information

A two-colour or ratio pyrometer which measures temperature from the ratio of spectral radiances emitted by the object at two different wavelengths, is calibrated for grey bodies and gives correct readings for grey and black bodies. If the utilized wavelengths are placed within the visible range of the radiation spectrum, the name of two-colour pyrometer is precisely correct. Nevertheless the same name is used for pyrometers working outside the visible spectrum range. The working principle of two-colour pyrometers, in which the ratio of spectral radiances at two wavelengths is estimated by a human eye, is thoroughly discussed in the papers of Forsythe (1973), Haase (1933), Naeser (1933, 1935/36) and Schmidt (1924/25).

A simplified diagram of a two-colour pyrometer given in Figure 7.36 shows that it is composed of a lens, eyepiece and a two-colour filter, in most cases red/green. The observer adjusts the filter position so that the target to be measured appears to be grey. This position corresponds to equal spectral radiances or spectral brightnesses, as they are felt by a human eye, in two supplementing colours. With increasing target temperature, the percentage of green colour radiation increases while the red one decreases, so that each temperature corresponds to a definite filter position. The measured temperature, called the colour temperature, T_c, can be read from the pointer position on a scale. A colour slider (Figure 7.37), which is based on a similar working principle is the simplest of all pyrometers. The target to be measured is observed through a two-colour graded sliding filter. At a position, where the target turns grey, the measured temperature is read directly on a scale. The error limit of the colour slider is about ± 20 to ± 30 K over its measurement range of 1200 to 2000 °C.

Modern two-colour pyrometers are mostly automatic with the human eye replaced by photoelectric detectors.

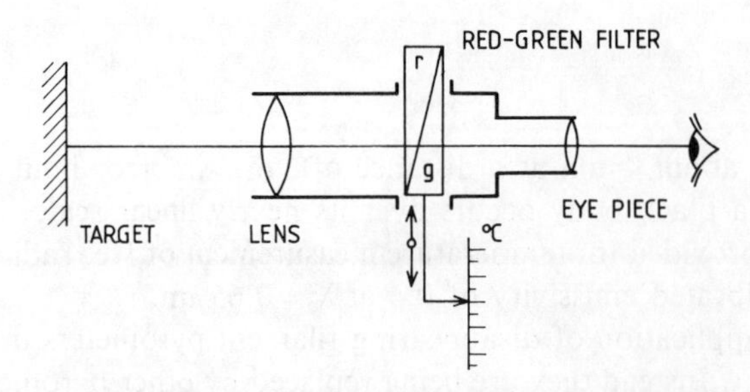

Figure 7.36
Principle of two-colour pyrometer.

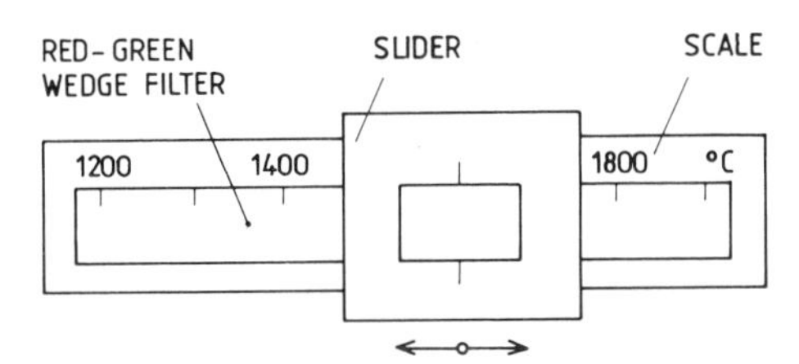

Figure 7.37
Colour slider.

7.6.2 *Scale defining equation*

Following Wien's law given in Equation (7.9), the spectral radiant intensity, W_{λ_1} emitted at the wavelength, λ_1, and at the temperature T_t by a body with emissivity ϵ_{λ_1} is given by:

$$W_{\lambda_1} = c_1 \epsilon_{\lambda_1} \lambda_1^{-5} \mathrm{e}^{-c_2/\lambda_1 T_t} \tag{7.82a}$$

and similarly, at the wavelength λ_2 it is:

$$W_{\lambda_2} = c_1 \epsilon_{\lambda_2} \lambda_2^{-5} \mathrm{e}^{-c_2/\lambda_2 T_t} \tag{7.82b}$$

According to Equation (7.62), the two equations above are also proportional to the spectral radiances L_{λ_1} and L_{λ_2}. The ratio of the spectral radiant intensities, or of the spectral radiances, at the wavelengths λ_1 and λ_2 is thus:

$$\frac{W_{\lambda_1}}{W_{\lambda_2}} = \frac{\epsilon_{\lambda_1}}{\epsilon_{\lambda_2}} \left(\frac{\lambda_2}{\lambda_1}\right)^5 \exp\left[\frac{c_2}{T_t}\left(\frac{1}{\lambda_2} - \frac{1}{\lambda_1}\right)\right] \tag{7.82}$$

Two-colour pyrometers give correct readings for grey bodies, whose emissivities are independent of wavelength ($\epsilon_\lambda = \text{const}$). Equation (7.82) will then be:

$$\frac{W_{\lambda_1}}{W_{\lambda_2}} = \left(\frac{\lambda_2}{\lambda_1}\right)^5 \exp\left[\frac{c_2}{T_t}\left(\frac{1}{\lambda_2} - \frac{1}{\lambda_1}\right)\right] \tag{7.83}$$

Depending upon pyrometer design, assume further that both wavelengths λ_1 and λ_2, are constant. Equation (7.83) can then be simplified to:

$$\frac{W_{\lambda_1}}{W_{\lambda_2}} = A\, \mathrm{e}^{B/T_t} \tag{7.84}$$

with the constants A and B, having the values:

$$A = \left(\frac{\lambda_2}{\lambda_1}\right)^5; \qquad B = c_2\left(\frac{1}{\lambda_2} - \frac{1}{\lambda_1}\right) \tag{7.85}$$

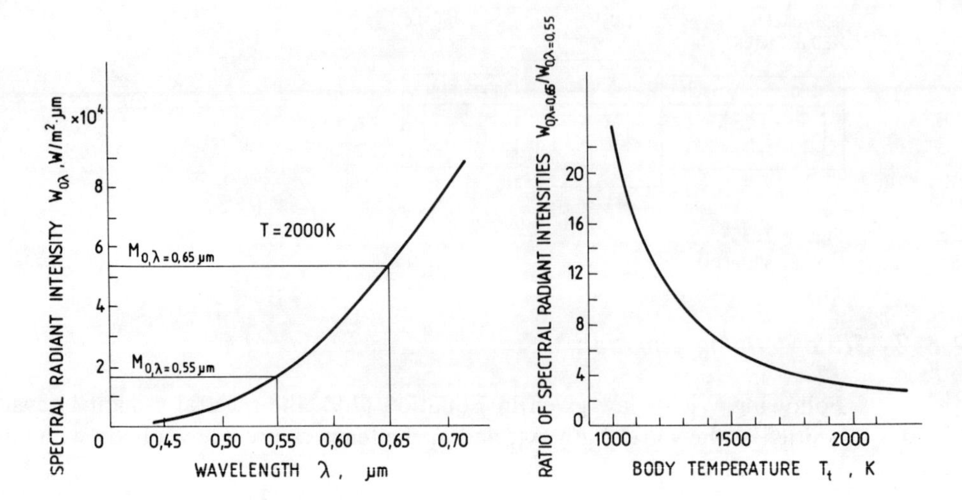

Figure 7.38
Two-colour pyrometer at black-body. (a) Spectral radiant intensity $W_{o\lambda}$ versus wavelength at 2000 K. (b) Ratio of spectral radiant intensities in red ($W_{o\lambda=0.65}$) and green ($W_{o\lambda=0.55}$) colour versus true temperature of radiating body T_t.

As can be seen from Equation (7.84), the ratio of spectral radiant intensities at λ_1 and λ_2 of any grey body, is an explicit function of the temperature T_t of the body and thus can be used for temperature measurement. In most two-colour pyrometers a red filter of effective wavelength $\lambda_1 = 0.65\ \mu$m and a green one of $\lambda_2 = 0.55\ \mu$m are used.

Figure 7.38(a) which shows part of the curve illustrating Planck's law of Equation (7.7) at a temperature of 2000 K, indicates the spectral radiant intensities at the two chosen wavelengths $\lambda_1 = 0.65\ \mu$m and $\lambda_2 = 0.55\ \mu$m. The ratio of the spectral radiant intensities at these wavelengths which has been calculated for different temperatures, is displayed in Figure 7.38(b). It can be seen that as the temperature T_t increases the ratio $W_{o,\lambda_1}/W_{o,\lambda_2}$ decreases. This shown limitation of sensitivity which is similar to the limitation in sensitivity exhibited by the human eye to colour variations, causes a reduction in the precision of temperature measurement also occurring with increasing target temperature. At the most commonly used wavelengths given above, the upper application limit of two-colour pyrometers is 2200 °C while the lower limit, resulting from the sensitivity of the human eye, is 700 °C.

The colour temperature of all black and grey bodies equals their true temperature. Care should be taken that two-colour pyrometers are not used by colour-blind operators. A detailed theory of two-colour pyrometers is given by Ruffino (1975).

7.6.3 Temperature measurement of non-grey bodies

Non-black and non-grey bodies also called selectively radiating bodies are characterized by the wavelength dependence of their spectral emissivity. If their temperature is measured by a two-colour pyrometer it is called the colour temperature, T_c. Ribaud *et al.* (1959) defined the colour temperature, T_c, of a body as that temperature of the body corresponding to the temperature of a black body, where the ratio of its radiant intensities at the wavelengths λ_1 and λ_2 equals the ratio of the radiant intensities of the body, whose temperature is to be measured at the same wavelengths.

According to that definition, and using Wien's law of Equation (7.9), it is apparent that:

$$\frac{\epsilon_{\lambda_1} e^{-c_2/\lambda_1 T_t}}{\epsilon_{\lambda_2} e^{-c_2/\lambda_2 T_t}} = \frac{e^{-c_2/\lambda_1 T_c}}{e^{-c_2/\lambda_2 T_c}} \tag{7.86}$$

where T_t is the true temperature of the body.

Taking logarithms and rewriting Equation (7.86), becomes:

$$\frac{1}{T_t} - \frac{1}{T_c} = \frac{\ln(\epsilon_{\lambda_1}/\epsilon_{\lambda_2})}{c_2[(1/\lambda_1) - (1/\lambda_2)]}$$

or finally:

$$T_c = \left(\frac{1}{T_t} - \frac{\ln(\epsilon_{\lambda_1}/\epsilon_{\lambda_2})}{c_2[(1/\lambda_1) - (1/\lambda_2)]}\right)^{-1} \tag{7.87}$$

From Equation (7.87) it is seen that when $\epsilon_{\lambda_1} \neq \epsilon_{\lambda_2}$ the indicated colour temperature, T_c, differs from the true value, T_t. This difference, $\Delta T = T_c - T_t$, depends on the ratio $\epsilon_{\lambda_1}/\epsilon_{\lambda_2}$ of the emissivities as well as on the chosen wavelengths λ_1 and λ_2. As the emissivity of metals increases for shorter wavelengths, their indicated colour temperatures are higher than their true temperatures. If ϵ_{λ_1} and ϵ_{λ_2} are known, the corresponding corrections to a colour temperature readings may be calculated.

Numerical example

The true temperature, T_t, of molybdenum was 1600 K, and its emissivities were $\epsilon_{\lambda_1} = 0.43$ and $\epsilon_{\lambda_2} = 0.45$ at $\lambda_1 = 0.65\ \mu m$, $\lambda_2 = 0.55\ \mu m$. Calculate the indicated colour temperature.

Solution: From Equation (7.87):

$$T_c = \left(\frac{1}{1600} - \frac{\ln(0.43/0.45)}{14\,388[(1/0.65) - (1/0.55)]}\right)^{-1} = 1629\ \text{K}.$$

A necessary correction to the readings is $\Delta T = T_t - T_c = 1600 - 1629 = -29$ K.

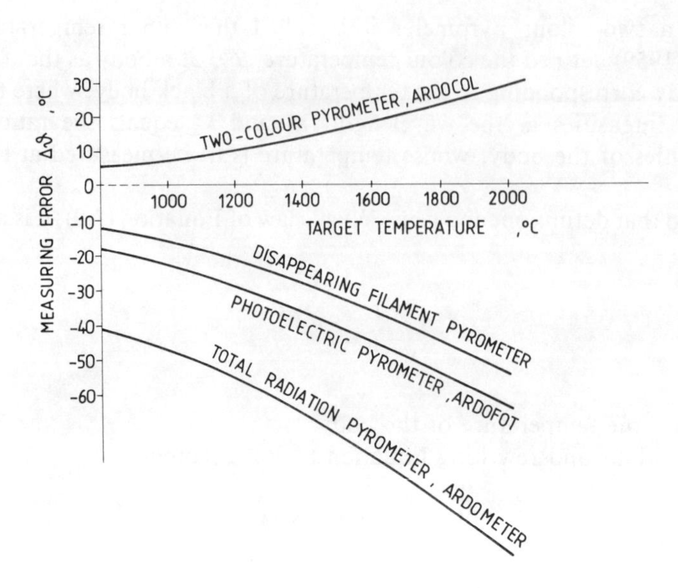

Figure 7.39
Measuring errors of different pyrometers versus target temperature at $\epsilon = 0.8$. (Courtesy of Siemens AG, Germany.)

For comparison, Figure 7.39 presents the measuring errors of different pyrometers in measuring the temperature of metals of $\epsilon = 0.8$, displayed against the object temperature. It is clearly seen that only two-colour pyrometers (such as the Ardocol instrument from Siemens AG, Germany) exhibit a positive error.

In most cases the errors, caused by small variations of the effective wavelengths as a function of the temperature of the filter are negligibly small (Ruffino, 1975).

7.6.4 Review of construction

To eliminate the influence of subjective colour estimation by different observers in two-colour automatic pyrometers, the eye of the observer is replaced by photoelectric detectors. The principles of automatic two-colour pyrometers are displayed in Figure 7.40. As shown in Figure 7.40(a) a single photoelectric detector D is irradiated alternately through a rotating disk having filters F_1 and F_2 of the effective wavelengths λ_{e1} and λ_{e2}. Applying a single detector for the comparison of two radiant intensities, helps to achieve a high stability of the readings. The system is based on a null balance principle, where one of the two radiant intensities is attenuated by the additional filter F_3, moved into the view-field by a servo-motor. Depending on which of the two radiant intensities is dominant, the phase of the detector output signal is changed by π. After amplification in a phase sensitive amplifier it alters the rotational direction of the motor which moves the filter F_3 so that a null balance is reached. In the state of optical balance the pointer, coupled with the movements of the filter F_3, indicates the measured temperature.

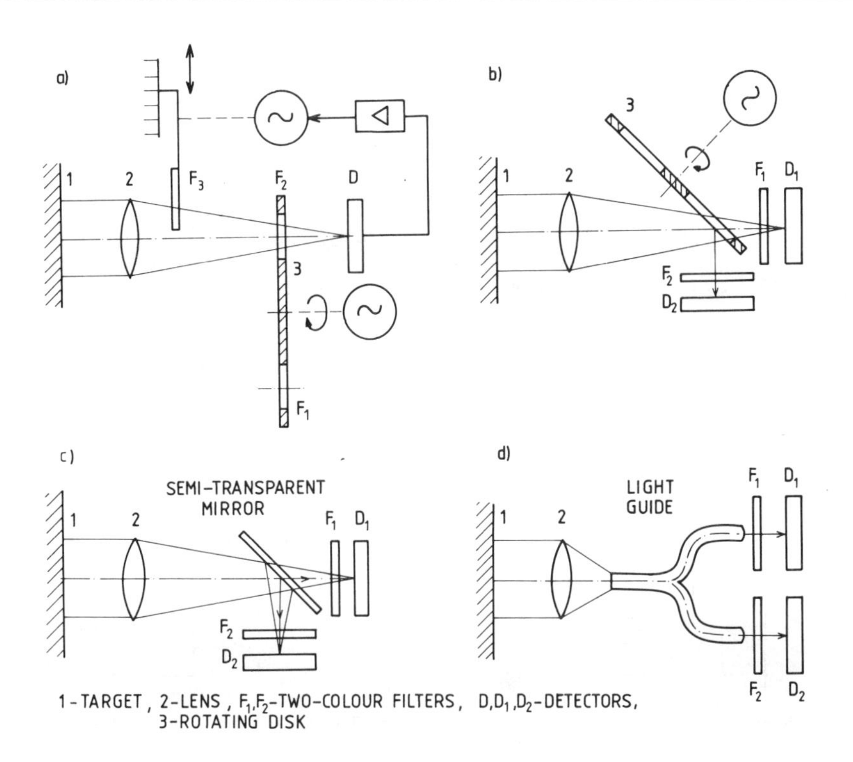

Figure 7.40
Principles of automatic two-colour pyrometers. F_1, F_2, two-colour filters; D, D_1, D_2, radiation detectors.

In a dual-wavelength radiometer from Wilkinson (USA), both radiant fluxes at λ_{e1} and λ_{e2} are directly amplified before a ratio output signal is formed.

In the system shown in Figure 7.40(b), the radiant flux which is either let through or reflected from the rotating disk with apertures falls alternately on the two detectors D_1 and D_2. Further electronic signal conditioning in a bridge circuit, with an adjustable resistor set by a servometer, gives the measured temperature as indicated by a pointer coupled with the slider position of the resistor. Other methods of splitting the incoming radiation in two channels, which may be done either by a semi-transparent mirror or by a bifurcated light guide are presented in Figure 7.40(c) and Figure 7.40(d). The design of the Ardocol pyrometer from Siemens AG (Germany), based on the principle shown in Figure 7.40(c), has the optical system given in Figure 7.41. Radiation from the target is concentrated on the light guide by a lens before falling on a semi-transparent indium phosphide filter, which allows the transmission of any radiation with a wavelength over 1 μm, while reflecting that with wavelengths under 1 μm. Split into these two parts, the radiation falls on the detectors D_1 and D_2. Superposition of the spectral sensitivities of the Si diode detectors as well as of the transmissivity and reflectivity of the InP filter results in selecting effective pyrometer wavelengths of $\lambda_{e1} = 0.888\ \mu$m and $\lambda_{e2} = 1.034\ \mu$m.

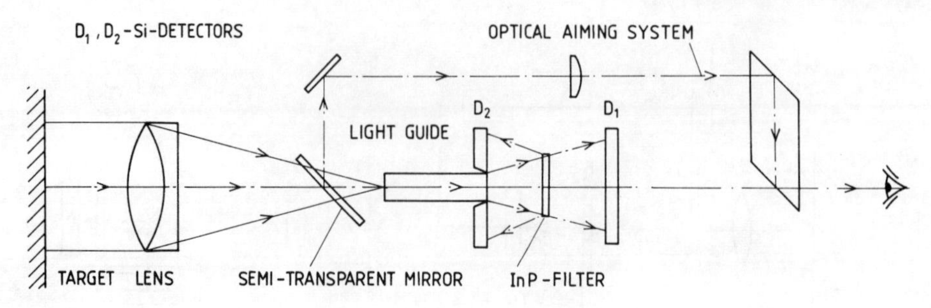

Figure 7.41
Optical system of two-colour pyrometer Ardocol. (Courtesy of Siemens AG, Germany.)

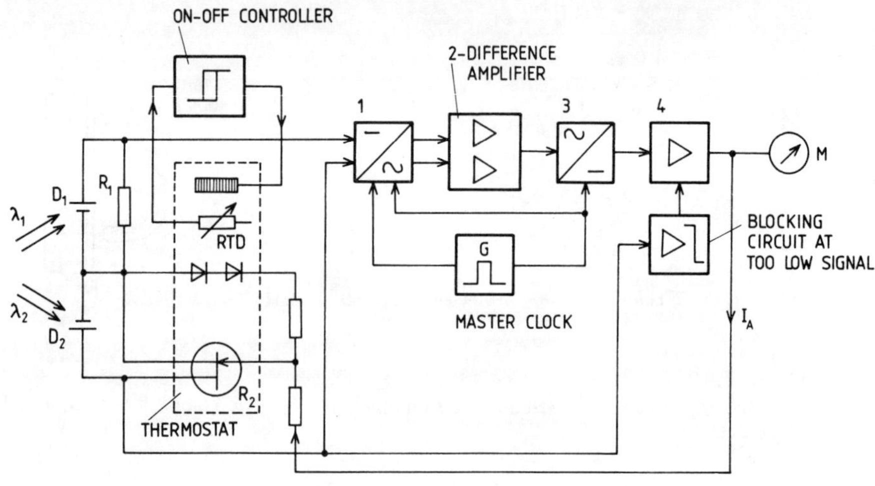

Figure 7.42
Electric system of two-colour pyrometer Ardocol. (Courtesy of Siemens AG, Germany.)

In Figure 7.42 which presents the electric circuit of the pyrometer, the resistor R_1 is used to set the lower limit of the measurement range. If this limit is exceeded, the detector D_1 delivers a higher voltage $V_{\lambda 1}$ than the voltage $V_{\lambda 2}$ of detector D_2 so that $V_{\lambda 1} > V_{\lambda 2}$. The bridge off-balance voltage which is changed into an a.c. voltage in the inverter 1 followed by amplification in the difference amplifier 2 and demodulation in 3, controls the output amplifier 4. Output current, I_A, which controls the resistance, R_2, of a field effect transistor, increases, until the voltage across the detectors D_1 and D_2 are equal. Balance of the bridge is reached when:

$$I_A = R_1 \frac{I_{\lambda 1}}{I_{\lambda 2}}$$

where $I_{\lambda 1}$ and $I_{\lambda 2}$ are the currents of both detectors. Thus the current I_A which is proportional to the ratio of both radiant intensities may be expressed as:

$$I_A = K \frac{W_{\lambda 1}}{W_{\lambda 2}}$$

where K is a constructional constant. The field effect transistor and the linearizing diodes are confined in a thermostat whose temperature is kept constant using an on–off controller and a resistance thermometer detector RTD. A special circuit blocks the pyrometer indications in the case when a signal which is too low comes from the target. The pyrometer which has a standardized output signal of 4 to 20 mA or 0 to 20 mA has an application range of 700 to 3100 °C, with measurement errors for grey bodies about $\pm 1\%$ to $\pm 1.5\%$ FSD. Depending on the measurement range, the distance ratio is 10 to 75. Modern, automatic two-colour pyrometers also ensure correct readings in the case of very low radiant intensities. For example, the Modline R Pyrometer from Ircon Inc. (USA) will tolerate a reduction of more than 95% in radiant intensity without error. Thus the pyrometer can cope with a small target area, low target emissivity or dust and smoke in the air. A special circuit warns the observer that the reading is incorrect if the radiant intensity becomes too low.

The PB60 two-colour pyrometer from Keller GmbH (Germany) which utilizes radiation at the wavelengths $\lambda_{e1} = 0.95\ \mu m$ and $\lambda_{e2} = 105\ \mu m$, is also produced as a fibre-optic pyrometer.

A unique hybrid design combining single and dual wavelength signals to cope with emissivity variations is a feature of the True Set 12000 pyrometer of Williamson (USA). It is produced especially for the temperature measurement of aluminium surface.

7.7 OUTPUT SIGNAL CONDITIONING

Different special circuits which are produced for the conditioning of pyrometer output signals may be built into the pyrometer or made as separate units.

A peak picker is used for holding the highest instantaneous temperature values, occurring as a stochastic function of time. In the working principle of the system, presented in Figure 7.43, the true temperature is tracked continuously without delay and then held with a slow, adjustable decay rate. A recorder connected to the system, follows the upper envelope of temperature values which change with time. In most practical cases, these are apparent temperature variations, resulting from any surface inhomogeneities of a moving object or from changes in radiation owing to absorption

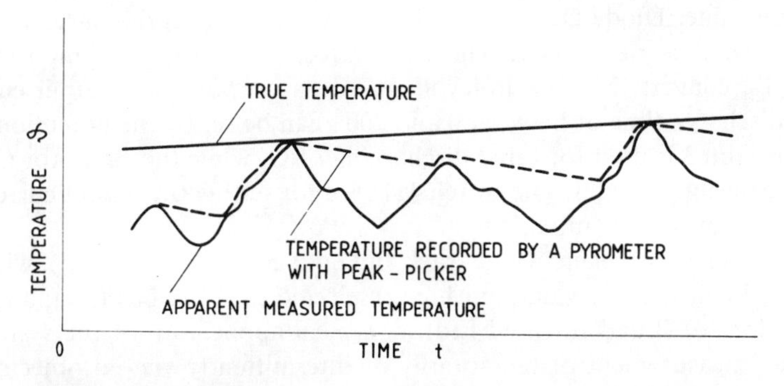

Figure 7.43
Principle of peak-picker.

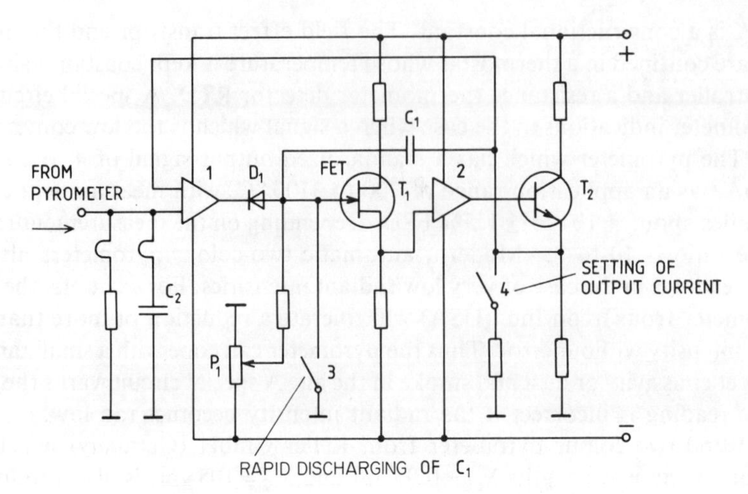

Figure 7.44
Basic diagram of a peak-picker. (Courtesy of Siemens AG, Germany)

by dust or smoke in the air between the target and the pyrometer. Typical applications of a peak picker occur in the temperature measurement or recording of:

- a succession of small parts to be viewed with spacing between them,
- moving surfaces, covered by oxides or scale,
- targets temporarily covered by dust, steam or smoke,
- non-moving inhomogeneous surfaces.

In this last application, the pyrometer sighting direction is oscillated periodically, scanning the area whose temperature is to be measured and picking the points of highest temperature. The target spot size should be as small as possible to pick out and hold only the highest and not the average temperature.

The schematic diagram of a peak picker by Siemens AG (Germany) which is shown in Figure 7.44, has its pyrometer output signal charging, through the amplifier 1, the capacitor C_1 to a voltage proportional to the instantaneous measured temperature value. Diode D_1 prevents the capacitor C_1 from discharging when the pyrometer signal is decreasing. The field effect transistor T_1, amplifier 2 and transistor T_2, convert the capacitor voltage into a proportional output current. A capacitor discharge time of between 0 and 200 s can be set by the potentiometer P_1. The push button 3 is used for rapid discharging of C_1 while the capacitor C_2 is used for filtering the input signal. The switch 4 is used for setting the output current range at 0 to 20 mA or 4 to 20 mA.

There are also peak pickers with a track and hold action (Figure 7.45), which trace the measured temperature values during preselected time periods (a) and then hold it for a period (b) (Land Infrared Ltd). A typical application of the system is the temperature measurement of temporarily or intermittently viewed objects.

A valley picker which exhibits a complementary reverse action by holding the lowest readings is similarly constructed.

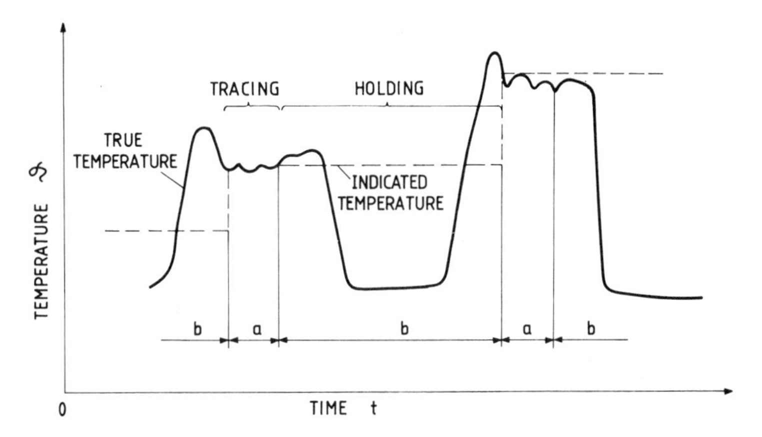

Figure 7.45
Principle of peak-picker with track and hold action.

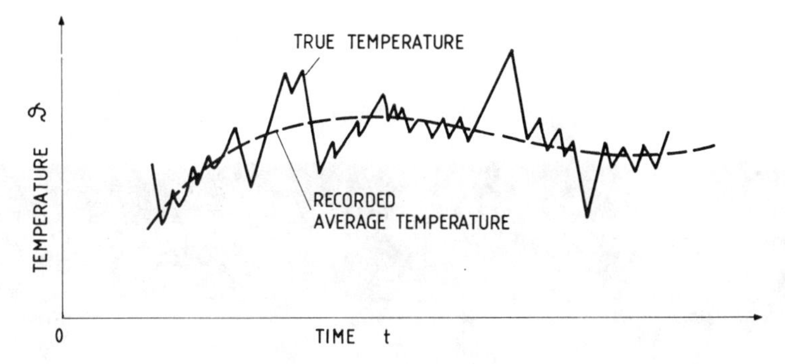

Figure 7.46
Principle of pyrometer signal averager.

An averager presents the mean value of a smoothed varying signal as the desired output. It is used for measuring, or more often recording, the average values of periodically or stochastically fluctuating temperatures, as shown in Figure 7.46. Photoelectric pyrometers with their extremely short response times, can follow these rapid temperature variations so that in theory their measurement or recording is possible, but of no practical use. Analog indication would be extremely difficult while digital indication is impossible. An average value which is easy to handle, can be formed with an averaging time set in the range from some seconds to 200 s. A schematic diagram of an averager by Siemens AG (Germany) is shown in Figure 7.47. The pyrometer output signal which is amplified in amplifier 1 is then converted into a linear output current in a complementary transistor pair TS. The value of this current in charging the capacitor C_1, sets the average time by the adjustment of the potentiometer P_1. This average time can also be set in the range from 0 to 30 s or from 20 to 150 s after the capacitor C_2 is additionally switched in. An output amplifier 2 converts the average signal into 0 to 20 mA or 4 to 20 mA output current,

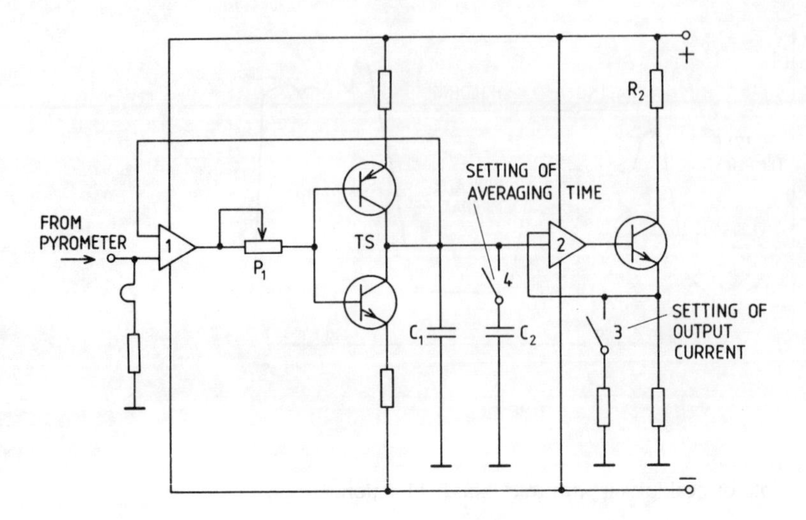

Figure 7.47
Basic diagram of an averager. (Courtesy of Siemens AG, Germany.)

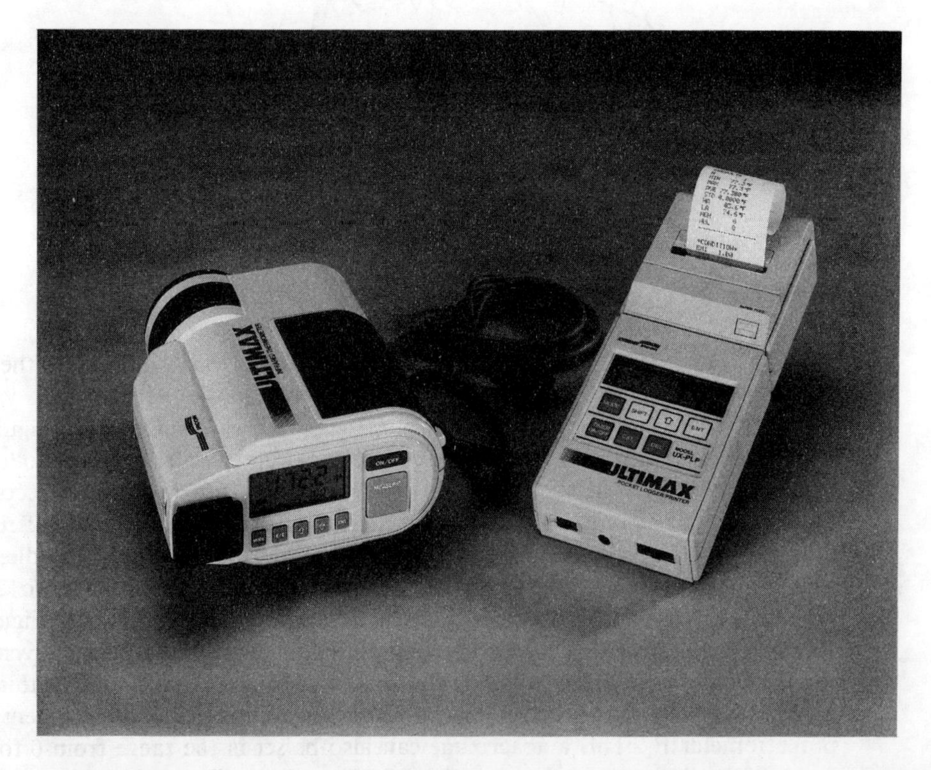

Figure 7.48
Portable pyrometer Ultimax with data logger. (Courtesy of Ircon Inc., USA.)

depending on the position of switch 3. For very rapid temperature variations an averager can be replaced by an averaging filter.

Many of the special functions descibed above can also be contained in a portable, hand-held pyrometer such as the Ultimax instrument by Ircon Inc. (USA). This truly versatile microprocess-based pyrometer illustrated in Figure 7.48, is produced in ten different series, equipped with either a thermopile or silicon photodiode detector. A short survey of the different series available is given in Table 7.5.

An accuracy of $\pm 1\%$ (± 1 digit) can be obtained when the pyrometer's distance ratio is confined, between 25 and 250, while the emissivity can be adjusted digitally between 0.1 and 1. The pyrometer disposes of the functions of a peak picker, delay, valley picker, real temperature, which are all easily programmed. A large LCD display at the rear of the pyrometer is duplicated in the field of view. The pyrometer which has a linearized analog output of 0 to 1 V d.c. or a digital output, is battery powered with a through-the-lens reflex focus. A programmable data logger printer also shown in Figure 7.48 which can log or store up to 1000 temperature readings can be supplied as an option. It comprises high and low limit alarms, data collection intervals and on-demand print out with the stored data transferrable to a computer.

It should be mentioned that the Ultimax UX-10 operating at 0.65 μm replaces the former disappearing filament pyrometer which is no longer manufactured. This enables the user to set the emissivity of various materials correctly, based upon the very large amount of data, collected and published over the many years when disappearing filament pyrometers were still in common use.

For temperature supervising or temperature scanning of large surfaces such as metal, or glass sheets, paper or textile tapes, metal blocks, furnace housing, and so on, sampling pyrometers with oscillating pyrometer heads are used (Land Pyrometers Ltd, undated). Another solution presents a stationary pyrometer with oscillating optics. In a pyrometer from Maurer GmbH (Germany) the possibility of scanning surfaces up to 2 m long, from a distance of 2 m, is offered. The same pyrometer may also be used for the temperature measurement of a moving wire. In this case the sighting point oscillations are only a few millimetres in amplitude with a peak picker retaining the maximum temperature reading of each oscillation.

7.8 PYROMETRIC MEASUREMENTS THROUGH ABSORBING MEDIA

The presence of absorbing media, like gases, smoke, dust or water vapour, in the sight path of a pyrometer may influence its readings. This influence can be caused by absorption of the incident radiation or by its reflection from dust particles which can also emit radiation themselves. Figure 7.21 shows both the absorbing wavelengths and the absorption free atmospheric windows where correctly designed pyrometers should operate. Some gases such as N_2, O_2 and H do not absorb the infrared radiation in practice, whereas others like CO_2, SO_2, NH_3 and water vapour are characterized by heavily absorbing wavelength bands. In these bands, following Kirchhoff's law, the gases mentioned above exhibit high emissivities. While the position of the absorbing bands of the gases depends on both temperature and

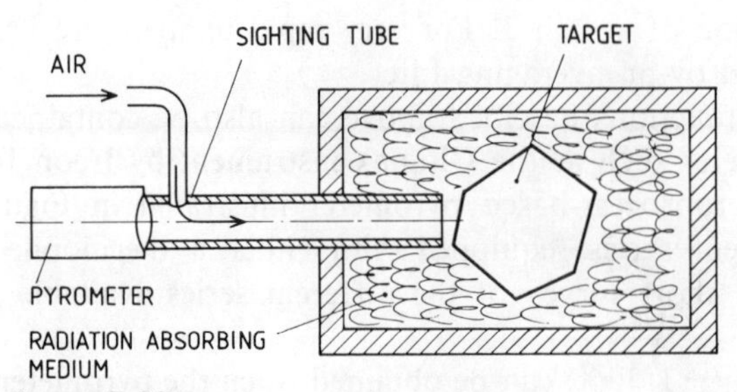

Figure 7.49
Total radiation pyrometer in temperature measurement in presence of absorbing media.

pressure, their absorptivity depends upon the thickness of the absorbing layer (Hackforth, 1960).

The principal absorbing bands of CO_2 and H_2O are:

CO_2: 2.4–3 μm; 4–4.8 μm; 12.5–16.5 μm.
H_2O: 1.3–1.5 μm; 1.7–2 μm; 2.2–3 μm; 4.8–8.5 μm; 12–30 μm.

Gas mixtures have a rather complicated character to their spectral dependence. More details are presented in Gröber *et al.* (1957) and Warnke (1972).

The absorption of dust, smoke and water mist depend on the size of their particles. If the particles are larger than the wavelengths of the pyrometer working band, a uniform radiation absorption can be observed. No measurement errors are observed when using two-colour pyrometers. If the particles are smaller than the wavelengths of the pyrometer working band, a selective radiation absorption which is very difficult to foresee, can be observed. Lotzer (1976) asserts that smaller measurement errors arise with a shorter pyrometer working wavelength band.

In the case when the temperature of absorbing gases in the sight path is equal to the temperature of the object to be measured, the absorbed radiant power is equal to the emitted one according to Kirchhoff's law of Equation (7.15). No measuring errors of any kind will be observed in that case.

Total radiation pyrometers are sensitive to the presence of polyatomic gases, combustion gases, smoke and water mist, which may cause measurement errors even as high as 100 K according to Lieneweg (1964a). To reduce these errors the pyrometer should be placed as close as possible to the target or an air stream applied to disperse dust and fumes from the pyrometer sight path (Figure 7.49). The same air can also be used for cooling the pyrometer head. In really difficult working conditions, Lotzer (1976) advised that the pyrometer should be mounted no further than about 1.5 m from the target.

Disappearing filament pyrometers do not exhibit any measurement errors in the presence of gases, which at $\lambda_e = 0.65\ \mu$m do not absorb thermal radiation. However, they are strongly influenced by the presence of dust with measurement errors per 1 m target distance amounting to:

0.7 K/m in steel or iron works with average air impurity,
2 K/m in steel or iron works with heavy air impurity,
5–15 K/m in foundries and in the neighbourhood of Martin furnaces.

Photoelectric pyrometers are very well adapted to operate in the presence of absorbing gases, because by choosing a convenient effective wavelength, their influence can easily be avoided. Nevertheless dust and smoke can influence the readings.

Two-colour pyrometers are practically immune to the influence of dust, smoke and water mist. In addition, those working at the effective wavelengths 0.55/0.65 μm, are not influenced by polyatomic gases. Other two-colour pyrometers exhibit measurement errors, when the absorptivities at the two applied wavelengths are different.

A simple method to check if any measurement errors are to be expected, is to repeat the same measurement from the shortest and longest possible distance. Both should be identical.

As an example of the influence of radiation absorption by water mist and water vapour, the results obtained in a rolling mill are relevant. Measuring the temperature $\vartheta_t = 1150$ °C, the readings of a total radiation pyrometer were 147 °C low, those of a photoelectric pyrometer with a Si diode were 76 °C low and a two-colour pyrometer gave exact readings.

7.9 PYROMETRIC MEASUREMENTS OF NON-HOMOGENEOUS SURFACES

In the case of hot-rolling of metal sheets or blocks, for example, it is quite common for the surface, whose temperature is to be measured, to be partially covered by an oxide layer. As the oxides are rapidly cooling, the resulting readings are too low. Let the percentage of surface covered by oxide be known, as well as the temperature of the oxides, ϑ_1, and of the metal, ϑ_2, with $\vartheta_1 < \vartheta_2$, while the corresponding emissivities are ϵ_1 and ϵ_2. Based on the known characteristic of any pyrometer, its readings can be determined.

As photoelectric band and spectral pyrometers exhibit higher steepness of the scale characteristic as a function of temperature (Figure 7.33), their readings depend to a lesser extent on the temperature ϑ_2, than those of total radiation pyrometers. Two-colour pyrometers are even less sensitive to temperature differences.

Comparative results of the temperature measurement of a surface, with 25% covered by oxides, are presented in Figure 7.50 following work reported by Lieneweg (1975). The true surface temperature is ϑ_1 and the oxide's temperature is ϑ_2.

In acquiring these results the three different Siemens AG pyrometers used were the Ardometer total radiation pyrometer, the Ardofot photoelectric pyrometer and the Ardocol two-colour pyrometer. All three pyrometers gave negative errors. The true temperature ϑ_1 of the free radiating, uncovered parts of the surface can be properly measured by a pyrometer with peak picker, described in §7.4.6, which periodically scans the surface.

7.10 SUMMARY OF THE PROPERTIES AND APPLICATIONS OF PYROMETERS

Total radiation pyrometers with quartz or glass lenses are used for continuous, industrial temperature measurement and recording. When measuring the temperature

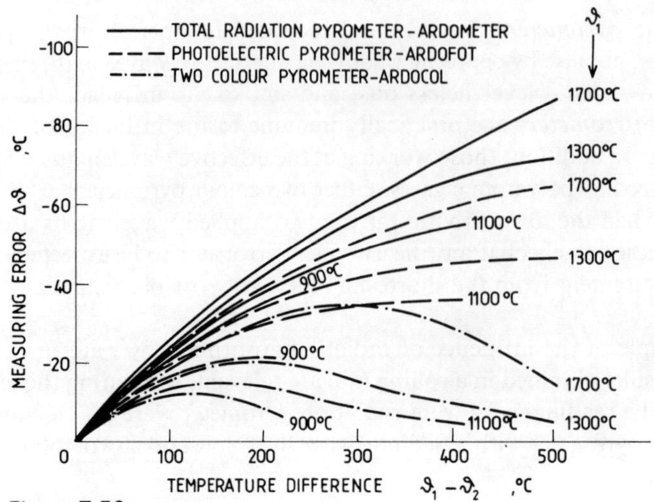

Figure 7.50
Errors $\Delta\vartheta$ of temperature measurement at a non-homogeneous surface (75%—ϑ_1, 25%—oxides of ϑ_2) by different pyrometers. (Courtesy of Siemens AG, Germany.)

Table 7.5
The Ultimax series of pyrometers surveyed. (Courtesy of Ircon Inc. USA.)

Series	Temperature span (°C)	Detector and operating wavelength	Applications
10	900–3000	Silicon 0.65 μm	Iron and steel, molten glass, semiconductor manufacture
20	600–3000	Silicon 0.96–1.05 μm	Steel mills, foundries, forging, induction heating, glass, semiconductors
40/41/42	−50–1000	Thermopile 8–13 μm	General purpose low temperature, plastics, paper, textiles, food, agriculture
50	400–1300	Thermopile 3.7–5.2 μm	Furnace and kiln maintenance
70/71	300–2800	Thermopile 4.8–5.2 μm	Refractory and ceramic production, glass, bricks, porcelain, quartz
80/81	50–400	Thermopile 7.5–8.5 μm	Plastics, plastic film, low temperature glass applications

of non-black bodies, measurement conditions should be arranged so that the target properties are as near those of a black body as possible. The apparent 'black' temperature could also be used as a reference value in mass production. Pyrometers are often used for temperature measurement in electric furnaces.

Total radiation pyrometers with fluorite or KRS.5 lenses and mirrors are mostly used for low temperature measurements of non-metallic surfaces like paper, textile, plastics or rubber. These materials can be regarded as black bodies for all practical purposes.

Table 7.6
Optical pyrometers—properties and average technical data (extremal values in parentheses).

Pyrometer type	Optical system	Radiation detector	Spectral band (μm)	Temperature range (°C)	Error	$t_{0.98}$* (s)	Distance ratio (l/d)	Output	Emissivity influence notes
Total radiation	Lens Mirror Fibre optic	Thermopile Pyroelectric Bolometer	$0.4 \div 10$[†] $8 \div 14$	$0 \div 2000$ (lens) $-50 \div 500$ (mirror)	$\pm 1 \div \pm 1.5\%$ at $\epsilon = 1$	$0.5 \div 4$ (0.015)	$4 \div 50$ (200)	Analog (mV, mA) Digital BCD	Great for metals, small for non-metals
Photo-electric	Lens Fibre optic	Photocon-ductor Photodiode Photovoltaic cell Vacuum photocell	To be chosen following the application	$250 \div 3000$	$\pm 0.5 \div \pm 1\%$ at $\epsilon = 1$	$0.02 \div 0.3$	$10 \div 300$ (700)	Analog, often linearized Rarely digital	Small. Specialized types built for different applications.
Dis-appearing filament	Lens	Observer's eye	0.65	$700 \div 2000$ (3500)	± 0.6–$\pm 2\%$ at $\epsilon = 1$	—	$20 \div 300$ (5000)	—	Great, but smaller than at total radiation pyrometers. Calculation of corrections possible for known $\epsilon_{0.65}$.
Two-colour	Lens	Observer's eye	0.55/0.65	$700 \div 2200$	± 10 °C	—	$10 \div 75$	—	No influence for grey bodies; small—for most real, non-grey bodies. Correct readings also for re-duced radiant intensity.
		Photoelectric	0.88/1.03 0.95/1.05 and others	$700 \div 3100$	$\pm 1\%$	0.04		Analog, linearized	

*Does not cover inertia of indicating instrument.
[†]Narrower spectral band if filters are applied.

Table 7.7
Survey of general purpose and dedicated purpose pyrometers produced by Ircon Inc. (USA).

Series	Type	Temperature range (°C)	Wavelength (μm)	Applications
2000	General purpose	500–3000	0.9	Metals, semiconductors
R	Two colour	700–3500	0.95 and 1.05	Varying emissivity, air impurities, metal, melting, wires, vacuum furnaces
G	General purpose	250–1100	1.64	Metal, average temperatures
6000	General purpose	80–500	2–2.6	Versatile, opaque plastics, rubber, metals
4000	General purpose	0–500	8–14	Specially for non-metals, paper, food, textiles, IR-heated objects
Q	Flames	800–2400	4.5	Flames, burners, chemical reactors, glass melting, drying
7000	Glass	50–1300	4.8–5.2	Glass surface temperatures
3400	Plastic films	0–800	3.43	Transparent plastic films of CH-absorption band, paint coatings
8000	Plastic films	20–400	7.9	Thin clear films of polyester and fluorocarbon, textiles

Photoelectric pyrometers are often applied in the presence of water vapour or absorbing gases, where total radiation pyrometers cannot be used. They also present the only realistic method for measuring either rapidly changing temperatures or the temperature of moving bodies. With a proper choice of their effective wavelength, photoelectric pyrometers can be adapted to the spectral properties of the target and of the surrounding atmosphere. Systems of output signal conditioning extend the application range of these pyrometers.

Disappearing filament pyrometers, portable and hand operated, are used for occasional measurements. As their precision is high they are often employed in industry for reference measurement and for the verification of other instruments. The possibility of calculating corrections, enables their use for measuring the temperature of non-black bodies. As their necessary target area must be small and their application range is high, they are especially useful in laboratory measurements.

Two-colour pyrometers are used for temperature measurement in the presence of absorbing gases, dust, smoke and vapour, where no other pyrometers are applicable, and also for the temperature measurement of non-black bodies of varying emissivity. For continuous measurements automatic two-colour pyrometers are appropriate. Hand-operated pyrometers are often used for reference measurements of bodies with unknown emissivity. They are also used for estimating the emissivities of different bodies.

A concise specification of the properties and technical data of pyrometers is given in Table 7.6. Many pyrometer manufacturers produce a whole range of different pyrometer types. For example, Ircon Inc., USA, offers the general purpose and dedicated purpose pyrometers of Modline Series summarized in Table 7.7. Nearly all of these applications are also covered by different models of the Ultimax portable, single lens reflex pyrometer also produced by Ircon Inc. (Table 7.5).

REFERENCES

Ackerman, S. (1962) Notes on design and performances of two-color pyrometer. *Temperature: Its Measurement and Control in Science and Industry*, Vol. 3, Part 2. Reinhold Publ. Co., New York, pp. 849–857.

Baker, H. D., Ryder, B. A. and Baker, N. A. (1953) *Temperature Measurement in Engineering*, Vol. 1, John Wiley and Sons, New York.

Bonnell, D. W. *et al.* (1972) The emissivities of liquid metals at their fusion temperatures. *Temperature: Its Measurement and Control in Science and Industry*, Vol. 4, Part 1. Instrument Society of America, Pittsburgh, pp. 483–488.

Brenden, B. B. (1962) An infrared ratio pyrometer. *Temperature: Its Measurement and Control in Science and Industry*, Vol. 3, Part 2. Reinhold Publ. Co., New York, pp. 429–433.

Considine, D. *et al.* (1957) *Process Instruments and Controls Handbook*, McGraw-Hill, New York.

Engel, F. (1974) *Temperaturmessungen mit Strahlungspyrometern*, Verlag Technik, Berlin.

Féry C. R. (1902) La mesure des températures élevées et la loi de Stephan. *Comptes Rendus de l'Academie des Sciences*, **155**, 977–980.

Féry, C. R. (1908) Photométrie-Rendement optique de quelques luminaires. *Le Journal de Physique et la Radium*, **7**, 632–640.

Forsythe, W. E. (1923) Color match and spectral distribution. *J. Opt. Soc. Am.*, **7**, 1115–1122.

Forsythe, W. E. (1941) Optical pyrometry. *Temperature: Its Measurement and Control in Science and Industry*, Reinhold Publ. Co., New York, pp. 1115–1131.

Foster, Ch. E. (1910) New radiation pyrometer. *Trans. Amer. Electr. Soc.*, 223–227.

Gerber, D. (1979) Berührungslose Temperaturmessung mit Pyrometern. *Technik*, **34**(11), 605–607.

Griffith, E. (1974) *Methods of Measuring Temperature*, C. Griffin, London.

Gröber, H., Erk, S. and Grigull, U. (1963) *Die Grundgesätze der Wärmeübertragung*, Springer Verlag, Berlin.

Haase, G. (1933) Farb-Pyrometrie. *ATM*, **214-2**(10), T133–T134.

Hackforth, H. L. (1960) *Infrared Radiation*, McGraw-Hill, New York.

Harmer, A. L. (1982) Principles of optical fibre sensors and instrumentation. *Measurement and Control*, **15**(4), 143–150.

Harrison, T. R. (1960) *Radiation Pyrometry and Its Underlying Principles of Radiant Heat Transfer*. John Wiley and Sons, New York.

Hecht, G. J. (1962) Two-wavelength near infrared pyrometer. *Temperature: Its Measurement and Control in Science and Industry*, Vol. 3, Part 2, Reinhold Publ. Co., New York, pp. 407–417.

Heimann, W. and Mester, U. (1975) Non contact determination of temperatures by measuring the infrared radiation emitted from the surface of a target. *Temperature Measurement*, Conference Series No. 26, The Institute of Physics, London, 1975, pp. 219–237.

Heinisch, R. P. (1972) The emittance of black body cavities. *Temperature: Its Measurement and Control in Science and Industry*, Vol. 4, Part 1. Instrument Society of America, Pittsburgh, pp. 435–448.

Henning, F. (1951) *Temperaturmessung*, J. A. Barth, Leipzig.

Hoffmann, G. (1981) Pyroelectric radiation detectors in the low temperature pyrometry. *Feingerätetechnik*, **30**, 25–27.

Holborn, L. and Kurlbaum, F. (1902) Über ein optisches Pyrometer. *Annalen der Phys.* **10**, 225–241.

Jakob, M. (1957) *Heat Transfer*, Vol. 1, John Wiley and Sons, New York.

Kelsall, D. (1963) An automatic emissivity-compensated radiation pyrometer. *J. Sci. Instrum.* **40**(1), 1–14.

Larsen, B. M. and Shenk, W. E. (1941) Temperature measurement with blocking layer photocells. *Temperature: Its Measurement and Control in Science and Industry*. Reinhold Publ. Co., New York, pp. 1150–1158.

Leclerc, G. (1976) *New Methods of Temperature Measurement*. Ircon Inc., USA.

Lieneweg, F. (1964a) Vergleichende Untersuchungen aber Temperaturmessungen mit Strahlungspyrometern. *Siemens-Z*, No. 38, pp. 5.

Lieneweg, F. (1964b) Fehler und Einflüsse bei der optischen Temperaturmessung mit Gesamtstrahlungs-Teilstrahlungs- und Farbpyrometern. *Arch. f. Eisenhüttenwesen*, **35**(12), 1145–1150.

Lieneweg, F. (1975) *Handbuch, Technische Temperaturmessung*. F. Vieweg, Braunschweig.

Lieneweg, F. and Menge, K. (1963) Das Farbpyrometer Ardocol. *Siemens-Z*, No. 12, pp. 821.

Lotzer, W. (1976) Probleme der berührungslosen Temperaturmessung. Fachberichte Hüttenpraxis. *Metallverarbeitung*, **14**, No. 9/10.

Mesures, Regulation, Automatisme (1981) Numére special, **46**, 4.

Murray, T. P. (1967) Polaradiometer—a new instrument for temperature measurement. *Rev. Sci. Instr.*, **38**, 791.

Murray, T. P. (1972) The polarized radiation method of radiation thermometry. *Temperature: Its Measurement and Control in Science and Industry*, Vol. 4, Part 1. Instrument Society of America, Pittsburgh, pp. 619–626.

Musbah A. and Zazabili, N. (1975) Berührungslose Temperaturmessung mit Infrarot-Strahlungspyrometern im Hochtemperaturbereich. *Siemens Fachberichte*, **49**(11), 896–902.

Naeser, G. (1930) Die Emissionsvermögen von flüssigen Eisenlegierungen. *Mitt. K. Wilh. Inst. Einsenforsch*, **12**, 365–372.

Naeser, G. (1935/36) Kombiniertes Farbpyrometer mit Vergleichslampe. *Arch. f. Eisenhüttenwesen*, **7**(9), 483–485.

Pepperhoff, W. (1960) Optische Pyrometrie im polarisiertem Licht. *Zeitsch. angew. Physik*, **12**, 168.

Reynolds, P. M. (1961) Emissivity errors of infra-red pyrometers in relation to spectral response. *Brit. J. of Appl. Phys.*, **11**(8), 401–405.

Ribaud, G. *et al.* (1959) *Etudes de pyrométrie pratique*, Eyrolles, Paris.

Righini, F., Rosso, A. and Ruffino G. (1972) Temperature dependence of effective wavelength in optical pyrometry. *Temperature: Its Measurement and Control in Science and Industry*, Vol. 4, Part 1. Instrument Society of America, Pittsburgh, pp. 413–424.

Rohsenow, W. M. and Harnett, J. P. (1973) *Handbook of Heat Transfer*, McGraw-Hill, New York.

Ruffino, G. (1975) Increasing precision of two-colour pyrometry. *Temperature Measurement*. Conference Series No. 26, The Institute of Physics, London 1975, pp. 264–272.

Schmidt, H. (1924/25) Über die Grundzüge der Farbpyrometrie. *Mitt, K. Wilh. Inst. Eisenforsch*, **12**, No. 6.

Siemens AG. (1987) *Strahlungs pyrometer mit temperaturlinearem Stromausgang*. Karlsruhe.

Tantzen, D. and Zureich, J. (1980) Temperaturen-Messen und-Regeln beim Hochfrequenz-Schweissen. *Bänder, Bleche, Röhre*, **9**, 381–382.

Thwing, C. H. and Burton, Ph. P. (1908) A new radiation pyrometer. *Journal of Franklin Inst.*, pp. 363–370.

Tingwaldt, C. (1960) Ein einfaches optischpyrometrisches Verfahren zur direkten Ermittlung wahrer Temperaturen glühender Metalle. *Z. Metallkunde*, **51**, 116–120.

Toyota, H., Yamada, T. and Nariai, Y. (1972) Improved radiation pyrometry for automatic emissivity compensation. *Temperature: Its Measurement and Control in Science and Industry*, Vol. 4, Part 1. Instrument Society of America, Pittsburgh, pp. 611–628.

Walter, L. (1981) Problems associated with the reduction of the influence of emissivity in contactless temperature measurement. *Temperature Measurement in Industry and Science*. First Symposium of IMEKO TC 12 Committee, Czechoslovak Scientific and Technical Society, Praha, 1981, pp. 89–94.

Warnke, G. F. (1972) Commercial pyrometers. *Temperature: Its Measurement and Control in Science and Industry*, Vol. 4, Part 1. Instrument Society of America, Pittsburgh, pp. 503–518.

Weichert L. *et al.* (1976) *Temperaturmessung in der Technik—Grundlagen und Praxis*, Lexika Verlag, Grafenau.

De Witt, D. P. and Hernicz, R. S. (1972) Theory and measurement of emittance properties for radiation thermometry applications. *Temperature: Its Measurement and Control in Science and Industry*, Vol. 4, Part 1. Instrument Society of America, Pittsburgh, pp. 459–482.

De Witt, D. P. and Kunz, H. (1972) Theory and technique for surface temperature determinations by measuring the radiance temperatures and the absorptance ratio for two wavelengths. *Temperature: Its Measurement and Control in Science and Industry*, Vol. 4, Part 1. Instrument Society of America, Pittsburgh, pp. 1164–1187.

Worthing, A. G. (1941) Temperature radiation emissivities and emittances. *Temperature: Its Measurement and Control in Science and Industry*, Reinhold Publ. Co., New York, pp. 1164–1187.

8 Conditioners, Transmitters, Indicators, Loggers and Recorders for Temperature Signals

8.1 INTRODUCTION

All of the items mentioned in the title of this chapter, which have special variants designed for applications in temperature measurements, will be described. Many of the instruments, which have a similar structure irrespective of the thermometer they are designed for, can be adapted to match any particular sensor either by switching or by exchanging plug-in modules.

For convenience it is important to draw a distinction between those items which are functionally and structurally close to the sensors. Conditioners and transmitters belong within this group. Instruments for indicating, logging or recording purposes are generally referred to as data presentation elements. These functional distinctions are drawn on the basis of how closely the element under consideration is structurally linked to the sensing function (McGhee *et al.*, 1986; McGhee and Henderson, 1989).

8.2 DATA ACQUISITION AND CONDITIONING FOR TEMPERATURE SIGNALS

8.2.1 Model structures for data acquisition

Most pieces of equipment which are called signal transmitters, perform more than the transmission function. They usually also include the functions of data conditioning, acquisition and conversion. The widespread application of micro-electronics, which has obscured the distinction between these diverse functions, has also led to dramatic changes in how these functions are now performed. Although single or multiple signal systems are possible, single signals are quite common in many temperature/thermal systems.

Data acquisition systems may have two main structures which are both included in Figure 8.1. One system has parallel conditioning and serial sampling. In the other, the sampling is taken from the common path and inserted into either all conditioning

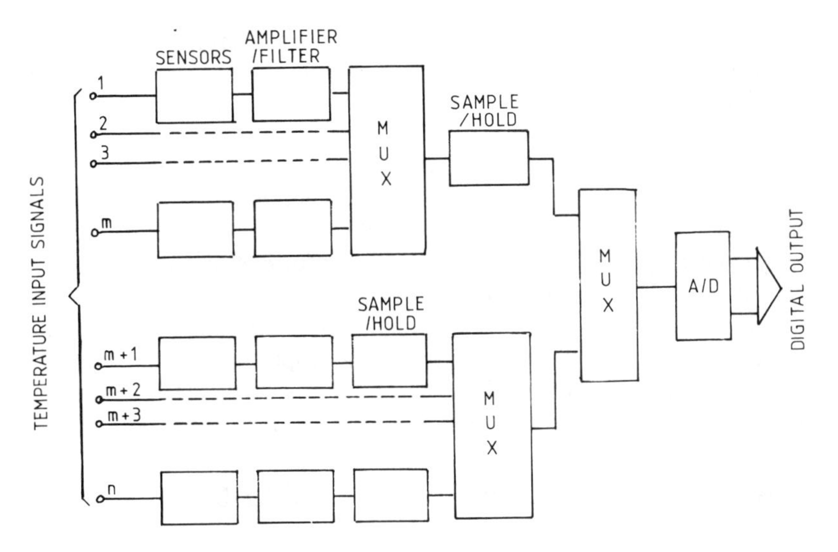

Figure 8.1
Block diagram of a complex data acquisition system.

paths or into a selected number of them. Recognizable functions are circuit completion and conditioning, multiplexing, pre-amplification/filtering, sample-and-hold, possible post-amplification with gain control, followed by analogue-digital conversion.

8.2.2 *Conditioning and amplification of temperature sensor signals*

The resistive modulating group of sensors, which must be supplied from an auxiliary or support energy source, may also require conditioning to remove bias. A classification of resistive sensor conditioning (Henderson and McGhee, 1990) indicates that auxiliary energy may be supplied from constant voltage or constant current sources. Bias removal is also accounted for by this means. Voltage sources, which may be used for conditioning purposes, use battery cells or rectified a.c. supplies regulated by integrated circuit voltage regulators based upon compensated zener diode or band-gap reference types. They are widely available as separate chip units or as an additional facility on multiple function chips. Bias removal, using deflection bridge circuit methods, may employ voltage or current sources with any one of the appropriate forms of multiple wire compensation techniques. As these bridges introduce non-linearity on their own account, any non-linearity of the sensor itself will be compounded. Constant current conditioners, which are based upon the seminal IC current mirror due to Widlar, now take more sophisticated forms (Hart and Barker, 1977).

The signal levels available from modern electrical temperature sensors, although not too low to be of immediate use, are frequently amplified using high gain instrumentation amplifiers. This amplification makes them less sensitive to interference and also permits more versatile handling of the signal. Many significant developments, which have taken place recently in the design and manufacture of integrated circuit operational amplifiers, include effective isolation barriers, chopper stabilization, auto-zeroing, reduced noise and improved transient performance. Input offset voltage and current are the two most common temperature dependent parameters of all operational amplifiers (van Putten, 1988). Special design methods are now being used to minimize the on-chip sensitivity to common mode temperature variations using common mode feedback. In most systems, where drift compensation is not possible, special care is taken to reduce errors from the temperature dependence of these two parameters. They are especially undesirable in temperature measurement systems.

A good example of a specialized IC element for conditioning thermocouple signals is the AD594/595 thermocouple amplifier by Analog Devices, USA. It provides a high level linearized output signal with automatic reference junction compensation as well as precalibration to match J- or K-type thermocouples. A resolution of 0.5 °C is accompanied with a sensitivity as high as 10 mV/°C. Using the AD7571 analogue-to-digital converter allows the design of a high resolution digital thermometer (Reidy, 1985).

8.3 TRANSMITTERS OF TEMPERATURE SIGNALS

Signal interfacing methods cover transmission in serial or parallel mode. Serial mode transmission is appropriate for both analogue and digital signals. In the case of analogue signals the interfacing, specified in BS 5836, is usually current loop of 4–20 mA or 0–20 mA, which are the conventional forms used in the process industry. With the increasing importance of digital communication methods and their commensurate economy in data intensive environments, there is a tendency to incorporate digital technology in modern complex process systems. As the twisted-pair cabling for current loop operation is suitable for serial digital transmission, it is feasible to upgrade from current loop operation to the digital method without the need for complete re-cabling. It seems likely that analogue transmitters will be displaced by digital types in the near future.

The technology of serial digital transmission, which is described by Lange (1991) and Maine (1986), for example, is well developed. In laboratory or small workshop installations data rates rarely exceed 20 kbits/s over distances up to about 15 metres. The Electronic Industry Association Standard, EIA-232D, which was formerly well known as RS-232C (Watson, 1988; Stephens, 1989), is probably the best choice for such applications. This standard is broadly compatible with CCITT V-24, V-28 and IS 2110. When faster transmission rates between more remote locations are required, then a more appropriate data communication interface is likely to be met in the specification of RS-449 (Weissberger, 1982) and its accompanying electrical specifications RS-422 and RS-423. This standard, which allows compatible operation with EIA-232D, is especially applicable if there is a need to upgrade an existing EIA-232D system. More recently, proposals to develop the serial interfacing Fieldbus system for complex manufacturing automation indicate a promising future. This interfacing technology has a flexible hierarchical structure (Gater, 1988; Warrior and Cobb, 1989; InstMC, 1990). As it allows the application of data communication techniques in factory installations, it will be important in the manufacture of products which require thermal treatments.

Parallel interfacing, as well as other methods (Dubbeldam *et al.*, 1989), are suitable for small scale laboratory installations, such as calibration environments. When many

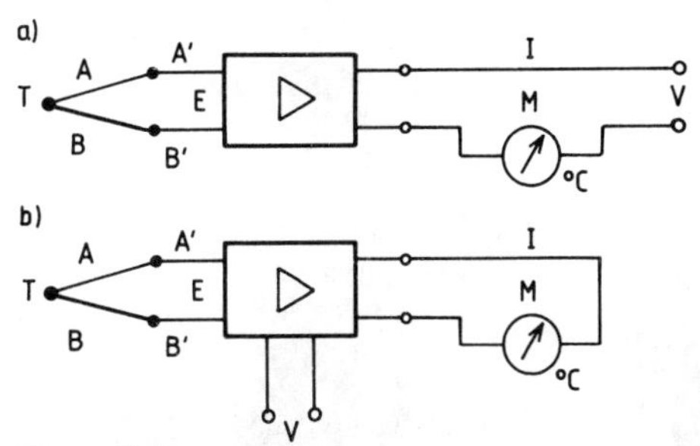

Figure 8.2
Thermometric thermometer with (a) two-wire and (b) four-wire transmitter.

different kinds of instruments, which may transmit data as 'talkers', receive data as 'listeners' or perform both functions, are employed, they usually have widely differing communicating abilities. A parallel interface which is compatible with this diversity through a highly structured well disciplined control is embodied in the IEEE-488 interface standard (Maine, 1986; Lange, 1991).

Process transmitters, similar to those based upon BS 5836, which still represent the most widely used forms, may be two-wire or four-wire transmitters (Figure 8.2). Depending on design, transmitters may be rack mounted, as in control rooms, field mounted (Bozarth and Hurd, 1972), or sensor head mounted. Field and head mounted transmitters commonly endure the most arduous field conditions while head mounted types additionally include reference junction compensation for thermocouple inputs.

Most thermocouple transmitters are based on the principle of variable current potentiometers (§4.8.4). A schematic diagram of such a unit, which is given in Figure 8.3, shows that the output signal is proportional to the measured e.m.f., E_x. The voltage drop across the resistor, R_c, caused by the current I_c, compensates the measured e.m.f., E_x, so that the input voltage, V_x, of the operational amplifier is:

$$V_x = E_x - I_c R_c \tag{8.1}$$

Defining the amplifier gain as:

$$K = \frac{I_c}{V_x} \tag{8.2}$$

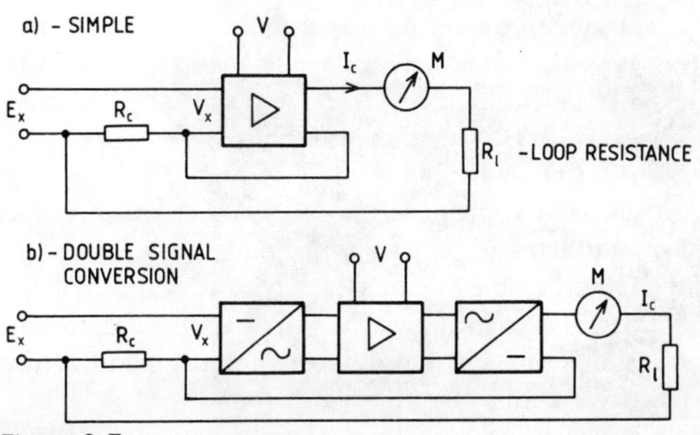

Figure 8.3
Four-wire thermocouple transmitter.

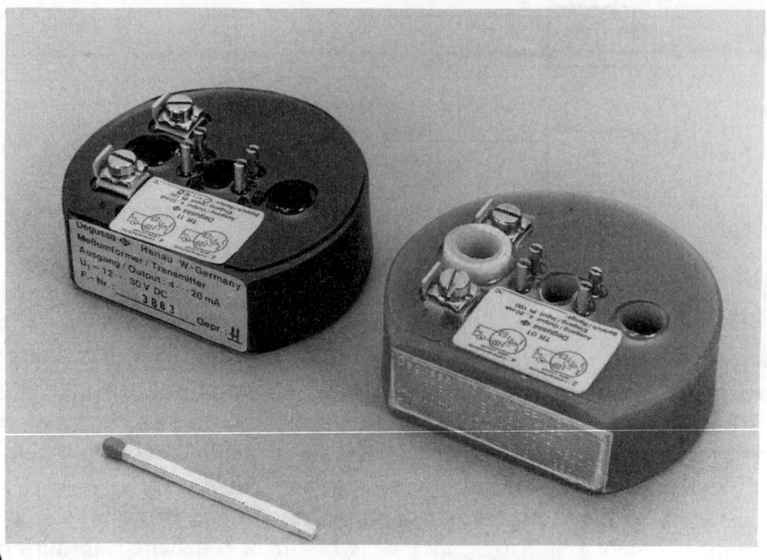

(a)

(b)

Figure 8.4
Two-wire transmitters. (a) General view, (b) mounted in a sensor head. (Courtesy of Degussa AG, Germany.)

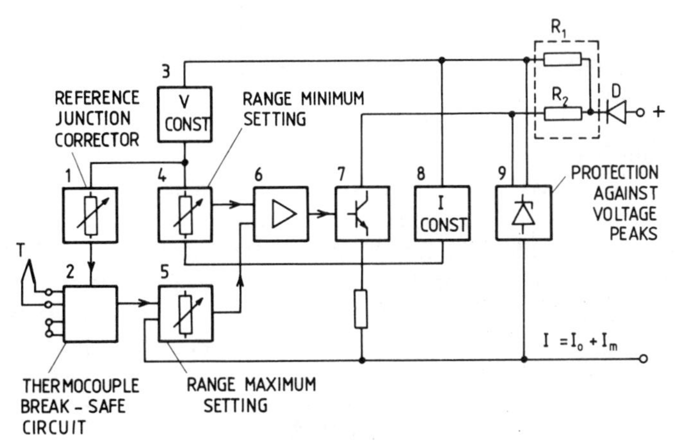

Figure 8.5
Block diagram of a two-wire TT01 transmitter for thermocouples. (Courtesy of Degussa AG, Germany.)

from Equations (8.1) and (8.2) one gets

$$E_x = I_c\left(R_c + \frac{1}{K}\right) \tag{8.3}$$

Assuming a high amplifier gain, K, Equation (8.3) can be rewritten as:

$$I_c \approx \frac{E_x}{R_c} \tag{8.4}$$

The compensating output current, I_c, is thus directly proportional to the measured e.m.f., E_x.

To eliminate the influence of the amplifier zero drift, there are also transmitters, as shown in Figure 8.3(b), with double signal conversion, having all three blocks forming an integrated circuit. As an example the epoxy encapsulated, head mounted two-wire transmitter from Degussa AG (Germany) (1988), which has the external view shown in Figure 8.4, has a diameter of 62 mm and is 20 mm thick. This transmitter, with the schematic diagram given in Figure 8.5, is primarily intended for use with thermocouple types NiCr–NiAl, Fe–CuNi and Pt10Rh–Pt. It is also compatible with all other standardized thermocouples as an option. Amplification of the thermocouple signal in 6 is followed by conversion in 7 into an output current, proportional to the measured e.m.f. The transmitter's measuring current, I, is composed of two components so that:

$$I = I_0 + I_m \tag{8.5}$$

The current, I_0, in most cases 4 mA in the 4 to 20 mA systems, which is used to supply the necessary power of two-wire systems to the transmitter, is kept constant by a current stabilizer 8. The component, I_m, is a function of the measured temperature. An advantage of 4 to 20 mA systems is that a transmitter failure, corresponding to the occurrence of zero output current, cannot be mistaken for a zero measured temperature value. The reference junction corrector 1, adjusted to match the thermocouple characteristic, simultaneously feeds the thermocouple break safe circuit 2. In the case of a thermocouple break the output current either falls to zero or increases over the 20 mA high limit, to a value of 35 mA in most transmitters. The voltage stabilizer 3 and the resistors of the circuit 4 determine the minimum measuring range, whereas the circuit 5 is used to set its maximum value. Both output current components flow through the diode, D, protecting it against any reversal of supply voltage, while protection for the transmitter against any voltage peaks is provided by the other zener diode 9. The transmitter, which also has an electronic current limiting circuit, is supplied with 12 to 30 V d.c. for a loop resistance, R_l, at the output. Knowing the supply voltage, V_s, allows R_l to be calculated from:

$$R_l \leqslant \frac{V_s - 12\ \mathrm{V}}{20\ \mathrm{mA}} \tag{8.6}$$

Accuracy is better than 0.2% of the upper current value (20 mA) of the transmitter at $V_s = 20$ V, in an ambient temperature of 20 °C. The operating temperature range is −25 to +90 °C for normal error limits or −50 to +110 °C for double error limits.

In the operation of different measuring instruments, a deciding factor is their ability to be used in hazardous areas such as occur in the chemical and petrochemical industries, in mines and also in quarries. Different techniques, developed for coping with these conditions, include enclosing in oil, high pressure inert gas or sand. In all cases it is essential to ensure that no explosion provoking spark may occur in the hazardous area. The technology which is associated with the design of safe instruments for use in hazardous environments is now well known as intrinsically safe design. A circuit is intrinsically safe when any spark which occurs either in normal operation or under fault conditions, cannot possibly cause thermal ignition towards explosion of a given atmosphere. In the described transmitter the resistors, R_1 and R_2, render it intrinsically safe.

The increasingly pervasive use of electronic instrumentation for temperature measurement in potentially hazardous environments demands the use of intrinsically safe equipment. Detailed consideration of this topic will not be given in this book. From the wide selection of literature in the field the reader is referred to Hutcheon (1989a, 1989b), IEE (1983), InstMC (1986), MTL (1983), Tortoishell (1990).

A transmitter type TR01 of Degussa for resistance thermometers is shown in Figure 8.6. As in the TT01 for thermoelectric signals (Figure 8.5), the temperature dependent resistance variations of the RTD are converted into an output current, linearized in current source 1. The input signal is fed through the temperature range module 2 and the amplifier 3 to the output current converter 4. A component of the current, I_0, smaller than 4 mA is set by the current stabilizer 5. Other parts are as in the TT01 thermocouple transmitter.

Two-wire, three-wire or four-wire connections are used for circuit completion between the RTD and transmitter. Although using a two-wire connection between the RTD and transmitter is a convenient method of circuit completion for head-mounted transmitters, it does not give precise measurements. It also imposes a limit upon the loop resistance variations. As three-wire and four-wire connections between the RTD and the transmitter ensure the compensation of loop resistance variations they are normally used in the case of larger distances between the sensor and the transmitter.

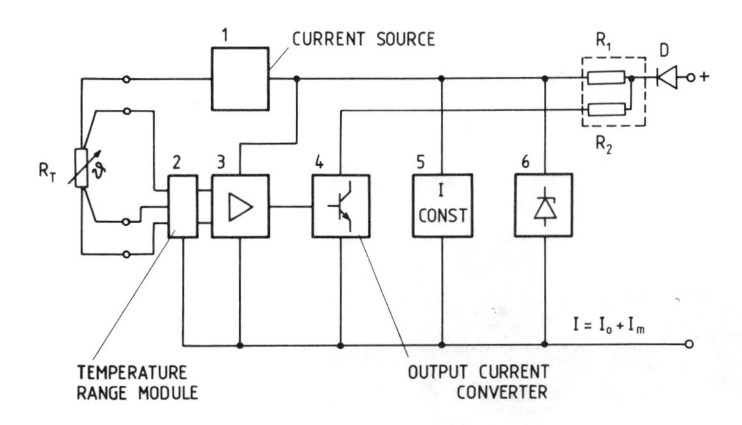

Figure 8.6
Block diagram of a two-wire transmitter TR01 for RTDs. (Courtesy of Degussa AG, Germany.)

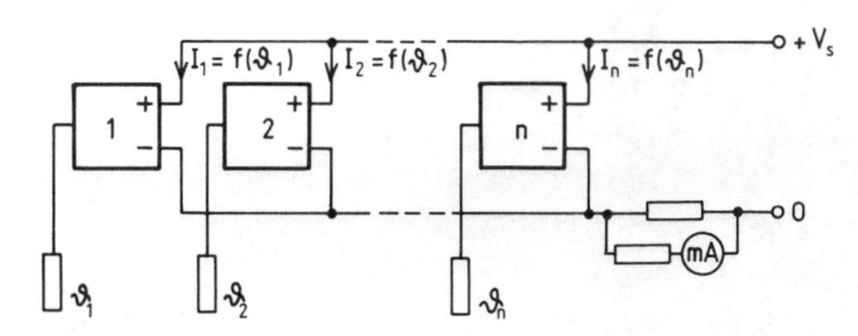

Figure 8.7
Measurement of mean temperature value by using *n* sensors and transmitters.

Transmitters can also be used for measuring the mean value of the temperature of *n* temperature sensors, as shown in Figure 8.7, provided galvanic isolation can be ensured between transmitters. This isolation may be achieved using transmitters with double signal conversion. Figure 8.8 illustrates transmitters with additionally built in analogue or digital indication. There are a number of advantages of using transmitters as additional elements in data logging or recording compared with direct logging or recording. In general, the signals available from transmitters are of a sufficient strength to give them immunity from electromagnetic interference even over the longer transmission distances (up to some kilometres) which they also allow. They may also provide good common-mode rejection through the use of isolation barriers. As they can withstand fairly high temperatures they allow economic use of compensating and reference junction devices in the case of thermocouple thermometers. Linearization of RTD signals can be accomplished with relative ease. Because of the low levels of energy stored by them they are eminently suitable for use in hazardous areas. Finally, their signals may be easily interfaced to computers.

An example of a temperature transmitter for pyrometers is the Modline 4 from Ircon Inc. (USA) shown in Figure 8.9. This transmitter, which is placed in the housing of a photoelectric spectral pyrometer of the Modline series, requires a supply voltage of 16 to 40 V d.c. giving an output signal of 4 to 20 mA, linear with measured temperature. A digital potentiometer enables the setting of the relevant value of

Figure 8.8
Temperature transmitter with a digital temperature display. (Courtesy of Hartmann & Braun AG, Germany.)

emissivity to between 0.1 and 0.99. The two-wire system used provides a robust signal, which is not influenced by environmental disturbances, allowing the transmitter to operate in the ambient temperature range from 0 to 60 °C. Its measuring ranges are from 0 to 250 °C up to 300 to 1300 °C.

Further development of temperature transmitters has led to what are now generically called intelligent transmitters described by Vincent (1987). For example, the 3044 Temperature Transmitter by Rosemount Ltd, shown in Figure 8.10, can accept inputs from a variety of temperature sensors such as Pt-RTD or different thermocouples (B, E, J, K, N, R, S, T-types) without changing any modules or altering any links or switches. This microprocessor based transmitter can be programmed by a remote transmitter interface which also enables the measuring ranges to be changed. The output signal can be a linear or any other function of measured temperature. Galvanic separation of input, output and earth is an important feature of the transmitter. An accuracy of + 1% of the reading over the entire working range of the sensor is claimed for the transmitter which is resistant to any airborne radio-frequency interference (RFI) owing to the included effective filtering. This last quality is very important in view of the increasing use of radio transceivers for on-plant personnel communication.

The schematic diagram of the transmitter is given in Figure 8.11. Selection of a sensor type causes the analogue switch network to gate the relevant sensor signal. The same network also gates the reference resistor, R_r, or the reference voltage, V_r,

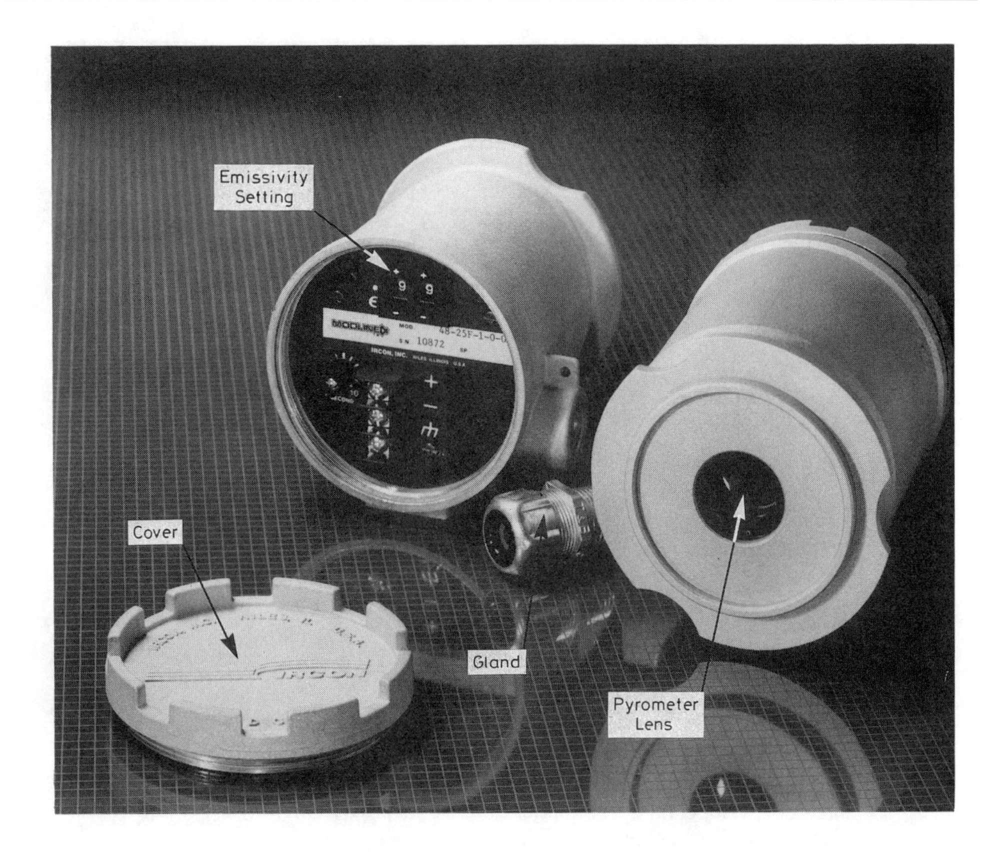

Figure 8.9
Temperature transmitter, Modline 4, built into the housing of a pyrometer. (Courtesy of Ircon Inc., USA.)

to the output circuit, which conditions the sensor signal. After conversion by the voltage/frequency converter, the frequency signal is passed through an isolating transformer to the microprocessor. Between samples from the sensor, the resistor, R_r, voltage, V_r, and short circuit are checked to calculate the circuit drift owing to ambient temperature variations or ageing of elements. The microprocessor which then corrects these influences, also compensates for the influence of changing reference junction temperature, if a thermocouple is used. Either a digital or an analogue 4 to 20 mA d.c. output signal is provided by the microprocessor. For computer interfacing, the 3044 transmitter can also communicate digitally over the 4–20 mA lines using a communications protocol based on the Bell 202 standard. Modern temperature transmitters are very reliable, reaching up to 12 000 working hours which is about eight years.

It is to be expected that digital temperature transmitters will replace the 4–20 mA standard in the foreseeable future. A review of transmitters, produced at present, may be found in 'Markt und Technik' (1987).

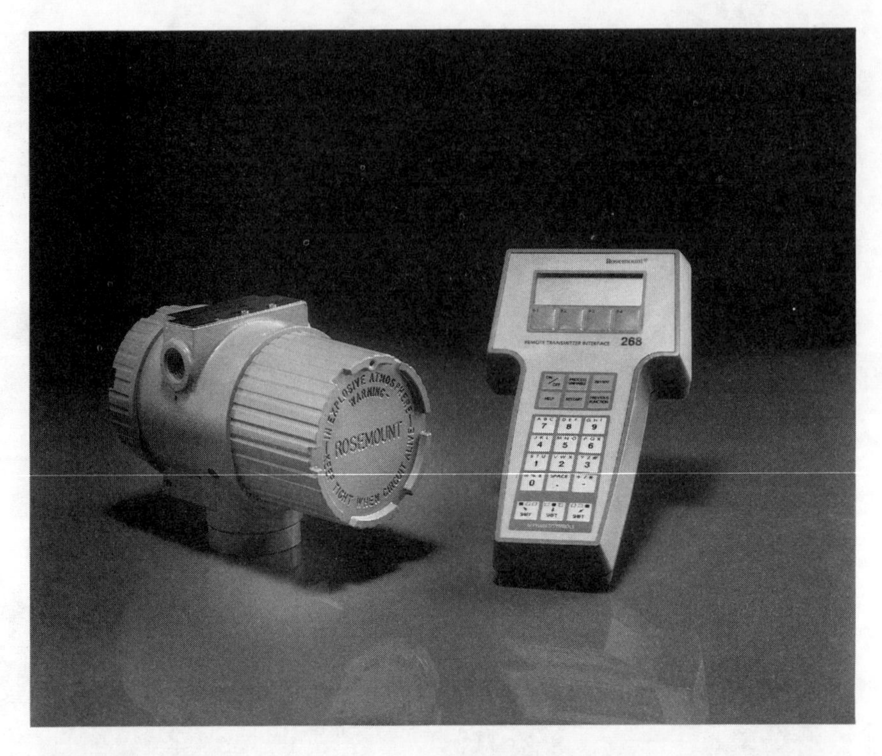

Figure 8.10
Intelligent Temperature Transmitter type 3044 programmed by a Remote Transmitter Interface. (Courtesy of Rosemount Ltd, USA.)

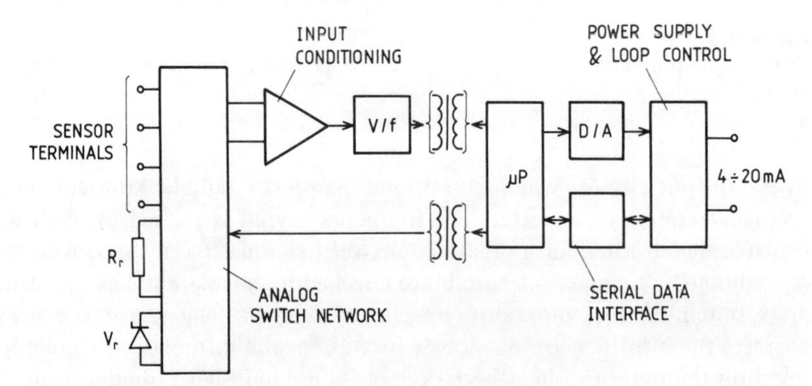

Figure 8.11
Block diagram of 3044 Transmitter shown in Figure 8.10. (Courtesy of Rosemount Ltd, USA.)

8.4 DIGITAL INDICATING INSTRUMENTS

Digital temperature indicators, which include single chip meters (Wyre, 1989), are becoming more and more popular because of their decreasing cost and increasing accuracy, reliability and versatility. Although they may use a variety of indicator technologies (Bylander, 1979), the most widely used is LED technology. Panel

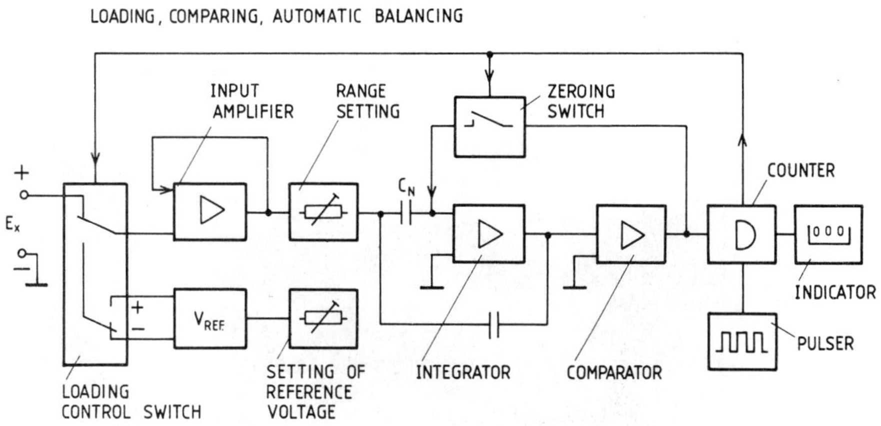

LOADING, COMPARING, AUTOMATIC BALANCING

INPUT AMPLIFIER RANGE SETTING ZEROING SWITCH COUNTER C_N E_x INDICATOR V_{REF} PULSER LOADING CONTROL SWITCH SETTING OF REFERENCE VOLTAGE INTEGRATOR COMPARATOR

Figure 8.12
Block diagram of a digital indicator based on the dual-slope or double integration principle. (Courtesy of Siemens AG, Germany.)

mounted digital indicators, designed for either discrete panel mounting or stacking with other similar units, require a panel cut-out with dimensions to DI standards ranging from 48 × 24 mm to 144 × 72 mm.

Distant reading is allowed by large 90-mm character indicators with cut-outs of, for example, 600 × 217 mm (Juchheim GmbH, Keller Pyro GmbH, Germany). It is usual to employ double integration or dual slope analogue-to-digital conversion methods for thermal or temperature signals. This conversion technique ensures very efficient attenuation of disturbances at the mains frequency or its harmonics (Lange, 1987). Multiple slope techniques, which may compensate for drift, are used less frequently. However, when high precision voltage measurements are required as part of a temperature measuring instrument, other high precision types are appropriate (Chenhall, 1987). Fuller consideration of analogue-to-digital conversion appears in many sources (Loriferne, 1982; Seitzer *et al.*, 1983). McGhee and Henderson (1989) have applied the science of classification to the grouping of analogue-to-digital converters.

The typical layout of a digital temperature indicator, shown in Figure 8.12, is an example of a Siemens AG (Germany) indicator. During the first integration lasting for a fixed time, the integrator charges to a value proportional to the measured value of temperature. A run-down phase then follows during which the charge decreases until the comparator reads a zero voltage at the integrator output. Thus the duration of this discharging phase is directly proportional to the measured value of temperature. Meanwhile, during this run-down phase a counter, whose initial value is zero, counts up clock generator pulses. The number, finally stored after completion of the run-down, is then converted into a BCD signal which drives a seven-segment display element. Before each measuring phase the zeroing of the analogue-to-digital converter is checked. If the integrator output is not zero the comparator generates an offset voltage across the capacitor, C_N, which compensates for any zero deviation. The digital temperature indicator in Figure 8.13, which is based on this principle is a

Figure 8.13
Typical digital temperature indicator. (Courtesy of Siemens AG, Germany.)

Siemens AG 144×72 mm 3½ or 4½-digit limit monitor with 14-mm high characters. It is designed for Pt-100 Ω resistance sensors or for thermocouples types J, K or S, with input linearization. The display starts to pulsate when a thermocouple break occurs. LED indicated alarms are triggered by values occurring outside the range between preset maximum and minimum values. These are adjustable over the whole range of the instrument, which also has a digital BCD output. Typical data for average temperature indicators are given in Table 8.1.

A new class of *portable*, small size *digital temperature indicator* has recently become popular. The technical data of these indicators are comparable with those of stationary indicators. An example is the family of instant action thermometers by Testoterm GmbH (Germany). The 7010 type, shown in Figure 8.14, is a microprocessor based, battery or mains operating indicator for Pt-100 Ω resistance detectors. A similar 9010 type is intended for K, or J-type thermocouples or NTC thermistors. The 7010 instrument is designed for two independent temperature inputs with simultaneous indication of both sensed values, or their difference, as well as minimum and maximum values together with store and hold functions. Its accuracy ranges from ±0.1 to ±0.2 °C (±1 digit) with a rate of 1 measurement per second. Two surface probes, an air probe, an immersion probe and a penetration probe, shown from left to right in Figure 8.14, represent the range of different temperature sensors which may be supplied. An optional plug-in logger for storage and printing, also shown in Figure 8.14, has a storage capacity of up to 2500 measured values. The measured

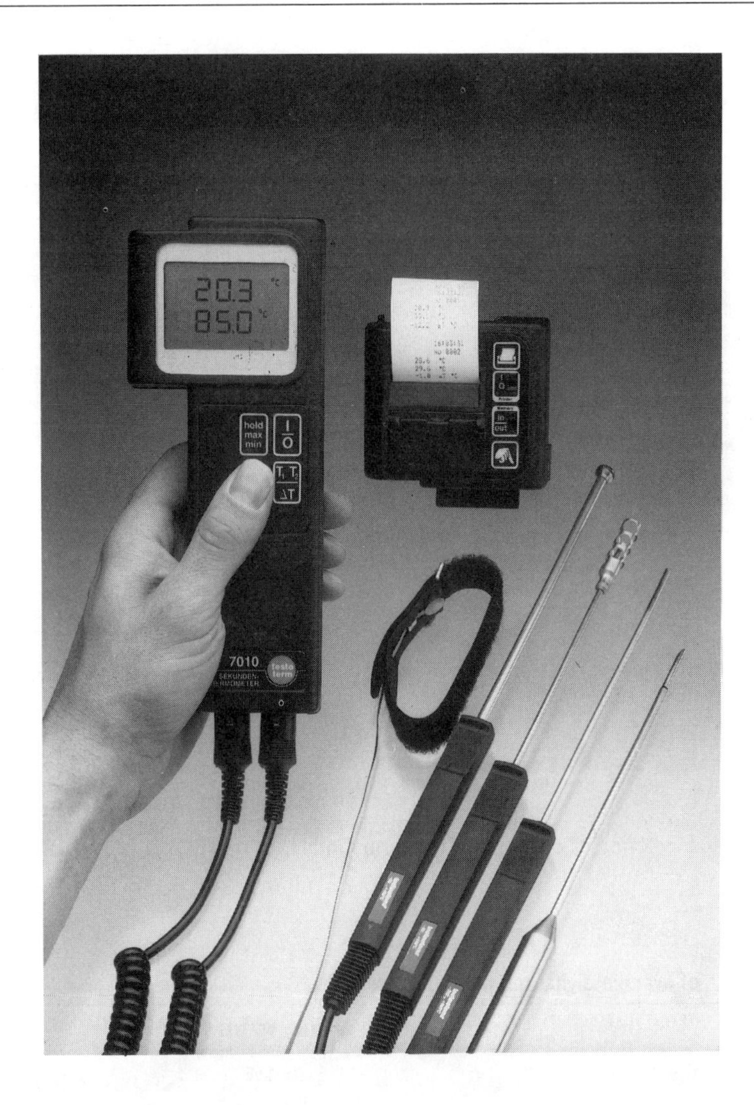

Figure 8.14
Portable digital temperature indicator with printer 7010 and temperature probes. (Courtesy of Testoterm GmbH, Germany.)

data, which can be recalled from memory at any time, can also be indicated and printed individually. An indicator, 'Pyrotherm 8500', of the same size, is intended for a pyrometer sensing head. One of the smallest digital indicators by 'Testoterm', shown in Figure 8.15 which has the two temperature ranges, -10 to $+110$ °C or -40 to $+50$ °C, is used in laboratories, food technology, refrigeration and air conditioning. A small size solar digital indicator is also produced by 'Testoterm' GmbH.

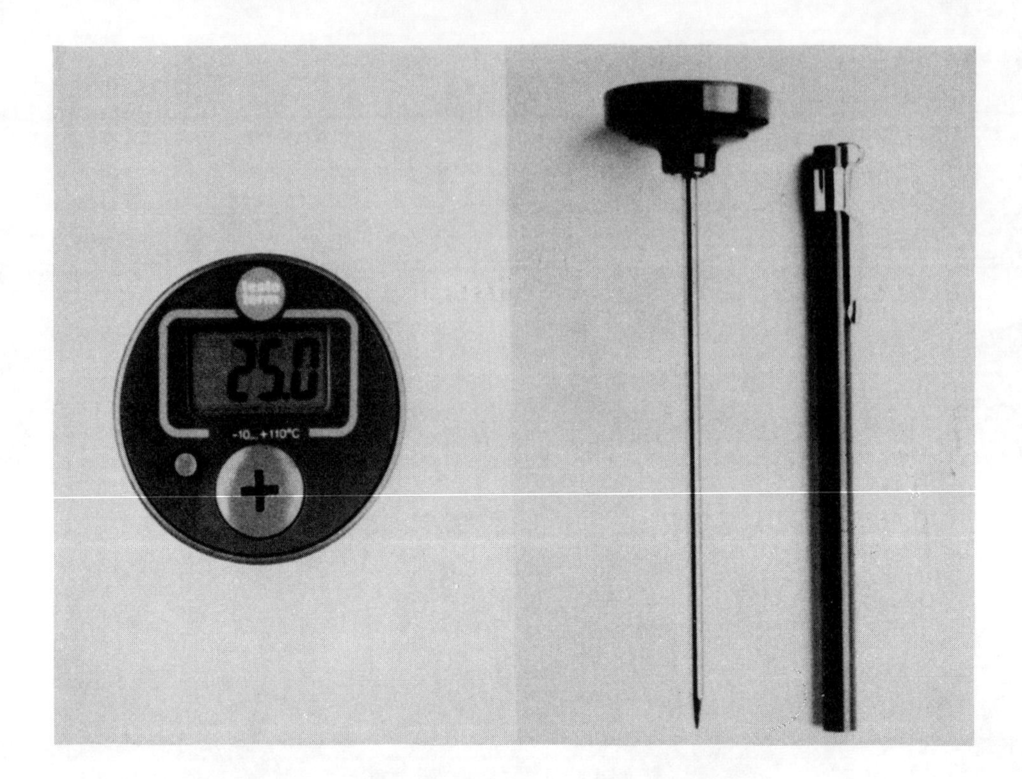

Figure 8.15
Small-size digital indicator. (Courtesy of Testoterm GmbH, Germany.)

Table 8.1
Technical data of average digital temperature indicators.

Parameter	Value, status
Number of digits	3½ to 4½
Digit height	14 mm
Measuring rate	2.5 to 4 per sec
Accuracy	±0.5% FSD + 1 digit
Resolution	0.1 to 1 °C
Ambient temperature drift	0.015 to 0.030%/°C
Operating temperature range	0 to 55 °C
Power supply	110/220 V a.c. ± 10% 50–60 Hz or 24 V ± 25% d.c.
Power consumption	2.5 to 8 W
Analogue output	4 to 20 mA d.c. or rarely 0 to 20 mA d.c.
Automatic polarity switching	Existing
External point shift	Existing
Thermocouple break or short-circuit indication	All digits blanked or pulsating
Input linearization	Optional
Reference junction compensation	Existing in thermocouple thermometers

A single purpose dedicated, digital microprocessor based indicator-controller Maxiline, which is offered by Ircon Inc. (USA) (Figure 8.16), can handle either one or two pyrometric sensing heads, having quite different temperature ranges and operating wavelengths. As well as operating as part of a digital process control system or with a computer, it also has a full range of analogue outputs. A unique feature is its continuous self-diagnostics and continuous autocalibration. Relevant LEDs indicate both OK and fault status or even if the properties of the head are changed. The readings are displayed in 20-mm high digits with an LED bar graph included for analogue indication of varying temperatures. All other signal conditioning possibilities, as described in §7.7, are also provided.

8.5 DATA LOGGERS

Data loggers and *scanners*, specially designed for temperature measurement, now present particularly versatile instruments. They can accept input signals from many different temperature sensors. The microprocessor based Model 740 system scanning thermometer from Keithley Instruments Inc. (USA) (Figure 8.17), which is a typical example, includes linearization of the seven most common thermocouples (types J, K, T, E, R, S and B), as well as a mV range. Mixing of these in any combination is also a feature of the Model 740. Its scanner card allows scanning of 1 to 9 channels at a rate of 20 channels/s. Using additional 705 Scanners the system can be expanded up to 81 channels. The built-in non-volatile data memory enables 100 single channel locations to be stored, or up to 81 different channels with channel number and time of day. This microprocessor based instrument incorporates eight LED digits of 12-mm characters in its display of decimal point, polarity, measured value, memory location, overflow, open thermocouple warning, IEEE-bus status, channel number with corresponding thermocouple type or mV indication. The indication accuracy is 0.02% ($+3\,\mu$V) for mV range and ± 0.4 to ± 1 °C for thermocouple ranges. Battery back-up maintains data in memory for up to one week.

A still more versatile and more expansive family of data acquisition and logging systems is offered by Hewlett-Packard (USA). They are operated by small hand-held or stationary computers. For instance, the HP 3056DL data logger (Figure 8.18) gives a choice of 17 separate inputs such as 2-wire RTD, 4-wire RTD, 2.2 kΩ thermistor, J, K, T, E, R, S-type thermocouples and so on. Each input is selected by pressing a key on the incorporated HP85 personal computer. Any other linearization equations or Basic subroutines can be generated to extend the flexibility even further. A monitor with graphics software helps in displaying the logged results. Adaptive data logging allows the scanning rate to be changed in preset out-of-limit conditions. A built-in printer, serving both data printing and hard-copy graphics, is one of the many options which can widen the application range of the system. The basic resolution of the analogue-to-digital converter is 5½ digits. In the case when the sampling rate is increased to 30 readings per second up to 56 channels can be monitored with a corresponding reduction in resolution to only 3½ digits. The normal sampling rate

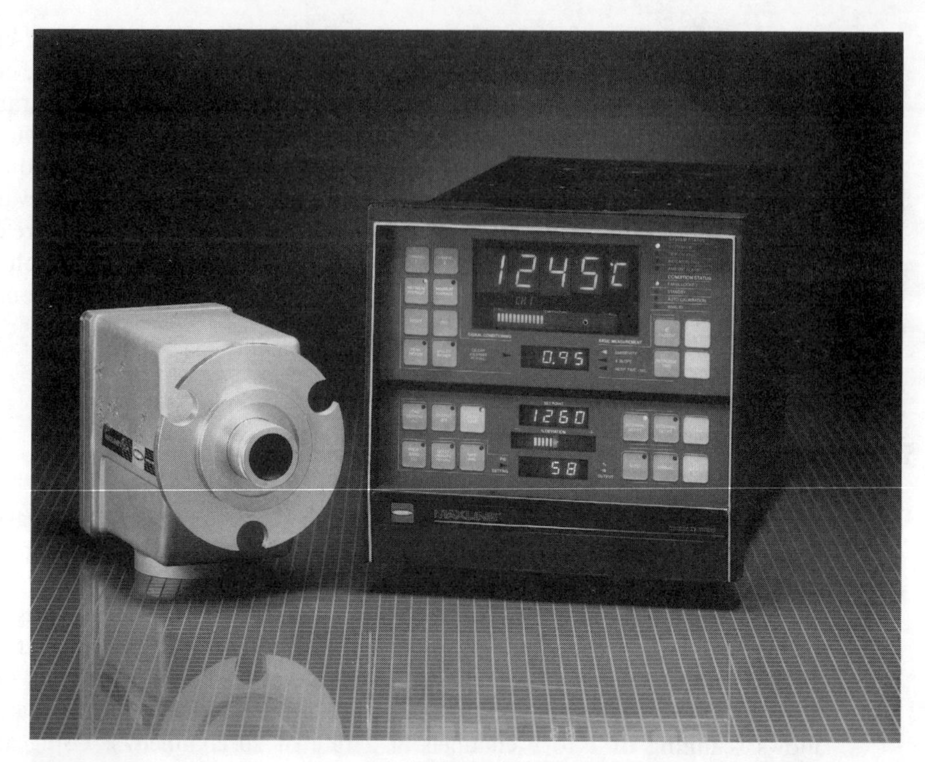

Figure 8.16
Maxiline type of microprocessor based indicator—controller, for pyrometers. (Courtesy of Ircon Inc., USA.)

Figure 8.17
Model 740 System Scanning Thermometer. (Courtesy of Keithley Instruments Inc., USA.)

Figure 8.18
Data logger HP3056DL. (Courtesy of Hewlett Packard, USA.)

for thermocouples is 0.9 readings per second with a long-term accuracy of about 1.5 °C including reference junction.

For very large plants, with up to 1000 or even 4000 monitored inputs, special computerized, universal systems, such as the Magna 2000 Monitoring System by Leeds & Northrup International (USA), can be used. This system comprises a central unit with colour monitor display and hermetically sealed reed relays for switching. Up to four operator consoles, with LED display, show the channel number, measured variable and point status. A 21 Mbyte, 5¼ inch hard disk is provided for data storage in the system which is designed for thermocouple and RTD inputs or for mV, V and mA as well. Recorder compatible analogue or digital output is also available.

A microprocessor based data logger for special applications is a 'testastor' by Testoterm (Germany) produced in two main types. Testastor 9500 (Figure 8.19(a)) which is designed for type K or S thermocouples, can operate in the ambient temperature range up to 140 °C for a time up to 6 h. This totally self-contained battery-operated apparatus can store up to a total of 1980 measured values in 1 to 4 channels. Any scanning rate between 4 s to 1 h can be set for one measuring cycle. Testastor, which can be used for temperature monitoring in sterilization or pasteurization processes or in any industrial processes, may be stationary or may move along with the charge in continuous processes. A special desk computer with a printer can be supplied for logger programming and read-out as shown in Figure 8.19(a). Time or temperature may be used to trigger the measuring cycle and rate. The same logger with additional super-heat protection can be used up to 1300 °C. A similar flat logger, the 'testastor 9510' (Figure 8.19(b)), operating up to 250 °C, has four channels and can store up to 1980 measuring values. Three different measuring rates can be programmed in advance, the shortest one being 0.25 s per channel.

Figure 8.19(a)
Heat withstanding data logger 'testastor.' (a) 9500-type, with programming and read-out computer, (b) 9510-type for four channels. (Courtesy of Testoterm GmbH, Germany.)

8.6 ANALOGUE RECORDERS

8.6.1 Introduction

Temperature recording by directly visible traces using analogue recorders is preferred to digital recording in many research or industrial applications. The main reason for this is the 'at-a-glance', direct and easy visibility allowed by them. The possibilities of observing trends and gradients and of quickly comparing temperatures at some chosen points is also commendable. Easy, compact and economic data storage are other benefits, although the inability to reconstruct the signal is a disadvantage. In spite of the dramatic development of digital techniques, analogue recorders using technology which is quite old are and will continue to be widely used.

The characteristic parameters of analogue recorders are listed in the CEI Publication No. 484 (CEI, 1974). Those which describe the recorder dynamics, are:

- 95% time (also 98% and 99% time),
- transference,

Figure 8.19(b)
Caption opposite

- maximum pen velocity,
- maximum pen acceleration,
- maximum overshoot for a step input,
- response time for a step input.

When using temperature sensors of very low thermal inertia, the main sources of all dynamic errors in recording are due to the dynamic lag or inertia of the recorders themselves. For this reason, analogue recorders will be described according to their ability to cope with the recording of quickly changing temperatures. In paper chart recorders this also means the ranges over which chart speed may be increased. Chart recorders are reviewed by Amrein (1980), Morris (1980) and Wilkinson (1988).

8.6.2 Non-electric recorders

Non-electric recorders are based on manometric thermometers (§3.4). Owing to their high deflecting torque the recorder pen can be fastened directly to the pointer, assuring continuous recording. In actual practice circular chart recorders, with one chart 'rotation' every 12 or 24 h, have a typical measuring range of -80 to $+600$ °C to a recording accuracy of $\pm 2\%$ FSD. Manometric thermometer recorders are very robust and vibration resistant. If the paper chart is moved by a spring clock mechanism they are then especially suitable for use in hazardous areas.

8.6.3 Electric deflection type recorders

Deflection type printing recorders, used in industry to record slowly changing temperatures, are based on millivoltmeter or quotient instruments. Their small deflecting torques do not allow the recording pen to be directly attached to the pointer. They are usually manufactured as 1, 2, 3, 4, 6 or 12 point recorders to the working principle explained in Figure 8.20. The pointer, which is periodically pressed to the inked ribbon by a tapper bar, marks coloured dots on a moving paper chart. In switching from one measuring point to another the inked ribbons move synchronously, so that each point is matched with a different colour at a printing rate of 10 to 30 s. Chart speed settings between 20 to 1200 mm/h are accompanied by a recorder accuracy of 1.5 to 2.5%, over a chart width of 100 to 150 mm. Similar recorders with an electronic input amplifier, allow a continuously writing pen to be directly attached to the pointer.

8.6.4 Potentiometric recorders

Automatic potentiometric recorders for thermocouples and pyrometers as well as automatically balanced bridges for resistance sensors, comprise the most important family of analogue recorders. Some of them have been briefly described already in

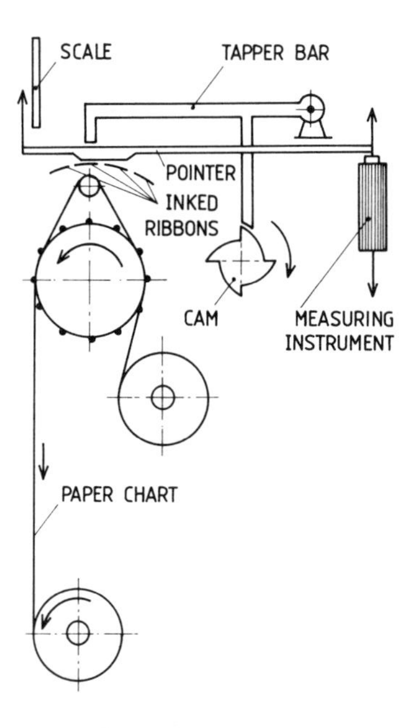

Figure 8.20
Working principle of deflection type printing recorder.

§4.8 and §5.7. The high driving torque of the balancing servo-motor enables the recording pen to be moved directly. Miniaturization of the balancing servo-system, allowing several independent systems to be built in one housing, is responsible for the existence of multichannel recorders.

This type of recorder may also be classified by the different recording methods depending on the pen speed.

At speeds of 2.7×10^{-4} to 500 mm/s, pen systems used may be:

1. Ink pens with writing tips made of steel, glass or sapphire. The writing tips, requiring frequent cleaning, may tear the paper chart and flow out during paper stops.
2. Ink pens with plastic tips that have to be exchanged about every 35-km line length. They flow out during paper stops.
3. Ink pens with plastic microcapillary tip, which require to be renewed every 10-km line length, are leak-proof.
4. Ball-point pen.
5. Ink-jet printing.

At speeds up to 50 m/s other systems are:

1. Spark recording on metallized paper (Figure 8.21) by 5 to 20 V between the pen-point and the paper. Reversing the polarity may result in colour change.
2. Thermal recording by a pen with a heated point on thermo-sensitive paper.

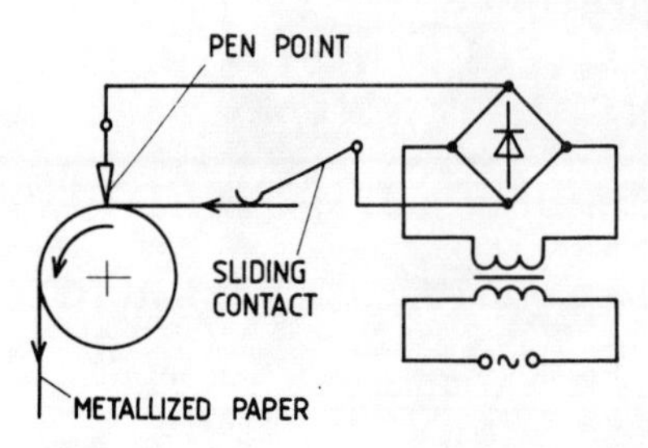

Figure 8.21
Spark recording on metallized paper.

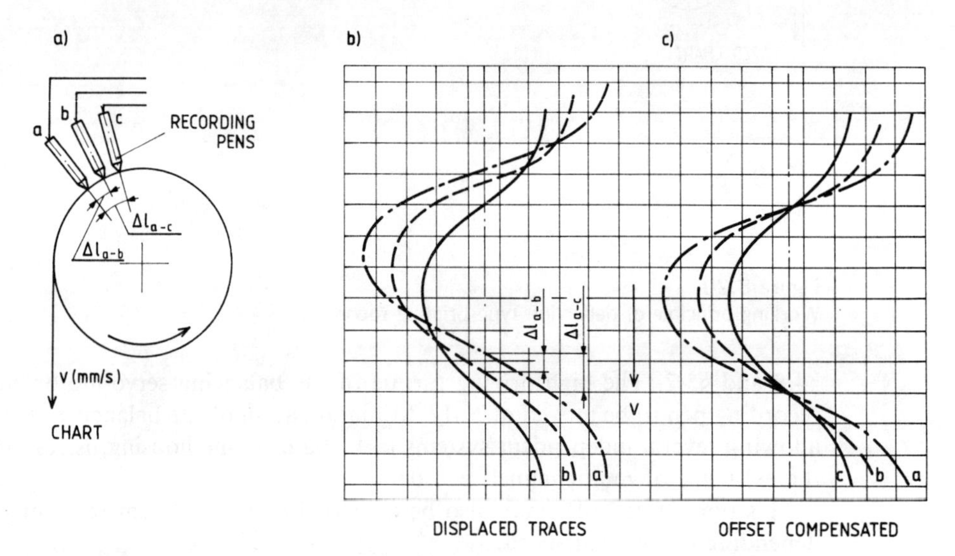

Figure 8.22
Offset compensation in a three-pan recorder.

Figure 8.22 shows a geometrical arrangement of three pens a, b and c, which have to be relatively displaced to allow them to pass each other. The recording trace of pen a lags behind that of pen b by the distance Δl_{a-b} mm and by Δl_{a-c} mm behind the recording trace of pen c. To compensate for these lags the input signals to pens b and c have to be delayed by Δt_{a-b} and Δt_{a-c}, respectively given by

$$\left.\begin{aligned} \Delta t_{a-b} &= \frac{\Delta l_{a-b}}{v} \\ \Delta t_{a-c} &= \frac{\Delta l_{a-c}}{v} \end{aligned}\right\} \tag{8.7}$$

where Δl is the linear shift in mm and v is the chart speed in mm/s.

(a)

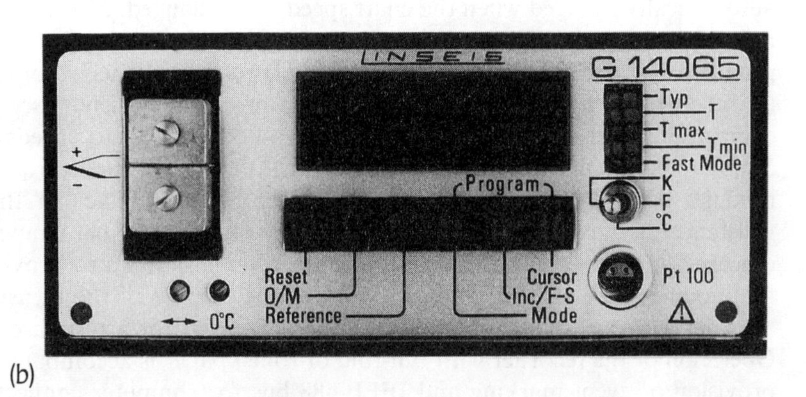

(b)

Figure 8.23
(a) L2065-Series, 6-channel, pen-offset compensated, microprocessor based potentiometric recorder, (b) an exchangeable module with a digital display for L2065-Series recorder. (Courtesy of Linseis GmbH, Germany.)

Figure 8.24
Microprocessor based circular chart recorder, type PX105. (Courtesy of ABB Kent-Taylor, UK.)

For example, when $\Delta l_{a-b} = 3$ mm, and $v = 10$ mm/s:

$\Delta t_{a-b} = 3/10 = 0.3$ s
$\Delta t_{a-c} = 6/10 = 0.6$ s.

The necessary microprocessor controlled time lags of Equation (8.7) are automatically changed when the chart speed, v, is changed. Modern potentiometric recorders equipped with pen offset compensation, similar to the L 2065-series recorders produced by Linseis GmbH (Germany), may have 2 to 6 channels recording on 250 mm (10 in) chart width (Figure 8.23). A Z-80 microprocessor, which is used for pen offset compensation, also controls the selection of 44 different chart speeds ranging from 5 cm/s to 0.1 cm/h. A modular construction allows a wide choice from among many modules which may be linearized or not for different kinds of thermocouples, resistance thermometers and thermistors. The initial and final temperature ranges are programmable, with reference junction compensation also provided. Some of the modules may also be equipped with digital display of the measured value. A recorder accuracy of 0.25% of FSD is possible even for a response time of 0.3 s. Operation of the recorder with fan-fold or rolled paper is accompanied by optional provision of event marking and IEEE-488 bus for computer connection.

Circular chart recorders, used mainly in cases where an easily stored visual record is required, are less popular. A microprocessor based recorder PX105 by ABB Kent-Taylor (Figure 8.24) is a good example. The chart, with rotation speed selectable between 1 hour and seven days in one hour steps, is driven by a microprocessor

Figure 8.25
Flat-bed recorder, LS52 Series. (Courtesy of Linseis GmbH, Germany.)

controlled stepper motor. A basic provision of 1, 2 or 3 pens is accompanied by indication, and alarm or control facilities. Input ranging, linearization, alarms, display and control outputs are also controlled by the microprocessor. Eight different standardized thermocouple inputs and resistance thermometer Pt 100 input are possible. Automatic reference junction compensation is provided. The overall accuracy of the recorder is $\pm 0.25\%$ of maximum span. The input value and units of measurement for each channel are displayed sequentially on a 20 character, single line display situated above the chart. One or two channels can be filtered with PID control, having a 0 to 20 mA analogue output.

Programming of the instrument is carried out by the operation of a sequence of tactile membrane switches: three scroll switches in conjunction with 'Up', 'Down', 'Decimal Point' and 'Enter' switches. An additional switch is used for raising or lowering the pens.

A number of flat-bed recorders, such as XY (Crook, 1980), X_1X_2Y or Xt, which are widely used in laboratories also have temperature ranges. Some of them, like the LS52 series of recorders by Linseis GmbH (Germany) shown in Figure 8.25, have interchangeable input modules for use with a number of standardized thermocouples, Pt-100 Ω sensors and thermistors. They may also have up to six channels with the same number of differently coloured independent pens. The chart, driven by stepping motor, may be set to speeds between 1000 mm/min to 10 mm/h. Chart reverse or advance is a standard feature. The recorder accuracy is 0.25% of FSD with a response time of 0.3 s.

Table 8.2
Technical data of potentiometric recorders.

Parameter	Average value	Extremal value
Indication error (% of FSD)	±0.3	±0.2
Non-linearity error (% of FSD)	±0.2	±0.1
Recording error (% of FSD)	±0.5	±0.3
Reproducibility (% of FSD)	±0.2	±0.1
Minimum span (mV)	2	0.1
Response time $t_{0.98}$ FSD (s)	1	0.16
Maximum chart speed (mm/s)	5	50
Maximum pen speed (cm/s)		250*
Channels number in printing recorders	12	24
Channels number in continuous line recorders	1÷2	6
Printing rate, time per point (s)	3÷30	1
Maximum source resistance	1÷10 kΩ	5 MΩ
Input impedance (MΩ)	2	70
Zero drift (V/K)	1	0.5
Ambient temperature limits (°C)	+10÷+40	−10÷+55
Zero suppression (in % of span)	100	500

*For X–Y recorders as slewing speed.

Table 8.2 presents typical average technical data and also best performance possibilities for potentiometric recorders.

8.7 HYBRID RECORDERS

A more frequent application of universal multi-channel hybrid analogue-digital data recorders is a generally observed trend (Anon, 1982). They offer many standardized temperature inputs for thermocouples, RTDs and pyrometers. The microprocessor based Speedomax 25000 Series, hybrid multipoint recorder by Leeds & Northrup (USA), which is a typical example (Figure 8.26), offers simultaneous data logging and alarm monitoring as well as analogue and/or digital recording. It can also display and record processes with up to 135 analogue or discrete input/output data points in various combinations. The basic recorder has a capacity of 60 analogue inputs which can be expanded by an additional 75 using an optional unit. It also provides a powerful calculation ability.

Inputs can be chosen from ten different standardized thermocouples in different temperature ranges, in °C or °F, with reference junction compensation, RTDs in °C and different pyrometric inputs in °C or °F.

Recording utilizes a 9-pin, 4-colour printing mechanism on a paper chart at a velocity which may be freely chosen between 1 cm/h and 200 cm/h. A preset alarm can activate a high-speed mode. Two modes of recording are available. The first allows a continuous trace of temperature with time using dot-filled lines. The second is an alphanumeric mode which permits the presentation of data in the form of tabular lists. Alternate or mixed modes may be chosen over preset time intervals. Many other features are included in the recording facilities. For example, scale printing, in up

Figure 8.26
Microprocessor based, hybrid, analogue-digital recorder Speedomax 25000. (Courtesy of Leeds & Northrup, USA.)

to 135 separate scales, may be accompanied with freely selected colouring from the four different coloured pens. These colours may be chosen to apply to any group of variables. A zoom scale possibility, analogous to optical zooming, allows expansion of a narrow scale segment to full scale to be chosen. Choosing to print such a zoom record is also under the control of the operator.

In addition to analogue and/or digital recording the operator may select summary reports to be written in the right or left margins or across the full chart width. Available reports are configuration or programming functions of the recorder, listing of selected analogue values, alarms and diagnostics as well as individual events, listing of discrete events and many other functions.

Among the many displays, selected by button, there are bar graph or pointer types, with an appropriate analogue scale for measured values, rate of change by arrows and alphanumeric display of selectable points, updated every 1 s to 4 min. All points can be displayed sequentially or in groups. Alarms and recorder status indication can also be displayed.

There are two keyboards available for the recorder. A simplified version is built into the front of the instrument, as shown in the illustration, while the other is hand-held. An IBM-format keyboard can also be used.

Speedomax 25000, which provides a large suite of built-in self-diagnostic capabilities, requires a code before an operator can gain access. Within the operating range of -5 to $+50$ °C, the accuracy of the recorder for thermocouples is $\pm 0.1\%$

of readings, with additional 0.3 °C reference junction error. The accuracy for RTDs is ±0.15% + 0.3 °C. Inclusion of a powerful mathematical package within Speedomax 25000 allows new, pseudo-data points to be calculated from other data points, or mathematical expressions. These pseudo-points can be used, processed, displayed and recorded in the same way as any other analogue values.

REFERENCES

Amrein, O. (1980) Stand und Trend der Registriertechnik bei elektrischen Betriebslinienschreibern. *messen + prüfen*, **16**(10), 712–718.

Anon. (1980) Schreiber Bell & Howell. *messen + prüfen*, **16**(3), 120–122.

Anon. (1982) Mindestens ein Multifunktionsrecorder wenn nicht noch mehr. *messen + prüfen*, **18**(6), 366–367.

Bozarth, T. B. and Hurd, E. (1972) Field mounted temperature transmitters their design and economic savings. *Temperature: Its Measurement and Control in Science and Industry*, Vol. 4, Part 1, Instrument Society of America, Pittsburgh, pp. 1339–1344.

Bylander, E. G. (1979) *Electronic Displays*. Texas Instruments Electronics Series, McGraw-Hill, New York.

C.E.I. (1974) *Appareil de mesure d'action indirecte*. Publication 484, Commission Electrotechnique Internationale, Paris.

Chenhall, H. (1987) Seeking the ultimate in A/D precision. *Electronic Prod Des*, Dec, pp. 39–42.

Crook, C. R. (1980) Dynamische Genauigkeit analoger XY-Schreiber. *messen + prüfen*, **16**(10), 694–702.

Degussa Messtechnik (1988) Transmitters for connection head and field mounting. *Catalog 8710E.*

Dubbeldam, J. F., de Groot, M. J. and Bloembergen, P. (1989) Computer aided thermometry—a complete approach. Proc. IMEKO TC-12 Symposium, *Microprocessors in Temperature and Thermal Measurement*, Lodz, Poland, pp. 23–30.

Gater, C. (1988) Fieldbus. *Meas. and Contr.*, **21**(2), 37–41.

Hart, B. L. and Barker, R. W. J. (1977) The design of constant current sources. *Electronic Eng.*, June, 85–88.

Henderson, I. A. and McGhee, J. (1990) A taxonomy of temperature measuring instruments. *Temperature and Thermal Measurement in Industry and Science* Proc. 4th IMEKO Symposium TEMPMEKO 90, Helsinki, pp. 400–405.

Hutcheon, I. C. (1989a) Intrinsic safety rules OK for process instrumentation. *Meas. and Contr.*, **22**(4), 108–116.

Hutcheon, I. C. (1989b) Intrinsic safety rules OK for process instrumentation. *Meas. and Contr.*, **22**(5), 149–156.

InstMC (1986) Special issue on instrumentation in hazardous areas. *Meas. and Contr*, **19**(7).

InstMC (1990) M and C feature on fieldbus. *Meas. and Contr.*, **23**(2).

I.E.E. (1983) Special feature: instrumentation for hazardous areas. *Electronics and Power*, **29**(4).

Lang, T. T. (1987) *Electronics of Measuring Systems* (English language edition edited by J. McGhee), John Wiley and Sons, Chichester.

Lang T. T. (1991) *Computerised Instrumentation* (English language edition edited by J. McGhee) John Wiley and Sons, Chichester.

Loriferne, B. (1982) *Analog-digital and Digital-analog Conversion*, John Wiley and Sons, Chichester.

McGhee, J. and Henderson, I. A. (1989) Holistic perception in measurement and control: applying keys adapted from classical taxonomy, in Linkens, D. A. and Atherton, D. P. (eds), *Trends in Control and Measurement Education*, IFAC Proc. Series, 1989, No 5. pp. 31–36.

McGhee, J., Henderson, I. A. and Sankowski, D. (1986) Functions and structures in measuring systems: a systems engineering context for instrumentation. *Measurement*, **4**(3), 111–119.
Maine, A. C. (1986) *Interfacing Standards for Computers*. Institution of Electrical and Electronics Incorporated Engineers Monograph.
Markt und Technik (1987) Anbieterübersicht-Messverstärker, No 1/2, pp. 99/102.
Morris, H. N. (1980) Strip-chart recorders keep pace with technology. *Control Eng*., **27**(4), 63–68.
MTL (1983) A user's guide to intrinsic safety. *Application Note AN9003*, Measurement Technology Ltd, Luton, England.
Putten, A. F. P. van (1988) *Electronic Measurement Systems*. Prentice-Hall, New York.
Reidy, J. (1985) Temperature measurement system to 10-bit resolution using the AD7571 and the AD594/595. *Application Note*, Analog Devices Ltd, Norwood, Mass., USA.
Ruhnau, K. and Klopfel, K. (1980) Neue Temperatur-messumformer für die Verfahrens-und Energietechnik. *messen + prüfen*, **16**(11), 691–692.
Seitzer, D., Pretzl, G. and Hamdy, N. A. (1983) *Electronic Analogue-to-digital Converters*, John Wiley and Sons, Chichester.
Stephens, G. (1989) Designers's guide to RS-232 standard. *Electronics and Wireless World*, April, pp. 340–342.
Tortoishell, G. (1990) Intrinsic safety—the second edition. *Meas. and Contr*., **23**(3), 81–84.
Vincent, P. (1987) The design of modern industrial temperature transmitters. *Meas. and Contr*., **20**(6), 29–31.
Warrior, J. and Cobb, J. (1989) Structure and flexibility for fieldbus messaging. *Meas. and Contr*., **22**(10), 292–294.
Watson, D. (1988) Update on revisions to RS-232 Standard. *Electronic Prod. Des*., January, pp. 57–60.
Weissberger, A. J. (1982) Upgrade data communication with an RS-449 interface. *EDN*, Feb 17. pp. 167–176.
Wilkinson, J. (1988) Chart recorders. *Contr. and Instr*., **20**(8), pp. 27–31.
Wyre, S. (1989) Single-chip digital panel meters. *Electronics and Wireless World*, April, pp. 347–351.

9 Dynamic Temperature Measurement

9.1 GENERAL INFORMATION

The term *dynamic temperature measurement* covers all measurements during which thermal transients occur in a sensor irrespective of whether the transient is caused by temperature variations in the medium, whose temperature is to be measured, or in the temperature sensor, itself. Thus, unavoidable dynamic errors occur during the measurement of any temperatures, changing with time. Errors also arise during the temperature measurement of a medium at a constant temperature using a temperature sensor immersed in the medium. Determination of the dynamic errors of a thermometer, requires knowledge of its dynamic properties. In many non-electric thermometers where the sensor and indicator form one inseparable unit, the dynamic properties to be described must refer to the whole device. By contrast, as electric thermometer sensors form a separate unit only the sensor needs to be described dynamically.

Electric thermometers are mostly applied when it is essential to know the dynamic error so that it can be taken into consideration. Consequently, the dynamic parameters of electric sensors will be the main topic for discussion in this chapter. It must be stressed, that dynamic errors in temperature measurement are principally caused by the sensor. For this reason, any influence of the dynamic properties of indicating instruments may be neglected in most cases.

Knowledge of the dynamic properties of a temperature sensor is necessary, among others, in the following cases:

- to determine the necessary immersion time, while measuring a constant medium temperature,
- to determine the dynamic errors while measuring temperatures changing with time,
- to compare the dynamic properties of different temperature sensors, so that the one best suited for a specific application, may be chosen,
- to determine the true temperature variations of temperatures changing in time by correcting known indicated values,
- to describe the dynamics of a sensor when it is part of a closed loop temperature control system as described by Michalski and Eckersdorf (1987),
- to choose the type and optimum settings of a corrector of dynamic errors.

As the problem of the dynamics of temperature sensors is approached in different ways, the number of existing references concerned with these dynamics is very large. Consequently only really concise principles are presented in this chapter.

9.1.1 *Transfer function of a temperature sensor*

The dynamic properties of a temperature sensor can be described by the differential equation:

$$F[y^n(t), y^{n-1}(t) \ldots y(t), \vartheta^m(t), \vartheta^{m-1}(t),, \ldots, \vartheta(t)] = 0 \tag{9.1}$$

where $y(t) \ldots y^n(t)$ are the sensor output signal and its time derivatives, and $\vartheta(t) \ldots \vartheta^m(t)$ are the measured temperature and its time derivatives.

In the case when the dynamic behaviour is linear, Equation (9.1) becomes:

$$\sum_{i=0}^{n} a_i \frac{d^i y(t)}{dt^i} = \sum_{j=0}^{m} b_j \frac{d^j \vartheta(t)}{dt^j} \tag{9.2}$$

where a_i and b_j are constant coefficients and $m < n$.

For describing and presenting the dynamic properties of a temperature sensor its transfer function, $G_T(s)$, which can be used, is the ratio of the Laplace transform of the sensor output signal, $y(s)$, to the Laplace transform of the measured temperature signal, $\vartheta(s)$, when the initial conditions are zero so that:

$$G_T(s) = \frac{y(s)}{\vartheta(s)} \tag{9.3}$$

Considering Equation (9.2), the transfer function can then be given by:

$$G_T(s) = \frac{\sum_{j=0}^{m} b_j s^j}{\sum_{i=0}^{n} a_i s^i} = \frac{L(s)}{M(s)} \tag{9.4}$$

The operational transfer function $G_T(s)$ is thus presented as the ratio of the polynomials $L(s)$ and $M(s)$. Poles of the transfer function are the roots of the equation $M(s) = 0$ and the zeros of the transfer function are the roots of the equation $L(s) = 0$. If the temperature input signal, $\vartheta(s)$, is known, a knowledge of the transfer function, $G_T(s)$, of a temperature sensor, enables its output signal to be determined as a function of time (Doetsch, 1961).

Each temperature sensor may be regarded as composed of the thermal and electrical conversion stages (Figure 9.1). In the thermal conversion stage, the temperature, ϑ, of the medium, whose temperature is being measured, is converted into the sensor's temperature, ϑ_T. The sensor's temperature, ϑ_T, is converted into the electrical output signal y (e.g. thermal e.m.f.) in the electrical conversion stage. This second conversion

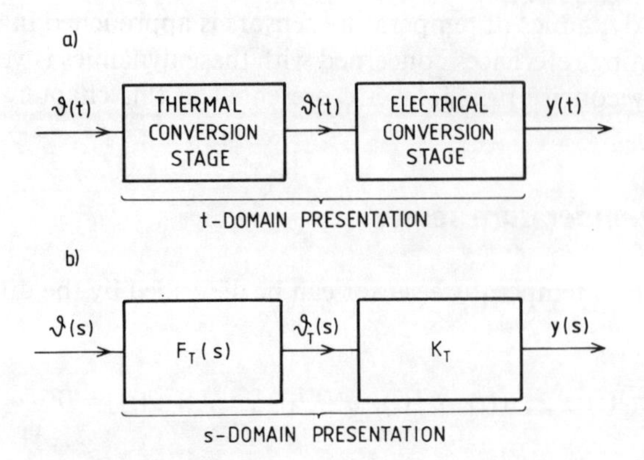

Figure 9.1
Block diagram of a temperature sensor.

stage has a purely static character. Thus, the sensor transfer function $G_T(s)$ of Equation (9.3) can be expressed as a product of the transfer function of the thermal conversion stage, $F_T(s)$, and of the coefficient K_T, representing the properties of the electrical conversion stage, and called the sensor gain so that:

$$G_T(s) = K_T F_T(s) \tag{9.5}$$

where:

$$K_T = \frac{dy}{d\vartheta} \tag{9.5a}$$

$$F_T(s) = \frac{\vartheta_T(s)}{\vartheta(s)} \tag{9.5b}$$

This approach to the presentation of sensor dynamics, makes it possible to limit further discussions of the dynamics to those of the thermal conversion stage.

In the case of steady-state periodic variations of measured termperature, the frequency response, $G_T(j\omega)$, of the sensor may be considered instead of the sensor transfer function, $G_T(s)$. The sensor frequency-response is the ratio of the phasor values of the output signal $y(j\omega)$ to the phasor value of the variable component of the sinusoidally changing measured temperature $\vartheta(j\omega)$:

$$G_T(j\omega) = \frac{y(j\omega)}{\vartheta(j\omega)} \tag{9.6}$$

where ω is the angular frequency and $j = \sqrt{-1}$

The frequency response of a sensor can be obtained by substituting $j\omega$ in Equation (9.3) in the place of the operator s. Consequently from Equation (9.4) the frequency response becomes:

$$G_{\mathrm{T}}(j\omega)=\frac{\sum_{j=0}^{m} b_{\mathrm{j}}(j\omega)^{j}}{\sum_{i=0}^{n} a_{i}(j\omega)^{i}}=\frac{L(j\omega)}{M(j\omega)} \tag{9.7}$$

In a similar fashion, the transfer function of Equation (9.5) can be rewritten as the frequency response function:

$$G_{\mathrm{T}}(j\omega)=K_{\mathrm{T}}F_{\mathrm{T}}(j\omega) \tag{9.8}$$

where K_{T} is the sensor gain, and $F_{\mathrm{T}}(j\omega)$ is the frequency response of the thermal stage.

Another way of expressing the frequency response of the sensor is:

$$G_{\mathrm{T}}(j\omega)=K_{\mathrm{T}}[P(\omega)+jQ(\omega)] \tag{9.9}$$

where

$$P=\mathrm{Re}\{F_{\mathrm{T}}(j\omega)\} \tag{9.9a}$$

$$Q=\mathrm{Im}\{F_{\mathrm{T}}(j\omega)\} \tag{9.9b}$$

or also

$$G_{\mathrm{T}}(j\omega)=\frac{\Delta y(\omega)}{\Delta\vartheta}\exp[j\varphi(\omega)] \tag{9.10}$$

where Δy is the amplitude of the output signal, $y(j\omega)$, $\Delta\vartheta$ is the amplitude of the first harmonic of measured temperature, $\vartheta(j\omega)$, and $\varphi(\omega)$ is the phase shift between $y(j\omega)$ and $\vartheta(j\omega)$.

Equations (9.9) and (9.10) are related as:

$$\frac{\Delta y(\omega)}{\Delta\vartheta}=|G_{\mathrm{T}}(j\omega)|=K_{\mathrm{T}}\sqrt{P^{2}(\omega)+Q^{2}(\omega)} \tag{9.11}$$

$$\varphi(\omega)=\arg G_{\mathrm{T}}(j\omega)=-\arctan\frac{Q(\omega)}{P(\omega)} \tag{9.12}$$

$\Delta y(\omega)/\Delta\vartheta$ of Equation (9.11) and $\varphi(\omega)$ of Equation (9.12) are respectively called the amplitude and phase characteristics of a temperature sensor. Sometimes it is more convenient to use the amplitude of the sensor temperature $\Delta\vartheta_{\mathrm{T}}(\omega)$ instead of the amplitude of the output signal $\Delta y(\omega)$,

The ratio:

$$\frac{\Delta\vartheta_{\mathrm{T}}(\omega)}{\Delta\vartheta}=\sqrt{P^{2}(\omega)+Q^{2}(\omega)}=|F_{\mathrm{T}}(j\omega)| \tag{9.13}$$

is called the amplitude characteristic of the thermal conversion stage of the temperature sensor.

9.1.2 Dynamic errors

The dynamic error can be defined as the difference between the sensor temperature, $\vartheta_T(t)$, and the same temperature, measured by another sensor, free of any thermal inertia but exhibiting the same static errors. Consequently the dynamic error is that part of the systematic error which varies with time. To simplify the problem, assume that the static error equals zero. The dynamic error is then defined as (Figure 9.2(a)):

$$\Delta\vartheta_{dyn}(t) = \vartheta_T(t) - \vartheta(t) \tag{9.14}$$

Measuring the stepwise changing temperature, the relative dynamic error is defined as follows (Figure 9.3(a)):

$$\delta\vartheta_{dyn}(t) = \frac{\Delta\vartheta_{dyn}(t)}{\Delta\vartheta} = \frac{\vartheta_T(t) - \vartheta(t)}{\Delta\vartheta} \tag{9.15}$$

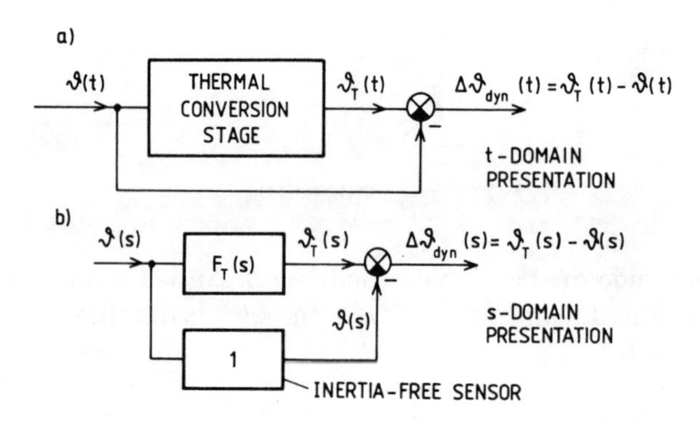

Figure 9.2
Definition of dynamic error.

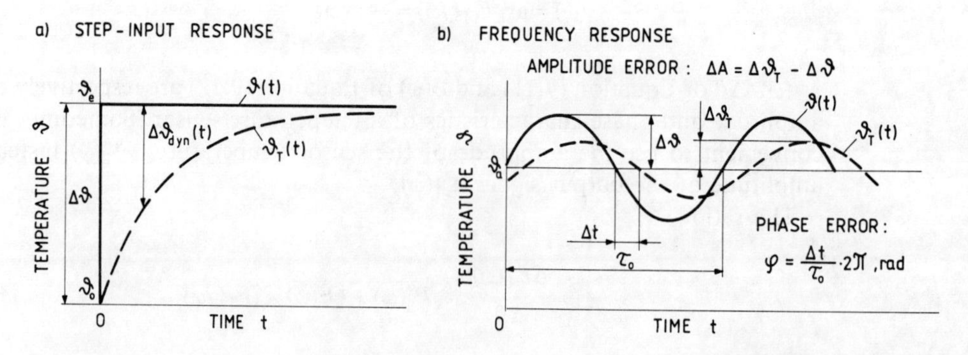

Figure 9.3
Time-display of dynamic errors.

Closely connected with the relative dynamic error, is the response time, t_r, after which the relative dynamic error does not exceed a certain value. For instance at $t \geq t_{r,5\%}$ the absolute value of the relative error is $|\delta\vartheta_{dyn}| < 5\%$.

The dynamic error can be also defined for other, non-periodic temperature variations. Its value is then related mostly to the maximum change of the measured temperature.

In dynamic temperature measurement Hofmann (1976) asserts that it is necessary to determine the dynamic errors in two cases. The first occurs when the measured, indicated temperature and the sensor's dynamic properties are known. Another occurs when the medium temperature, the input, is known as a function of time as well as the dynamic properties of the sensor. In both cases it is convenient to represent the dynamic error as a Laplace transform (Figure 9.2(b)) with the form:

$$\Delta\vartheta_{dyn}(s) = \vartheta_T(s) - \vartheta(s) \tag{9.16}$$

In accordance with the definition of the transfer function, $F_T(s)$, of the thermal conversion stage given in Equation (9.5b) as:

$$F_T(s) = \frac{\vartheta_T(s)}{\vartheta(s)} \tag{9.17}$$

which is valid for zero initial conditions, it follows that:

$$\Delta\vartheta_{dyn}(s) = \vartheta_T(s)\left[1 - \frac{1}{F_T(s)}\right] \tag{9.18}$$

or

$$\Delta\vartheta_{dyn}(s) = \vartheta(s)\,[F_T(s) - 1] \tag{9.19}$$

By applying the inverse Laplace-transform, the same value as a function of time can be determined as

$$\Delta\vartheta_{dyn}(t) = \mathscr{L}^{-1}\{\Delta\vartheta_{dyn}(s)\} \tag{9.20}$$

The definitions of the dynamic error given above can be used when the temperature $\vartheta(t)$ is an elementary function of time.

According to Michalski and others (1981), when measuring the sinusoidally changing temperature $\vartheta(\omega t) = \Delta\vartheta \sin\omega t + \vartheta_p$, the dynamic errors as shown in Figure 9.3(b), are composed of two parts:

1. Amplitude error, ΔA, is given by the difference of the amplitudes of the sensor temperature $\Delta\vartheta_T$ and of its true value $\Delta\vartheta$ so that

$$\Delta A = \Delta\vartheta_T - \Delta\vartheta \tag{9.21}$$

or the relative error, δA, which is related to the amplitude of the measured temperature may be written as:

$$\delta A = \frac{\Delta A}{\Delta \vartheta} = \frac{\Delta \vartheta_T - \Delta \vartheta}{\Delta \vartheta} \tag{9.22}$$

It is also given by the amplitude ratio:

$$\eta = \frac{\Delta \vartheta_T}{\Delta \vartheta} \tag{9.23}$$

2. Phase error is defined as the phase shift between the sensor temperature $\vartheta_T(t)$ and the measured temperature $\vartheta(t)$.

The cut-off frequency, f_c, of a temperature sensor defines a frequency below which the amplitude error does not exceed a given value (e.g. at $f < f_{c,5\%}, |\delta A| < 5\%$) when a state of stationary oscillations exists.

From Equations (9.12) and (9.13) when the frequency response $F_T(j\omega)$ of the sensor is known the amplitude and phase errors can also be derived as:

$$\Delta A = \Delta \vartheta_T \left[1 - \frac{1}{|F_T(j\omega)|} \right] \tag{9.24}$$

or

$$\Delta A = \Delta \vartheta \left[|F_T(j\omega)| - 1 \right] \tag{9.25}$$

with the relative amplitude error given by:

$$\delta \mathrm{A} = [F_T(j\omega) - 1] \tag{9.26}$$

The amplitude ratio is then seen to be:

$$\eta = |F_T(j\omega)| \tag{9.27}$$

with the phase shift:

$$\varphi = -\arctan \frac{\mathrm{Im}\{F_T(j\omega)\}}{\mathrm{Re}\{F_T(j\omega)\}} \tag{9.28}$$

9.2 IDEALIZED SENSOR

The existence of an idealized temperature sensor, designed as a homogeneous cylinder made of a material having infinitely great thermal conductivity λ, will be assumed.

Let this sensor have the mass m, specific heat c and a surface area A for heat exchange with its surroundings. During the temperature measurement of a liquid or gaseous medium, the sensor is completely immersed, so that it does not exchange heat with any other medium with a different temperature. As an example, any electric temperature sensor can be taken, provided that it is connected with the indicating instrument by extremely thin wires. It is further assumed, that the thermal capacity, mc, of the sensor is negligibly small compared with the total thermal capacity of the medium and that the heat transfer coefficient, α, between the sensor and the medium is constant.

9.2.1 Transfer function

To set up the differential equation, which describes the sensor's dynamics and thus its transfer function, the method of heat balances will be used. Assume, at the time $t=0^-$, an infinitesimally small time before zero, that the sensor is in a steady state, with its temperature equal to the ambient temperature $\vartheta_T=\vartheta_a$. At $t=0^+$ immerse the sensor in the medium at temperature ϑ, higher than the ambient temperature ($\vartheta>\vartheta_a$). For the temperature excess over the given reference value introduce the notation Θ. In this book Θ is also simply referred to as temperature. The initial conditions at $t=0^-$ are given by:

$$\Theta_T=\vartheta_T-\vartheta_a=0$$

$$\Theta=\vartheta-\vartheta_a>0$$

According to Newton's law, when the sensor is immersed in the medium the heat transferred to the sensor in the time interval dt will be

$$dQ=\alpha A(\Theta-\Theta_T)\,dt \tag{9.29}$$

where α is the heat transfer coefficient between sensor and medium and A is the heat exchange area.

The heat stored in the sensor is:

$$dQ=mc\,d\Theta_T \tag{9.30}$$

where m is the mass of the sensor, and c is the specific heat of the sensor material.

From Equations (9.29) and (9.30), it follows that

$$\alpha A(\Theta-\Theta_T)\,dt=mc\,d\Theta_T \tag{9.31}$$

or

$$\frac{mc}{A\alpha}\frac{d\Theta_T}{dt}+\Theta_T=\Theta \tag{9.32}$$

Introducing the notation:

$$\frac{mc}{A\alpha} = N_T \tag{9.32a}$$

Equation (9.32) can be expressed as:

$$N_T \frac{d\Theta_T}{dt} + \Theta_T = \Theta \tag{9.33}$$

N_T, which is called the sensor time constant at the given heat transfer conditions ($\alpha = \text{const}$), is expressed in time units. Equation (9.33) corresponds to Equation (9.2) with $a_0 = 1$, $a_1 = N_T$, $b_0 = 1$. The sensor output signal $y(t)$ is replaced by $\Theta_T(t)$ and the medium temperature $\vartheta(t)$ by the excess in temperature $\Theta(t)$. Taking the Laplace transform Equation (9.33) becomes:

$$N_T s \Theta_T(s) + \Theta_T(s) = \Theta(s) \tag{9.33a}$$

where s is the Laplace operator. Defining the transfer function of the thermal stage of the sensor as:

$$F_T(s) = \frac{\Theta_T(s)}{\Theta(s)} \tag{9.34}$$

Equation (9.33a) becomes

$$F_T(s) = \frac{1}{1 + sN_T} \tag{9.35}$$

so that the sensor transfer function is:

$$G_T(s) = \frac{y(s)}{\Theta(s)} \tag{9.36}$$

or

$$G_T(s) = K_T F_T(s) = K_T \frac{1}{1 + sN_T} \tag{9.37}$$

The frequency response of the thermal stage of an idealized sensor may be written as:

$$F_T(j\omega) = \frac{\Theta_T(j\omega)}{\Theta(j\omega)} = \frac{1}{1 + j\omega N_T} \tag{9.38}$$

and of the sensor as a whole:

$$G_T(j\omega) = \frac{y(j\omega)}{\Theta(j\omega)} = K_T \frac{1}{1 + j\omega N_T} \tag{9.39}$$

Equation (9.39) shows that the transfer function of an idealized temperature sensor is that of a first order inertia.

9.2.2 *Step response*

The step temperature response from an initial temperature ϑ_b to a final temperature ϑ_e is to be measured by an idealized temperature sensor. In practice this case corresponds to the immersion of a temperature sensor with a temperature ϑ_b into a medium with a temperature ϑ_e, which is written mathematically as:

$$\vartheta(t) = \begin{cases} \vartheta_b & \text{for } t \leq 0 \\ \vartheta_e & \text{for } t > 0 \end{cases}$$

or

$$\vartheta(t) = (\vartheta_e - \vartheta_b)\mathbf{1}(t) + \vartheta_b \qquad (\mathbf{1}(t) \text{ is a unit step function at } t = 0) \tag{9.40}$$

As it is necessary to obtain zero initial conditions for the Laplace transform, the excess of temperature $\Theta = \vartheta - \vartheta_b$, will be used, to obtain:

$$\vartheta(t) = \Theta_e \mathbf{1}(t) \tag{9.40a}$$

$$\Theta(s) = \frac{\Theta_e}{s} \tag{9.41}$$

From (9.34) it follows that:

$$\Theta_T(s) = \Theta(s) F_T(s) \tag{9.42}$$

Inserting $F_T(s)$ from (9.35) and $\Theta(s)$ from Equation (9.41), into Equation (9.42), the Laplace transform of the sensor temperature will be:

$$\Theta_T(s) = \frac{\Theta_e}{s(1 + sN_T)} \tag{9.43}$$

After the inverse-transformation in accordance with Doetsch (1961) the temperature–time dependence will be:

$$\Theta_T(t) = \mathcal{L}^{-1}\{\Theta_T(s)\} \tag{9.44}$$

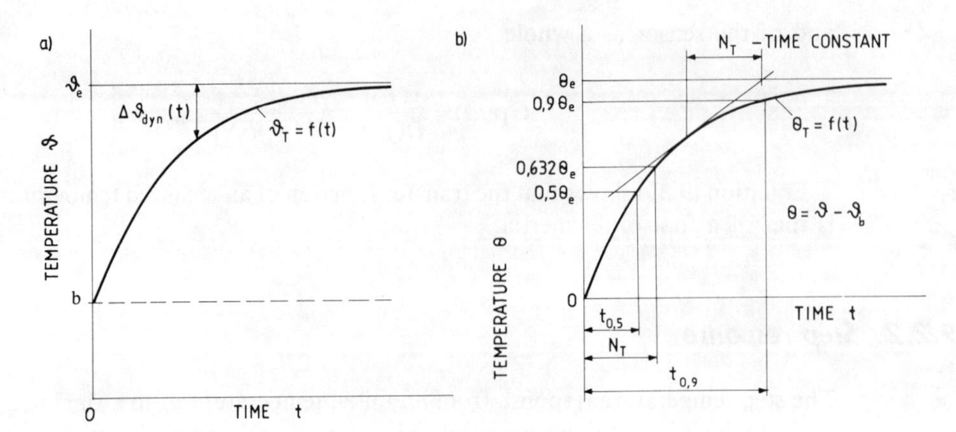

Figure 9.4
Step-input response of an idealized sensor.

$$\Theta_T(t) = \Theta_e(1 - e^{-t/N_T}) \tag{9.45}$$

The final result is then:

$$\vartheta_T(t) = (\vartheta_e - \vartheta_b)(1 - e^{-t/N_T}) + \vartheta_b \tag{9.46}$$

From Equations (9.45) and (9.46) it follows that the step input response of an idealized temperature sensor is an exponential curve, having the time constant N_T as shown in Figure 9.4(b). From this curve, the time constant N_T can be found in a graphical way from the tangent to the curve $\Theta_T = f(t)$ at any point, or as the time after which $\Theta_T = 0.632\Theta_e$.

Also the half-value time or 50% rise time $t_{0.5}$, which is the time when $\Theta_T = 0.5\Theta_e$, can be used to determine the time constant. From Equation (9.45) at $t = t_{0.5}$, it is clear that:

$$0.5\Theta_e = \Theta_e(1 - e^{-t_{0.5}/N_T})$$

After some transformation the time constant is obtained as:

$$N_T = \frac{1}{\ln 2} t_{0.5} = \frac{1}{0.693} t_{0.5} \tag{9.47}$$

The nine-tenth value time, $t_{0.9}$, is also a characteristic time of the step temperature response. From the exponential function it can be shown that:

$$\frac{t_{0.9}}{t_{0.5}} = 3.32 \tag{9.48}$$

From Equation (9.14) the dynamic error as represented in Figure 9.4(a) will be:

Table 9.1
Dynamic error $\delta\vartheta_{dyn}$ as a function of the immersion time t (N_T is the sensor time constant).

Time t	$0.623N_T$	N_T	$3N_T$	$3.9N_T$	$4.6N_T$	$6.9N_T$	$9.2N_T$
$\delta\vartheta_{dyn}$	-0.5	$-\frac{1}{e}$	-0.05	-0.02	-0.01	-0.001	-0.0001

$$\Delta\vartheta_{dyn}(t) = \vartheta_T(t) - \vartheta_e = \Theta_T(t) - \Theta_e = -\Theta_e \, e^{-t/N_T} \tag{9.49}$$

with the relative dynamic error given by:

$$\delta\vartheta_{dyn}(t) = \frac{\vartheta_T(t) - \vartheta_e}{\vartheta_e - \vartheta_b} = \frac{\Theta_T(t) - \Theta_e}{\Theta_e} = -e^{-t/N_T} \tag{9.50}$$

The dependence of $\delta\vartheta_{dyn}$ upon the immersion time t is given in Table 9.1. This relative error as a function of time, tends to zero in an exponential manner.

Numerical example

The medium temperature, $\vartheta_e = 220$ °C, is to be measured by a temperature sensor with a time constant, $N_T = 30$ s, and an initial ambient sensor temperature of $\vartheta_b = 20$ °C. How long should the immersion time be, to ensure that the indication error is less than 2 °C?

Solution: $|\Delta\vartheta_{dyn}| \leq 2$ °C corresponds to:

$$|\delta\vartheta_{dyn}| \leq \frac{2}{220 - 20} = 0.01$$

From Table 9.1 for $\delta\vartheta_{dyn} = -0.01$, the necessary immersion time is $t = 4.6N_T$, so that: $t > 4.6N_T = 4.6 \times 30 \text{ s} = 138 \text{ s}$

9.2.3 Ramp response

Assume that the measured temperature is a linear function of time of the form:

$$\vartheta(t) = \vartheta_b + kt \tag{9.51}$$

which corresponds to

$$\Theta(t) = kt \tag{9.52}$$

The Laplace transform of Equation (9.52) is

$$\Theta(s) = \frac{k}{s^2} \tag{9.53}$$

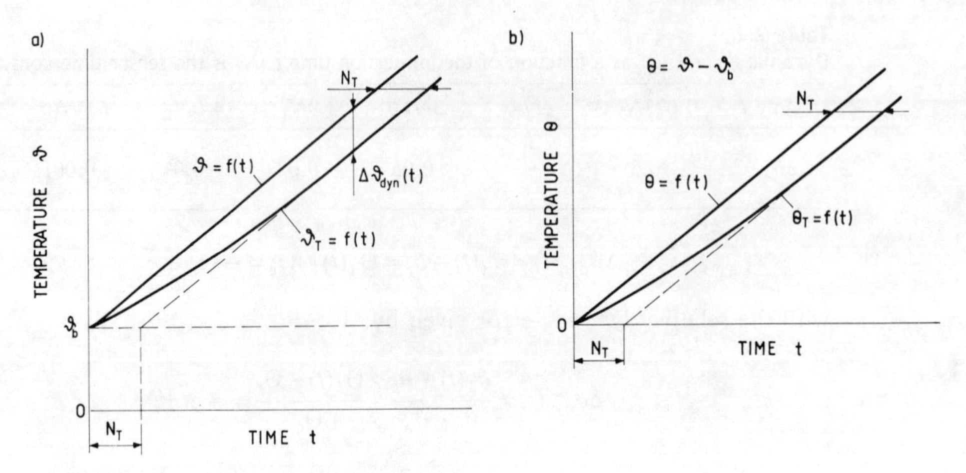

Figure 9.5
Ramp-input response of an idealized sensor.

Inserting $F(s)$ from Equation (9.35) and $\Theta(s)$ from Equation (9.53) into Equation (9.42), gives the Laplace transform of the temperature excess as:

$$\Theta_T(s) = \frac{k}{s^2} \frac{1}{1 + sN_T} \tag{9.54}$$

The inverse Laplace transformation of Equation (9.54) gives the sensor temperature as a function of time in the form

$$\Theta_T(t) = k[t - N_T(1 - e^{-t/N_T})] \tag{9.55}$$

which also leads to

$$\vartheta_T(t) = k[t - N_T(1 - e^{-t/N_T})] + \vartheta_b \tag{9.56}$$

The functions for $\Theta_T(t)$ and $\vartheta_T(t)$ given in Equations (9.55) and (9.56) are shown in Figure 9.5. The e^{-t/N_T} decreases as t increases so that when $t \gg N_T$ as $e^{-t/N_T} \ll 1$ it follows that simplified formulae which can be used, are:

$$\Theta_T(t) = k(t - N_T) \tag{9.57}$$

$$\vartheta_T(t) = k(\mathrm{t} - N_T) + \vartheta_b \tag{9.58}$$

Equations (9.57) and (9.58) show that the sensor temperature lags behind the measured temperature by a time which is equal to the value of the time constant. The dynamic error in Figure 9.5(a) which conforms to Equation (9.14) when $t \gg N_T$, becomes equal to:

$$\Delta\vartheta_{dyn}(t) = \vartheta_T(t) - \vartheta(\mathrm{t}) = -kN_T \tag{9.59}$$

From Equation (9.59) it follows that, for $t \gg N_T$, a linearly changing temperature is measured with a constant, dynamic error, proportional to the sensor time constant and also to the time-rate of change of the temperature. The measurement of temperature linearly changing in a known manner can be used to determine the time constant of a sensor (Figure 9.5(b)).

9.2.4 Exponential response

In practice it is often necessary to measure temperatures which are exponentially changing in accordance with:

$$\vartheta(t) = (\vartheta_e - \vartheta_b)(1 - e^{-t/N}) + \vartheta_b \tag{9.60}$$

which may also be otherwise written:

$$\Theta(t) = \Theta_e(1 - e^{-t/N}) \tag{9.61}$$

The Laplace transform of Equation (9.61) is:

$$\Theta(s) = \frac{\Theta_e}{s(1 + sN)} \tag{9.62}$$

where N is the time constant of the measured temperature variations.

Inserting $F_T(s)$ from Equation (9.35) and $\Theta(s)$ from Equation (9.62) into Equation (9.42), gives the sensor temperature excess as:

$$\Theta_T(s) = \frac{\Theta_e}{s(1 + sN)(1 + sN_T)} \tag{9.63}$$

Applying an inverse Laplace transformation to Equation (9.63), the time domain temperature excess is obtained in the form:

$$\Theta_T(t) = \Theta_e\left[1 - \frac{1}{N - N_T}(N\,e^{-t/N} - N_T e^{-t/N_T})\right] \tag{9.64}$$

which may also be written in the alternate form:

$$\vartheta_T(t) = (\vartheta_e - \vartheta_b)\left[1 - \frac{1}{N - N_T}(N\,e^{-t/N} - N_T e^{-t/N_T})\right] + \vartheta_b \tag{9.65}$$

The functions (9.64) and (9.65) are displayed in Figure 9.6. In most cases $N \gg N_T$ so that for $t \gg N_T$ simplified versions of Equations (9.64) and (9.65) are:

$$\Theta_T(t) = \Theta_e\left(1 - \frac{N}{N - N_T}e^{-t/N}\right) \tag{9.66}$$

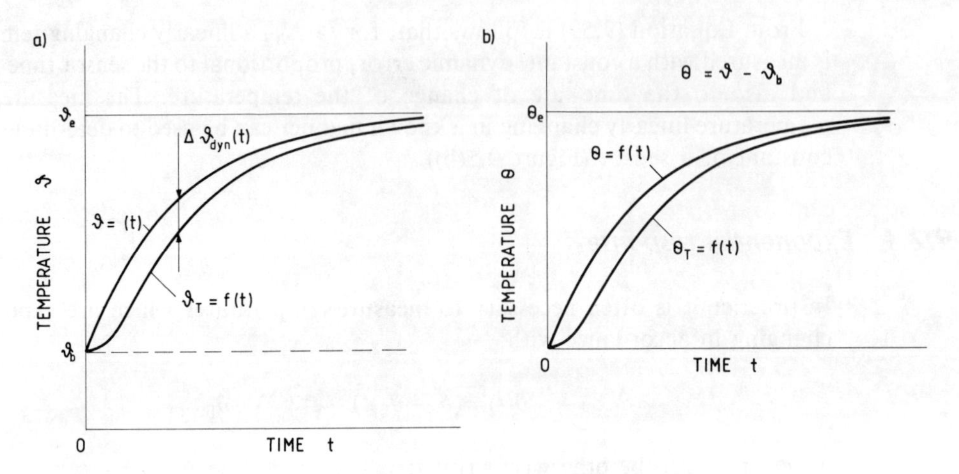

Figure 9.6
Exponential-input response of an idealized sensor.

$$\vartheta_T(t) = (\vartheta_e - \vartheta_b)\left(1 - \frac{N}{N - N_T} e^{-t/N}\right) + \vartheta_b \tag{9.67}$$

Assuming that $N \gg N_T$ and $t \gg N_T$, the dynamic errors corresponding to Equations (9.14) and (9.15) are given by:

$$\Delta\vartheta_{dyn}(t) = \vartheta_T(t) - \vartheta(t) = -\Theta_e \frac{N_T}{N - N_T} e^{-t/N} \simeq -\Theta_e \frac{N_T}{N} e^{-t/N} \tag{9.68}$$

$$\delta\vartheta_{dyn}(t) = -\frac{N_T}{N - N_T} e^{-t/N} \simeq -\frac{N_T}{N} e^{-t/N} \tag{9.69}$$

When measuring an exponentially changing temperature, Equation (9.69) shows that the dynamic error decreases exponentially with increasing values of time.

9.2.5 Sinusoidal response

Let the measured temperature be a sinusoidal function of time:

$$\vartheta(t) = \Delta\vartheta \sin \omega t + \vartheta_b \tag{9.70}$$

where $\Delta\vartheta$ is the amplitude of the temperature oscillations and ω is the angular frequency.

Assuming the notation $\Theta = \vartheta - \vartheta_b$, Equation (9.70) becomes:

$$\Theta(t) = \Delta\Theta \sin \omega t \tag{9.71}$$

In a continuous state of temperature oscillations, the sensor temperature is also a sinusoidal function of time. As the sensor temperature may be characterized by an amplitude $\Delta\vartheta_T$ and phase-shift φ, it is given by:

$$\Theta_T(t) = \Delta\Theta_T(\omega)\sin[\omega t + \varphi(\omega)] \tag{9.72}$$

or

$$\vartheta_T(t) = \Delta\vartheta_T(\omega)\sin[\omega t + \varphi(\omega)] + \vartheta_b \tag{9.73}$$

From Equation (9.13):

$$\frac{\Delta\vartheta_T(\omega)}{\Delta\vartheta} = |F_T(j\omega)|$$

Inserting $F_T(j\omega)$ from Equation (9.38), one gets:

$$\Delta\vartheta_T(\omega) = \Delta\vartheta\left|\frac{1}{1 + j\omega N_T}\right| = \Delta\vartheta\frac{1}{\sqrt{1 + (\omega N_T)^2}} \tag{9.74}$$

From Equation (9.12) it follows that the phase angle is:

$$\varphi = -\arctan \omega N_T \tag{9.75}$$

The measured and the sensor temperatures in a continuous state of oscillations, for given ω, are shown in Figure 9.7. Equations (9.24) to (9.27) give amplitude errors while the phase error is given by Equation (9.75). Both of these errors increase with increasing frequency, ω, and with increasing values of the sensor time-constant, N_T.

Numerical example

A sinusoidally varying medium temperature was measured by a temperature sensor of $N_T = 30$ s. The period of temperature oscillations was $\tau_0 = 100$ s, while the amplitude of the sensor temperature $\Delta\vartheta_T = 5$ °C. Determine the true amplitude of the temperature of the medium.

Solution: From Equation (9.74), the amplitude of the medium temperature may be calculated to obtain:

$$\Delta\vartheta = \Delta\vartheta_T\sqrt{1 + (\omega N_T)^2} = 10.65 \text{ °C}$$

9.2.6 Periodic, non-sinusoidal response

Let the measured medium temperature be a periodical triangular function of time, as shown in Figure 9.8. This function can be presented as the sum of harmonics:

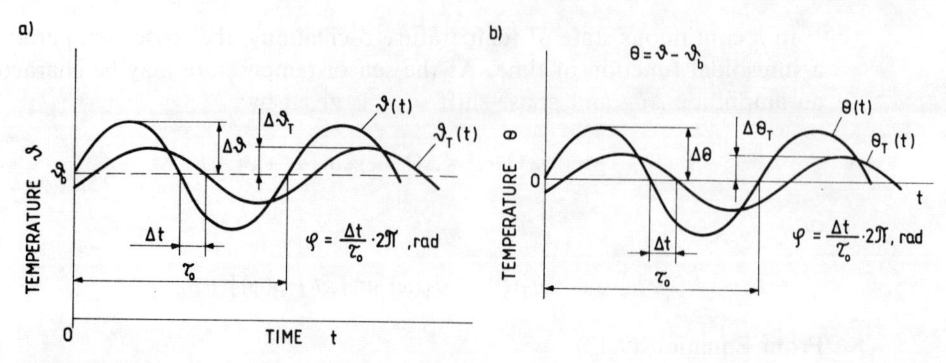

Figure 9.7
Sinusoidal-input response of an idealized sensor.

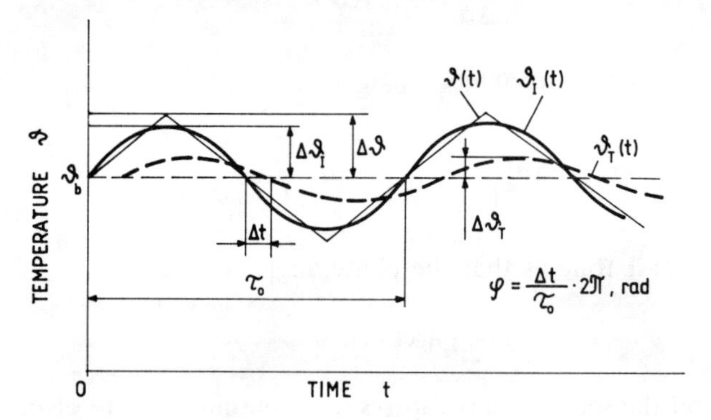

Figure 9.8
Periodic, non-sinusoidal input response of an idealized sensor.

$$\vartheta(t) = \frac{8}{\pi^2}\Delta\vartheta\left(\sin\omega t - \frac{\sin 3\omega t}{9} + \frac{\sin 5\omega t}{25} + \ . \ . \ .\right) + \vartheta_b \tag{9.76}$$

As the higher order harmonics are more strongly attenuated by the sensor inertia it is sufficient for approximate calculations to limit consideration to the first harmonic. Since the sensor is linear the principle of superposition applies so that the amplitude of the oscillations of the sensor temperature is then:

$$\Delta\vartheta_{T,I} \simeq \Delta\vartheta\frac{8}{\pi^2}\frac{1}{\sqrt{1+(\omega N_T)^2}} \tag{9.77}$$

From Equation (9.75) the phase shift of the first harmonic of the sensor temperature, relative to the fundamental harmonic of the oscillating temperature is then:

$$\varphi_1 = -\arctan\omega N_T$$

The sensor response to other non-sinusoidal periodic variations of the temperature can be assessed in a similar way.

9.3 REAL SENSORS

The dynamic properties of real industrial sensors differ from those of the idealized sensors described in §9.2. Materials and specific structures as well as working conditions, which must be considered in detail for the description of real sensor behaviour will be dealt with in §§9.3.1 and 9.3.2.

9.3.1 *Sensor design*

The temperature indications of an idealized sensor depend upon the average temperature of its whole mass. As it was assumed that the thermal conductivity, λ, of the material was infinitely high, the sensor temperature was the same all over its volume.

In real industrial sensors, which are mostly cylindrical, the sensitive part of the sensor which is relevant for the measurement may be:

- the whole mass, as in mercury-in-glass thermometers, neglecting the extremely thin glass layer (Figure 9.9(a)),
- the surface of the sensor as in bare resistance temperature detectors with the resistance wire wound on the surface (Figure 9.9(b)),
- the centre of the cross-section as in sheathed thermocouple sensors (Figure 9.9(c)).

Figure 9.9 also presents the corresponding step input temperature responses for all of the cases described above. The first curve (a) is most similar to the step input response of an idealized sensor. As proposed by Lieneweg (1975) all of the sensors can be classified by the ratio of their response times, $t_{0.9}/t_{0.5}$ so that for:

(a) sensors with volumetric response, $t_{0.9}/t_{0.5} \approx 3.32$,
(b) sensors with surfacial response, $t_{0.9}/t_{0.5} > 3.32$,
(c) sensors with central response, $t_{0.9}/t_{0.5} < 3.32$.

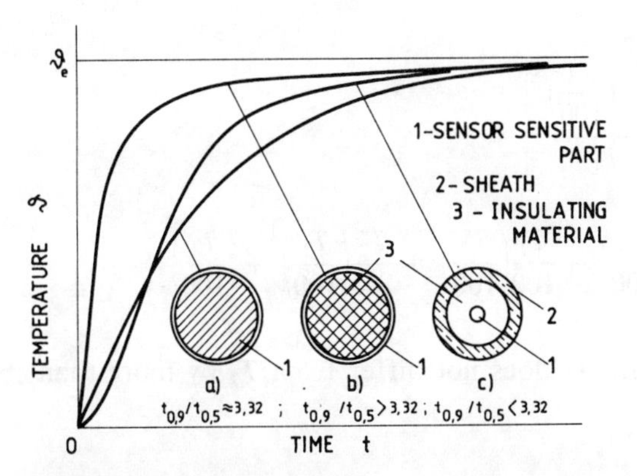

Figure 9.9
Step-input responses of real temperature sensors.

Describing a sensor as belonging to any one of the above three groups, depends on its step response which itself is determined not only by its design, but also by the working temperature and heat transfer conditions.

9.3.2 Changing heat transfer coefficient

In the derivation of the differential Equation (9.33), describing the dynamic properties of an idealized sensor, it was assumed that both the sensor time constant and the heat transfer coefficient between the sensor surface and the surrounding medium were constant and temperature independent. In reality, as the overall heat transfer coefficient, including convection, conduction and radiation, is a function of the medium temperature as well as of the instantaneous sensor temperature, so also the time constant of a sensor will vary.

These problems, are especially apparent while measuring temperatures due to a predominantly radiative heat exchange, as occurs inside chamber furnaces working above 600 °C (Hackforth, 1960; Michalski, 1966). The radiant heat flux between the chamber walls at temperature T_2 and the temperature sensors at T_1, is given by Equation (7.20a) as:

$$\Phi_{12}=A_1\epsilon_1 C_0\left[\left(\frac{T_2}{100}\right)^4-\left(\frac{T_1}{100}\right)^4\right]$$

where A_1 is the sensor heat exchange surface and ϵ_1 is the sensor emissivity.

This formula which is valid for the walls surface $A_2 > 3A_1$, can be rewritten as:

$$\Phi_{12}=A_1\alpha_r(T_2-T_1) \tag{9.78}$$

where α_r is the radiant heat transfer coefficient given by:

$$\alpha_r=\epsilon_1 C_0\left[\frac{\left(\frac{T_2}{100}\right)^4-\left(\frac{T_1}{100}\right)^4}{T_2-T_1}\right]$$

$$=\epsilon_1 C_0 10^{-2}\left[\left(\frac{T_3}{100}\right)^3+\frac{T_1}{100}\left(\frac{T_2}{100}\right)^2+\frac{T_2}{100}\left(\frac{T_1}{100}\right)^2+\left(\frac{T_1}{100}\right)^3\right] \tag{9.79}$$

If the sensor temperature T_1 does not differ from T_2 by more than $\pm 10\%$ T_2, corresponding to:

$$0.9<\frac{T_1}{T_2}<1.1$$

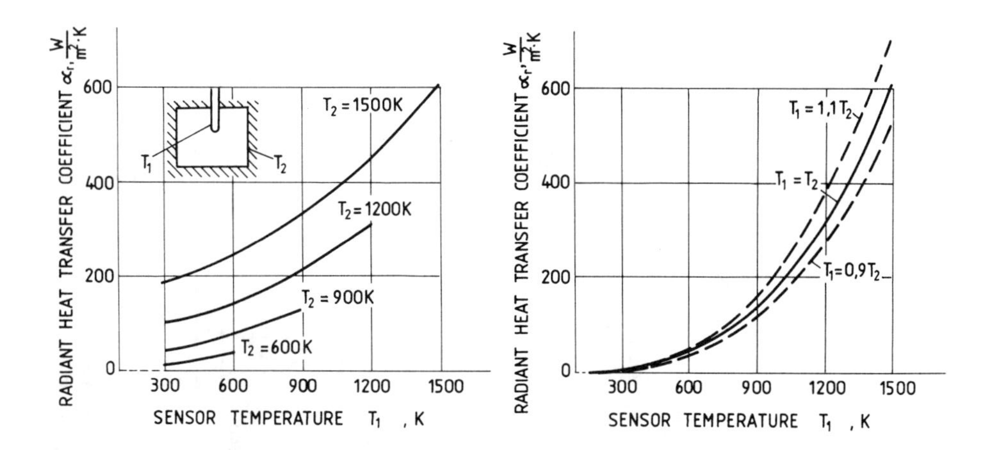

Figure 9.10
Radiant heat transfer coefficient α_r versus sensor temperature T_1, for a sensor of $\epsilon_1 = 0.8$, at given temperature T_2 of furnace walls.

then Equation (9.79) can be replaced by the following approximate dependence of accuracy to within 1%.

$$\alpha_r \approx \epsilon_1 C_0 0.5 \times 10^{-2} \left(\frac{T_1}{100} + \frac{T_2}{100} \right)^3 \tag{9.80}$$

According to Eijkman (1955) and Lienewig (1975) the coefficient α_r increases as the temperatures of the furnace walls and the sensor increase (Figure 9.10), achieving far higher values than the convective heat exchange coefficient α_k. The temperature dependence of α_r causes a considerable difference between the step responses of bare thermocouples in an electric furnace and the exponential curves. Considering their high λ-values, though, they closely approach ideal sensors.

9.3.3 Equivalent transfer function

The application of the concept of the transfer function to real temperature sensors which is only possible under some simplifying assumptions, leads to the idea of an equivalent transfer function. Some necessary simplifying assumptions are:

1. The dynamic properties of the sensor are linearized within the given temperature range.
2. A mean time constant is used as a value to describe the bidirectional heat flow between the sensor and the surrounding medium as described by Skoczowski (1982).
3. The sensor is represented by a lumped parameter model.

Table 9.2
Equivalent transfer functions of electrical temperature sensors.

Equivalent transfer function		Examples of sensor design
$G_T(s) = K_T \dfrac{1}{1 + sN_T}$	(9.81)	THERMOCOUPLES; SHEATH; MI-THERMOCOUPLES; d < 3mm, α - SMALL; d - ANY, α - SMALL
$G_T(s) = K_T \dfrac{1}{(1 + sN_{T1})(1 + sN_{T2})}$	(9.82)	THERMOCOUPLES; SHEATHS; MI-THERMOCOUPLE; PRESSURE SHEATH; d; RESISTANCE THERMOMETER; SHEATH; RTD
$G_T(s) = K_T \dfrac{e^{-sL}}{1 + sN_T}$	(9.83)	
$G_T(s) = K_T \dfrac{e^{-sL}}{(1 + sN_{T1})(1 + sN_{T2})}$	(9.84)	
$G_T(s) = K_T \dfrac{1 + sN_{T3}}{(1 + sN_{T1})(1 + sN_{T2})}$	(9.85)	MI-THERMOCOUPLES; d < 4mm, α - MEDIUM; d ⩾ 4mm, α - SMALL; d < 4mm, α - MEDIUM & BIG; d ⩾ 6mm, α - MEDIUM; BARE RTD; INSULATION TUBE; RESISTANCE WIRE

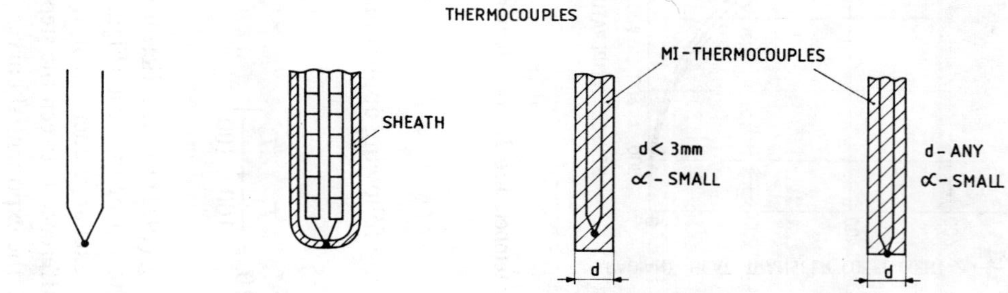

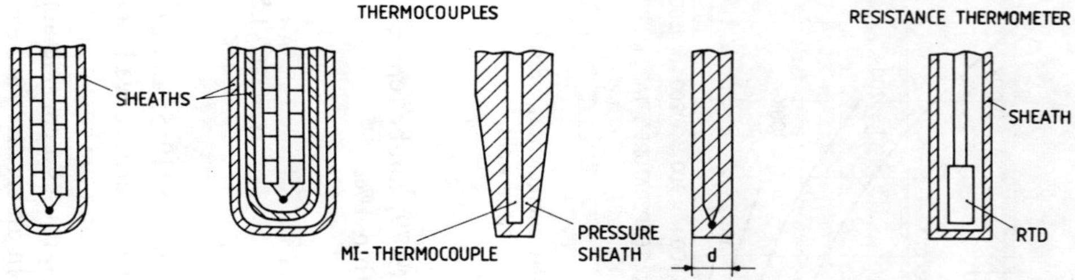

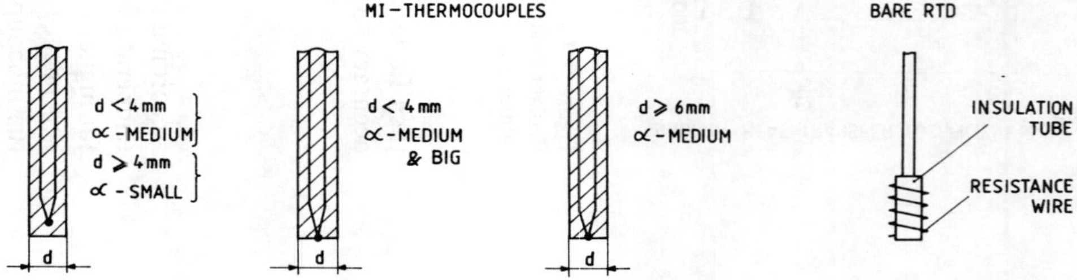

The most commonly used sensor transfer functions are given in Table 9.2. These models, which take the design of the sensor and the heat transfer conditions into account, have been considered by Bliek and Fay (1979), Bliek *et al.* (1978), Eckersdorf (1980), Hofmann (1965, 1966, 1967b), Rubin and Feldman (1968), Schwarze (1964) and Souksounov (1970). The majority of industrial thermocouples and resistance thermometer sensors may be simulated by the second order system given by Equation (9.82) in Table 9.2. Low inertia sensors are usually represented by a first order system model in accordance with formula (9.81) in Table 9.2.

9.4 CALCULATION OF DYNAMIC PROPERTIES OF REAL SENSORS

9.4.1 Homogeneous sensors

It is assumed that the temperature sensor can be represented by a homogeneous cylinder of thermal conductivity λ, specific density ρ and specific heat c. The sensor is totally immersed in the medium whose temperature is to be measured. Its thermal capacity is infinitesimally small compared with that of the medium, while the heat transfer coefficient between the sensor and the medium, as well as the sensor parameters, are constant. Internal heat sources are also assumed to be non-existent in the sensor. Under the above assumptions the Fourier differential equation of an infinitesimally long cylinder will be (Hoffman, 1976; Jakob, 1957, 1958):

$$\frac{\partial\Theta(r,t)}{\partial t}=a\left[\frac{\partial^2\Theta(r,t)}{\partial r^2}+\frac{1}{r}\frac{\partial\Theta(r,t)}{\partial r}\right] \tag{9.86}$$

where a, the thermal diffusivity of the cylinder material, is equal to $\lambda/\rho c$.

For real temperature sensors, the solution of this equation follows that for the boundary condition of the third kind which have the general form:

$$\frac{\partial\Theta(R,t)}{\partial r}=-\frac{\alpha}{\lambda}[\Theta(R,t)-\Theta_e] \tag{9.86a}$$

$$\frac{\partial\Theta(0,t)}{\partial r}=0 \tag{9.86b}$$

and also with zero initial conditions given by:

$$\Theta(r,0)=0 \tag{9.86c}$$

This solution, considered by Hofmann (1976), gives the temperature at any chosen point of the cylinder at a distance r from the cylinder axis ($0<r<R$) for a step change of the ambient temperature from 0 to Θ_e

$$\frac{\Theta(r,t)}{\Theta_e}=1-\sum_{\mu_n=1}^{\infty}\frac{2J_1(\mu_n)}{\mu_n[J_0^2(\mu_n)+J_1^2(\mu_n)]}J_0\left(\mu_n\frac{r}{R}\right)e^{-\mu_n^2\mathrm{Fo}} \tag{9.87}$$

where μ_n are the roots of the characteristic equation:

$$\mu_n J_1(\mu_n) = \mathrm{Bi} J_0(\mu_n) \tag{9.87a}$$

$J_0(\mu_n)$,$J_1(\mu_n)$ are Bessel functions of the first kind of zero and first order respectively, Fo the Fourier number is equal to at/R^2 and Bi the Biot number is equal to $\alpha R/\lambda$.

From Equation (9.87), putting $r=R$, the relative step input response at the cylinder surface is obtained as:

$$h_{r=R}(\mathrm{Fo}) = \frac{\Theta(R,t)}{\Theta_e} = 1 - \sum_{n=1}^{\infty} \frac{2\mathrm{Bi}}{\mu_n^2 + \mathrm{Bi}^2} e^{-\mu_n^2 \mathrm{Fo}} \tag{9.88}$$

At the cylinder axis, corresponding to $r=0$, the solution is:

$$h_{r=0}(\mathrm{Fo}) = \frac{\Theta(0,t)}{\Theta_e} = 1 - \sum_{n=1}^{\infty} \frac{2\mathrm{Bi}}{J_0(\mu_n)(\mu_n^2 + \mathrm{Bi}^2)} e^{-\mu_n^2 \mathrm{Fo}} \tag{9.89}$$

The functions given in Equations (9.88) and (9.89) can also be presented as a sum of exponential time functions. When $r=R$, such a relation is:

$$h_{r=R}(t) = 1 - \sum_{n=1}^{\infty} A_{r=R,n}\, e^{-t/N_n} \tag{9.90}$$

where

$$A_{r=R,n} = \frac{2\mathrm{Bi}}{\mu_n^2 + \mathrm{Bi}^2} \tag{9.90a}$$

$$N_n = \frac{K}{\mu_n^2} \tag{9.90b}$$

$$K = \frac{R^2}{a} \tag{9.90c}$$

When $r=0$, the relation is:

$$h_{r=0}(t) = 1 - \sum_{n=1}^{\infty} A_{r=0,n} e^{-t/N_n} \tag{9.91}$$

where

$$A_{r=0,n} = \frac{2\mathrm{Bi}}{J_0(\mu_n)(\mu_n^2 + \mathrm{Bi}^2)} \tag{9.91a}$$

The roots μ_n of the characteristic Equation (9.87a) can be found in a graphical way, with the assistance of the graphs by Hofmann (1976), or numerically, using previously computed mathematical tables (Gröber *et al.*, 1963). Graphical display of the functions in Equations (9.88) and (9.91) are also given in Hofmann (1976).

Following the well known rules of the Laplace transformation, Michalski and others (1981) show that it is easy to find the transfer function $G(s)$ from the relative step response $h(t)$ using:

$$G(s) = s\mathscr{L}\{h(t)\} \tag{9.92}$$

The transformation in Equation (9.92) when applied to the function (9.90), allows the establishment of the transfer function on the cylinder surface as:

$$G_{r=R}(s) = s\mathscr{L}\{h_{r=R}(t)\} = 1 - s\sum_{n=1}^{\infty} A_{r=R,n}N_n\frac{1}{1+sN_n} \tag{9.93}$$

In the same way the transfer function corresponding to Equation (9.91) on the cylinder axis is

$$G_{r=0}(s) = s\mathscr{L}\{h_{r=0}(t)\} = 1 - s\sum_{n=1}^{\infty} A_{r=0,n}N_n\frac{1}{1+sN_n} \tag{9.94}$$

Detailed calculations of the approximation errors in the step-input response and in the transfer functions for the surface and axis of an infinitely long cylinder (9.90), (9.91), (9.93), (9.94) as a function of the number of applied elements n and of the Biot number will be found in Hofmann (1976). For small Biot numbers ($B_i \approx 10^{-2}$) even with $n = 1$, Hofmann (1976) shows that the approximation errors are well below 1%. With increasing Biot number, the necessary n number also increases. Nevertheless, if somewhat larger errors are accepted near the start of a step response, a value for $n = 1$ can be used even for larger Bi values. These somewhat larger errors fall below 1% after a time corresponding to $\text{Fo} \geq 0.03$.

The theory presented above as developed for a simple sensor can also be extended to multi-layer sensors as shown by Bernhard (1978), Hofmann (1976), Lieneweg (1938a, 1938b, 1941, 1962) and Yarishev (1967).

For some sensors it is also possible to use other simple models such as that of a sphere or plate. The corresponding solutions of the Fourier equation for boundary conditions of the third kind are given in Gröber *et al.*. (1963), Hofmann (1976) and Jakob (1957, 1958).

9.4.2 Multi-layer sensors

To determine the dynamic properties of real sensors, first it is necessary to set up their equivalent circuits. As temperature sensors with tubular sheaths are widely used in practice, most publications concern this type of sensors. The majority of authors, such as Caldwell *et al.* (1959), Eijkman (1955), Eijkman and Verhagen (1958), Meyer-Witting (1959) and Yarishev (1967), use equivalent circuits based on a second order system, as shown in Figure 9.11. The thermal capacity m_1c_1 of the sheath is represented by a capacitor C_1 and that of the sensitive part of the sensor, m_2c_2 by

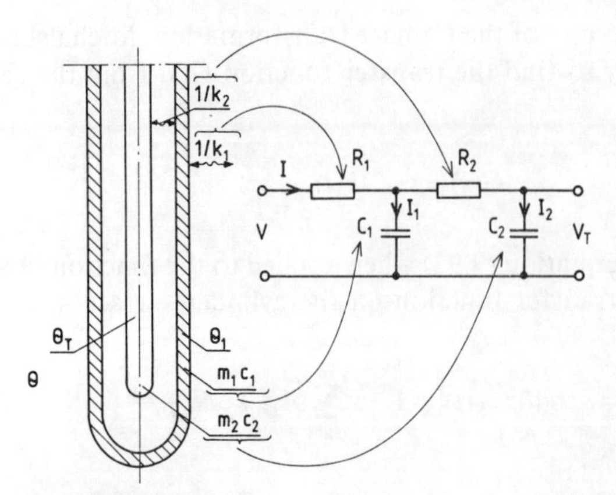

Figure 9.11
Real temperature sensor and its electric analogue circuit

C_2. The thermal resistances between the sheath and its environment $1/k_1$ and between the sheath and the temperature sensitive part, $1/k_2$ are similarly represented by the resistors R_1 and R_2 respectively. All these capacitances and resistances are given as relative values per unit of the sensor length provided that the sensor is sufficiently long to neglect any heat exchange along its length. The thermal resistance across the sheath wall has also been neglected. Element pairs of the analogous quantities of the thermal and electric models, are given in Table 9.3 (Gröber *et al.*, 1963). The input voltage, V, corresponds to the measured mean temperature, Θ, the output voltage, V_T, to the temperature, Θ_T, of the sensor's sensitive part, so that the transfer function of the analogue circuit (Figure 9.11) is:

$$G(s) = \frac{V_T(s)}{V(s)} = \frac{1}{s^2 R_1 C_1 R_2 C_2 + s(R_1 C_1 + R_2 C_2 + R_1 C_2) + 1} \tag{9.95}$$

Denoting

$$R_1 C_1 = N_1 \tag{9.95a}$$

$$R_2 C_2 = N_2 \tag{9.95b}$$

$$N_1 N_2 = N^2 \text{ and } N = \sqrt{N_1 N_2} \tag{9.95c}$$

and

$$N_1 + N_2 + R_1 C_2 = 2\xi N$$

$$\xi = \frac{1}{2} \frac{N_1 + N_2 + R_1 C_2}{\sqrt{N_1 N_2}} \tag{9.95d}$$

allows Equation (9.95) to be written in a standard second order form as follows:

Table 9.3
The analogy between electric and thermal systems.

Thermal system	Electric system	Scale factor
Temperature over ambient $\Theta\,[°C]$	Voltage $V\,[V]$	of voltage $K_V = \frac{V}{\Theta}\,[V/K]$
Thermal resistance $\frac{1}{k}\,[K/W]$	Resistance $R\,[\Omega]$	of resistance $K_R = Rk\,[\Omega W/K]$
Thermal capacitance $mc\,[J/K]$	Capacitance $C\,[F]$	of capacitance $K_c = C/mc\,[FK/J]$
Time t [s]	Model time $t_{el}\,[s]$	of time $K_t = t_{el}/t\,[s/s]$

$$G(s) = \frac{1}{s^2N^2 + s2\xi N + 1} \tag{9.96}$$

The step response of the circuit as given by Doetsch (1961) is then

$$h(t) = \mathscr{L}^{-1}\left\{\frac{1}{s}G(s)\right\}$$

$$= 1 + \frac{e^{-\xi t/N}\sin\left[(\sqrt{1-\xi^2}t/N) - \Phi\right]}{\sqrt{1-\xi^2}} \tag{9.97}$$

where

$$\Phi = \arctan\frac{\sqrt{1-\xi^2}}{-\xi} \tag{9.97a}$$

To obtain the transfer function of a sensor or its step response, the electric quantities in Equations (9.96) and (9.97) should be replaced by the analogous thermal quantities given in Table 9.3.

As this rather complicated solution is of only limited practical use, a temperature sensor can be modelled for simulation purposes by two, non-interacting RC-elements. Such a model will be used for the experimental determination of the step responses of a temperature sensor. Since the analytical method is not very precise it is rarely applied.

9.5 EXPERIMENTAL DETERMINATION OF DYNAMIC PROPERTIES OF REAL SENSORS

9.5.1 Classification of methods

Dynamic properties of real sensors which are mostly determined experimentally should be found in the same environment and at the same temperature in which the sensor will be used. Any recorder used for this purpose, should have as small a response time as possible, at least five times shorter than the sensor's response time.

At low temperatures, below about 100 °C, where heat exchange by radiation is negligibly small, the working conditions of the sensor can be described explicitly by the convective heat exchange coefficient α_k. Most measuring methods are based on measurements in either flowing water or air. At temperatures over 600 °C, in conditions of free convection, in practice there is only radiant heat exchange, characterized by the source temperature ϑ_e as the final temperature. In all other cases it is necessary to precisely specify the heat exchange conditions, by giving the values of the heat exchange coefficient α and of the final temperature ϑ_e.

In most publications such as those of Higgins and Keim (1954), Hofmann (1976), Huhnke (1973), Jakob (1958), and Kondratiev (1947), the experimental determination of the sensor dynamic properties is performed at low temperatures where the methods and instruments are well known. These will be discussed in Section 9.5.2. Only a few publications deal with the temperature range over 600 °C (Rubin and Feldman, 1968; Bernhard, 1978; Eckersdorf, 1980). Section 9.5.3 is reserved for the presentation of these methods.

The classification of experimental methods which is presented in Figure 9.12 mostly concern industrial sensors.

Determination of the dynamics of thin bare thermocouples and of thin resistance wires is allowed by using direct heating from an electric current flowing in them (Petit *et al.*, 1982). The dynamics of resistance thermometers and thermocouples can now also be determined in this way, as for example in nuclear power stations (Kerlin, *et al.*, 1978; Bernhard and Noack, 1981; Kerlin *et al.*, 1981; Eckersdorf, 1986; Michalski and Eckersdorf, 1990). This method permits the determination of sensor dynamics *in situ*.

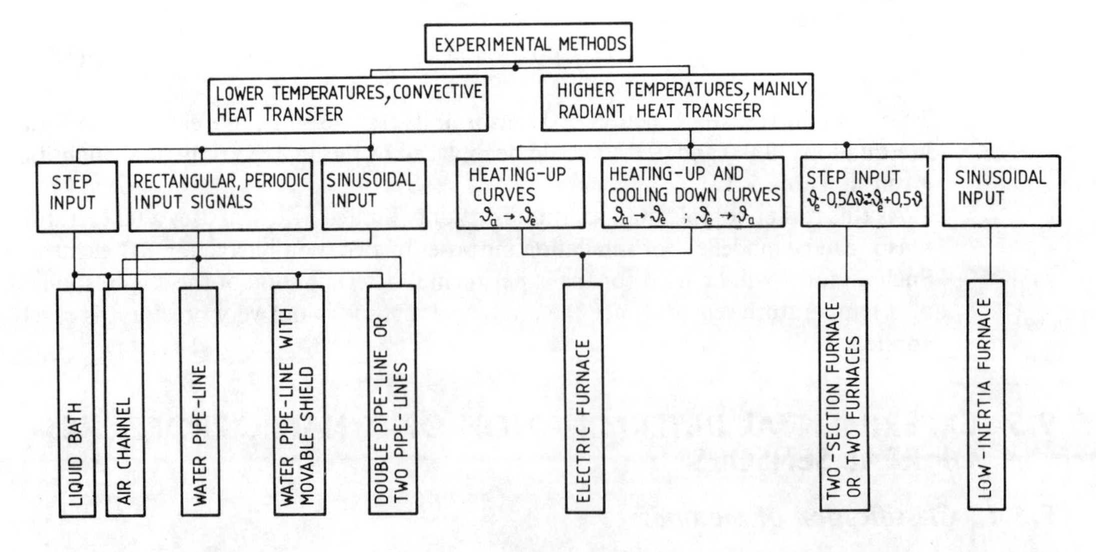

Figure 9.12
Classification of methods for the experimental determination of sensor dynamics.

9.5.2 *Convective heat transfer*

Step-input method

The simplest and most popular method of measuring the step response of a temperature sensor is to immerse it in well-stirred water in a thermostat, whose temperature is 20 to 60 °C above the ambient. Berger and Balko (1972) advise the mechanization of the immersion procedure.

The heat transfer coefficient α between the sensor and the medium is of some thousands of W/m^2 °C. This method gives explicit and reproducible measuring conditions, because any variations of the high α-values, which occur, do not exert a significant influence on the sensor dynamics (Kondratiev, 1947; Yarishev, 1967). For example, Kondratiev (1947) applied a step input to the sensor, by transferring the sensor from well stirred water at 30 to 35 °C to another tank with well stirred water at 15 to 20 °C. From this he derived the so called index of thermal inertia, ϵ, whose values can be found in a graphical way, by displaying the step-input response in the half logarithmic scale as given in Figure 9.13. On the linear part of the diagram, two temperatures Θ_1 and Θ_2, which are as far away from each other as possible, are marked so allowing the corresponding times t_1 and t_2 to be determined. The ϵ-value is then calculated as

$$\epsilon = \frac{t_2 - t_1}{\ln\Theta_1 - \ln\Theta_2} \tag{9.98}$$

This index ϵ which is a certain measure of the sensor inertia, may be used for comparing the dynamics of different sensors in reference conditions (well stirred water at $\vartheta \approx 25$ °C). This index also introduced into Soviet Standards (GOST 6616-61, 6651-59), offers no possibility of calculating the dynamic properties of sensors in other conditions or at other temperatures. A strictly physical interpretation of the index

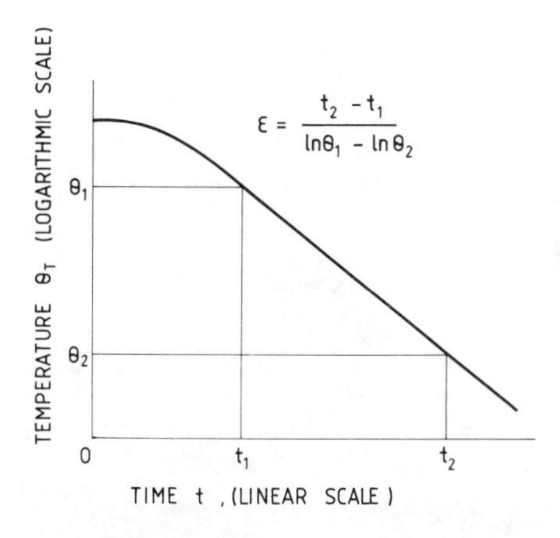

Figure 9.13
Determination of an index of thermal inertia ε.

ϵ indicates that it is the largest of some time constants of a sensor at infinitely high heat transfer coefficient on the sensor surface.

To measure sensor dynamics at different α-values, some special channels and pipe-lines are constructed, in which a gas or liquid (mostly air and water) flows at a changing, adjustable velocity. This velocity is measured in order to enable a comparison of obtained results.

Hofmann (1976) describes an air channel in which the flow velocity may be adjusted between 0 and 20 m/s. It is intended for sensors, having sheath diameter, D, up to 10 mm and an immersion length up to 100 mm. The highest achievable α values were 150 W/m^2 °C, for $D = 10$ mm, and 400 W/m^2 °C, for $D = 1$ mm. Chohan and Natour (1988) describe an air channel, which was designed for sensor testing at air velocities up to 2000 m/s.

In the water pipe-lines described by Hofmann (1976) and Huhnke (1973), the water velocity is adjustable between 0.1 and 2 m/s. The α-values at the lowest water flow velocity are comparable with those at the highest air flow velocity, while the α-values for wet air approach the values for liquid (Chohan and Natour, 1988).

Measurements in air and water proceed in a similar way. The tested sensor is washed by hot water coming from a pipe inside a shield, as shown in Figure 9.14. Protection of the sensor from contact with the flowing medium is ensured by the shield. The hot-water temperature is 10 °C above the medium temperature. At a given moment, the springy shield is pushed in the direction of the flowing medium and simultaneously the hot-water flow stops. A deciding factor is the step input duration, which has to be shorter than 0.1 of the half-value time ($t_{0.5}$) of the tested sensor. In the described stand a mechanical device ensures that the step time ($t_{0.95}$) is shorter than 20 to 100 ms.

Temperature step change can also be realized using a low-inertia mesh heating element. Such a mesh is mostly used for gas heating, because in liquid heating far greater heating power would be needed. A principal diagram of an air-channel described by Huhnke (1973) is shown in Figure 9.15. The thin wire mesh heating element is placed parallel to the tested temperature sensor and perpendicular to the

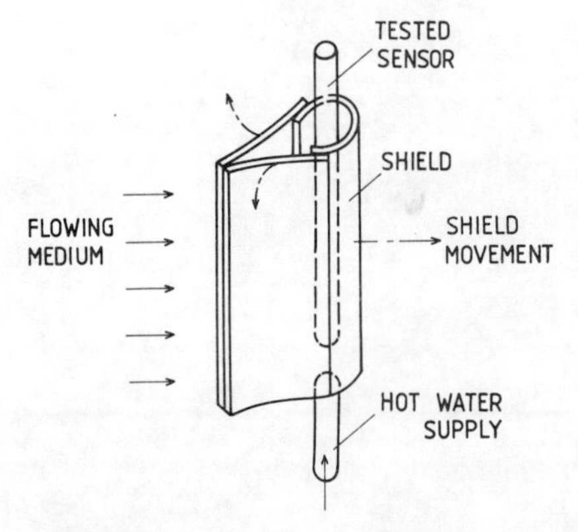

Figure 9.14
Pipe-line with movable shield.

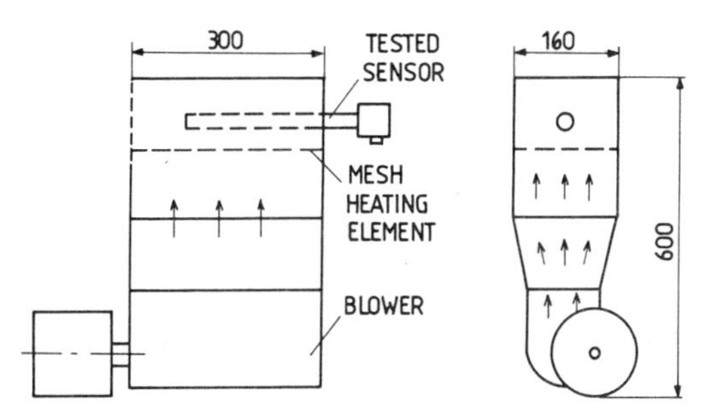

Figure 9.15
Air-channel with mesh heating element.

direction of the air stream. Step change of the temperature of the flowing medium is achieved by switching the heating power on and off. In most cases the temperature changes of about 10 °C which are applied, should be precisely regarded rather as exponential temperature changes characterized by a time-constant N (Huhnke, 1973). The value of the time constant can be calculated from:

$$\mathrm{N} \simeq K \frac{d^{5/3}}{v^{1/3}} \tag{9.99}$$

where K is an experimentally determined coefficient, d is the wire diameter, m, and v is the air velocity, m/s. For $d = 0.02$ to 0.05 and $v = 1$ to 15 m/s, the K value is 1.3×10^6.

For example, at an air velocity of $v = 2$ to 15 m/s, the time constant of the heating wires of $d = 0.02$ mm, is $N = 16$ to 7 ms, corresponding to a response time $t_{0.95} = 50$ to 20 ms. Application of a low inertia heating element, also makes it possible to generate other temperature test signals, especially sinusoidal ones.

The method of interpreting the recorded step-input response of a temperature sensor depends upon its character and upon the necessary precision. Directly from the recorded step-input response it is possible to determine the $t_{0.5}$ or $t_{0.9}$ times described in §9.2.2 or the response time, t_r, of §9.1.2. Some of the most popular methods of presenting the sensor dynamics, which are based upon the ratio $t_{0.9}/t_{0.5}$ will be described.

Sensors with volumetric response ($t_{0.9}/t_{0.5} \approx 3.32$) are presented as first order inertia elements, characterized by the time-constant N_T. This time-constant N_T can be found graphically as shown in Figure 9.4(b) or calculated from the half-value time $t_{0.5}$ given in Equation (9.47). Sometimes, instead of recording the whole step-response it is sufficient to measure the time-constant or only the half-value time. Figure 9.16 shows the block representation and the equivalent analogue electric circuit of a sensor which is regarded as a first order inertia element with a transfer function (Table 9.2 and Equation (9.81)) given by:

$$G_T(s) = K_T \frac{1}{1 + sN_T} \tag{9.100}$$

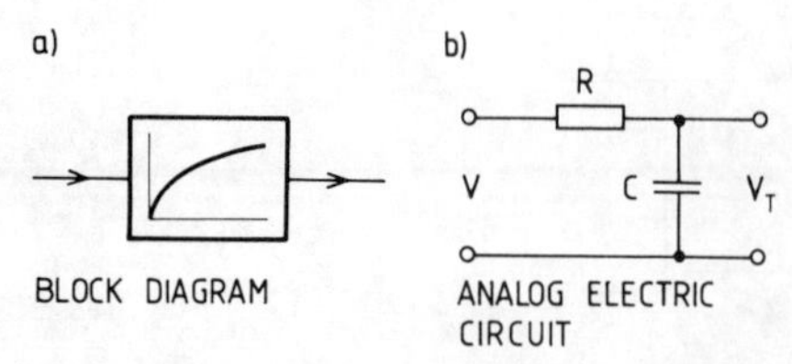

Figure 9.16
Temperature sensor with volumetric response as a first-order inertia element.

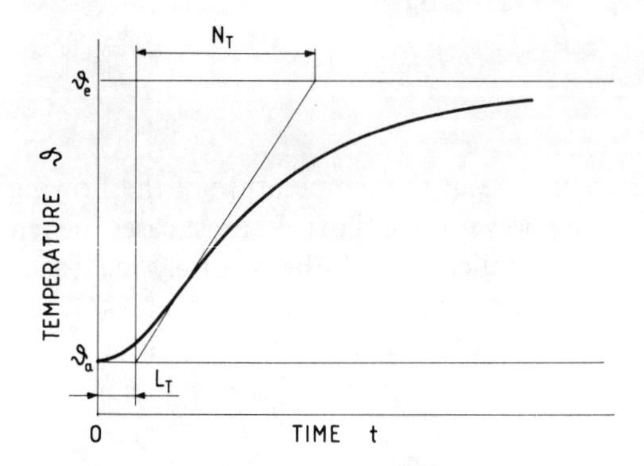

Figure 9.17
Step-input response of a temperature sensor with central response.

The relevant differential equation, and the responses to different input-signals are identical to those of the idealized sensor described in §9.2.

Sensors with central response ($t_{0.9}/t_{0.5} < 3.32$) are characterized by s-shaped, step-response curves. Their dynamics can be approximately presented, by one time constant, N_T, found from the measured half-value time. Another way is based on a recorded step-response as given in Figure 9.17. The tangent to this curve, cuts off the equivalent lag L_T on the line $\vartheta = \vartheta_a$ and the time constant N_T on the line $\vartheta = \vartheta_e$. This sensor model consisting of a series connection of a first order inertia element and a pure lag, is rarely used. Its transfer function, shown in Table 9.2, and given in Equation (9.83), is:

$$G_T(s) = K_T \frac{e^{-sL_T}}{1 + sN_T} \tag{9.101}$$

and its step-input response is

$$\Theta_T(t) = 0 \qquad \text{in } t < L_T \tag{9.102a}$$

$$\Theta_T(t) = \Theta_e(1 - e^{-(t - L_T)/N_T}) \qquad \text{in } t \geq L_T \tag{9.102b}$$

A second-order inertia approach, which is a series connection of two first-order inertia elements, characterized by two time-constants, N_{T1} and N_{T2}, is more popular. Its block diagram and equivalent electric circuit, are presented in Figure 9.18.

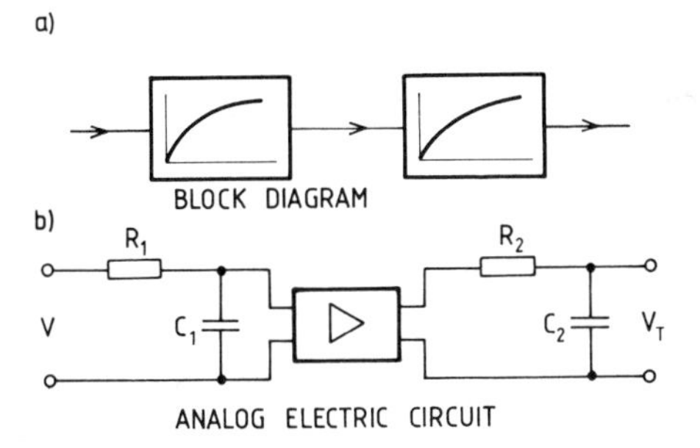

Figure 9.18
Temperature sensor with central response as a second order inertia element.

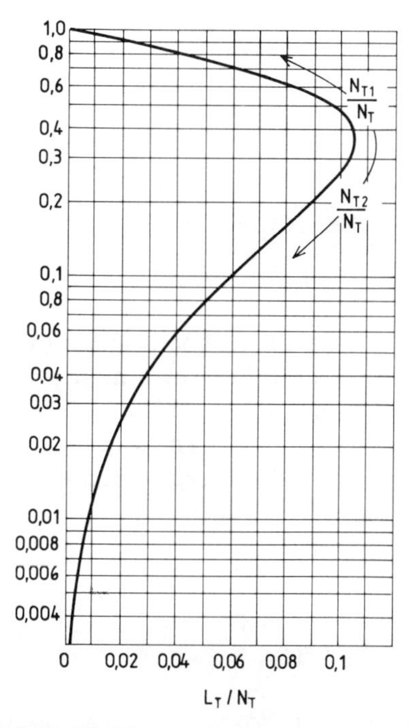

Figure 9.19
Diagram for determining the time-constants N_{T1} and N_{T2} when L_T/N_T ratio is known. L_T and N_T to be found as in Figure 9.17.

The approximate transfer function of this sensor, following Table 9.2 and Equation (9.82), is:

$$G_T(s) = K_T \frac{1}{(1 + sN_{T1})(1 + sN_{T2})}$$

and its step-input response is given by:

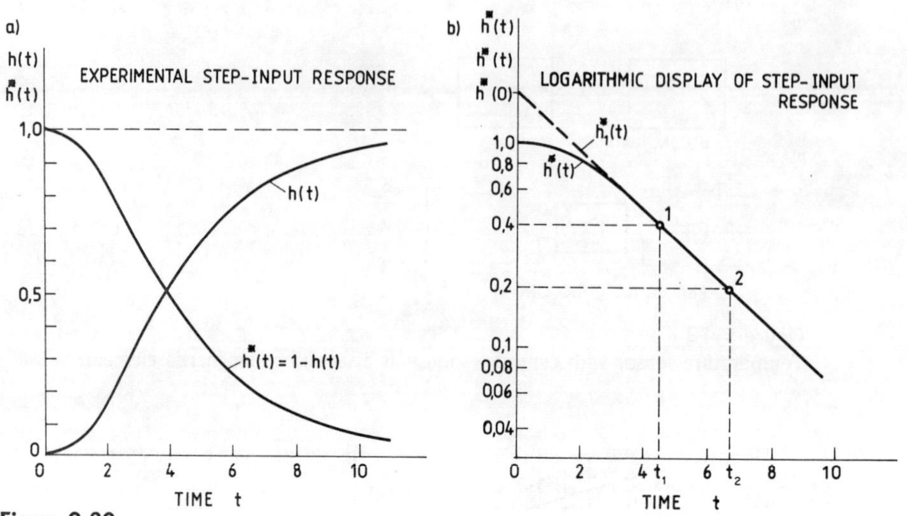

Figure 9.20
Logarithmic method of determining the time constants N_{T1} and N_{T2}

$$\Theta_T(t) = \Theta_e \left[1 - \frac{N_{T1}}{N_{T1} - N_{T2}} e^{-t/N_{T1}} + \frac{N_{T2}}{N_{T1} - N_{T2}} e^{-t/N_{T2}} \right] \tag{9.103}$$

The values of N_{T1} and N_{T2} can be found from the Kühne diagram (Kühne, 1957) (Figure 9.19), when the ratio L_T/N_T (Figure 9.17) is known. A more precise method of determining N_{T1} and N_{T2}, when $N_{T1} \gg N_{T2}$, is the logarithmic technique shown by Yarishev (1967) in Figure 9.20 described as follows. The step-input response of Equation (9.103), when presented as the relative unit-step response:

$$h(t) = \frac{\vartheta_T(t)}{\vartheta_e} \tag{9.104}$$

allows a definition of the function given by $h^*(t)$:

$$h^*(t) = 1 - h(t) \tag{9.105}$$

When $t \gg N_{T2}, h^*(t)$ will be transformed into $h_1^*(t)$:

$$h_1^*(t) = \frac{N_{T1}}{N_{T1} - N_{T2}} e^{-t/N_{T1}} \tag{9.106}$$

whose logarithm presents the straight line:

$$\ln h_1^*(t) = \ln \left(\frac{N_{T1}}{N_{T1} - N_{T2}} \right) - \frac{t}{N_{T1}} \tag{9.107}$$

As shown in Figure 9.20(b) two points which are chosen on the straight line in $\ln h_1^*(t) = f(t)$ correspond to the moments t_1 and t_2. The time constant N_{T1}, can then be found from the formula

$$N_{T1} = \frac{t_2 - t_1}{\ln h_1^*(t_1) - \ln h_1^*(t_2)} \tag{9.108}$$

To determine the time-constant N_{T2}, the value of $h_1^*(0)$ should be read from Figure 9.20(b). From Equation (9.106) at $t = 0$, it follows:

$$h_1^*(0) = \frac{N_{T1}}{N_{T1} - N_{T2}} \tag{9.109}$$

Eventually:

$$N_{T2} = \frac{h_1^*(0) - 1}{h_1^*(0)} N_{T1} \tag{9.110}$$

The time-constant N_{T2} can also be found in a similar way as N_{T1}, by presenting the function $h_2^*(t) = h_1^*(t) - h^*(t)$ in semi-logarithmic scale. This method is then especially advised to check if the second-order inertia system gives a sufficiently precise model of the sensor. If the function $h_2^*(t)$ is not a straight line, the sensor should be approximated by a higher order system.

Sensors with surfacial response ($t_{0.9}/t_{0.5} > 3.32$) are mostly approximated by first-order inertia systems, described by one time-constant N_T (Figure 9.16). If a more precise approximation is necessary, a transfer function as given in Equation (9.85) from Table 9.2 is used:

$$G_T(s) = K_T \frac{1 + sN_{T3}}{(1 + sN_{T1})(1 + sN_{T2})}$$

The step-response of this type of sensor is given by

$$\Theta_T(t) = \Theta_e \left[1 - \frac{N_{T1} - N_{T3}}{N_{T1} - N_{T2}} e^{-t/N_{T1}} - \frac{N_{T2} - N_{T3}}{N_{T2} - N_{T1}} e^{-t/N_{T2}} \right] \tag{9.111}$$

A method to find the time-constants N_{T1}, N_{T2} and N_{T3}, based on the recorded step-input response, is described in Hofmann (1976) and Yarishev (1967).

Rectangular, periodic input signals

The step-input response for the determination of temperature sensor dynamics, is easy to implement, but it does not give sufficiently high precision for low-inertia sensors with high heat-transfer coefficients. For this type of sensor, periodic input signals are advised. Rectangular periodic input signals are easy to generate, but rather time consuming in the interpretation of the results. More details of this method can be found in Bliek and Fay (1979), Bliek *et al.* (1978), Huhnke (1973) and Woschni (1980).

Compactly coded, rectangular type multifrequency binary signals, called Strathclyde compact MBS (Henderson *et al.*, 1987; McGhee *et al.*, 1986, 1987), have also been used for the immersion testing of temperature sensors (McGhee *et al.*, 1989, 1990, 1991).

Sinusoidal input signals

Application of sinusoidal input signals for determining sensor dynamics is not very popular so far, as it requires rather complicated instrumentation and time consuming measurements. The design of a special channel with low-inertia heating elements described by Huhnke (1973) and shown in Figure 9.15, decisively simplifies the application of this method. A simplified diagram of the heating element power control, shown in Figure 9.21, arranges a sum of the output signals from the sinusoidal signal generator and the biasing constant d.c. voltage source which activate the final control element to change the heating power. The constant voltage component and the amplitude of the sinusoidally changing voltage are adjusted as a function of the air velocity. Measurement of the air temperature is usually by a thin bare wire thermocouple or a resistance temperature sensor of Pt wire. The diameter of the Pt wire can be calculated from Equation (9.99) or determined experimentally as described by Higgins and Keim (1954) and Huhnke (1973). Recording of the temperatures, ϑ, of the air and of the tested sensor, ϑ_T, at different frequencies of the sinusoidal temperature oscillations allows a determination of the frequency response of the sensor mostly in the form of a Bode diagram like that shown in Figure 9.22 (Michalski *et al.*, 1981).

An experimentally determined frequency response of a temperature sensor is the most precise way of presenting its dynamics, because no simplifying assumptions of any kind are needed and at the same time all existing nonlinearities are taken into consideration. If the sensor is approximated by a first order inertia element it is easy to find its time constant from the corner frequency ω_c as $N_T \approx 1/\omega_c$ (Figure 9.22).

Presentation of the frequency response of a sensor by a Bode diagram is especially convenient in the synthesis of an automatic temperature control system. (Michalski and Eckersdorf, 1987).

The frequency response is very useful in conceiving the best sensor model and its characteristic parameters (Michalski *et al.*, 1981).

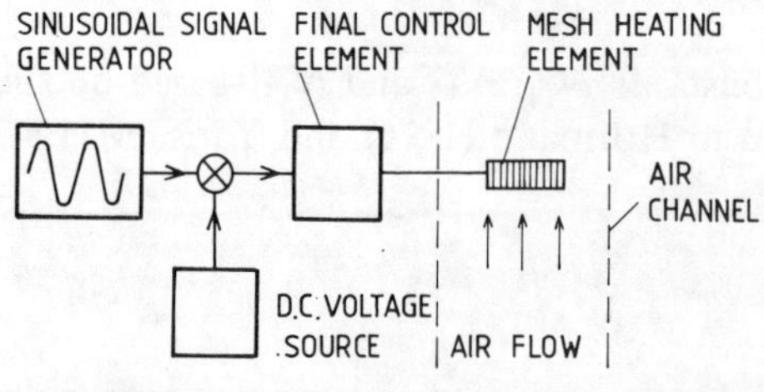

Figure 9.21
Power control of a mesh heating element in an air-channel.

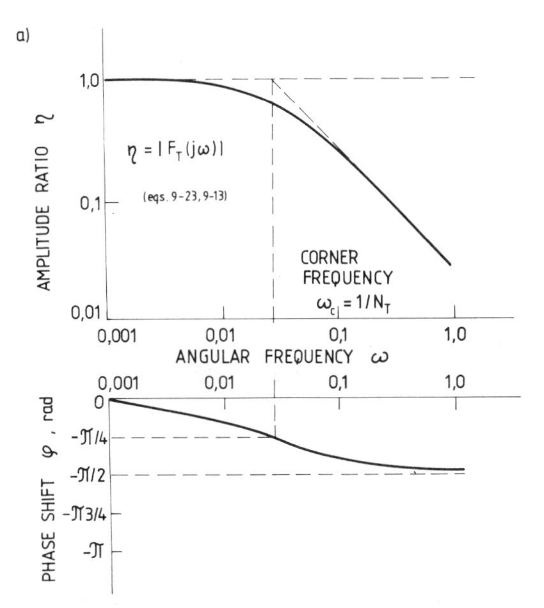

Figure 9.22
Bode diagram display of the frequency response of a temperature sensor.

9.5.3. *Radiant heat transfer*

Heating-up curve method

The heating-up curve of a temperature sensor can be recorded by placing the sensor in a furnace chamber at the temperature ϑ_e. When used to measure sensor dynamics it is known as heating-up curve method. To give explicit and reproducible measuring conditions, the furnace should be constructed so that the emissivity of the furnace chamber is as near to unity as possible, that its thermal capacity be far greater than that of the sensor, and that the furnace is heated to a uniform temperature over a zone much longer than the sensor immersion depth. In addition the furnace temperature has to be precisely controlled in the range from 500 °C up to the maximum temperature of application of the sensor.

In that temperature range, the sensor cannot be approached as a linear element and therefore it should not be described by a transfer function in the manner given in §§9.3.2. and 9.3.3. The experimental heating-up curve of a sensor, can only be used for really rough, approximate estimation of the sensor dynamics. Commonly the $t_{0.5}$ and $t_{0.9}$ times, are used with their values referred to the final value of the temperature ϑ_e of the heating-up curve.

Heating-up and cooling-down curves method

For temperature sensors exhibiting volumetric response ($t_{0.9}/t_{0.5}<3.32$) an easy way to determine their equivalent transfer function based on the heating-up curve and the first part of the cooling-down curve is known as the heating-up and cooling down curves method (Eckersdorf and Michalski, 1984). A furnace, similar to that for the heating-up curve method described above, should be used for the method which is based on the step response Equation (9.103). In this equation two functions which can be distinguished are:

$$f_1(t/N_{T1}) = \frac{N_{T1}}{N_{T1} - N_{T2}} e^{-t/N_{T1}}$$

$$f_2(t/N_{T2}) = \frac{N_{T2}}{N_{T1} - N_{T2}} e^{-t/N_{T2}}$$

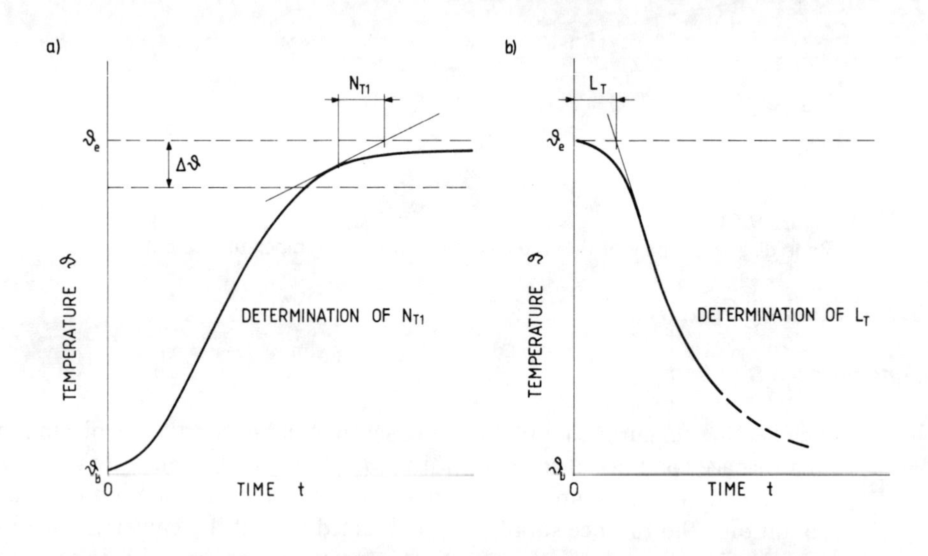

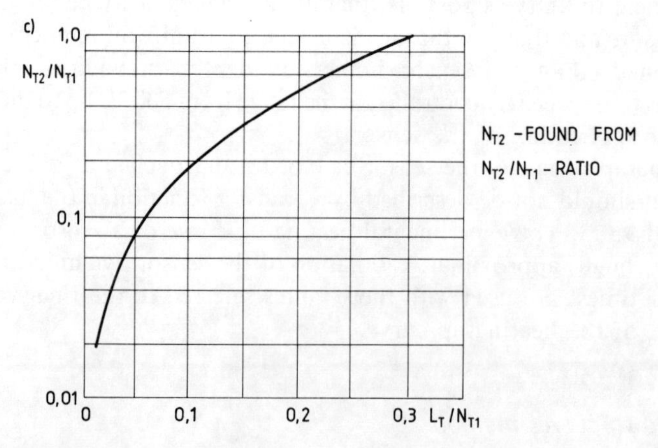

Figure 9.23
Determination of time constants N_{T1} and N_{T2} of a sensor by the method of heating-up (a) and cooling-down (b) curves.

Assuming that $N_{T2}<0.5N_{T1}$, the function f_2 decreases with time far quicker than f_1. After $t>1.5N_{T1}$, since $f_2(t/N_{T2})<f_1(t/N_{T1})$, it follows that the latter part of the heating-up curve is only determined by the bigger time-constant N_{T1}. In the temperature interval, $\Delta\vartheta$, below 100 °C, the sensor can be regarded as a linear element so that the graphically found time-constant N_{T1} (Figure 9.23(a)) has a correct value, relevant to the final temperature value ϑ_e.

From the cooling-down curve of Figure 9.23(b) the equivalent lag, L_T, of the sensor may be found. This is then used to find its second time-constant N_{T2} as shown in Figure 9.23(c). The diagram $(N_{T2}/N_{T1})=f(L_T/N_{T1})$ is based on the data from Figure 9.19. The determined values of N_{T1} and N_{T2} are referred to the temperature ϑ_e. It has been proved by Eckersdorf and Michalski (1984) that the proposed method is sufficiently precise for those industrial sensors typically applied in electric furnaces in the temperature range from 600 °C to 1200 °C.

Step-input response method

To precisely realize a step-input at a given temperature level for use in the step-input response method, the sensor should be transferred, as quickly as possible, from temperature $(\vartheta_e-0.5\Delta\vartheta)$, to temperature $(\vartheta_e+\Delta\vartheta)$. Each time, the sensor should be immersed up to its normal immersion depth. To increase the measurement precision the step increase and the step decrease of temperature should be repeated twice, taking the average values as the final result. The value of temperature step $\Delta\vartheta$ is of utmost importance. Too big a step is not allowed, to ensure that the sensor response remains mainly linear, while a step which is too small may result in large random errors. In practice the $\Delta\vartheta$ value should be about 60 to 100 °C. Either two electric furnaces, standing side-by-side or one two-section furnace can be used. In the first case, in transferring the sensor from one furnace to another, the temperature versus time is far from a pure step-input, thus resulting in additional errors (Figure 9.24). In the second case, although the step-input generation assures a much higher precision, its

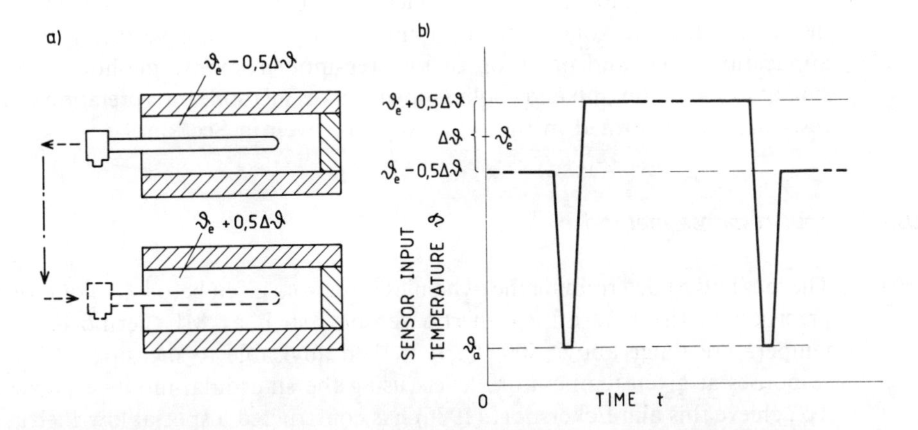

Figure 9.24
Realization of step-input using two furnaces. (a) sensor transfer, (b) sensor input temperature versus time.

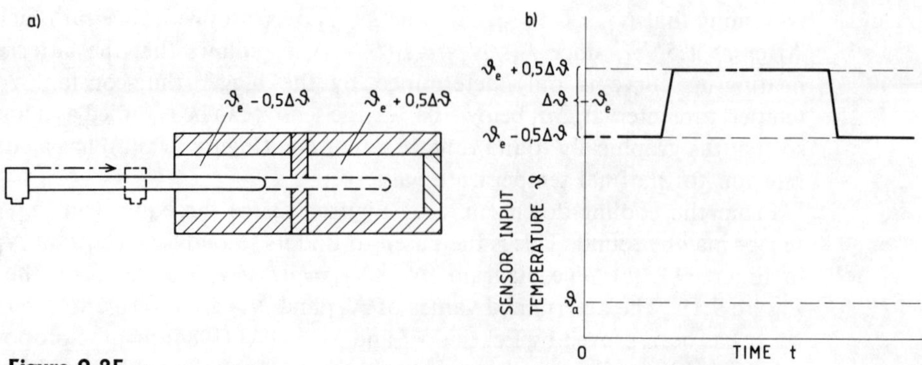

Figure 9.25
Realization of step-input using one, two-section furnace (a) sensor transfer, (b) sensor input temperature versus time.

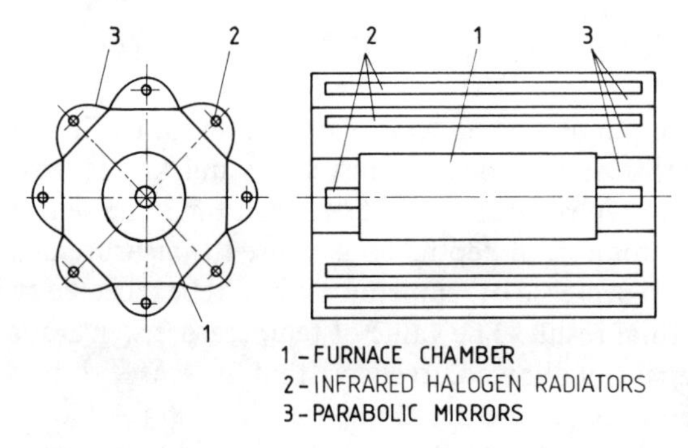

Figure 9.26
Special electric furnace for measuring frequency response of low-inertia sensors in conditions of radiant heat transfer.

application range is restricted, depending upon the complexity of furnace construction (Figure 9.25). In particular, the furnace, which should have sufficiently long non-interacting heating zones, will be limited only to testing very long sensors. The application range and precision of the step-input response method of $\vartheta_e \pm \Delta\vartheta$ are similar to the heating-up and cooling-down curves method. Interpretation of measured responses is performed in the same ways as given in Section 9.5.2.

Sinusoidal-input response method

The methods of determining the dynamics of sensors given before do not give sufficient precision in the case of low-inertia sensors, such as MI thermocouples, in the temperature range above 500 °C. It is then advisable to measure their frequency responses at given temperature levels using the sinusoidal-input response method. To achieve this aim Eckersdorf (1986) has constructed a special low-inertia furnace, Figure 9.26, which enables the generation of sinusoidal temperature variations from 0.001 to 0.5 Hz in the range of from 600 to 1000 °C. The furnace chamber, 1, made of a temperature resistant steel tube with closed ends is heated by halogen radiators, 2, whose radiation is concentrated by parabolic mirrors, 3. Sinusoidal temperature variations are generated in a closed-loop system applying a generator, a temperature controller and a thyristor final control element, with thin thermocouples for temperature measurement and control welded to the chamber tube, 1. The temperature amplitudes for the measurements are 10 to 40 K. Normal presentation of the results is similar to that described in §9.5.2.

9.6 MIXED EXPERIMENTAL–CALCULATIVE METHOD FOR DETERMINING THE DYNAMIC PROPERTIES OF REAL SENSORS

Methods to be described in this section enable the dynamic properties of sensors to be determined for any surfacial heat-transfer coefficient, when experimental results at given α-values are available. These methods are applicable in low temperature ranges, where the α-values describe the working conditions of the sensor with sufficient precision.

To help with the practical application of the described methods, the average α-values for cylindrical sensors of diameter D in different media and at different temperatures are given in Table 9.4 (Hofmann, 1976; Yarishev, 1967).

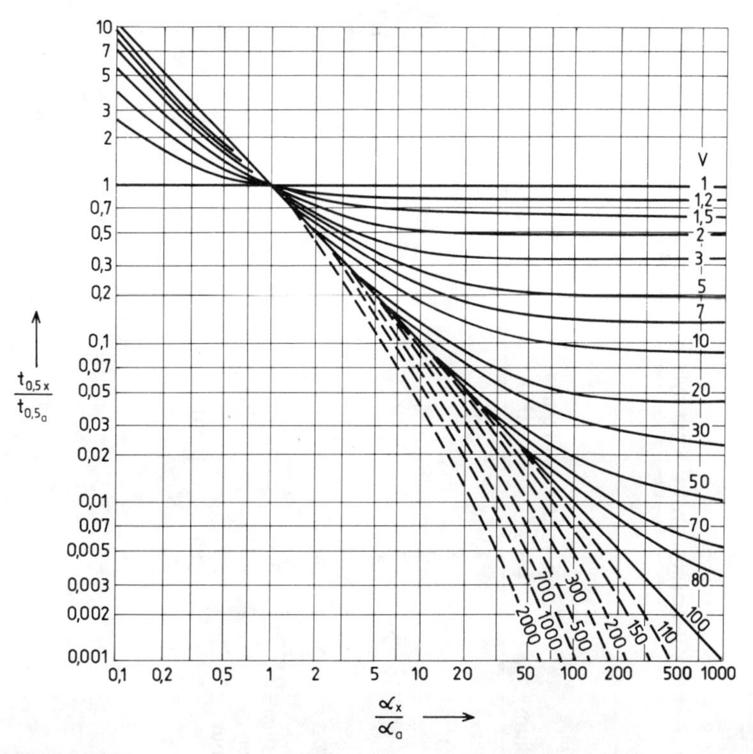

Figure 9.27
Diagram to determine the characteristic parameter V, following the Lieneweg method.

Table 9.4
Heat transfer coefficient α, W/m^2 K for cylindrical temperature sensors of diameter D, mm.

			Flow velocity (m/s)							
			0				2			
			$D=$				$D=$			
Medium	**Temperature (°C)**	**Presure (MPa)**	**1**	**4**	**10**	**20**	**1**	**4**	**10**	**20**
Water	30			840	700	580				
	50		2400	1390	1130	950				
	70			1900	1530	1390				
Air	50		37	16.0	11.3	10.2				
	100			20.0	14.7	12.7	140	49	42	38
	200			25.3	18.8	16.2				
Overheated water vapour	300	0.5								
	400	2								
	300	5								
	500	10								

			Flow velocity (m/s)								
			0			0.01			0.1		
			$D=$			$D=$			$D=$		
Medium	**Temperature (°C)**	**Presure (MPa)**	**4**	**10**	**20**	**4**	**10**	**20**	**4**	**10**	**20**
Oil	70		52	36	28	174	116	87	520	337	232

Medium	Temperature (°C)	Pressure (MPa)	Flow velocity (m/s) 5				10				20				50			
			D=				D=				D=				=			
			1	4	10	20	1	4	10	20	1	4	10	20	1	4	10	20
Water	30																	
	50			1100 to 11000														
	70																	
Air	50																	
	100			80	70	64	300	120	105	93		168	147	133	520	285	248	212
	200																	
Overheated	300	0.5		104	156	261		162	226	360		226	320	500		360	505	800
water	400	2		290	383	630		383	545	860		560	780	1200		870	1230	1950
vapour	300	5		755	1100	1800		1100	1570	2550		1570	2260	3600		2500	3550	4200
	500	10		930	1330	2030		1330	1850	2900		1850	2610	4070		2960	4100	6600

Condensing, saturated water vapour: 5800–10 500
Condensing, overheated water vapour: 460
Condensing liquids: 1750–35 000

9.6.1 Lieneweg method

The Lieneweg method (Lieneweg, 1975) is based on experimentally determined step-input responses of the tested sensor in two different media. Surfacial heat transfer coefficients α must be known in each case. In most cases the measurements are executed in water and air to find either the half-value times $t_{0.5,w}$ in water and $t_{0.5,a}$ in air or the time constant $N_{T,w}$ in water and $N_{T,a}$ in air.

Let the corresponding heat transfer coefficient values be α_w in a water medium, α_a in an air medium and α_x in any other medium. The characteristic parameter, V, of the sensor is found from Figure 9.27, knowing the ratio $t_{0.5,w}/t_{0.5,a}$ of the half-value times and the ratio α_w/α_a of the heat transfer coefficients. When α_x in the new condition is known, the new ratio $t_{0.5,x}/t_{0.5,a}$, and thus the sought value, $t_{0.5,x}$, can also be found from the ratio α_w/α_a and the value of V using Figure 9.27. In a similar way, the time constants or $t_{0.9}$ of the sensor can be determined.

Numerical example

The heat transfer coefficient, α_w, between a temperature sensor and a water medium has the value 1200 W/m² °C. Under these conditions the sensor has the time constant, $N_{T,w} = 10$ s. With an air medium the values were $\alpha_a = 15$ W/m² °C and $N_{T,a} = 100$ s. Find the time constant in the medium x when the heat transfer coefficient is $\alpha_x = 150$ W/m² °C.

Solution:

For $N_{T,w}/N_{T,a} \approx t_{0.5,w}/t_{0.5,a} = 0.1$ and for $\alpha_w/\alpha_a = 80$ then $V = 10$ from Figure 9.27.
For $\alpha_x/\alpha_a = 10$ and $V = 10$, it follows from Figure 9.27 that $t_{0.5,x}/t_{0.5,a} \approx N_{T,x}/N_{T,a} = 0.18$.
Finally $N_{T,x} = 0.18 \times 100 = 18$ s

9.6.2 Model method

The model method described by Hofmann (1976) is based on the similarity between the sensor's experimental step-input response and the model step-input response of an infinitely long solid-cylinder (see §9.4.1). Equality of the $t_{0.9}/t_{0.5}$ ratio is assumed as the similarity criterion of both functions. The method hinges upon the dependence of the ratio $t_{0.9}/t_{0.5}$ upon the Biot number, for points on the cylinder surface (Equation (9.90)) and in its axis (Equation (9.91)) as shown in Figure 9.28.

From the experimental step-input response of the sensor characterized by the heat transfer coefficient α, the ratio $t_{0.9}/t_{0.5}$ is determined in given measurement conditions. The corresponding value of the Biot number is read from Figure 9.28. For $t_{0.9}/t_{0.5} > 3.32$ the model step-input response at a Bi value read from Figure 9.29 for the cylinder surface, corresponds to the step-input response of the real sensor. For $t_{0.9}/t_{0.5} < 3.32$ the corresponding step-input response of the model is that of the cylinder axis.

Knowing the Bi value of the model function, the coefficients $A_{r=R,1}$, $A_{r=R,2}$, ..., $A_{r=R,n}$, $A_{r=0,n}$ are consecutively calculated from Equation (9.90a) or Equation

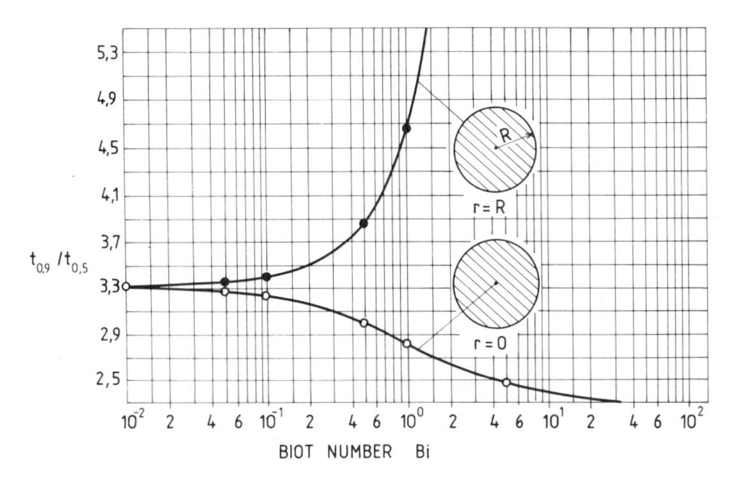

Figure 9.28
$t_{0.9}/t_{0.5}$ versus Biot number Bi display for a homogeneous, infinitely long, solid cylinder.

(9.91a). The N_1, N_2, . . . , N_n of Equation (9.90b) are then calculated finally yielding the step-input responses $h_{r=R}(t)$ of Equation (9.90) or $h_{r=0}(t)$ of Equation (9.91) and the transfer functions $G_{r=R}(s)$ of Equation (9.93) or $G_{r=0}(s)$ of Equation (9.94).

To determine the step-input response and the transfer function of the same sensor at different operating conditions for α_x, the new Biot number for the model function is calculated as:

$$\mathrm{Bi}_x = \mathrm{Bi}\frac{\alpha_x}{\alpha}$$

Any further calculations are completed as described above. To allow the calculations, some necessary data are given in Figure 9.29.

In the work of Hofmann (1976) the relevant computer algorithm can be found.

Numerical example

A dynamic response experiment on a temperature sensor yielded the data:

$t_{0.5} = 20$ s; $t_{0.9} = 60$ s; with $\alpha = 30\ \mathrm{W/m^2\,K}$.

Determine the transfer function for $\alpha = 30\ \mathrm{W/m^2\,K}$ and for $\alpha = 100\ \mathrm{W/m^2\,K}$.

Solution: From the response time ratio, $t_{0.9}/t_{0.5} = 60/20 = 3$ in conjunction with Figure 9.28 the Biot number is Bi = 0.55. The model step-input response is at the cylinder axis. From Figure 9.22(d), for Bi = 0.55 it follows that:

$$A_{r=0;1} = 1.1; \quad A_{r=0;2} = -0.18; \quad A_{r=0;3} = 0.08$$

From Figure 9.29(b) for Bi = 0.5 it follows that:

$$N_1/K = 10; \quad N_2/K = 0.06; \quad N_3/K = 0.02$$

Continued on page 298

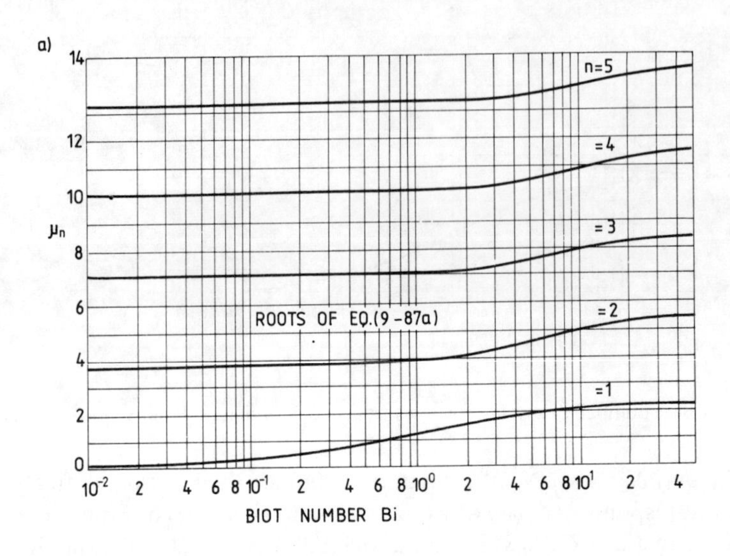

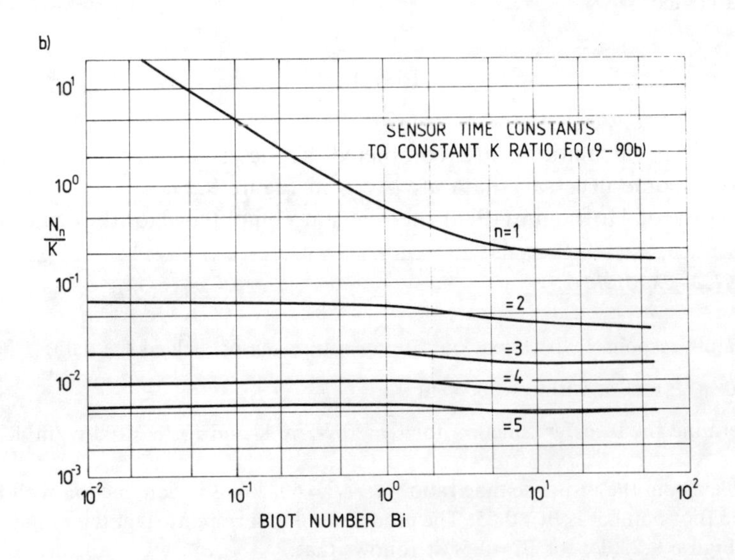

Figure 9.29
Coefficients from equations describing dynamic properties of a homogeneous, infinitely long, solid cylinder versus Biot number Bi.

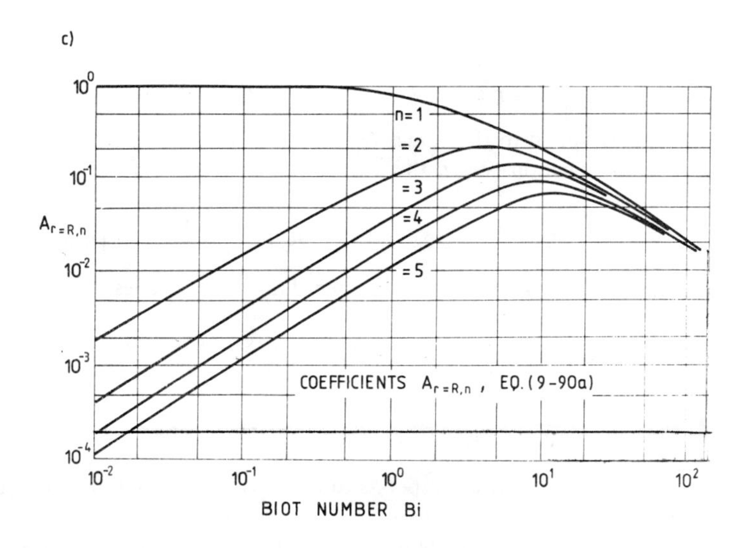

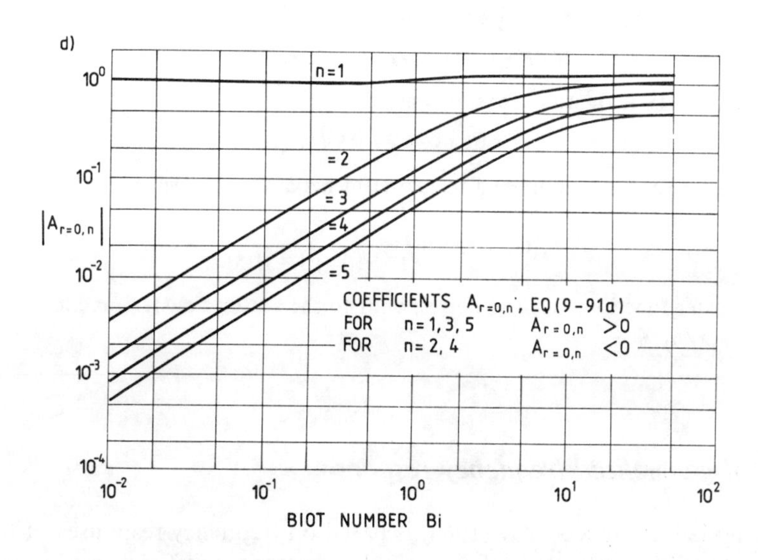

Figure 9.29
(continued)

Continued from page 296

To determine the constant K, the first time constant is calculated as:

$$N_1 = t_{0.5}/\ln(2A_{r=0;1}) = 20/\ln(2 \times 1.1) = 25.3 \text{ s}$$

using this value gives $K_1 = N_1/n_1 = 25.3/1.0 = 25.3$ s.

The other time constants are:

$$N_2 = Kn_2 = 25.3 \times 0.06 = 1.5 \text{ s}, \quad N_3 = Kn_3 = 25.3 \times 0.02 = 0.5 \text{ s}$$

where n_1, n_2, n_3 correspond to the values N_n/K from Figure 9.29(b).

Conforming with Equation (9.94), the transfer function will be

$$G_{r=0}(s) = 1 - s\left[A_{r=0;1}N_1\frac{1}{1+sN_1} + A_{r=0;2}N_2\frac{1}{1+sN_2} + A_{r=0;3}N_3\frac{1}{1+sN_3}\right]$$

Since $N_3 \ll N_1$, the above expression can be shortened to only two series elements so that:

$$G_{r=0}(s) = 1 - s\left[1.1 \times 25.3\frac{1}{1+s25.3} - 0.18 \times 15\frac{1}{1+s1.5}\right]$$

$$= \frac{1}{(1+s25.3)(1+s1.5)}$$

This transfer function, $G_{r=0}(s)$, corresponds to the sensor transfer function $F_T(s)$ for $\alpha = 30$ W/m^2 K.

For $\alpha' = 100$ W/m^2 K, the calculations proceed in a similar way. For $\text{Bi}' = \text{Bi}(\alpha'/\alpha) = 0.55(100/30) = 1.83$, it follows from Figure 9.29(d) that:

$$A_{r=0;1} = 1.37; \quad A_{r=0;2} = -0.44; \quad A_{r=0;3} = 0.23$$

From Figure 9.29(b) for Bi = 1.83 the values are:

$$N_1/K = 0.41; \; N_2/K = 0.06; \; N_3/K = 0.02$$

Inserting $K = 25.3$, the time constants are thus:

$$N_1 = 10.4 \text{ s}; \; N_2 = 1.5 \text{ s}; \; N_3 = 0.5 \text{ s}$$

The transfer function, shortened to two elements as before, will be:

$$G_{r=0}(s) = \frac{1}{(1+s10.4)(1+s1.5)}$$

This transfer function, $G_{r=0}(s)$, corresponds to the sensor transfer function $F_T(s)$ for $\alpha' = 100$ W/m^2 K.

9.6.3 Method of generalized frequency response

The method given by Kocurov (1963) is based on frequency responses of the tested sensor at two different values of surfacial heat transfer coefficient α. In the thermal conversion stage of the sensor given in Equation (9.5b), two elements can be distinguished. The first has a transfer function $F_1(s)$ dependent on the coefficient α while the second has a transfer function $F_2(s)$ dependent on the construction, materials and size of the sensor (Figure 9.30).

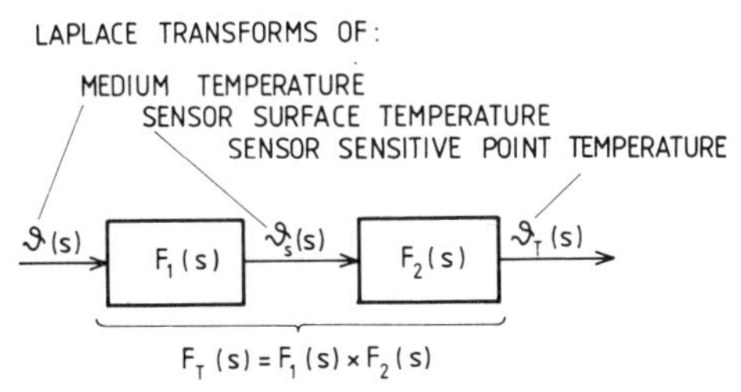

Figure 9.30
Equivalent block diagram of the thermal conversion stage of a temperature sensor.

It can be shown (Equation (9.34)) that a generalized frequency response of the thermal conversion stage for any α, is given by

$$F_{T,\alpha}(j\omega)=\frac{kF_{T,\alpha 1}(j\omega)F_2(j\omega)}{F_2(j\omega)+(k-1)F_{T,\alpha 1}(j\omega)} \tag{9.112}$$

where $k=\alpha/\alpha_1$, $F_{T,\alpha 1}(j\omega)$ is the experimental frequency response with $\alpha=\alpha_1$, and $F_2(j\omega)=F_{T,\alpha\to\infty}$ is the experimental frequency response at $\alpha\to\infty$.

Thus, by experimentally determining the sensor's frequency response at the known α_1 and at $\alpha\to\infty$ (as occurs in intensively mixed water), it is possible to find the frequency response $F_{T,\alpha}(j\omega)$ for any α-value.

All of the formulae given above are also valid in the s-domain.

9.7 DYNAMIC PROPERTIES OF CHOSEN SENSORS

Table 9.5 presents the characteristic parameters of some chosen temperature sensors. The parameters of MI thermocouples are given in Table 9.6 as well as in Figures 9.31 and 9.32.

The dynamic properties of thermocouple sensors, intended for application over 600 °C at atmospheric pressure, are displayed against temperature, in Figure 9.33. In this figure two time constants N_{T1} and N_{T2} or one equivalent time-constant N_{Te} versus temperature are shown for two different industrial sensors. Determination of the equivalent time constant N_{Te} is based on the measured half-value time $t_{0.5}$ in accordance with Eckersdorf and Michalski (1984). Experimental frequency responses for the same sensors obtained by the same authors are shown in Figure 9.34.

Frequency responses of MI thermocouples and of bare, unprotected thermocouples, experimentally determined as described in §9.5.3, are shown in Figure 9.35 and 9.36. Approximating the dynamics of MI thermocouples by the equivalent transfer functions of the first order inertia elements gave relevant time constants (Figure 9.37) which were found from the frequency responses, as $N_T=1/\omega_c$ as shown in Figure 9.22.

For rough estimation and comparison of the dynamics of different thermocouple sensors, data from Figure 9.38, can be used. The equivalent time constants N_T, have been found from the heating-up curves from ambient temperature to ϑ_e as $N_{Te}=t_{0.5}/0.693$.

Table 9.5
Characteristic dynamic parameters of temperature sensors determined from step-response from 20 to 40 °C (VDE/VDI—Richtlinien 5522, 1979).

Temperature sensors	Medium					
	Air at $v = 1$ m/s			Water at $v = 0.2$ m/s		
	$t_{0.5}$ (s)	$t_{0.9}$ (s)	$t_{0.9}/t_{0.5}$	$t_{0.5}$ (s)	$t_{0.9}$ (s)	$t_{0.9}/t_{0.5}$
I Liquid-in-glass thermometers						
Mercury, bulb φ6×12 mm	48	150	3.12	2.4	6.2	2.58
Mercury, bulb φ6×38 mm, steel sheath φ10 mm	121	350	2.89	23	77	3.35
Alcohol, bulb φ6×38 mm, steel sheath φ12 mm	200	500	2.50	81	251	3.10
II Manometric liquid-filled thermometers						
Bulb φ17 mm	180	600	3.33	4.4	12.5	2.84
Bulb φ17 mm, steel sheath φ22 mm	340	1090	3.21	72	224	3.10
III Thermocouple sensors applied at atmospheric pressure						
NiCr–NiAl chromium iron alloy sheath φ22 mm	390	1030	2.64	112	440	3.92
NiCr–NiAl, chromium iron alloy sheath φ15 mm	190	490	2.58	57	170	2.98
NiCr–NiAl, ceramic sheath φ15 mm	220	580	2.63	88	245	2.78
Pt10Rh–Pt, ceramic sheath φ15 mm	180	455	2.52	42	125	2.98
Pt10Rh–Pt, two ceramic sheaths φ15 mm, φ26 mm	235	570	2.42	85	230	2.70
IV Thermocouple sensors applied at medium and high pressure						
Cylindrical thermowell φ9 mm, thermocouple insert φ6 mm (max 4MPa, max 400 °C)	92	300	3.26	7	30	4.28
Cylindrical thermowell φ14 mm, thermocouple insert φ8 mm (max 10MPa, max 500 °C)	148	460	3.11	21	80	3.81
Conical thermowell φ24 mm, thermocouple insert φ6 mm (max 60MPa, max 540 °C)	225	800	3.55	7	30	4.28
Conical thermowell φ18 mm, MI thermocouple φ3 mm (max 16MPa, max 540 °C)	190	750	3.95	2	9.5	4.75
V Resistance thermometer detectors (RTD)						
Glass platinum RTD, φ4 mm	11.5	47	4.08	0.7	2.3	3.2
Glass platinum RTD, φ3 mm	8	28	3.50	0.7	2.0	2.8
Glass platinum RTD, φ1 mm	1.3	4.1	3.28	0.15	0.6	4.0
Ceramic platinum RTD, φ2.8 mm	10	39	3.90	0.45	1.6	3.5
Ceramic platinum RTD, φ1.5 mm	6	20	3.33	0.2	0.6	3.0
Ceramic nickel RTD, φ5.2 mm	21	67	3.19	2.3	10.5	4.5
Resistance thermometer insert φ6 mm	72	220	3.06	6.2	16	2.58
Resistance thermometer insert φ8 mm	115	335	2.31	19	55	2.8
VI Resistance thermometer sensors applied at medium and high pressure environment						
Cylindrical thermowell φ9 mm, RT-insert φ6 mm (max 4MPa, max 400°C)	140	390	2.78	30	80	2.6
Cylindrical thermowell φ14 mm, RT-insert φ8 mm (max 10MPa, max 500°C)	240	720	3.00	53	155	2.9
Conical thermowell φ24 mm, RT-insert φ6 mm (max 60MPa, max 540°C)	315	1070	3.40	38	110	2.9
VII Low inertia resistance thermometer sensors						
Metal sheath φ6 mm (max 500°C)	32 ÷ 40	118 ÷ 130		1.2 ÷ 2.1	4.5 ÷ 7.2	
Metal sheath φ3.5 mm, thinned down to φ2.3 mm (max 600°C)	4.5	14	3.11	0.35	1.3	3.7
High pressure cylindric thermowell φ12 mm thinned down to φ4.5 mm (max 10MPa, max 400°C)	80	280	3.50	6	20	3.3
High pressure conical thermowell φ24 mm, thinned down to φ4.5 mm (max 60MPa, max 400°C)	80	250	3.12	6	20	3.3

Table 9.6
Half-value time $t_{0.5}$, s, of MI thermocouples determined from step-response from 20 to 40 °C

Medium	Sheath diameter Φ (mm)					
	0.5	1	1.5	3	4.5	6
Water at $v = 0.2$m/s	0.06	0.15	0.21	1.2 (0.6)*	2.5	4 (4)*
Air at $v = 2$m/s	1.8	5	8	23 (26)*	37	60 (55)*

*Value for MI resistance sensor.

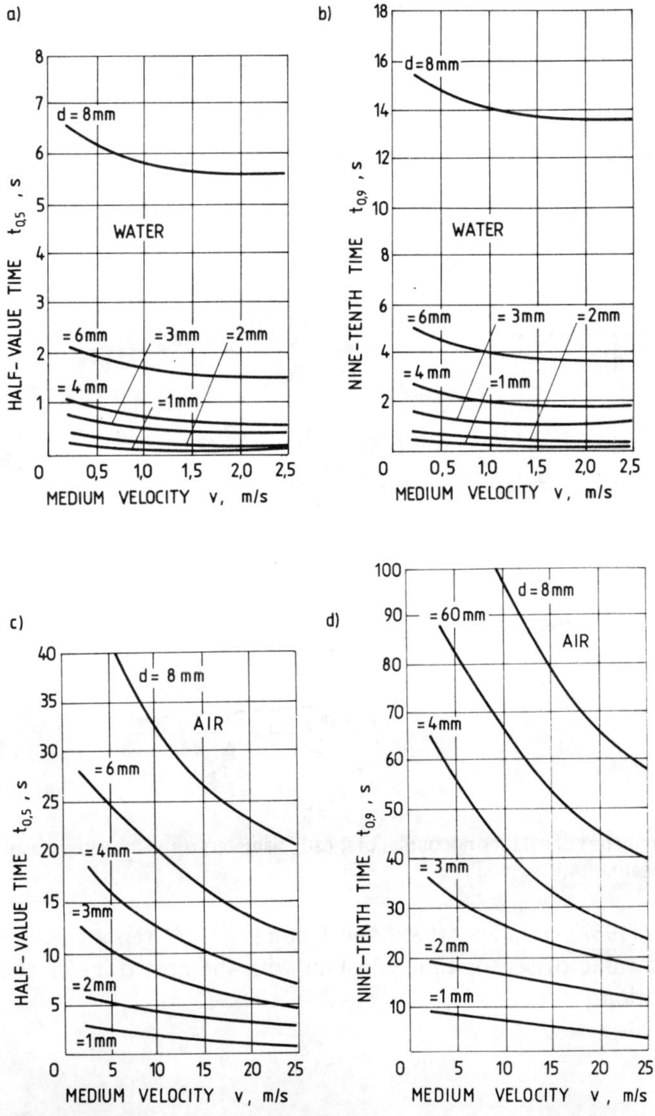

Figure 9.31
Characteristic parameters of MI thermocouples of sheath diameter d and insulated measuring junction.

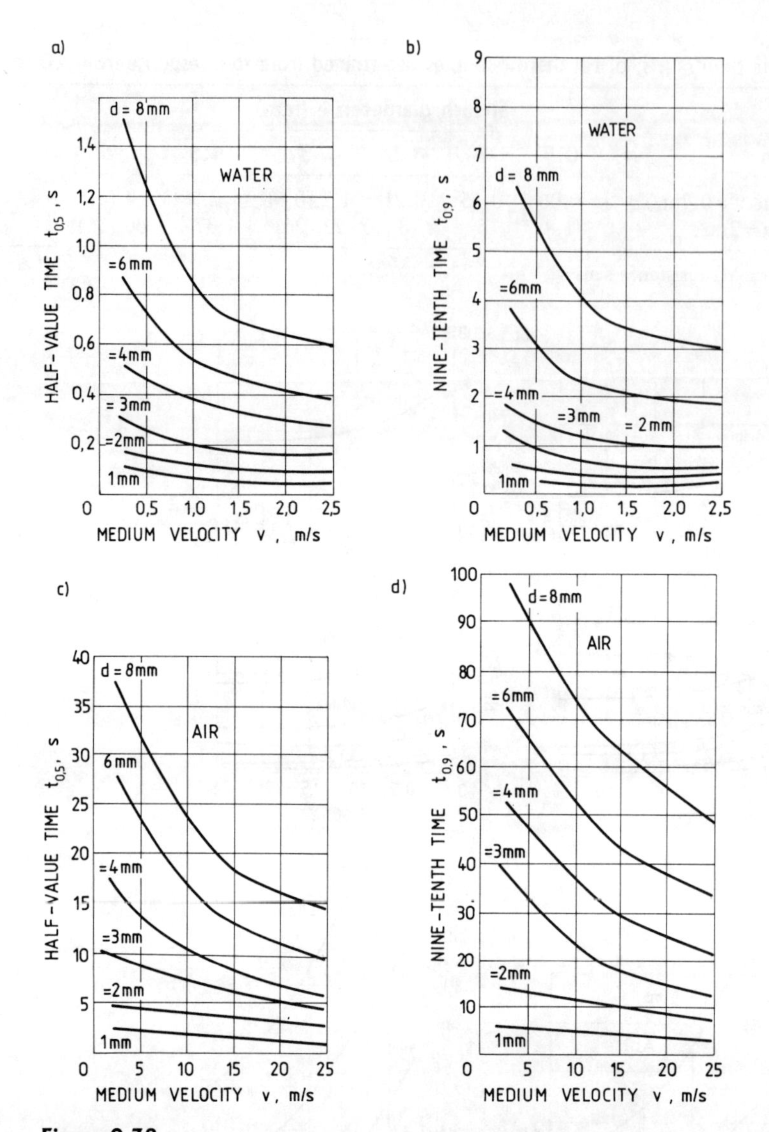

Figure 9.32
Characteristic parameters of MI thermocouples of sheath diameters *d* and insulated measuring junction welded to the sheath.

Kerlin *et al.* (1982), Pandey (1985) and Chohan (1986) report the possibility of correlating the value of sensor time constant with the heat transfer coefficient by the simple relation:

$$N_\alpha = C_1 + C_2/\alpha \tag{9.112a}$$

where C_1 and C_2 are constants depending upon the construction of the sensor and α is the heat transfer coefficient.

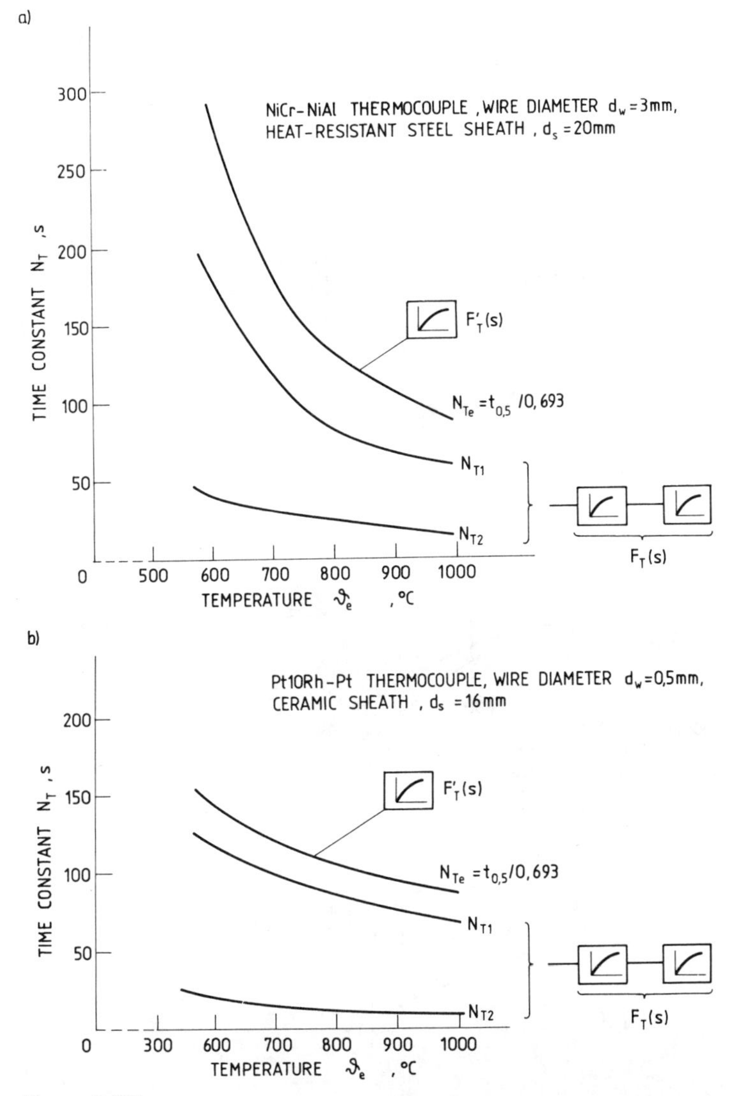

Figure 9.33
Characteristic parameters of thermocouples, measured in an electric chamber furnace in still air.

9.8 CORRECTION OF DYNAMIC ERRORS

9.8.1 *Principles of correction*

To eliminate the dynamic error of a temperature sensor, a corrector of the transfer function $G_C(s)$ shown in Figure 9.39, can be used.

The dynamic error will be zero if:

$$G_{TC}(s) = G_T(s)G_C(s) = K_T K_C \tag{9.113}$$

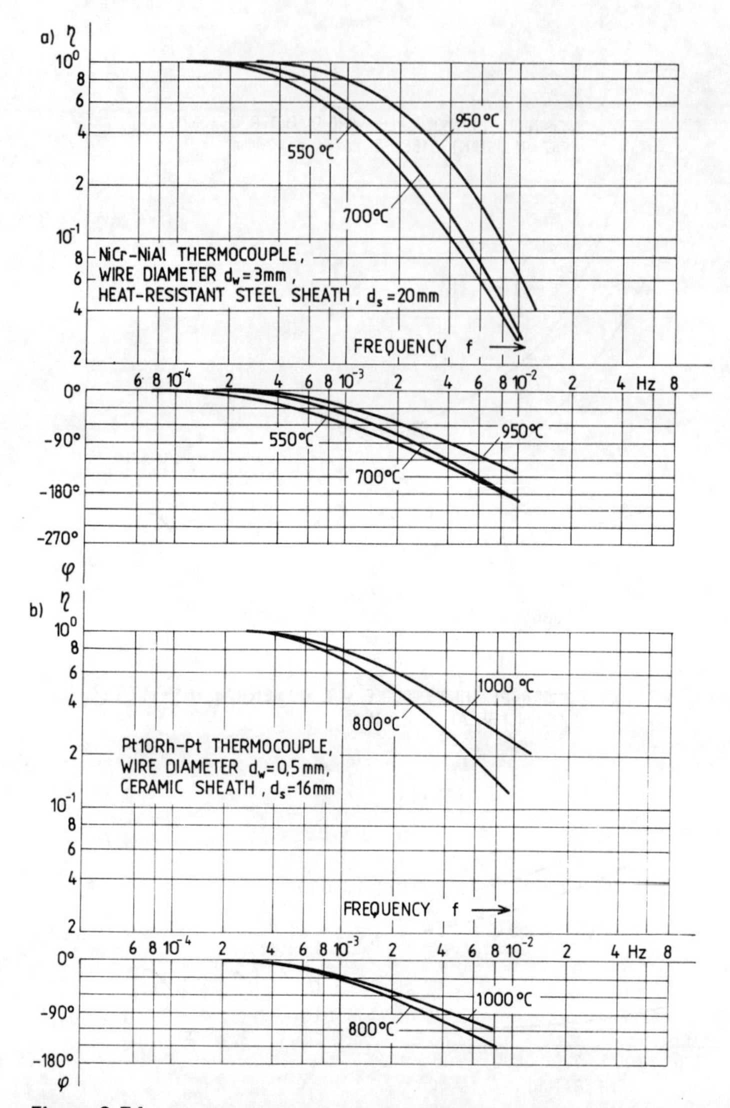

Figure 9.34
Frequency response of industrial thermocouples, measured in an electric chamber furnace in still air.

Thus the transfer function of a perfect corrector should be:

$$G_C(s) = \frac{K_T K_C}{G_T(s)} \tag{9.113a}$$

where K_C is the corrector gain. For a sensor, approximated by a first order inertia element having the transfer function given from Equation (9.81) in Table 9.2, the transfer function of the corrector should be:

$$G_C(s) = K_C(1 + sN_C) \tag{9.114}$$

where $N_C = N_T$.

In practice each corrector has some inertia, which has to be considered in the transfer function of Equation (9.114), finally yielding:

$$G_C(s) = K_C \frac{1 + sN_C}{1 + s\dfrac{N_C}{k}} \tag{9.115}$$

where k is the correction coefficient, $k > 1$.

Satisfying the condition $N_C = N_T$, the transfer function of a sensor-corrector set will be:

$$G_{TC}(s) = K_T K_C \frac{1}{1 + s\dfrac{N_T}{k}} = K_{TC} \frac{1}{1 + s\dfrac{N_T}{k}} \tag{9.116}$$

From the above equation it follows, that a corrected system, like the sensor alone, is also a first order inertia element, but with a time-constant k times smaller. In practice corrector types described by Equation (9.115) are also used to correct the dynamics of sensors described as second order inertia elements having a transfer function like Equation (9.82) or Equation (9.85). If $N_C = N_{T1}$, the 'sensor corrector' set approaches a double inertia sensor, whose larger time-constant N_{T1} is reduced k times.

The optimum corrector parameters N_C and k, depend on the type of sensor transfer function, the range of variations of sensor dynamics and the time dependence of the measured temperature.

Optimization criteria are usually based on the comparison of step-input response and the amplitude versus frequency characteristic of the sensor and of the sensor-corrector assembly. Further consideration of this question appears in Bainbridge and Kaltner (1984), Eckersdorf (1980), Hofmann (1967a, 1970, 1976), Kraus and Woschni (1979) and Michalski and Eckersdorf (1978). Performance indices, typically based upon the response time $t_{r,0.5\%}$ and the cut-off frequency $f_{0.5\%}$, are used to judge the correction efficiency.

Figure 9.40 presents the dynamic errors of the step-input response of a sensor, having the transfer function of Equation (9.81) connected to a corrector having the transfer function Equation (9.115), for different corrector parameters. For the same assembly, Figure 9.41 also presents the influence of corrector parameters on corrector performance indices, defined as the ratio of response times of sensor-corrector, $t_{r,0.5\%TC}$, to the response time of the sensor alone, $t_{r,0.5\%T}$. It is apparent that the deciding influence on assembly performance is the correct choice of the ratio of corrector time constant N_C to that of sensor N_T.

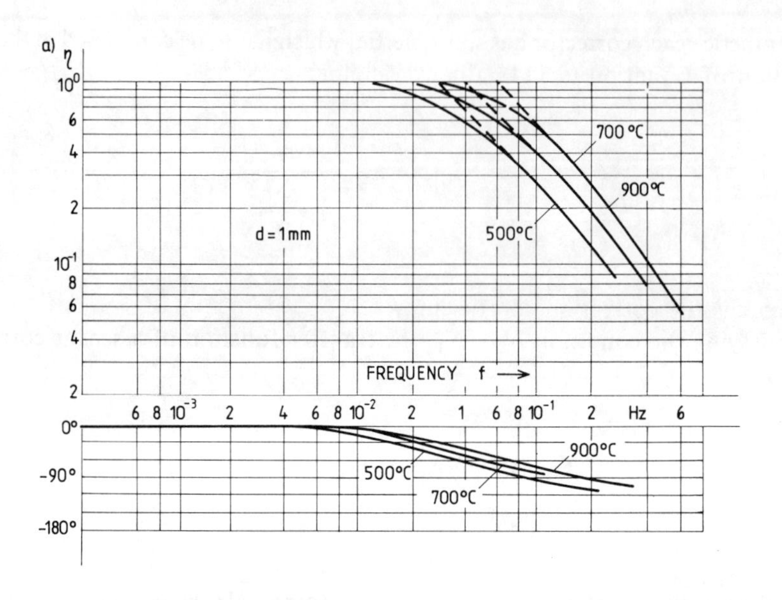

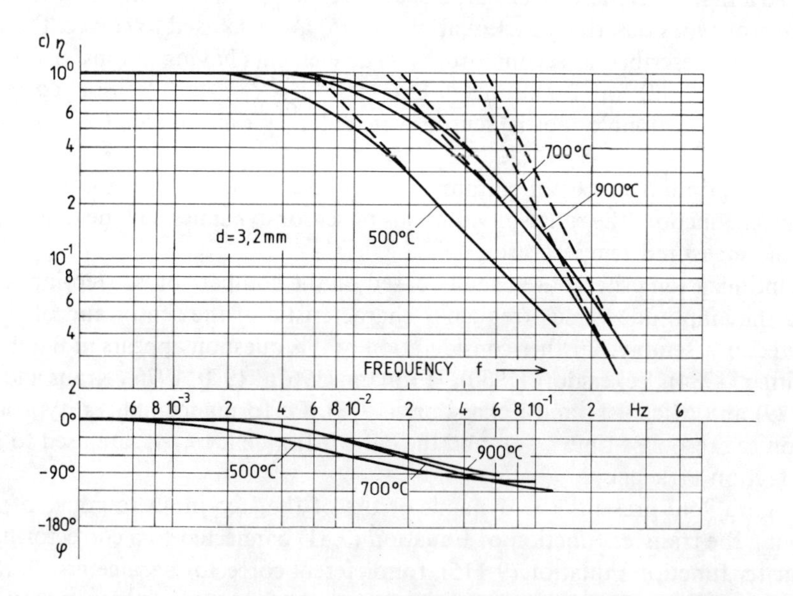

Figure 9.35
Frequency response of MI thermocouples of diameter *d*, with isolated measuring junction in a radiant heat transfer condition.

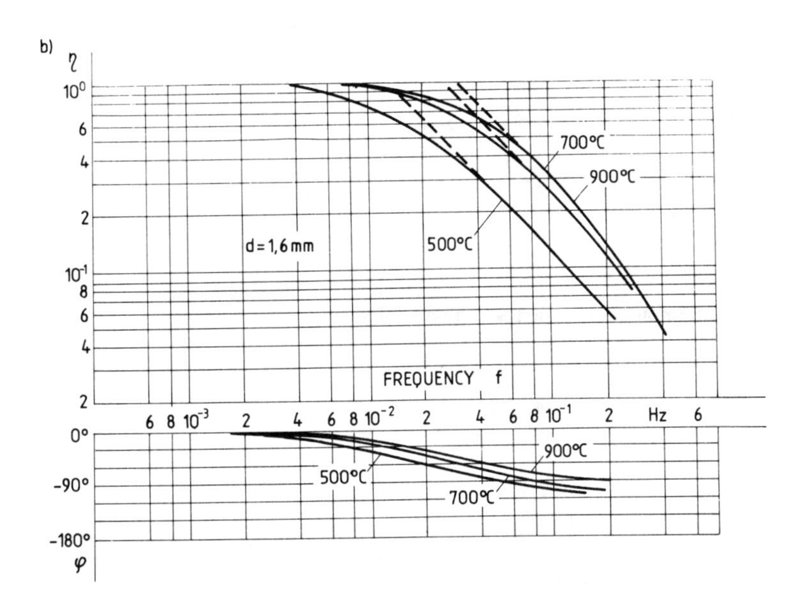

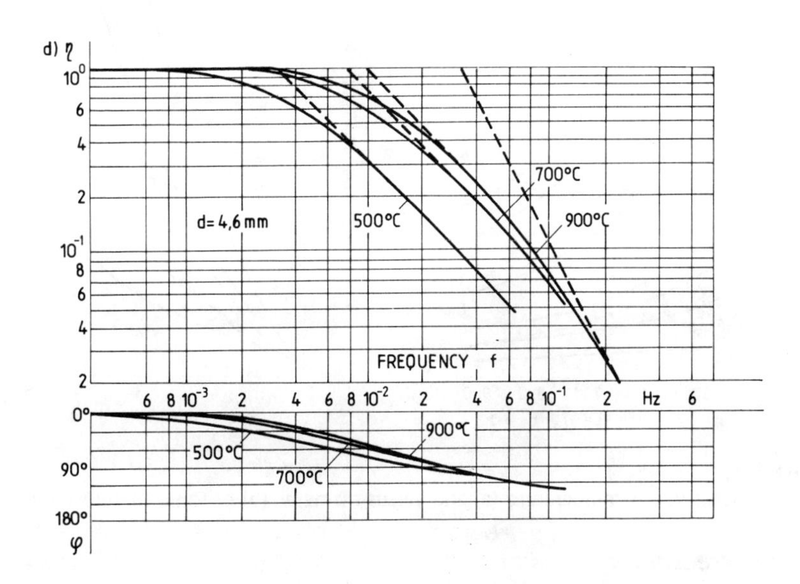

Figure 9.35
(continued)

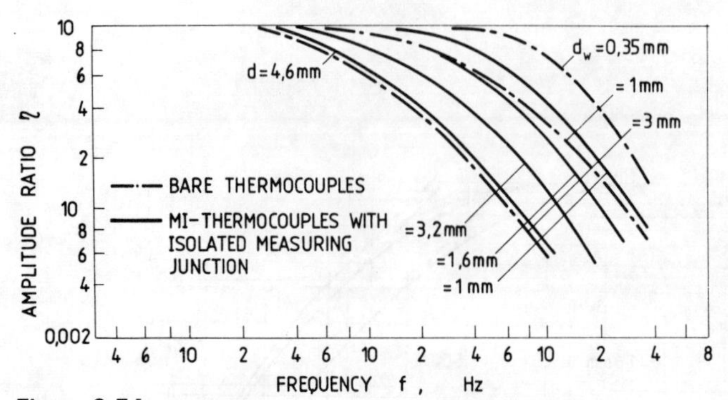

Figure 9.36
Comparison of amplitude frequency response of some thermocouples at 100 °C.

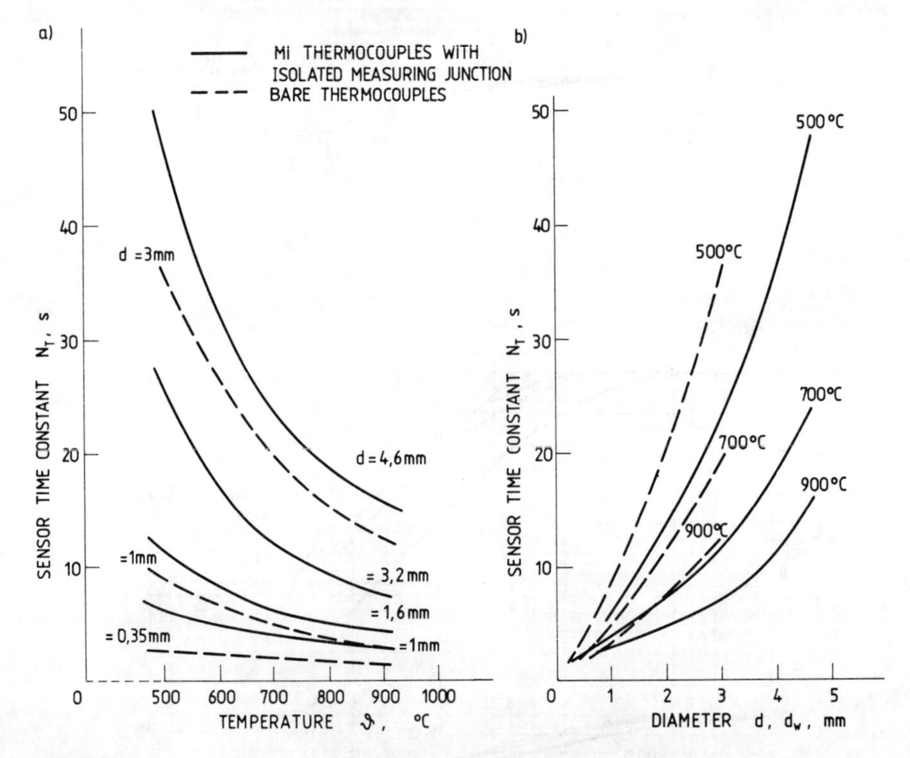

Figure 9.37
Time constants versus temperature (a) and versus diameter (b) of some thermocouples.

9.8.2 *Passive correctors*

The simplest passive correctors, having the transfer function of Equation (9.115) are shown in Figure 9.42. A characteristic feature of all passive correctors is their gain K_C which is the reciprocal of the correction coefficient k (Figure 9.42). It follows that the steady-state value of the corrector's output signal is k times smaller than

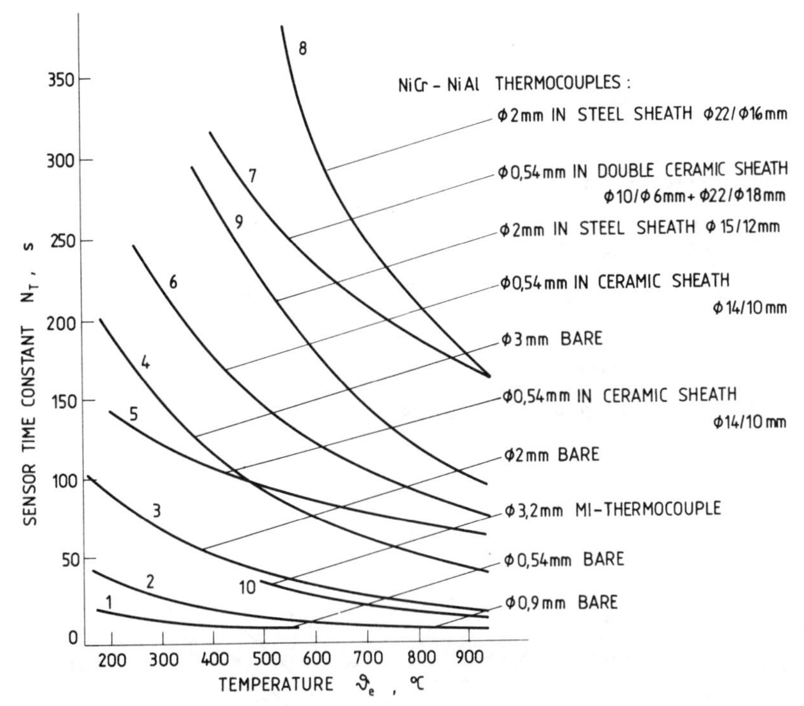

Figure 9.38
Approximate values of time-constants of NiCr–NiAl thermocouples in an electric chamber furnace, in still air. Time constants were found from a step input response from ambient to temperature ϑ_e.

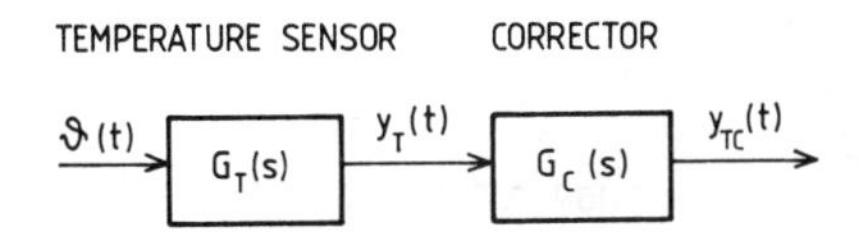

Figure 9.39
Block diagram of sensor-corrector circuit.

the steady-state value of the input signal. Thus, a preliminary amplification of the sensor output signal is necessary. In passive correctors it is thus difficult to get large values of the time-constant N_C because of the limited range of the values of R, L and C which may be used. As high values of correction coefficient, k, give increasing noise level, the value of k cannot be increased indefinitely.

In practice, Souksounov (1970) asserts that passive correctors can be applied, when $N_C < 5$ s and $k < 30$.

9.8.3 Active correctors

By combining proportional, derivative and integral elements in different configurations, based on operational amplifiers it is easy to design a corrector of

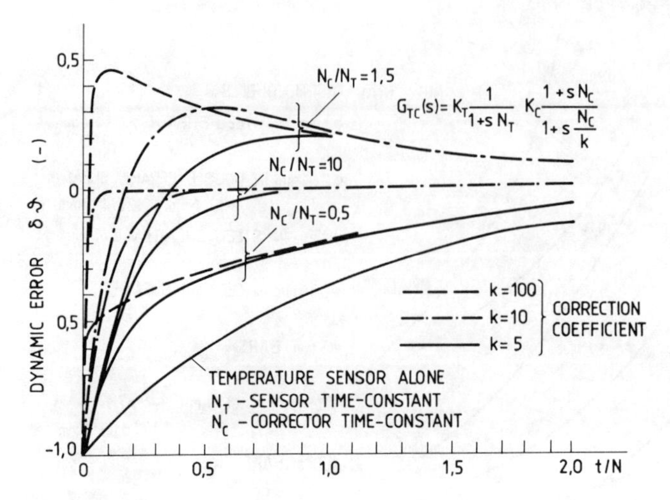

Figure 9.40
Dynamic errors of sensor-corrector set having a transfer function $G_{TC}(s)$, with step input.

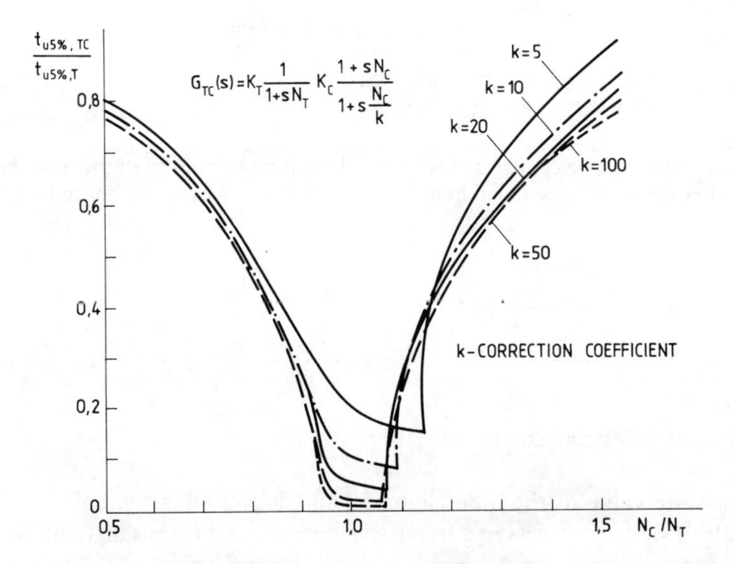

Figure 9.41
Ratio of 5% response times of sensor-corrector set to sensor alone versus the ratio of the corrector time-constant N_C to the sensor time-constant N_T.

any required transfer function. The same transfer function can be achieved from many different circuits. To characterize the usability of different circuits, and thus to help in a correct choice, two indices proposed by Hofmann (1967a, 1976) are used for the circuits described by Equation (9.115). The first is a sensitivity factor Q_1 written as:

$$Q_1 = |K_C| = f(k)$$

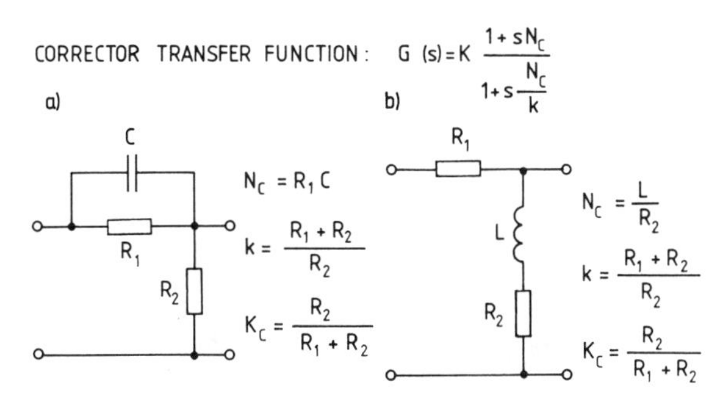

Figure 9.42
Types of passive correctors.

where K_C is the corrector gain and k is the correction coefficient while the second is a ratio of time-constants, Q_2, given by:

$$Q_2 = \frac{N_C}{N_{Ce}}$$

where N_C is the time constant in the corrector transfer function and N_{Ce} is the time constant of the applied derivative or inertia element.

For sensors with large time constants, correctors should have large values of Q_2. The value of the index Q_1 is of decisive importance if the sensor output signals are small when Q_1 should be as high as possible.

Some chosen corrector circuits, having transfer functions as in Equation (9.115) are presented in Figure 9.43 (Hofmann, 1976). More sophisticated circuits are described by Hofmann (1976), Praul and Hmurcik (1973) and Souksounov (1970).

In correctors, based on operational amplifiers, it is necessary to amplify the sensor output signal to the voltage levels needed by operational amplifiers. As active correctors based on operational amplifiers will also be influenced by amplifier noise, circuits which do not contain differentiating elements as in Figure 9.43(b), are preferred to those in Figure 9.43(a).

A detailed analysis of the quality factors of operational amplifiers, described by the influence of their noise, drift, internal resistance and dynamic behaviour is given in Souksounov (1970). In a similar way to the passive corrector, the maximum values of N_C and k are also limited in active types by the quality of the R and C elements and of the operational and measuring amplifiers. Nevertheless, the maximum achievable values of N_C and k of active correctors are much higher than in passive ones. For example, Souksounov (1970) states that some of them reach even $N_{C1} \approx 800$ s and $k \approx 500$.

9.8.4 Adaptive correctors

The performance of a corrector critically depends upon good matching of its parameters to the transfer function of the sensor. In changing measuring conditions

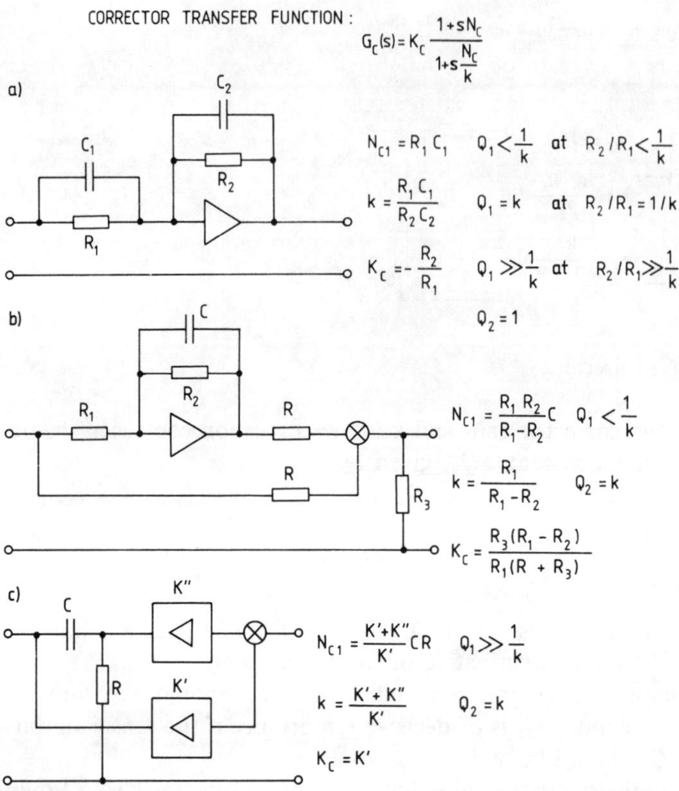

Figure 9.43
Types of active correctors.

the equivalent transfer function of the sensor also changes. Eckersdorf (1986) has shown that if the main time constant of the sensor N_T or N_{T1} respectively, changes by more than 30 to 40% in relation to the value for which the time constant of the corrector has been chosen, then the application of correctors with constant parameters is not advised. Under such changing conditions adaptive or self-tuning correctors should be used.

To conceive a corrector which adapts its parameters to changing sensor dynamics, it is necessary to continuously identify the sensor parameters during the measurements. Different ways of solving this problem have been presented by Souksounov (1970) and Zubov and Semenistyj (1973, 1974).

The principle of an adaptive corrector, in which two temperature sensors are used, is shown in Figure 9.44. An auxiliary sensor T_2 of N_{T2} is placed, side-by-side with the measuring sensor T_1, having the time-constant $N_{T1}(N_{T2}>N_{T1})$. Considering, that both sensors are subject to the same heat-transfer conditions to and from the surrounding medium it is valid to assume that:

$$\frac{N_{T2}}{N_{T1}} \simeq C = \text{constant} \tag{9.117}$$

It is also assumed, that the transfer functions of both sensors $G_{T1}(s)$ and $G_{T2}(s)$ are as in Equation (9.81) and that the transfer functions of the correctors $G_{C1}(s)$ and $G_{C2}(s)$ are as in Equation (9.115). Assume further, that the correction coefficient k_2 of

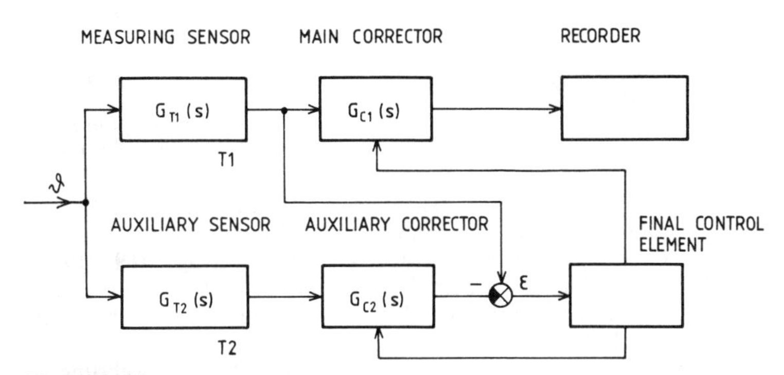

Figure 9.44
Block diagram of an adaptive corrector.

the auxiliary corrector is $k_2 = C$ and that the time constants of of the correctors are such, that $N_{C2}/N_{C1} = C$. It may be easily proven, that the condition of optimum correction $N_{C1} = N_{T1}$ is satisfied, when the difference signal ϵ equals zero. Assume a change of the main sensor's time constant from N_{T1} to N'_{T1} and that of the auxiliary sensor's from N_{T2} to N'_{T2}. The difference signal ϵ driving the final control element results in a new setting of both correctors until a new equilibrium state is reached. This state occurs when:

$$N'_{C1} = N'_{T1} \ (\text{at} \frac{N'_{C2}}{N_{C1}} = K)$$

As the adaptive correctors described above are rather complicated and expensive they are rarely applied in an industrial environment. However, they are easy to implement using microprocessor systems.

9.8.5 *Correction by computers*

Correcting action may be implemented, using computers or more frequently micro-computers and microprocessors. The correction principles, described in §9.8.1, enable algorithms based on the necessary corrector transfer function $G_C(s)$ of Equation (9.115), to be set up so that the output signal of the sensor-corrector assembly satisfies the equation:

$$y_{TC}(s) = y_T(s) G_C(s) \tag{9.118}$$

where $y_{TC}(s)$ is the Laplace transform of the output signal of the 'sensor corrector' assembly, $y_T(s)$ is the Laplace transform of the sensor output signal, as in Figure 9.39. To get the output signal $y_{TC}(t)$, it is necessary to perform the operation:

$$y_{TC}(t) = \mathscr{L}^{-1}\{y_T(s) G_C(s)\} \tag{9.119}$$

From the principles of operational calculus it follows that

$$y_{TC}(t) = \int_0^t y_T(t-\tau) g_C(\tau) \, d\tau \tag{9.120}$$

where $g(\tau)$ is the impulse response of corrector given by $g_C(\tau) = \mathscr{L}^{-1}\{G_C(s)\}$.

Equation (9.120) is the basis for setting up the necessary algorithm.

Application of this type of correction is especially justified when a computer used for data logging or signal conditioning (McGhee, 1989; Woschni, 1979) forms a part of the temperature measuring system.

REFERENCES

Bainbridge, B. L. and Kaltner, N. R. (1984) Application of transient response characterization and enhancement to thermal instrument. *Temperature Measurement Symposium*, Knoxville, TN, pp. 21.01–21.18.

Berger, R. L. and Balko, B. (1972) Thermal sensor coatings suitable for rapid response biomedical application. *Temperature: Its Measurement and Control in Science and Industry*, Vol. 4, Part 3, Instrument Society of America, Pittsburgh, pp. 2169–2192.

Bernhard, F. (1978) Der Einfluss der Betriebsbedingungen auf das dynamische Verhalten von Berührungsthermometern in fluidischen Medien. *Mereni teploty 1978 Praha*, Dum Techniky CSVTS, Praha, Pt II, pp. 60–71.

Bernhard, F. (1979) Umrechenbarkeit dynamischer Kennwarte von Berührungsthermometern auf andere Betriebsbedingungen. *m.s.r.*, **22**(8), 448–449.

Bernhard, F. and Noack, G. (1981) Possibilities and borders of the identification of dynamical parameters of thermometers under working conditions. First Symposium of the IMEKO TC-12 *Temperature Measurement in Industry and Science*, Karlovy Vary, pp. 219–231.

Bliek, L., and Fay, E. (1979) Ein Messverfahren zur einfachen Ermittlung des Ubertragungsverhaltens von Temperaturaufnehmern. *Techn. Messen*, **46**(7/8), 283–286, 291–292.

Bliek, L. Fay, E. and Gitt, W. (1978) Eine einfache Beschreibung für das dynamische Verhalten von Temperaturaufnehmern und von Temperaturmessystemen. *Mereni teploty 1978 Praha*, Dum Techniky CSVTS, Praha, pp. 102–109.

Caldwell, W. I., Coon, G. A. and Zoss, L. M. (1959) *Frequency Response for Process Control*, McGraw-Hill, New York.

Chohan, R. K. (1986) Response time correlation for industrial temperature sensors. *J. Phys. E: Sci., Instrum.*, **19**, 786–787.

Chohan, R. K. and Natour, M. (1988) Temperature sensor testing in air. *J. Phys. E: Sci., Instrum.*, **21**, 550–553.

DDR-Norme (1979) TGI 33208, 1979, Berührungsthermometer Beschreibungsverfahren für das dynamische Verhalten.

Doetsch, C. (1961) *Anleitung zum praktischen Gebrauch der Laplace-Transformation*, R. Oldenbourg, München.

Eckersdorf, K. (1980) Optimization of correctors for correction of dynamics of electric temperature sensors, PhD Thesis. Technical University Lodz (Polish).

Eckersdorf, K. (1986) Experimental determination of dynamic properties of temperature sensors installed in electroheat plants. *31 Int. Wiss Koll*, T. H. Illmenau, pp. 27–31.

Eckersdorf, K. and Michalski, L. (1979) Zur optimierung elektrischer Korrekturglieder bei dynamischen Temperaturmessugen. *24 Int. Wiss Koll*, T. H. Illmenau, Vol. 3, pp. 39–42.

Eckersdorf, K. and Michalski, L. (1984) Dynamics of the low inertia temperature sensors under the conditions of the radiant heat exchange. *Temperature Measurement in Industry, and Science*, TEMPMEKO-84, 2nd Symposium IMEKO TC-12, Kammer der Technik, Suhl, pp. 305–314.

Eijkman, E. G. J. (1955) Temperatuurmetingen, Fouten bij dynamische Temperatuurmetingen. *De Ingenieur*, **67**(2), 0.15–0.30

Eijkman, E. G. J. and Verhagen, D. M. (1958) Response and phase-lag of thermometers. *Process Control*, pp. 158–166.

Goodwin, W. M. (1945) Response time and lag of a thermometer element mounted in a protecting case. *Trans. AIEE*, **64**(9), 665–670.

Gröber, H., Erk, S. and Grigull, U. (1963) *Die Grudgesätze der Wärmeübertragung*. Springer-Verlag, Berlin.

Hackforth, H. L. (1960) *Infrared Radiation*, McGraw-Hill, New York.

Henderson, I. A., Ibrahim, A. A., McGhee, J. and Sankowski, D. (1987) Assembler generated binary sequences for process identification. *IFAC Proc Series, 1987, No. 7*, Pergamon Press, Oxford, pp. 77–82.

Higgins, S. P. and Keim, J. R. (1954) A thermal sine wave apparatus for testing industrial thermometers. *ASME*, Vol. 76, Paper 54-SA-20, pp. 1–9.

Hofmann, D. (1965) Meßdynamik elektrischer Industriethermometer. *m.s.r*, **8**(12), 407–410

Hofmann, D. (1966) Meßdynamik elektrischer Industriethermometer, *m.s.r*, **9**(1), 17–20.
Hofmann, D. (1967a) Zur elektrischer Korrektur des dynamischen Verhaltens von trägen Meßwandlern, *m.s.r*, **10**(1), 20–26.
Hofmann, D. (1967b) Zur Meßdynamik von elektrischen Berührungsthermometern. *IMEKO IV*, Warszawa, DDR-160.
Hofmann, D. (1970) Erhöhung der Genauigkeit von Meßsysteme mittels Korrektur. *IMEKO V*, Versailles, B-404.
Hofmann, D. (1976) *Dynamische Temperaturmessung*. Verlag Technik, Berlin.
Huhnke, D. (1973) Neue Apparaturen zur Aufnahme der Übergangsfunktion elektrischer Berührungsthermometern in strömenden Medien. *VDI-Berichte, 1987, Technische Temperaturmessung*, pp. 103–109.
Jakob, M. (1957) *Heat Transfer, Vol. 2*, John Wiley and Sons, New York.
Jakob, M. (1958) *Heat Transfer, Vol. 1* (6th edition), John Wiley and Sons, New York.
Kerlin, T. W., Miller, L. F. and Hashemian, N. M. (1978) In situ response time testing of platinum resistance thermometers *ISA Transactions*, **17**(4), 71–88.
Kerlin, T. W., Hashemian, N. M. and Peterson, K. M. (1981) Time response of temperature sensors. *ISA Transactions*, **20**(1), 65–67.
Kerlin, T. W., Hashemian, H. H., and Peterson, K. M. (1982) Response characteristics of temperature sensors installed in processes. *ACTA IMEKO, III*, North-Holland, pp. 95–103.
Kocurov, W. I. (1963) Determination of dynamic characteristics of temperature sensors. *Energetika*, **14**(5), 91–97 (in Russian).
Kondratiev, G. M. (1947) Universal method of determination of the index of thermal inertia. *Trudy WNIIM*, **4**, 62–83 (in Russian).
Kraus, M. and Woschni, F. G. (1979) *Meßinformationssysteme*, EB Verlag, Berlin.
Kühne, Ch. (1957) Über den Zusamenhang der Zeitkonstanten einer Regelstrecke 2. Ordnung mit den entsprechenden Tot- und Analaufzeiten, *Regelungstechnik*, **5**(6), 206–207.
Lieneweg, F. (1938a) Bestimmung der Anzeigeverzögerung von Thermometern. *ATM*, No. 2, pp. T16–T18.
Lieneweg, F. (1938b) Anzeigedämpfung von Thermometern bei zeitlichen Temperaturänderungen. *ATM*, **11**, T143–T144.
Lieneweg, F. (1941) Bestimmung der Anzeigeverzögerung von Thermometern. *ATM*, **6**, T79–T80.
Lieneweg, F. (1962) Die Übergangsfunktion beim Abkühlung und Erhitzen fester Körper in beliebigen Mitteln, einschliesslich der Anzeigeverzögerung von Thermometern. *Regelungstechnik*, **10**(4), 159–165.
Lieneweg, F. (1975) *Handbuch, Technische Temperaturmessung, F. Vieweg*, Braunschweig.
McGhee, J. (1989) The impact of electronics and information technology upon measurement and control in thermal systems. Keynote paper IMEKO TC-12, Symp. *Microprocessors in Temperature and Thermal Measurement*, Lodz, Poland.
McGhee, J., Fisher, G. and Henderson, I. A. (1987) A fast DFT algorithm for on-line MBS process identification. *Proc. CSCS7, 7th Int. Conf. Con. Sys. Comp. Sc.*, Vol. 1, Bucharest, pp. 111–119.
McGhee, J., Henderson, I. A., Ibrahim, A. A. and Sankowski, D. (1986) Identification: Systems, Structures, Signals, Similarities, Systems Science IX, Int Conf on Systems Science, Wroclaw, Poland.
McGhee, J., Henderson, I. A. and Jackowska-Strumillo, L. (1990) Identifying temperature sensors using Strathclyde compact multifrequency binary sequences, *Proc. CSCS 8, 8th Int. Conf. Cont. Sys. Comp. Sc.*, Bucharest, to be published.
McGhee, J., Henderson, I. A. and Mackie, S. (1991) Simulation of MBS identification of temperature sensors, *Proc. TEMPMEKO 90, 4th Symp. on Temp. and Therm. Meas. in Sci. and Ind.*, Helsinki, Finland, pp. 201–206.
McGhee, J., Sheppard, D. and Henderson, I. A. (1989) Immersion testing of temperature sensors using a microprocessor based MBS generator/analyser. *Proc. IMEKO TC-12 Symposium, Microprocessors in Temperature and Thermal Measurement*, Lodz, Poland, pp. 65–72.
Meyer-Witting, O. (1959) Berchnungsverfahren zur Bestimmung der Anzeigeverzögerung von Temperaturfühlern bei periodischen Temperaturänderungen, PhD thesis, T. H. Braunschweig.
Michalski, L. (1966) Temperatur eines Kammerofens. *Elektrotechniek (Netherlands)*, **44**(20), 466–471.

Michalski, L., and Eckersdorf, K. (1978) Electrical correction of dynamics of temperature sensors, having temperature dependent properties. *Meřeni Teploty*, Praha, 1978, Dum Techniky CSVTS, pp. 110–119.

Michalski, L. and Eckersdorf, K. (1987) Temperature sensors in a closed loop temperature control. TEMPMEKO 87, *Thermal and Temperature Measurement in Science and Industry*, Sheffield UK, pp. 127–136.

Michalski, L. and Eckersdorf, K. L. (1990) In situ determination of dynamics of temperature sensors, *Temperature and Thermal Measurement in Industry and Science*, Proc. 4th IMEKO Symposium, TEMPMEKO 90, Helsinki, pp. 193–200.

Michalski, L., Kuzminski, K. and Sadowski, J. (1981) *Temperature Control in Electroheat*, WNT, Warsaw (in Polish).

Pandey, D. K. (1985) Response time correlations for chromel-constantan thermocouples in flowing hot air. *J. Phys. E: Sci. Instrum.* **18**, 712–713.

Pelepecenko, I. P. (1969) Determination of time constant for resistance thermometer sensors by an A. C. method. *JVUS, Priborostroyenye*, **8**, 956–999.

Petit, C. *et al.* (1982) Frequency response of fine-wire thermocouple. *J. Phys. E: Sci, Instrum.*, **15**(17), 760–764.

Praul, S. H. and Hmurcik, L. V. (1973) Instantaneous temperature measurement. *Rev. Sci., Instrum.*, **14**(9), 1363–1364.

Reed, R. P. (1972) The transient response of embedded thin film temperature sensors. *Temperature: Its Measurement and Control in Science and Industry*, Vol. 4, Part 3, Instrument Society of America, Pittsburgh, pp. 2193–2196.

Rubin, G. and Feldman, I. (1968) Radiant heat exchange and dynamics of thermal processes in resistance furnaces. *VI. Int. Congress on Electroheat*, Brighton, paper 314.

Schwarze, G. (1964) Neue Erghebnisse zur Bestimmung der Zeitkonstanten im Zeitbereich, *m.s.r*, 7(1).

Sirs, J. A. (1961) Measurement of rapid temperature changes by thermocouples, *J. Sci. Instrum.* **38**, 489–490.

Skoczowski, S. (1982) Meßfehler verursacht durch die dynamische Nichtlinearität des Meßfühlers. *Techn. Messen*, **49**(1), 27–30.

Souksounov, V. E. (1970) Correcting elements in instruments for measuring non-stationary temperatures. *Biblioteka pro automatike*, No. 33, Ind, Energiya, Moscow (in Russian).

Taylor, H. R. and Navarro, H. A. (1987) A method to determine and compensate for the frequency response of a platinum resistance thermometer under practical conditions. TEMPMEKO-87, *Thermal and Temperature Measurement in Science and Industry*, Sheffield, UK, pp. 137–147.

VDE/VDI-Richtlinien 3511 (1964) Technische Temperaturmessungen.

VDE/VDI-Richtlinien 5522 (1979) Das Zeitverhalten von Berührungs-Thermometern.

Woschni, E. G. (1964) *Meßfehler bei dynamischen Messungen und Auswertung von Meßergebnissen*, Verlag Technik, Berlin.

Woschni, E. G. (1979a) Vergleich der statistischen Linearitäts und dynamischen Korrektur von Systemen mit Mikrorechnern, insbesondere in der Prozesstechnik. 24 *Int. Wiss. Koll. Ilmenau*, T. H. Ilmenau No. 3, pp. 55–58.

Woschni, E. G. (1979b) Beziehungern zwischen den verschiedenen Definitionen von dynamischen Fehlern in der Meßtechnik und Schlußfolgerungen für praktische Anwednung. *m.s.r*, **22**(5), 259–261.

Woschni, E. (1980) Die Meßtechnik als Teilgebiet der Informationstechnik *Techn., Messen* **47**(1) 7–13.

Yarishev, N. A. (1967) Theoretical principles of measurement of non-stationary temperatures. *Izd. Energiya*, Leningrad (in Russian).

Zubov, W. G. and Semenistyj, K. C. (1973) Active corrector for the correction of inertia of temperature sensors, *Priborostroyenye* **14/15** 107–109 (in Russian).

Zubov, W. G. and Semenistyj, K. C. (1974) Self tuned measuring system with dynamic correction of inertia of temperature sensors. *Odbor i Peredaca Inf.*, pp. 74–76 (in Russian).

65B Central Office, 24, Industrial platinum, resistance thermometers or sensor. *Standard Project of (C.E.I).*

10 Measurement of the Surface and Internal Temperature of Solid Bodies

10.1 INTRODUCTION

One of the most frequently encountered problems in temperature measurement is measuring the temperature of solid bodies on their surfaces in contact with a surrounding gas or liquid. This may be solved either by contact or non-contact (pyrometeric) methods or also by a 'semi-contact method', which has not been described so far. The non-contact or pyrometric methods are described in detail in Chapter 7. For a rough estimation of surface temperatures, temperature indicators, described in §3.5, are also used.

The problem of measuring the internal temperatures of solid bodies, presented in §10.6, is also similar to surface temperature measurement.

10.2 THEORY OF THE CONTACT METHOD

It is assumed that a solid body in contact with a surrounding gaseous medium, as shown in Figure 10.1, remains in a thermal steady-state. A surfacial heat source is placed inside the solid body whose true surface temperature ϑ_t is higher than the ambient gas temperature ϑ_a. The surface temperature ϑ_t is to be measured by a contact sensor which is a bare thermocouple having a flat-cut measuring junction, in contact with the surface. Under the influence of the contact sensor (Figure 10.1(b)), the original isotherms in the solid body (Figure 10.1(a)) are deformed. Assume that the heat transfer between the investigated surface and the surrounding gaseous medium takes place through convection and conduction, with the isotherms in the gas deformed in the surrounding of the sensor. The corresponding temperature distribution in the direction normal to the surface is shown in the lower part of Figure 10.1(c,d).

While making contact with the investigated surface, the sensor causes a more intense heat flow from the surface, resulting in a drop in its temperature from the original value ϑ_t to a new temperature ϑ' (Figure 10.1(d)). The temperature difference $\Delta\vartheta_1 = \vartheta' - \vartheta_t$ is called the first partial error of the measurement, caused by the deformation of the original temperature field. Between the flat cut measuring junction

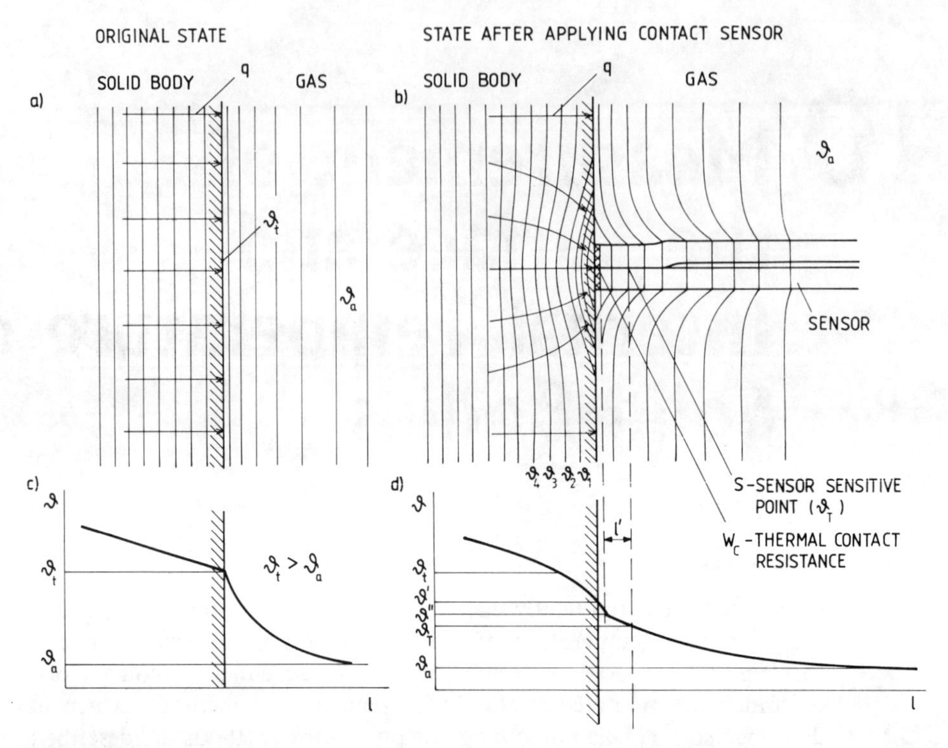

Figure 10.1
Surface temperature measurement of a solid body by a contact sensor. a, b, isotherms and heat flux density lines q; c, d, temperature distribution along direction normal to the surface; ϑ_t, original, true surface temperature; ϑ', surface temperature after applying the contact sensor; ϑ'', temperature at sensor frontal area; ϑ_T, temperature of sensor sensitive point.

of the thermocouple and the investigated surface there is always a thermal contact resistance, W_c, caused by a non-ideal contact. The temperature drop across this contact resistance $\Delta\vartheta_2 = \vartheta'' - \vartheta'$ is called the second partial error.

It is further assumed, as in all sensors, that there is also a sensitive point in a contact sensor determining the thermometer readings, ϑ_T. From Figure 10.1(b) this point, S, in a contact thermocouple is placed at the distance, l', from the investigated surface. The temperature ϑ_T at the point S differs from ϑ'' by a value $\Delta\vartheta_3 = \vartheta_T - \vartheta''$, called the third partial error, which depends on the sensor design. All of the differences $\Delta\vartheta_1$, $\Delta\vartheta_2$ and $\Delta\vartheta_3$ are systematic errors of the contact method of surface temperature measurement of a solid body in the thermal steady-state. To determine their values the temperature field of a solid body in contact with a sensor and the heat flux entering the sensor will be analysed. In §§10.2.1 and 10.2.2, the temperature excess, Θ, over ambient will be used.

10.2.1 *Disturbing temperature field*

The investigated temperature field of a solid body in contact with a sensor, according to Kulakov and Makarov (1969), can be regarded as a superposition of two fields.

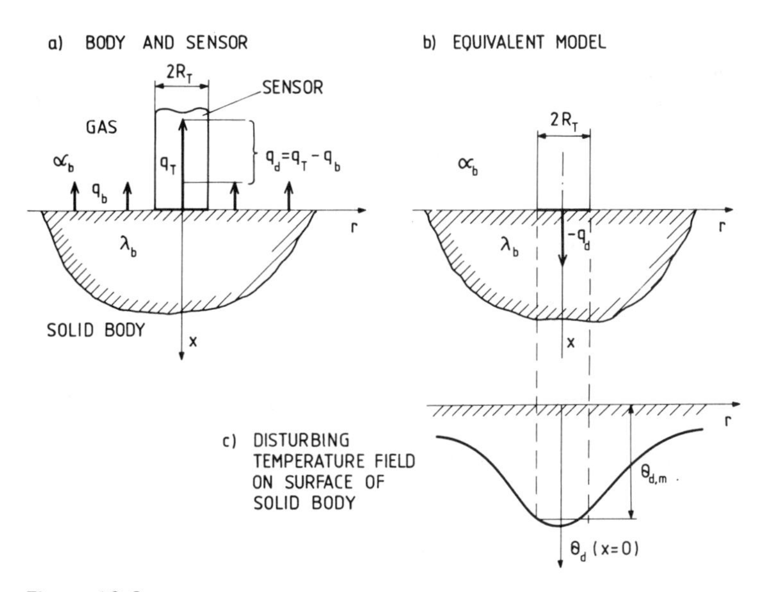

Figure 10.2
Disturbing temperature field on the surface of a semi-infinite body, resulting from the application of a contact temperature sensor.

Firstly, there is the original temperature field in the body without the sensor, described by $\Theta_b = f(x,r)$ and secondly, the disturbing temperature field $\Theta_d = f(x,r)$. The disturbing temperature field is caused by the disturbing heat flux, resulting from the difference between the density, q_T, of the heat flux conducted along the sensor and the density, q_b, of the heat flux transferred from the body to its ambient surrounding. This density, q_d, of disturbing heat flux is given by

$$q_d = q_T - q_b \tag{10.1}$$

Using the semi-infinite body in Figure 10.2 as an example, gives an explanation of the manner of determining the disturbing flux and also shows the surfacial temperature distribution. The medium value of disturbing temperature Θ_{dm} (Figure 10.2(c)) at the contact surface between sensor and body permits the first partial error $\Delta\vartheta_1$ to be determined.

The differential equation of heat conduction, describing the disturbing temperature field in a semi-infinite body, is

$$\frac{\partial^2 \Theta_d}{\partial r^2} + \frac{1}{r}\frac{\partial \Theta_d}{\partial r} + \frac{\partial^2 \Theta_d}{\partial x^2} = 0 \tag{10.2}$$

with the boundary conditions:

$$\left[\frac{\partial \Theta_d}{\partial x} - \frac{\alpha_b}{\lambda_b}\Theta_d\right]_{x=0} = -\frac{q_d}{\lambda_b}\bigg|_{r \leqslant R_T} \tag{10.3}$$

$$\Theta_d|_{x=0} < \infty; \Theta_d|_{x=\infty} = 0; \Theta_d|_{r=\infty} = 0$$

where λ_b is the thermal conductivity of the body, and α_b is the heat transfer coefficient at surface of the body.

Solution of Equation (10.2) with the boundary conditions of Equation (10.3) is:

$$\Theta_d = -\frac{q_d R_T}{\lambda_b} \int_0^{\infty} \frac{e^{-k_x v} I_1(v) I_0(k_r v)}{v+B} dv \tag{10.4}$$

where v is the variable of integration,

$$k_x = \frac{x}{R_T}, \quad k_r = \frac{r}{R_T}; \quad B = \frac{\alpha_b R_T}{\lambda_b}$$

$I_0(v)$ and $I_1(v)$ are Bessel functions of first kind, of zero and first order, respectively.

From Equation (10.4), let the value of the integral, a function of k_x, k_r and B, be described by $F(k_x, k_r, B)$, so that relation (10.4) becomes:

$$\Theta_d = -\frac{q_d R_T}{\lambda_b} F(k_x, k_r, B) \tag{10.5}$$

To help with the practical use of Equation (10.5) Kulakov and Makarov (1969) graphically display the values of the function F versus k_x in Figure 10.3 and versus k_r in Figure 10.4 as well as the mean value F_m of the function F at the contact surface between the sensor and body versus parameter B (Figure 10.5).

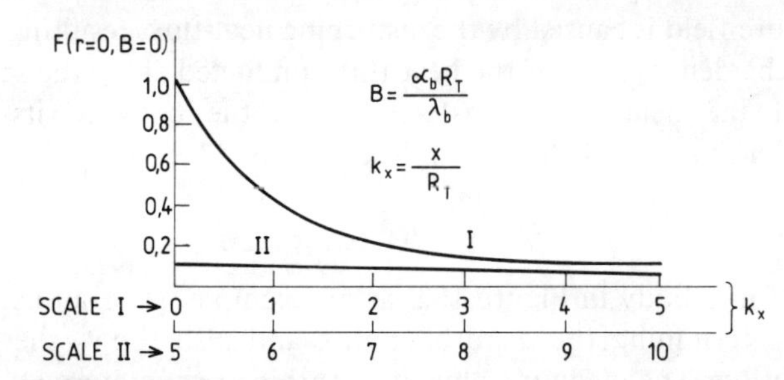

Figure 10.3
Function F from Equation (10.5) versus parameter k_x, for $r=0$, $B=0$.

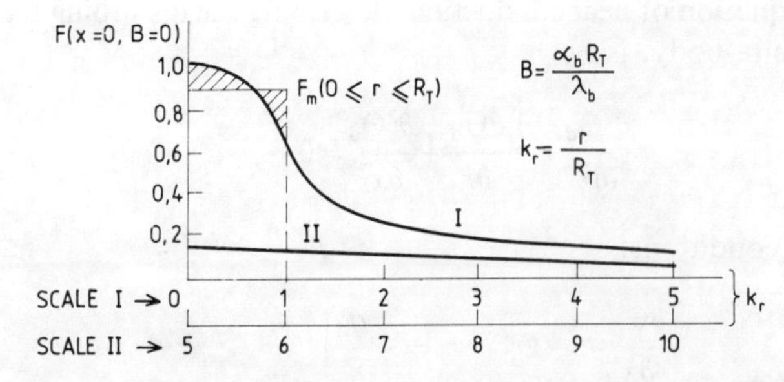

Figure 10.4
Function F from Equation (10.5) versus parameter k_r, for $x=0$, $B=0$.

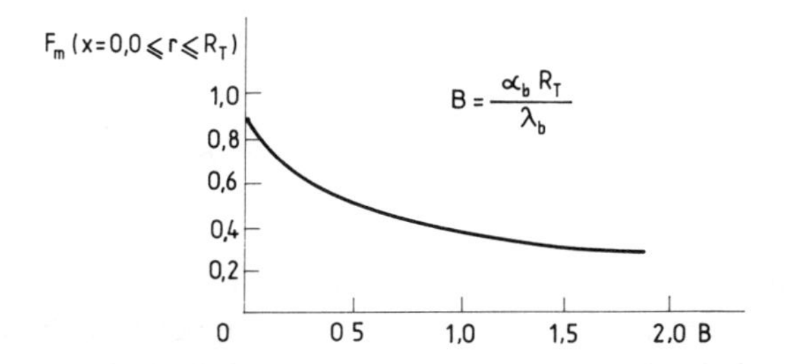

Figure 10.5
Mean value F_m of function F from Equation (10.5) versus parameter B, for $x=0$, $0\leqslant r\leqslant R_T$.

Figure 10.6 presents a case in which the disturbing heat flux of density, q_d, penetrates into an infinitely large plate of limited thickness, through a surface limited by a circle of radius R_T. The differential equation of heat conduction, characterizing the disturbing temperature field, is the same as that given in Equation (10.2) for the semi-infinite body with the boundary conditions:

$$\left.\begin{aligned} &\left[\frac{\partial \Theta_d}{\partial x}-\frac{\alpha_1}{\lambda_b}\Theta_d\right]_{x=0} = -\frac{q_d}{\lambda_b}\ \Bigg|\ {}_{r\leqslant R_T} \\ &\left[\frac{\partial \Theta_d}{\partial x}-\frac{\alpha_2}{\lambda_b}\Theta_d\right]_{x=l_b} = 0 \\ &\Theta_d|_{r=0}<\infty ; \Theta_d|_{r=\infty}=0 \end{aligned}\right\} \tag{10.6}$$

The solution of Equation (10.2) for the boundary conditions (10.6) is:

$$\Theta_d = -\frac{q_d R_T}{\lambda_b}\int_0^{\infty}\frac{(\nu+B_1)e^{[\nu(k_l-k_x)]}+(\nu-B_2)\,e^{[-\nu(k_l-k_x)]}}{(\nu+B_1)(\nu+B_2)\,e^{k_l\nu}-(\nu-B_1)(\nu-B_2)\,e^{-k_l\nu}}I_1(\nu)I_0(k_r^2)\,d\nu$$

$$= -\frac{q_d R_T}{\lambda_b}H(k_x,k_r,k_l,B_1,B_2) \tag{10.7}$$

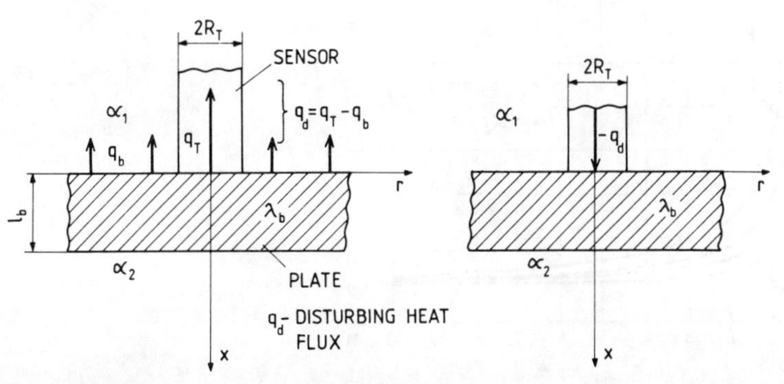

Figure 10.6
Infinitely large plate with a surface temperature sensor.

where:

$$k_x = \frac{x}{R_T}, \quad k_r = \frac{r}{R_T}, \quad k_l = \frac{l_b}{R_T}, \quad B_1 = \frac{\alpha_1 R_T}{\lambda_b}, \quad B_2 = \frac{\alpha_2 R_T}{\lambda_b}$$

To calculate the value of the first partial error, $\Delta\vartheta_1$, which equals the medium value of the disturbing temperature, $\Theta_{d,m}$, at the contact surface between the body and sensor, the function H from Equation (10.7) must be known for k_r varying in the range $0 \leqslant k_r \leqslant 1 (0 \leqslant r \leqslant R_T)$ and for $k_x = 0$.

As the function H depends upon several parameters it is difficult to display it graphically in a universal way. Figure 10.7 presents values of H, for $k_r = 0$ and $k_x = 0$, as a function of relative plate thickness k_l for some chosen values of $B_1 = B_2 = B$. This corresponds to a case when the heat transfer coefficients on both sides of the plate are the same ($\alpha_1 = \alpha_2 = \alpha$). Figure 10.7 permits the maximum value of disturbing temperature in the centre of the contact surface to be found. From the curves of $H = f(k_l)$, it follows that the disturbing temperature values decrease with increasing plate thickness. If the plate is sufficiently thick it can be regarded as a semi-infinite body and Equation (10.5) can be applied. Kulakov and Makarov (1969) discuss some bodies with finite dimensions and other shapes.

10.2.2 Heat flux entering the sensor

As follows from relation (10.1), determination of the density, q_d, of disturbing heat flux, requires knowledge of the density, q_T, of heat flux entering the sensor. Four simple sensor models will be considered (Figure 10.8), under some simplifying assumptions. It will be assumed that the heat transfer coefficients have constant values, that the temperature field in the cross-section of the rod or plate is uniform, with a uniform density of heat flux all over the front surface of the sensor as well. In addition it is assumed that the rods shown in Figure 10.8(b), Figure 10.8(c) and Figure 10.8(d) are infinitely long.

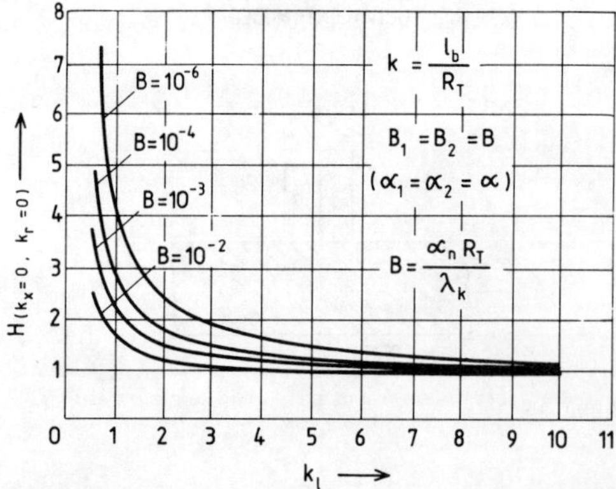

Figure 10.7
Function H from Equation (10.7) versus relative plate thickness k_l, for different B values.

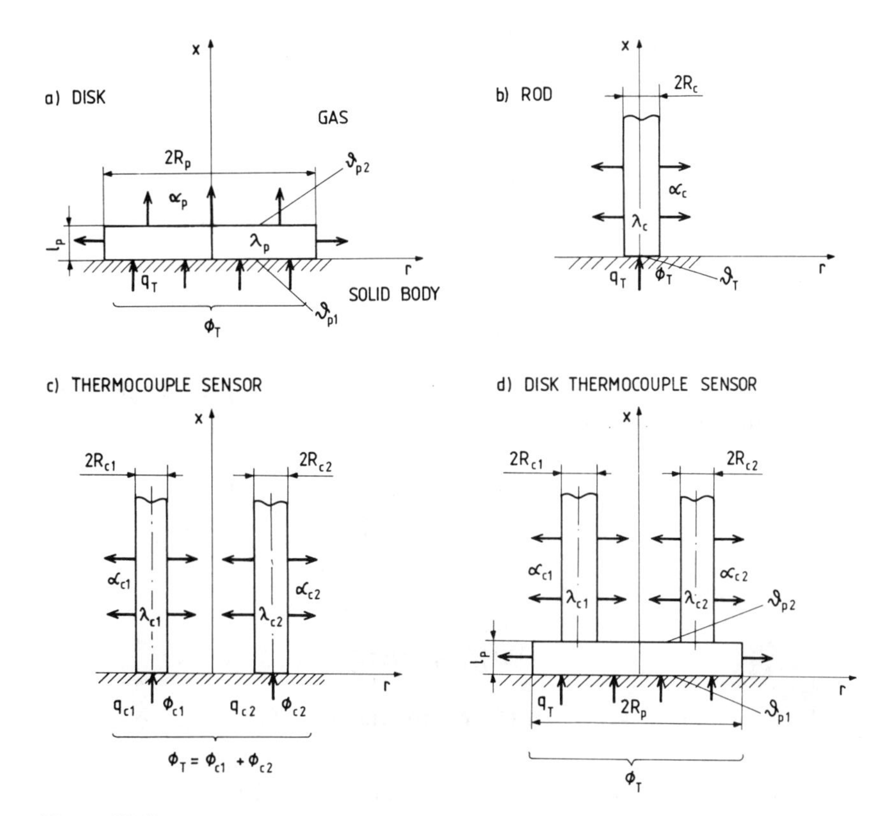

Figure 10.8
Sensor models for evaluating the heat flux Φ_T entering the sensor. q_T is the heat flux density.

For each model the heat flux, Φ_T, entering the sensor, will be determined as well as the heat flux density, q_T, at the sensor surface, A_T, in contact with the investigated body:

$$q_T = \frac{\Phi_T}{A_T} \tag{10.8}$$

Also to be determined is the thermal resistance, W_T, of the sensor, defined by Mackiewicz (1976a) in terms of, Θ_T, the temperature of the surface of the sensor in contact with the investigated body as,

$$W_T = \frac{\Theta_T}{\Phi_T} \tag{10.9}$$

To simplify the problem of the disk of Figure 10.8(a), it is assumed, that the disk (plate) only transfers heat to the environment, from its upper surface ($l_p \ll 2R_p$), and that $\vartheta_{p1} = \vartheta_{p2} = \vartheta_T$ corresponding to a thin disk, with a large value of thermal conductivity λ_p. The heat flux Φ_T is:

$$\Phi_T = \pi R_p^2 \alpha_p (\vartheta_T - \vartheta_a) = \pi R_p^2 \alpha_p \Theta_T \tag{10.10}$$

with the heat flux density given by:

$$q_T = \frac{\Phi_T}{\pi R_p^2} = \alpha_p \Theta_T \tag{10.11}$$

The thermal resistance of the disk, following Equation (10.9), is:

$$W_T = \frac{1}{\pi R_p^2 \alpha_p} \tag{10.12}$$

Rod—Figure 10.8(b). The differential equation describing the temperature distribution along the rod (conductor) of Figure 10.8(b) is:

$$\frac{d^2\Theta_c(x)}{dx^2} - \frac{2\alpha_c}{\lambda_c R_c}\Theta_c(x) = 0 \tag{10.13}$$

with the boundary conditions:

$$\Theta_c(x)|_{x=0} = \Theta_T; \Theta_c(x)|_{x=\infty} = 0 \tag{10.14}$$

The solution of Equation (10.13) for the boundary conditions given above is:

$$\Theta_c(x) = \Theta_T \exp\left(-\sqrt{\frac{2\alpha_c}{\lambda_c R_c}}x\right) \tag{10.15}$$

The heat flux, Φ_T, entering the sensor equals the total heat flux transferred to the environment from the side surface of the rod in accordance with:

$$\Phi_T = \int_0^\infty 2\pi R_c \alpha_c \Theta_c(x) dx \tag{10.16}$$

Substituting the expression for $\Theta_c(x)$ from Equation (10.15) into Equation (10.16), yields:

$$\Phi_T = \pi R_c \sqrt{2\alpha_c \lambda_c R_c} \Theta_T \tag{10.17}$$

The heat flux density, q_T, is then

$$q_T = \frac{\Phi_T}{\pi R_c^2} \tag{10.18a}$$

and finally:

$$q_T = \sqrt{\frac{2\alpha_c \lambda_c}{R_c}} \Theta_T \tag{10.18b}$$

Following the definition of Equation (10.9), the thermal resistance W_T of the rod is given by:

$$W_T = \frac{1}{\pi R_c \sqrt{2\alpha_c \lambda_c R_c}} \tag{10.19}$$

The double conductor thermocouple sensor of Figure 10.8(c) can be regarded as two independent rods from the model in Figure 10.8(b). In most cases it can be assumed that:

$$R_{c1} = R_{c2} = R_c; \quad \alpha_{c1} = \alpha_{c2} = \alpha_c; \quad \vartheta_{T1} = \vartheta_{T2} = \vartheta_T$$

and then the total heat flux, Φ_T, entering the sensor, is

$$\Phi_T = R_c \pi \sqrt{2\alpha_c R_c}(\sqrt{\lambda_{c1}} + \sqrt{\lambda_{c2}})\Theta_T \tag{10.20}$$

and the thermal resistance of the sensor, following Equation (10.9) will be:

$$W_T = \frac{1}{R_c \pi \sqrt{2\alpha_c \lambda_c R_c}(\sqrt{\lambda_{c1}} + \sqrt{\lambda_{c2}})} \tag{10.21}$$

The heat flux density of each conductor is given by Equation (10.18).

For the disk thermocouple sensor of Figure 10.8(d), the same simplifications as for the models of Figure 10.8(a) and 10.8(b) but with $R_c \ll R_p$ are assumed. The total heat flux, Φ_T, entering the sensor then equals the sum of the heat fluxes of both conductors and of the disk (plate):

$$\Phi_T = [\pi R_p^2 \alpha_p + \pi R_c \sqrt{2\alpha_c R_c}(\sqrt{\lambda_{c1}} + \sqrt{\lambda_{c2}})]\Theta_T \tag{10.22}$$

The appropriate heat flux density in this case is:

$$q_T = \frac{\Phi_T}{\pi R_p^2} \tag{10.23a}$$

or

$$q_T = \left[\alpha_p + \frac{R_c}{R_p^2}\sqrt{2\alpha_c R_c}(\sqrt{\lambda_{c1}} + \sqrt{\lambda_{c2}})\right]\Theta_T \tag{10.23b}$$

and the thermal resistance, W_T, conforming with Equation (10.9) is:

$$W_T = \frac{1}{\pi [R_p^2 \alpha_p + R_c \sqrt{2\alpha_c R_c}(\sqrt{\lambda_{c1}} + \sqrt{\lambda_{c2}})]} \tag{10.24}$$

10.2.3 Method errors

The *first partial error*, $\Delta\vartheta_1$, conforming with the definition from §10.2, is

$$\Delta\vartheta_1 = \vartheta' - \vartheta_t \tag{10.25}$$

This error equals the medium value of the disturbing temperature $\Theta_{d,m}$, at the contact surface between the sensor and the body. For a semi-infinite body, following Equation (10.5), the error, $\Delta\vartheta_1$, is described by

$$\Delta\vartheta_1 = \Theta_{d,m}\Big|_{\substack{x=0\\ 0\leqslant r\leqslant R_T}} = -\frac{q_d R_T}{\lambda_b} F_m\Big|_{\substack{x=0\\ 0\leqslant r\leqslant R_T}} \tag{10.26}$$

The density, q_d of disturbing heat flux in Equation (10.26), is calculated from Equation (10.1) as:

$$q_d = q_T - q_b \tag{10.27}$$

Determination of q_T for the four sensor models is described in Section 10.2.2. To simplify the problem it is assumed that $\vartheta_T = \vartheta'(W_c = 0)$. The value of q_b is calculated from $q_b = \alpha_b\Theta_T$ while the value of F_m in Equation (10.26) is found from Figure 10.5.

Calculation of the error $\Delta\vartheta_1$ is accomplished in a step by step iterative way or graphically. This will be explained in a numerical example.

To calculate these $\Delta\vartheta_1$ errors, a ready formula can also be used. This is derived for a semi-infinite body by Mackiewicz (1976a), taking into consideration the thermal contact resistance W_c in the manner

$$\Delta\vartheta_1 = \frac{(W_T + W_c)/W_a - 1}{1 + (\pi R_T \lambda_b / F_m)(W_T + W_c)}\Theta_t \tag{10.28}$$

where λ_b is the thermal conductivity of the investigated body, W_T is the thermal resistance of the sensor (Section 10.2.2), W_c is the thermal contact resistance between the sensor and the investigated body, W_a is the thermal resistance of heat transfer from the surface of the body to the environment, before applying the sensor, given by the relation

$$W_a = \frac{1}{\alpha_b R_T^2 \pi} \tag{10.29}$$

In a similar way, based on the theory of disturbing heat flux for an infinitely large plate, the error, $\Delta\vartheta_1$, can be found with the calculations based on Equation (10.7). The application range of this method is limited to the case, when the heat transfer coefficients on both plate surfaces are the same ($\alpha_1 = \alpha_2$) and the plate itself is composed of one layer only.

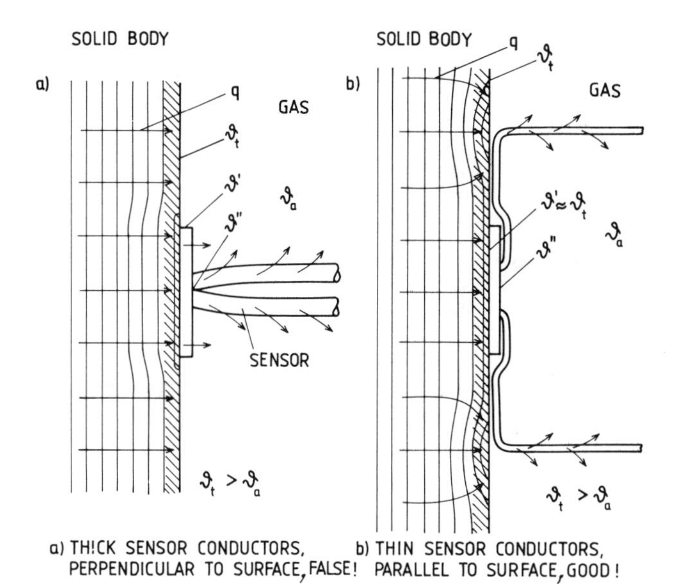

Figure 10.9
Surface temperature measurement of a solid body by a disk thermocouple.

There are different ways to reduce the first partial error $\Delta\vartheta_1$:

1. Error $\Delta\vartheta_1$ can be reduced by increasing the contact surface area between the sensor and the investigated body. In this way the heat flux density is diminished at the contact area and thus the deformation of the original temperature field is also reduced. One of the ways to achieve this is to apply an additional metal disk of high thermal conductivity (Figure 10.9(a)). The influence of the disk on the value of $\Delta\vartheta_1$ may be easily shown by comparing the value of the $\Delta\vartheta_1$ error calculated from the formula (10.26), expressed as:

$$\Delta\vartheta_1 = -\frac{(q_T - q_b)R_T}{\lambda_b} F_m \Bigg|_{\substack{x=0 \\ 0 \leqslant r \leqslant R_T}}$$

   First introduce q_T for a disk sensor (Equation (10.23)) and then for a twin conductor thermocouple sensor (Equation (10.18)). In the first case the contact area equals the disk area while the contact area equals the frontal cross-sectional area of both thermocouple conductors in the second case.
2. Reduction of the value of $\Delta\vartheta_1$ can be obtained by decreasing the heat flux conducted from the measuring point along the sensor or its conductors. For this purpose the thermocouple conductors should be as thin as possible and should be led in their primary part, parallel to the investigated surface which is along the isotherms. Any influence of the conductor radius, R_c, on the heat flux entering the sensor is seen from the relation of Equation (10.17).
3. Total elimination of the error, $\Delta\vartheta_1$, can be achieved by applying a thin disk sensor, made from material of high conductivity and having the same emissivity

as the investigated surface and fastened permanently to it. In such a solution the heat flux densities q_b and q_T are equal ($q_b = q_T$) as well as thermal resistances $W_a = W_T$; ($\alpha_b = \alpha_p$). For a disk fastened permanently to the surface the thermal contact resistance, between the sensor and the surface, is nearly zero ($W_c \approx 0$). From Equations (10.26) and (10.28) it follows that $\Delta\vartheta_1 \simeq 0$ in the case described above.

The *second partial error* $\Delta\vartheta_2$ is defined, conforming to Figure 10.1 as

$$\Delta\vartheta_2 = \vartheta'' - \vartheta' \tag{10.30}$$

This error results from the existence of a thermal contact resistance, W_c, between the sensor and the body. It can be regarded as a temperature drop across the resistance W_c under the influence of heat flux, Φ_T, entering the sensor. Thus, $\Delta\vartheta_2$ is given by

$$\Delta\vartheta_2 = -\Phi_T . W_c \tag{10.31}$$

In the case of a semi-infinite body $\Delta\vartheta_2$ is given by

$$\Delta\vartheta_2 = -\frac{(W_c/W_a) + (\pi R_T \lambda_b / F_m) W_c}{1 + (\pi R_T \lambda_b / F_m)(W_T + W_c)} \Theta_t \tag{10.32}$$

where all symbols are as in Equation (10.28).

It is rather difficult to calculate the thermal contact resistance, W_c, which depends upon many factors such as the smoothness and cleanliness of the surface, the force with which the sensor is pressed to the investigated surface, the elasticities of the sensor and the surface materials, and so on. Tye (1969) and Michalski (1978) show how it can be found experimentally as illustrated, for example, in Figure 10.10. Here the thermal contact resistance between a copper rod and a steel surface is displayed against the contact force at different contact temperatures ϑ_c. Based on these results

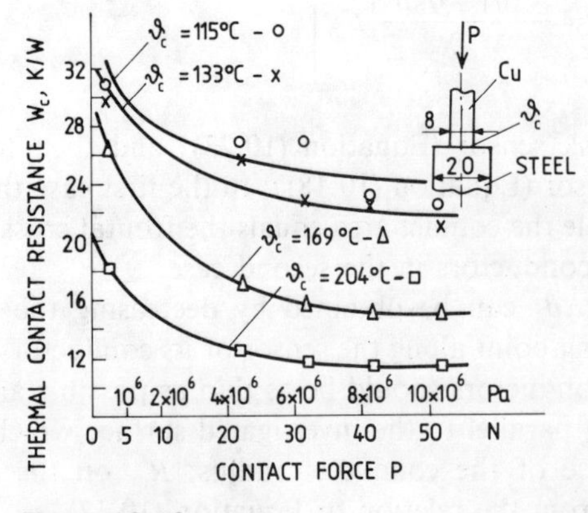

Figure 10.10
Thermal contact resistance, W_c, of copper-steel contact versus contact force P at different temperatures.

Michalski (1978) has advised the application of a force of about 30 N for copper-plate sensors of diameter 8 mm. Reduction of the thermal contact resistance is possible by cleaning the contacting surfaces to remove oxides, fats and other impurities followed by the application of a paste of high thermal conductivity (Tye, 1969; Michalski, 1978).

The *third partial error*, $\Delta\vartheta_3$, is defined, conforming to Figure 10.1 as

$$\Delta\vartheta_3 = \vartheta_T - \vartheta'' \qquad (10.33)$$

For its calculation, the temperature distribution along the sensor must be known. However, for really small distances between the sensor sensitive point and its front surface, the heat flux from the side surface of the sensor along the l' length can be neglected. Thus the third partial error is given by:

$$\Delta\vartheta_3 = -q_T \frac{l'}{\lambda_T} \qquad (10.34)$$

where λ_T is the thermal conductivity of the sensor material.

To reduce the third partial error, $\Delta\vartheta_3$, the distance l' (Figure 10.1) should be kept as small as possible. This can be achieved by using thin flat band-thermocouples, thin plates, of well-conducting materials or thermocouples having non-joined, pointed conductors (Figure 10.15(a)), which can only be used on metallic surfaces.

In contact surface temperature measurements, especially when using hand-held sensors, dynamic errors are also observed. These errors occur when the moment of taking the readings takes place before a sensor, touching the investigated surface, has reached a thermal steady state. This error can be eliminated by a sufficiently long contact time before the readings are taken. In practice, this time has to be below about 1 min, as for longer times it starts to be tiresome for the operator. Also the probability of ensuring the correct sensor position on the surface decreases with increasing time. It is possible to reduce dynamic errors by applying a peak-picker device of the type described in §7.7.

A theory of dynamic errors in contact temperature measurements has not been properly developed so far (Mackiewicz, 1971b; Znichenko, 1963).

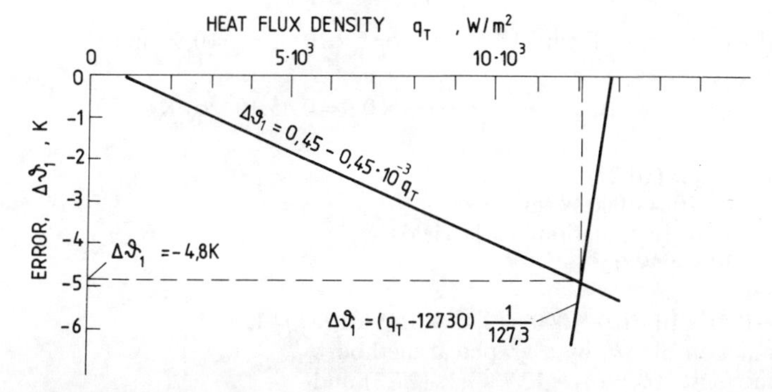

Figure 10.11
Auxiliary diagram for the numerical example.

Numerical example

Calculate the method error of contact temperature measurement of a chromium-nickel steel surface, using a disk thermocouple. It is assumed that the thermal contact resistance W_c between the sensor and investigated surface is null.

Data: original surface temperature $\vartheta_t = 120$ °C,
ambient temperature $\vartheta_a = 20$ °C
body thermal conductivity $\lambda_b = 10$W/m K,
heat transfer coefficient at body's surface
$\alpha_b = 10$ W/m^2 K

Sensor: NiCr–NiAl thermocouple, with copper disk of: $l_p = 1.5$ mm
$R_p = 7.5$ mm, $\lambda_p = 372$ W/m K
NiCr conductor: $R_{c1} = 1.5$ mm,
$\lambda_{c1} = 13$ W/m K.
NiAl conductor: $R_{c2} = 1.5$ mm
$\lambda_{c2} = 58$ W/m K
Heat transfer coefficient on disk surface: $\alpha_p = 10$ W/m^2 K.
Heat transfer coefficient on side surfaces of conductors $\alpha_c = 50$ W/m^2 K.

Solution:

1. Calculation of $q_T = f(\Delta\vartheta_1)$ conforming to Equation (10.23):

$$q_T = \left[10 + \frac{1.5\times10^{-3}}{(7.5\times10^{-3})^2}\sqrt{2\times50\times1.5\times10^{-3}}(\sqrt{13}+\sqrt{58})\right]\Theta_T = 127.3\ \Theta_T \text{W/m}^2$$

Assuming: $\Delta\vartheta_2 = 0(W_c = 0)$ and $\Delta\vartheta_3 = 0$, method error $\Delta\vartheta = \Delta\vartheta_1$, and consequently:

$\vartheta_T = \vartheta_t + \Delta\vartheta_1$ or $\Theta_T = \vartheta_T - \vartheta_a = (\vartheta_t - \vartheta_a) + \Delta\vartheta_1$
thus:
$q_T = 127.3[(120-20) + \Delta\vartheta_1] = 12\ 730 + 127.3.\Delta\vartheta_1$
and finally:

$$\Delta\vartheta_1 = (q_T - 12\ 730)\frac{1}{127.3}\text{K}$$

2. Calculation of $\Delta\vartheta_1 = f(q_T)$
From Equation (10.4):

$$B = \frac{10\times7.5\times10^{-3}}{15} = 5\times10^{-3}$$

From the diagram in Figure 10.5, for $B = 5\times10^{-3}$, $F_m \approx 0{,}9$, and thus,

$$\Delta\vartheta_1 = \frac{q_d.7.5\times10^{-3}}{15}\times0.9 = 0.45.10^{-3}q_d\ \text{K}$$

From Equation (10.27):
$q_b = 10(120-20) = 1000$ W/m^2
Substituting q_b in Equation (10.1) yields:
$q_d = (q_T - 1000)$; W/m^2
and then
$\Delta\vartheta_1 = -0.45\times10^{-3}(q - 1000) = 0.45 - 0.45\times10^{-3}q_T$ K

3. Determination of $\Delta\vartheta_1$ by a graphical method.
The functions: $\Delta\vartheta_1 = (q_T - 12\,730)(1/127.3)$ and
$\Delta\vartheta_1 = 0.45 - 0.45\times10^{-3}q_T$ are displayed in one diagram (Figure 10.11).

The intersection point of both lines determines the error $\Delta\vartheta_1 \approx -5$ K.

10.2.4 Influence of thermal properties of bodies on errors

The main physical property characterizing a body, whose temperature is to be measured, is its specific thermal conductivity λ_b. With increasing λ_b, the disturbing temperature, ϑ_d, decreases in agreement with Equation (10.5) and thus the first partial error, $\Delta\vartheta_1$, also decreases as in Equation (10.26). This dependence is not strictly inversely proportional, because in Equation (10.26) the function F_m also depends upon the thermal conductivity λ_b, hidden in the constant B (Figure 10.5). For low values of the heat transfer coefficient, α_b, at the surface of the body, and for normally applied dimensions of the end part of the sensor, the error $\Delta\vartheta_1$ is a hyperbolically decreasing function of thermal conductivity λ_b. This explains the difficulties encountered in the use of contact methods for the temperature measurement of non-metallic bodies whose thermal conductivity is many times smaller than that of metals.

The thermal conductivity λ_b of a body also influences the response time of the contact sensor, which is far longer for non-metals. A decisive influence is exercised by the size of the investigated body, as compared with the sensor dimensions, upon the precision of the readings.

The disturbing temperature values in a semi-infinite body also decreases in a direction perpendicular to the surface (x in Figure 10.3) as along it (r in Figure 10.4) as a function of distance from the contact point. At a depth x and a distance r, about five times greater than the radius R_T of the contacting area, the disturbing temperature falls below 10% of its maximum value. It follows that, any body having dimensions x and r more than five times greater than the radius of contacting area ($x > 5R_T$; $r > 5R_T$), can be regarded as a semi-infinite body. For any other bodies, or characteristic dimensions below those given above, the deformation of the original temperature field and so the errors $\Delta\vartheta_1$ will be far greater.

The surface state of the investigated body influences the thermal contact resistance, W_c, between sensor and body and thus also the precision of the readings. Errors, resulting from the presence of thermal contact resistance, often exceed all other partial errors.

10.3 SENSORS

10.3.1 Portable contact sensors

Portable contact sensors or probes, pressed to the investigated surface, give good readings, known as spot readings, when a thermal steady-state is reached. Many such thermometers have exchangeable measuring tips, adapted to the condition, shape and material of the surface (Figure 10.12). All of the sensor tips used should fit the surface in the best possible way.

Besides method errors (§10.2), sometimes large random errors may also occur. They are mostly caused by incorrect positioning of the sensor tip relative to the surface. The greater the sensor response time the higher is the probability of occurrence of random errors. To reduce them it is advisable to use sensors of low thermal inertia or to repeat the measurements several times. A marked improvement is obtained by

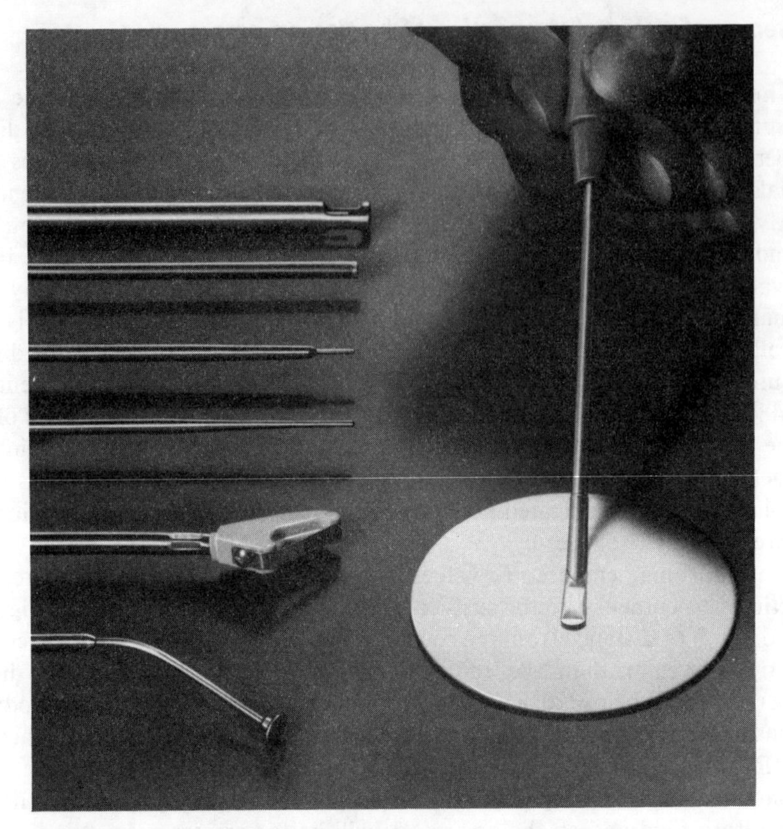

Figure 10.12
Portable contact sensors. From the top: 2 for gas, 2 for immersion and 2 for surface temperature measurement. (Courtesy of Heraeus GmbH, Germany.)

the application of a peak-picker device (§7.7), which is essential when using digital indicators. Without a peak-picker the readings of a digital instrument are a series of random, erroneous values.

Most surface sensors are thermocouples. For temperature measurement of smooth, flat surfaces, disk thermocouples (Figure 10.13) are mostly used. The thermocouple conductors are soldered into a copper (up to 400 °C) or silver (up to 600 °C) disk. A detailed discussion of its constructional parameters is given in §10.2.3. As an example, a thermocouple with a silver disk of diameter $d_p = 10$ mm and thickness $l_p = 1$ mm, having NiCr–NiA conductors of $d_c = 1$ mm, gives readings for metal surfaces which are about 10 to 20 K low. Its 98% response time $t_{0.98} \approx 10$ s. This type is of no use for non-metals. Application of heat conducting paste improves the precision by causing smaller $\Delta\vartheta_2$. A survey of special surface sensors applied in scientific instruments is given by Gatowski *et al.* (1989). A better solution for steel and other low-conductivity metals is ensured by flat, spiral, elastic tips (Figure 10.14), which are mostly NiCr–NiAl types, combining the merits of a thin, flat measuring junction (small $\Delta\vartheta_3$), with conductors led along isotherms (small $\Delta\vartheta_1$). They can be used up to about 500 °C, securing errors of about -5 to -12 K

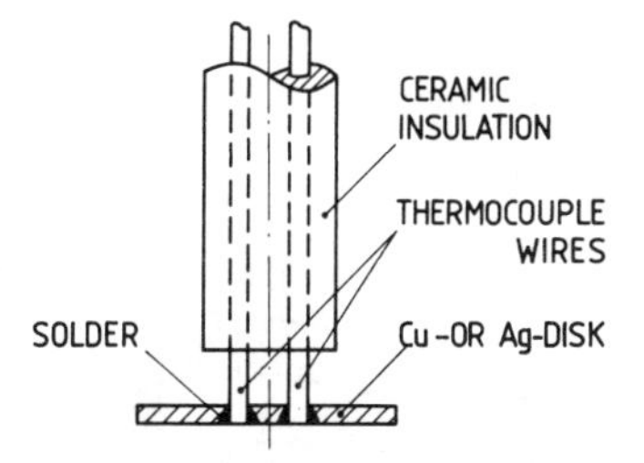

Figure 10.13
Disk thermocouple.

for metallic surfaces and -10 to -30 K for ceramics. In the case of 0.8 mm spiral thickness the $t_{0.98}$ response time is about 15 s for metals and about 2 min for ceramics.

Thermocouples with a non-soldered point tip are used (Figure 10.15(a)) to measure the temperature of clean non-ferrous metallic surfaces such as Al or Cu. The conductors, which are pressed with a force P to the surface, form the measuring junction with the body under measurement so eliminating the third partial error $\Delta\vartheta_3$. NiCr–NiAl and NiCr–NiCu thermocouples are used as they are the only types hard enough to remain sharp for long. Readings are a function of conductor diameter, the shape of the points and the pressing force against the surface, as shown in Figure 10.15(b) for NiCr–NiAl thermocouples. The necessary pressing force is secured by springs in the conductors' mounting. In the case of NiCr–NiAl conductors with $d_c = 3$ mm the $t_{0.98}$ response time is about 1 s.

For cylindrical surfaces bow-band thermocouples stretched across elastic yokes as in Figure 10.16, are used. In this way, a perfect contact with the cylindrical surface is assured. Due to the tangential position of the band, no deformation of the original temperature field occurs so that $\Delta\vartheta_1 \approx 0$. This thermocouple can be used for metallic

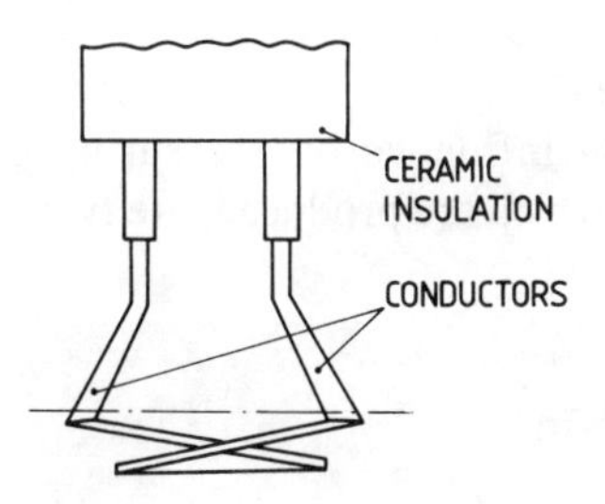

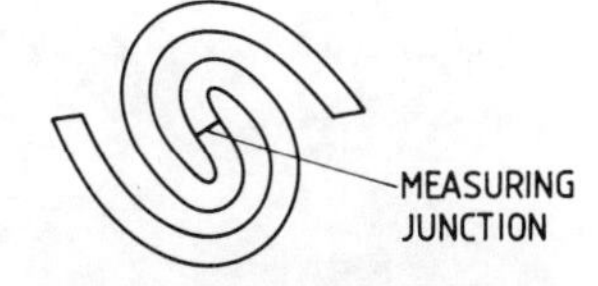

Figure 10.14
Flat-spiral thermocouple.

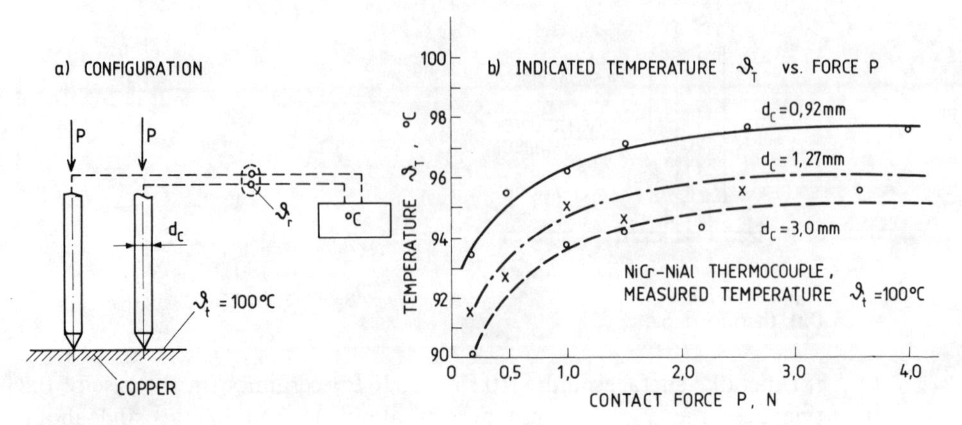

Figure 10.15
Point-contact tip thermocouple.

and non-metallic surfaces. For example, when measuring the temperature of a copper tube of 30-mm diameter, the overall error was about 2 K at 100 °C and the $t_{0.98}$ response time was about 3 s.

For flat non-metallic surfaces flat band thermocouples, like those in Figure 10.17, are used. With a similar construction as for the bow-band type, the band with the measuring junction is pressed to the surface by a non-metallic bar with a coil-spring.

Measuring errors of a few kelvins may be negative or positive. They are positive in the case when the band emissivity is smaller than the surface emissivity.

For temperature measurement of small metallic or non-metallic bodies of flat, concave or convex surfaces convex-band thermocouples (Figure 10.18) are used. They are produced among others by Omega Engineering Inc. (USA), or Ultrakust GmbH (Germany). The elastic, convex band thermocouple, Chromel-Konstantan or Chromel-Alumel is flattened, while touching the investigated surface, thus ensuring good thermal contact. This flattening is limited by a buffer, to prevent damage. The longitudinal incisions permit the shape of the thermocouple to adjust to any surface irregularities. Band thickness has a deciding influence on measuring errors. The different sizes of convex-band thermocouples, which are produced, are typically applied

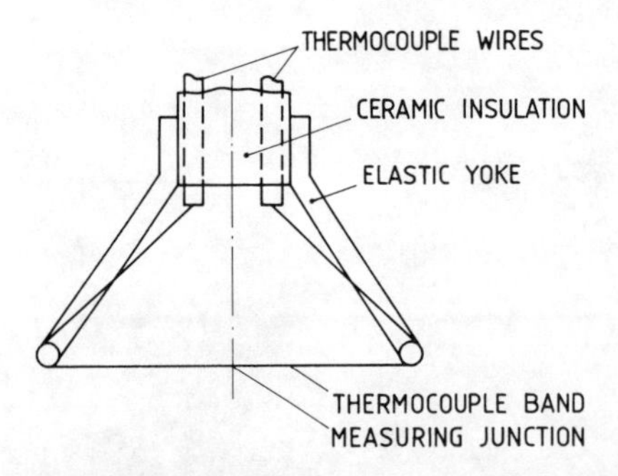

Figure 10.16
Bow band thermocouple.

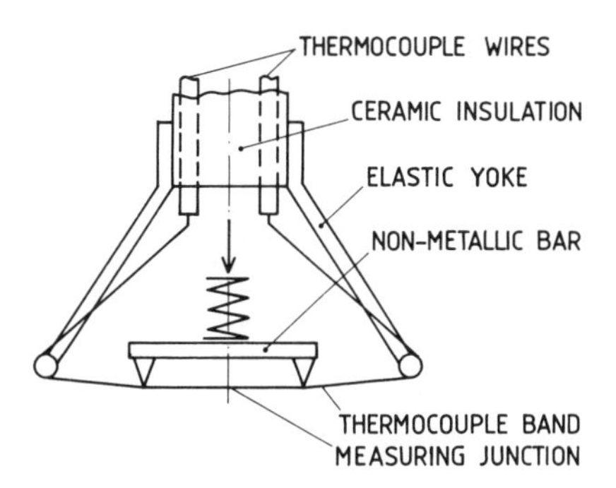

Figure 10.17
Flat band thermocouple.

in the temperature measurement of foundry forms, metal plates, walls, glass-ware, ball-bearings and so on, up to about 500 °C. The smallest sensors are used in temperature measurements of electronic components.

The $t_{0.9}$ response time of the Ultrakust sensor on metallic surfaces is $t_{0.9}=0.2$–0.5 s and on non-metals $t_{0.9}=3$–10 s.

Thermistors which are also used for surface contact measurements, are mounted in a small silver plate to intensify the heat transfer between the thermistor and the investigated surface. They are used mainly for metallic surfaces up to 250 °C and have short response times.

For less precise surface temperature measurement when mercury-in-glass thermometers are also used, they are mounted directly in special thermometric pockets or holes or in additional copper or brass metallic stands (Figure 10.19). A thermometer placed in a thermometric hole may give readings which differ from the surface temperature.

10.3.2 Fixed contact sensors

For stationary, continuous surface temperature measurement the temperature sensors are either pressed in a hole or glued or soldered to the surface. Their indications are

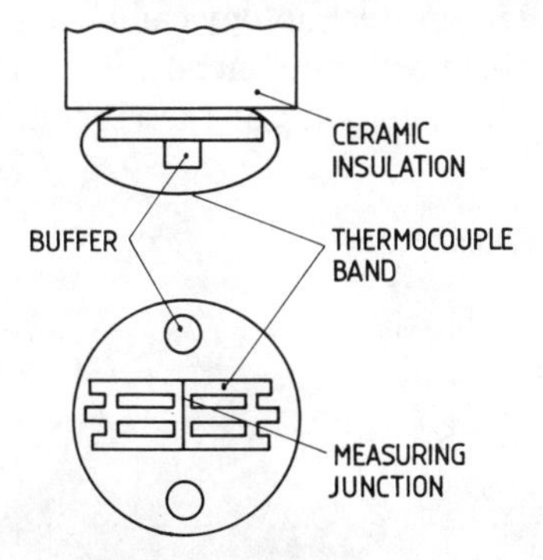

Figure 10.18
Convex band thermocouple.

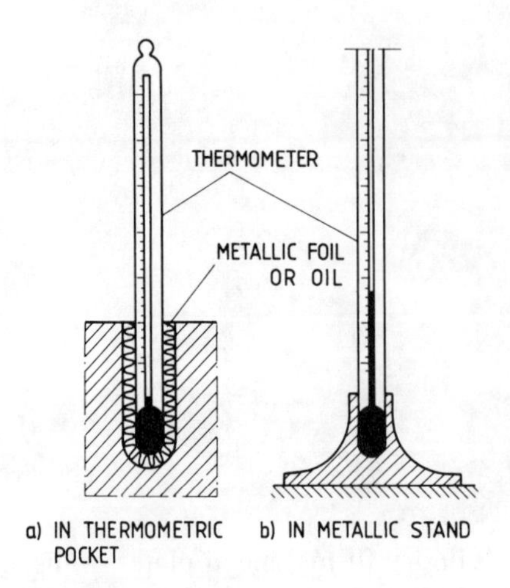

Figure 10.19
Mercury-in-glass thermometer for surface temperature measurement.

more precise than those of portable types, especially as there are no thermal contact resistances. Figure 10.20 presents a typical method of fixing thermocouple conductors in a hole by peening, using a special tool. Both conductors are peened separately to the surface metal which itself constitutes the measuring junction. For peening, a special tool (Figure 10.20(c)) with a conductor inserted through its axial hole, is hammered lightly. Following the law of the third metal (§4.1.2) and for a small distance between both conductors, the existence of the third metal does not change the readings. Leading the conductors along the metal surface, ensures that all three partial errors are negligibly small. Sometimes the thermocouple is soldered to metallic surfaces or glued to non-metallic ones by a refractory cement (Figure 10.21). A detailed description of the ways of fixing thermocouples is given by Baker *et al.* (1953).

With the development of thin-film technology thin-film thermocouples are also used. They are vacuum deposited on thin glass or ceramic laminate plate and glued to the investigated surface (Figure 10.22) (Moeller, 1963a,b; Browning and Hemphill, 1962; Kinzie, 1973; Bransier, 1975). Although this technology can be applied for all thermocouple materials, best results are obtained with pure metals. It should be taken into consideration that the thermoelectric characteristics of thermocouples of film

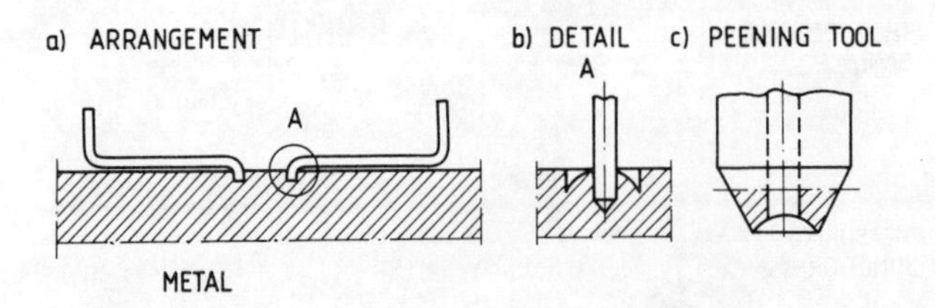

Figure 10.20
Peened thermocouple.

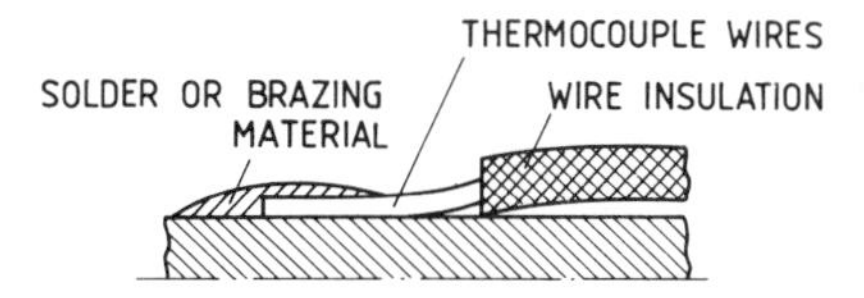

Figure 10.21
Soldered thermocouple.

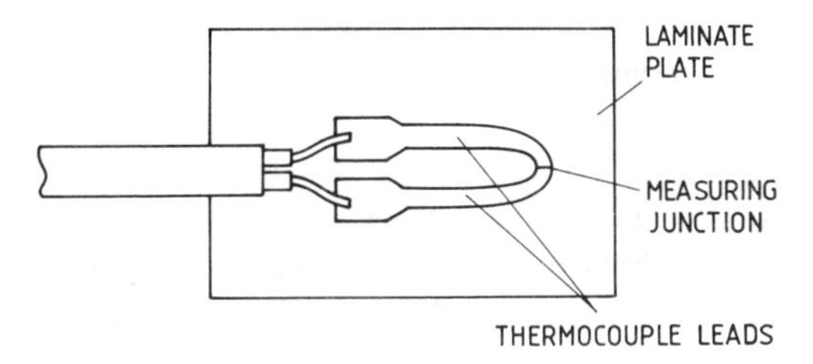

Figure 10.22
Thin-film thermocouple.

thickness below 0.25 μm can essentially deviate from those of the standardized ones (Bransier, 1975). Film-thermocouples, thin as they are, give very small measurement errors. They also have a very low thermal inertia, with low time constants of some milliseconds. Omega Engineering Inc., USA, offers three types of thin surface 'Cement-On' thermocouples in which the thermocouple is embedded between glass-reinforced polymer laminates intended for continuous application up to 260 °C and up to 370 °C for short duration. The overall sensor thickness is about 0.1 mm. Special epoxy based glues which are supplied with thermocouples, can be used on metals, ceramics, glass, plastics and paper.

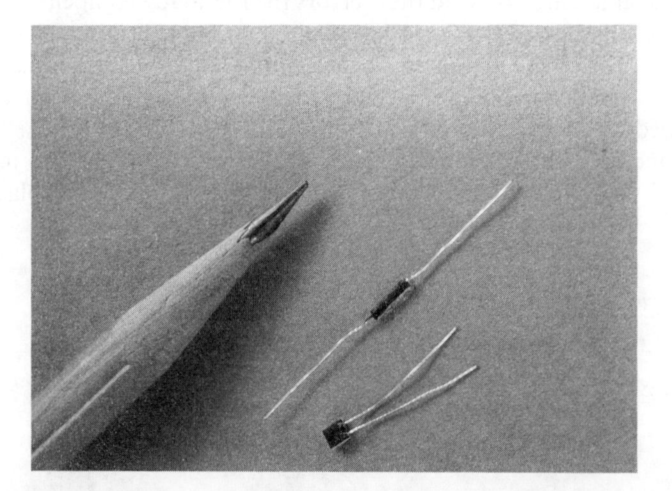

Figure 10.23
Thin-film resistance detectors. (Courtesy of Omega Engineering Inc., USA.)

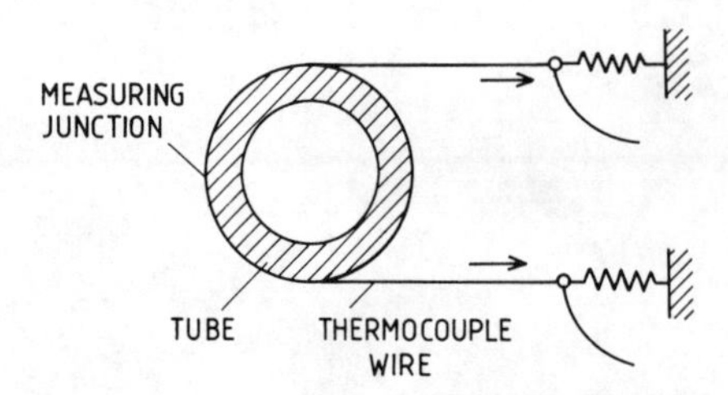

Figure 10.24
Thin-wire thermocouple for tubes and cylindrical surfaces.

Thin-film resistance detectors for surface temperature measurement which are also produced (§5.2.3) can be used from −20 up to 500 °C. They can be glued to the surface either directly or in a flat sheath which gives protection against mechanical damage (Figure 10.23). The size of a sensor produced by Omega Engineering by way of example measures $10.2 \times 3.2 \times 1$ mm.

The temperature of tubes and cylindrical surfaces is easily measured by thin-wire or thin-band thermocouples strung by springs as shown in Figure 10.24. Another way to apply resistance wire wound round a tube and then fixed by epoxy or silicon resin. Wire of Pt or Ni of diameter 0.03 to 0.1 mm are used as in Figure 10.25.

10.3.3 Thermally compensated sensors

In measuring surface temperature by portable sensors one of the error sources was the deformation of the original temperature field by the heat flux conducted from the surface along the sensor. The most common case is when the surface temperature is higher than the ambient. To avoid these errors the thermally compensated sensors which are used, are heated by an additional low-power heating element to a temperature near that of the measured surface. In most cases the thermal design is based on either the static method described by Mackiewicz (1976b), or on a dynamic method described by Mackiewicz (1976a).

In the static method, the thermocouple sensor (Figure 10.26) has two measuring junctions, called the main and auxiliary one. The auxiliary junction is placed some

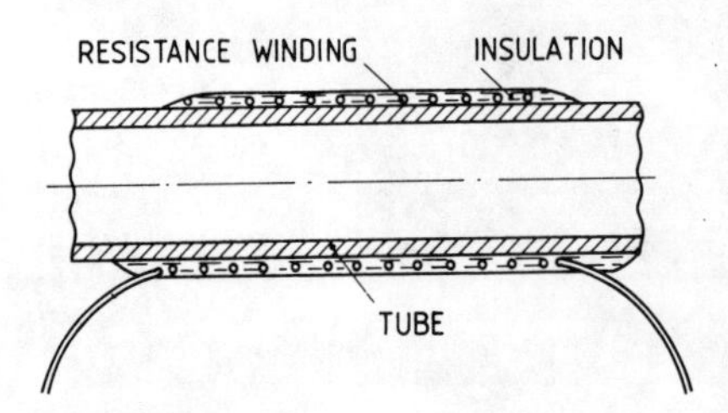

Figure 10.25
Wire resistance detector wound round a tube.

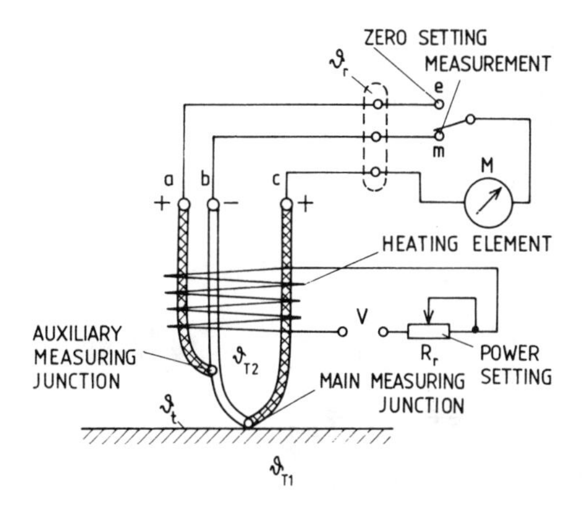

Figure 10.26
Thermally compensated thermocouple for the static method.

millimetres from the investigated surface. After contact is made, the sensor is heated by a small heating element, whose power is set by resistor, R_r, till the indications of both thermocouples ab and cb are equal. The difference of both thermoelectric forces is controlled by the measuring instrument M in position e of the switch. When this difference becomes zero no heat exchange between the sensor and surface takes place. The sensor temperature, ϑ_T, near to the original surface temperature, ϑ_t, is measured by the instrument M in position m of the switch.

Different configurations of thermally compensated sensors (Figure 10.27) described by Emelyanenko (1960) and Stamper (1963) can be represented by an equivalent cylindrical system as shown in Figure 10.28, which also presents the temperature distribution along the equivalent sensor. Mackiewicz (1976a) points out that at the moment the reading is taken the temperatures of the main and auxiliary measuring junctions are equal ($\vartheta_{T1} = \vartheta_{T2}$). Similarly, as in §10.2, there are also the three partial

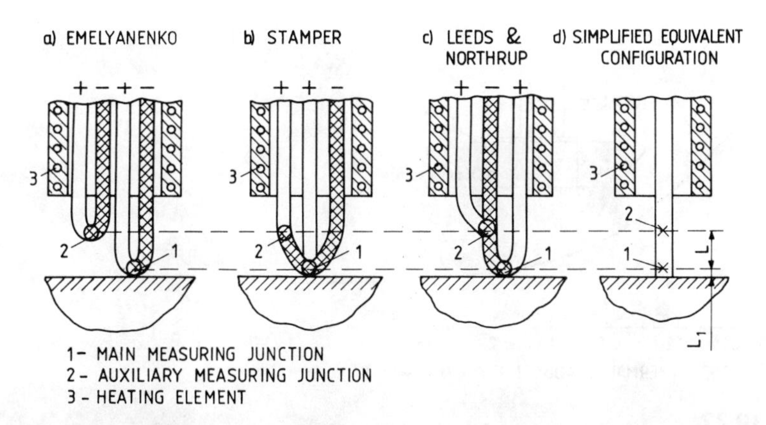

Figure 10.27
Different configurations of thermally compensated sensors.

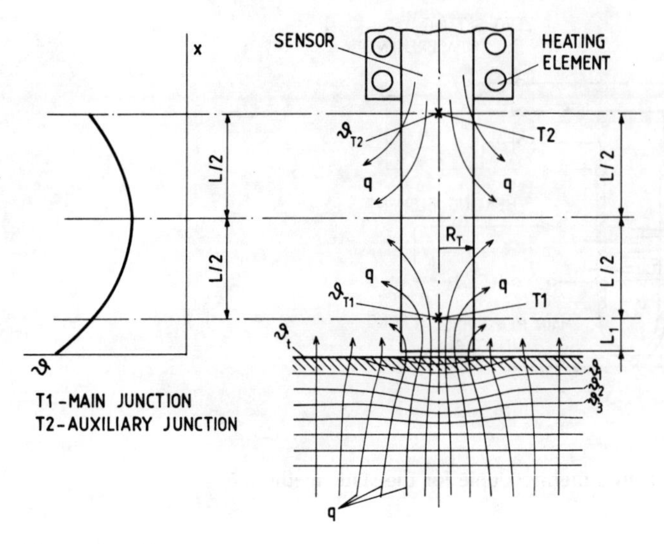

Figure 10.28
Thermally compensated thermocouple. Heat flux q and temperature ϑ distributions.

errors $\Delta\vartheta_1$, $\Delta\vartheta_2$ and $\Delta\vartheta_3$ present in this case. Error, $\Delta\vartheta_1$, is given by Equation (10.28) and, $\Delta\vartheta_3$, by Equation (10.32) in agreement with Mackiewicz (1976a). The thermal resistance of the sensor is given by the approximate relation:

$$W_T = \frac{1}{\alpha_T A_T} \tag{10.35}$$

where A_T is the cylindrical area of the sensor surface along the length $L/2$ as shown in Figure 10.28, and α_T is the heat transfer coefficient from the sensor surface to the ambient environment.

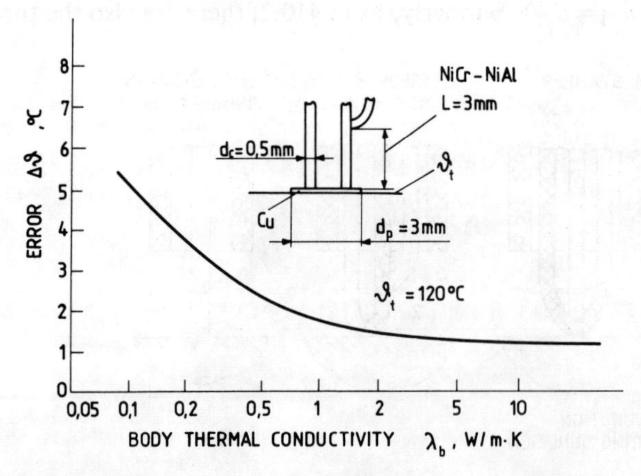

Figure 10.29
Measuring error $\Delta\vartheta$ versus thermal conductivity λ_b of the investigated body when using a thermally compensated sensor in the static method.

Although the error, $\Delta\vartheta_3$, is difficult to estimate, it is negligibly small assuming $L_1 \approx 0$ (see Figure 10.28). According to formulae (10.28), (10.32) and (10.35) as well as to practical experience, to reduce errors, thermally compensated thermocouples should have conductor diameters which are as small as possible. The thermal contact resistance, W_C, should also be kept small by polishing the contact surface and by applying heat conducting paste. Overall measuring errors, which can be positive or negative, may contain some additional errors caused by the limited sensitivity of the measuring instrument M (Figure 10.26). The errors which mainly depend upon the thermal conductivity, λ_b, of the body, whose temperature is to be measured, (Figure 10.29) (Mackiewicz, 1976a), decrease with increasing λ_b. It is clear that the described method is especially well adapted for temperature measurement of metals, where the overall errors are below ± 1 K.

The measuring systems of thermally compensated sensors are mostly automated with the power of the heating element automatically set so that the condition $\vartheta_{T1} = \vartheta_{T2}$ is fulfilled (Figure 10.26). Typically the measuring range is about 20 to 650 °C with the response time amounting to some seconds. Thermally compensated sensors based on the dynamic method described by Mackiewicz (1976b) are rarely used. In this method Sasaki and Kamada (1952) point out that the sensor is slowly heated up, or cooled down, by periodically pressing it against the investigated surface either mechanically or manually. From the recorded temperature versus time, the measured temperature may be found from a determination of the moment of thermal equilibrium.

10.3.4 Comparison of different sensors

A survey of results obtained using different types of contact sensors is given in this section for aluminium, steel and ceramic surfaces (Figures 10.30, 10.31, 10.32).

The types of sensors which were applied were:

NiCr–NiAl, silver disk (Figure 10.13), $d_p = 10$ mm, $l_p = 0.8$ mm,
NiCr–NiAl, spiral (Figure 10.14), spiral thickness 0.8 mm,
NiCr–NiAl, non-soldered, pointed (Figure 10.15), $d_c = 3$ mm,
NiCr–NiAl, flat band (Figure 10.17), band thickness 0.1 mm,
NiCr–NiAl thermally compensated by the static method (Figure 10.29).

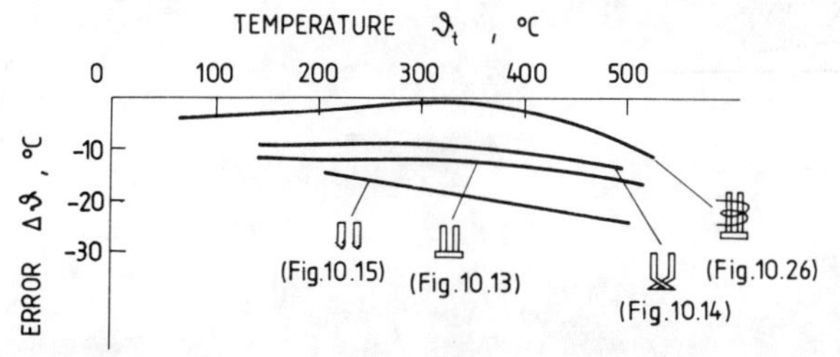

Figure 10.30
Errors $\Delta\vartheta$ of the temperature measurement of an aluminium surface versus body temperature ϑ_t.

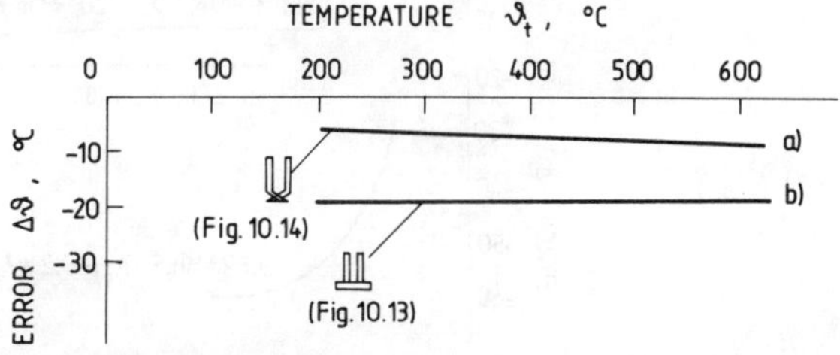

Figure 10.31
Errors $\Delta\vartheta$ of the temperature measurement of a clean steel surface versus body temperature ϑ_t.

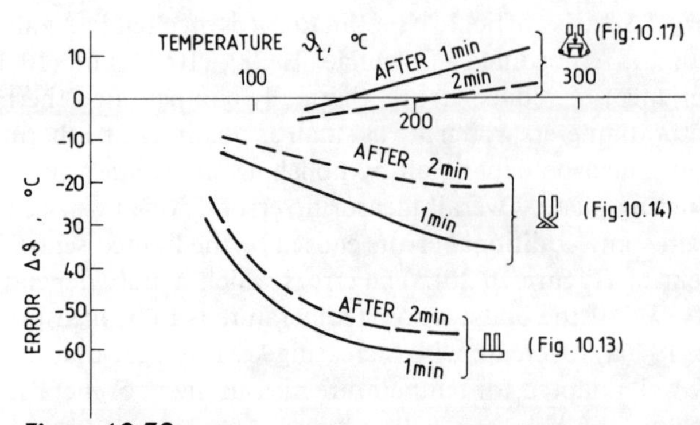

Figure 10.32
Errors $\Delta\vartheta$ of the temperature measurement of a ceramic surface versus body temperature ϑ_t after 1 and 2 min.

Measuring errors, applying a commonly used silver disk sensor of diameter d_p = 10 mm and l_p = 0.8 mm, are shown in Figure 10.33, and their dependence upon the thermal conductivity of the body is shown in Figure 10.34. These results agree with §10.2.4. Another set of measurements concern temperature measurements of a copper sheet 1 mm thick, in contact from below with boiling water (Table 10.1). All results in this section are average values of about a dozen readings taken by an experienced operator.

Table 10.1
Errors $\Delta\vartheta$ of temperature measurement of copper surface at ϑ_t = 100 °C.

Temperature sensor	Figure	Error (°C)
NiCr–NiAl, silver disk (d_p = 10 mm, l_p = 0.8 mm)	10.8(d) 11.13	−5
NiCr–NiAl, non-soldered, pointed d_c = 3 mm	10.15	−5
Thermistor		−7
Mercury-in-glass thermometer in metal tube	10.19(b)	−10
NiCr–NiAl thermally compensated by static method	10.29	−1.3

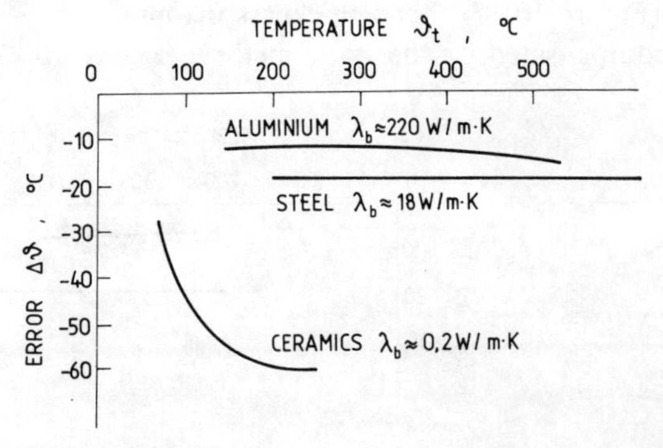

Figure 10.33
Errors $\Delta\vartheta$ of the temperature measurement of the surfaces of different materials, by a disk thermocouple, versus temperature ϑ_t.

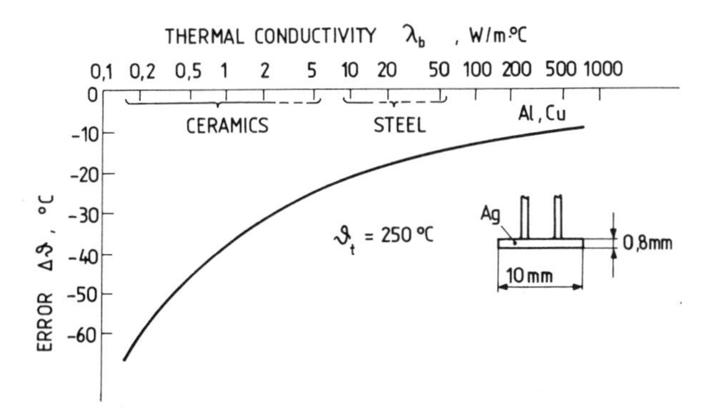

Figure 10.34
Errors $\Delta\vartheta$ of the temperature measurement of a semi-infinite body by a disk thermocouple versus body thermal conductivity λ_b at $\vartheta_t = 250$ °C.

10.4 QUASI-CONTACT METHOD

The quasi-contact method comprises the properties of both the pyrometric and the contact methods. A quasi-contact sensor which is shown in Figure 10.35, has a cup for its main part. This cup is a mirror of polished gold inside, which is placed at a distance smaller than 1 mm from the investigated surface. Thermal radiation emitted from the surface irradiates a thermopile radiation detector through a fluorite window. Although oxidized metal surfaces have an emissivity factor well below unit, the presence of the hemispherical golden mirror means that the whole system approaches black-body conditions. In this manner, the readings which are emissivity independent, are not influenced by any outside radiation due to the presence of the protecting hemisphere.

One of the drawbacks of the semi-contact method are the changed heat-transfer conditions from the estimated surface to the environment after applying the mirror. To eliminate this influence, the mirror is sometimes made with a reflectivity which is a bit lower, or readings are taken during a short time of application of the sphere (Drury *et al.*, 1951; Burton, 1953).

A measuring device by Land Infrared Ltd operates in the temperature ranges from 200 to 900 °C, 500 to 1300 °C and 100 to 400 °C. The sphere diameter is 50 mm and the response time of the device is about 5 to 6 s. The overall error for oxidized metals and non-metals is about ± 5 K. This instrument, which can also be used for concave surfaces with a radius of curvature over 50 mm, is typically applied for short-time temperature measurements of ingots, cylinders, foundry forms, furnace surfaces and so on. The same instrument can be used to measure surface emissivities. For this purpose, the mirror is temporarily covered by a black insert and comparative readings are taken. From the readings with and without the black insert, a set of diagrams permits the surface emissivity to be found.

Convective methods also belong to the class of semi-contact methods. The forced convection intensifies the heat transfer between the surface and the sensor.

(a)

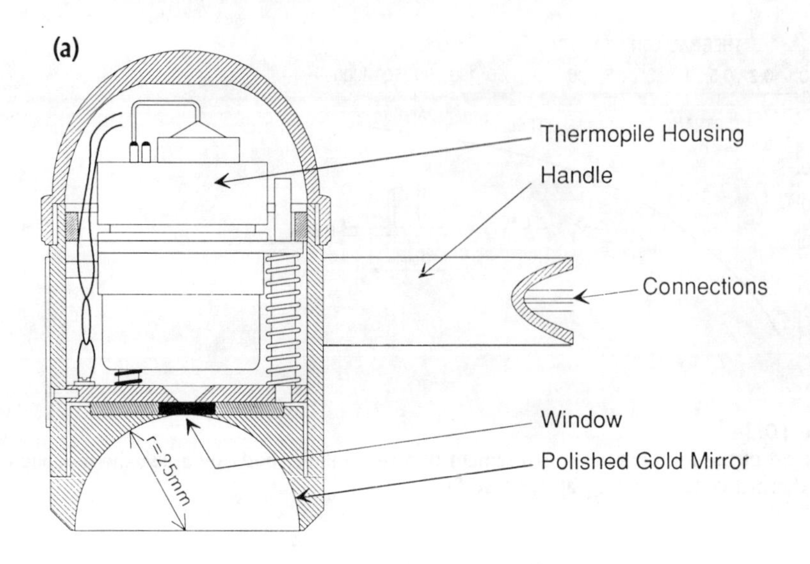

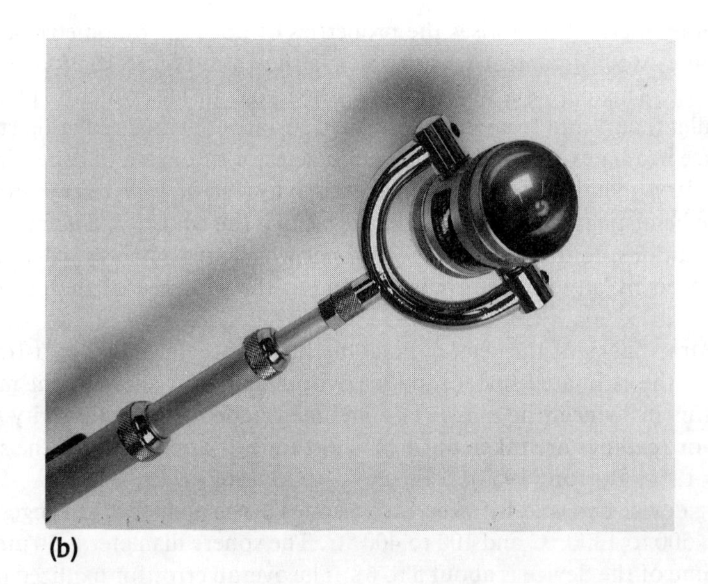

(b)

Figure 10.35
Quasi-contact sensor for surface temperature measurement. (a) Schematic diagram, (b) general view. (Courtesy of Land Infrared Ltd, UK.)

Although convective compensating methods help in attaining more precise readings, they need additional heating of the sensor. In both these methods, the sensor itself does not touch the investigated surface. These two methods are mainly applied in the temperature measurement of moving bodies. They will be described in §13.4.

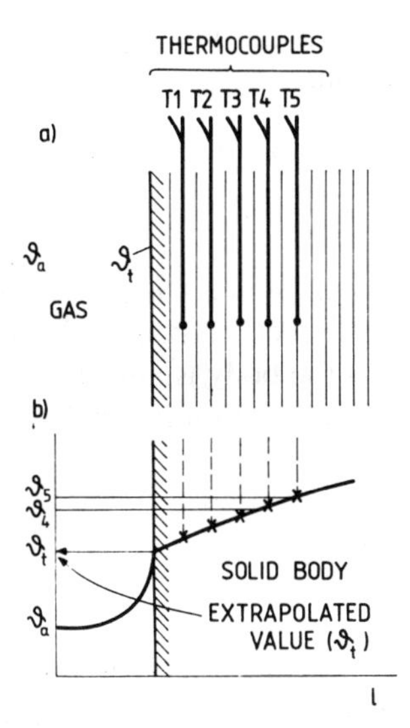

Figure 10.36
Extrapolation method of surface temperature measurement. (a) Arrangement of the sensors, (b) graphical extrapolation.

10.5 EXTRAPOLATION METHOD

The extrapolation method is one of the most precise ways to determine the surface temperature of solids (Figure 10.36). Thin thermocouples which are placed along the isotherms inside a solid body, are bare or mineral insulated thermocouples. In metals and semiconductors only MI thermocouples with insulated measuring junctions are used.

Based on the indications of particular thermocouples, the surface temperature, ϑ_t, can be determined, without any deformation of the original surface temperature distribution. Because of the necessity of introducing n thermocouples and drilling the holes, Wilcox and Rohsenow (1970) assert that the extrapolation method can only be used in some chosen cases, most commonly in some specialized measuring instruments. This is also the case in devices for the standardization of the contact sensor, or as pointed out by Michalski (1978), in devices for measuring the thermal contact resistance.

Whereas a simple graphical extrapolation is applied in sporadic measurements (Figure 10.36(b)), microcomputers are used in continuous measurements. Michalski and Borowik (1985) show that the most suitable method then uses polynomial extrapolation following the algorithm of Neville and Aitken (Stoer, 1972). In most technical problems the application of thermocouples is normally adequate.

In determining the temperature of rather thick bodies, which may be regarded as semi-infinite bodies, the two-sensor method of Yarishev and Minin (1969) is sufficiently precise. In this method, based on the indications ϑ_1 and ϑ_2 of two sensors, at the respective distances x_1 and x_2 from the investigated surface, the surface temperature is calculated from the relation:

$$\vartheta_t(t)=\vartheta_1(t)\left[\frac{x_2}{x_2-x_1}-\frac{x_2}{x_2-x_1}\sqrt[3]{\frac{\vartheta_2(T)}{\vartheta_1(t)}}\right]^3 \tag{10.36}$$

With a small range of temperature variations and small temperature differences within a solid body, such as plates of finite thickness, the thermal conductivity λ_b can be assumed constant so that the temperature drop in the body is also nearly linear. In this case, the simplest possible linear extrapolation, based on two measured values, has the form:

$$\vartheta_t(t)=\frac{\vartheta_1(t)x_2-\vartheta_2(t)x_1}{x_2-x_1} \tag{10.37}$$

Extrapolation methods which are used in principle for steady-state surface temperature determination, can only be rarely used for transients. Emelyanenko (1960) and Hollander (1959) illustrate their use to determine temperature transients of the edges of cutting tools.

The main error sources of extrapolation methods are due to thermal non-homogeneity and anisotropy of the solid body material and the deformation of the existing temperature field by the introduction of sensors inside the solid body. To avoid this last source of errors it is advisable to use thermocouples which are as thin as possible, introduced along the isotherms.

10.6 MEASUREMENT OF INTERNAL TEMPERATURE OF SOLID BODIES

Similarly as in the measurement of surface temperature the sensor which is introduced inside a solid body, deforms the original temperature field. It is assumed, as in normal practice, that the measurement is executed by a thermocouple, which for further analysis is replaced by an equivalent homogeneous cylinder.

Figure 10.37(a) presents the isotherms in a solid body in contact with a gaseous medium. To measure temperature at a depth l from the surface, a sensor T is inserted through a hole (Figure 10.37(b)). The sensitive point S of the sensor is at a distance l_1 from its end. Figure 10.37(b) presents the isotherms and heat-flux lines after-inserting the sensor T, assuming that its thermal conductivity is λ_T, which is greater than that of the body, λ_b.

The overall measurement error is composed of three partial errors:

1. $\Delta\vartheta_1=\vartheta_1'-\vartheta_t$ is the error resulting from the deformation of the original temperature field,
2. $\Delta\vartheta_2=\vartheta''-\vartheta'$ is the error resulting from the temperature drop across the thermal contact resistance, and across the air gap, together,
3. $\Delta\vartheta_3=\vartheta_T-\vartheta''$ is the error resulting from the temperature drop across the distance l_1 between the end of sensor and its sensitive point.

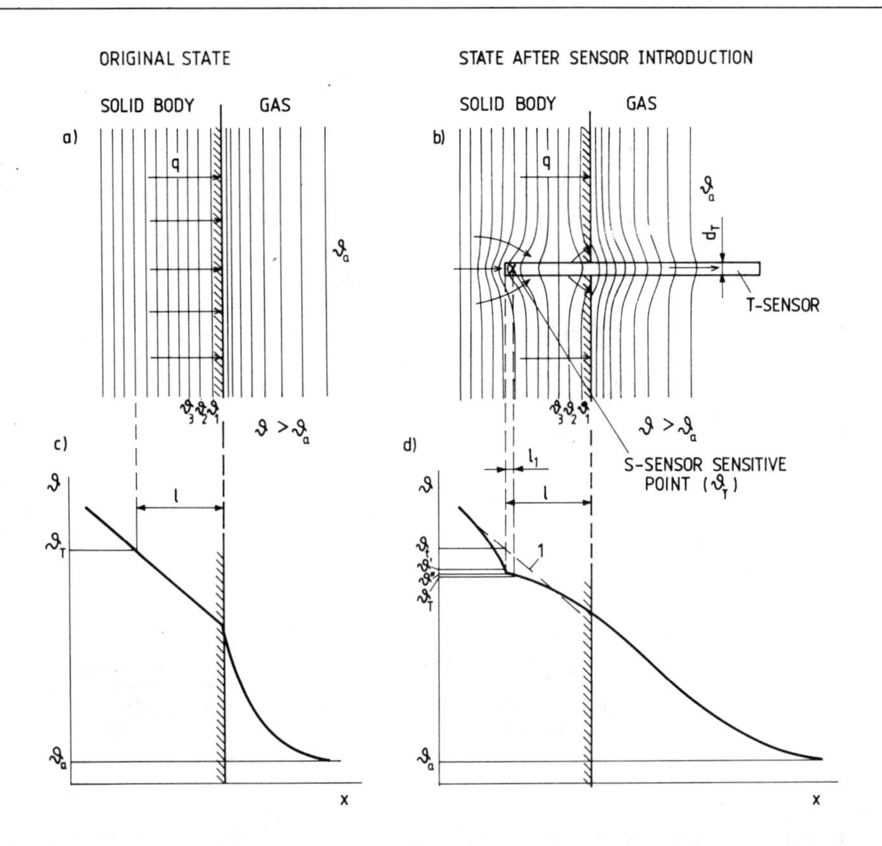

Figure 10.37
Internal temperature measurement in a solid body by a rod sensor T. a,b isotherms and heat-flux density lines q. c,d, temperature distribution along the direction, normal to the surface; ϑ_t, original, true body temperature at depth l from the surface; ϑ', temperature at the hole-bottom after inserting the sensor; ϑ'', temperature of the sensor end.

Concerning the above partial errors, the following comments are appropriate:

1. To reduce $\Delta\vartheta_1$, thin thermocouple conductors made of materials of low thermal conductivity should be used. The sensor should be inserted as deep as possible into the solid body. $\Delta\vartheta_1$ errors increase with increasing difference in the thermal conductivities of the solid body and the thermocouple materials. This is shown in Figure 10.38 which presents the $\Delta\vartheta_1$ errors versus the thermal conductivity of the investigated body of λ_b for four different insertion depths. Bare Cu–CuNi thermocouples in varnish insulation were used. The measured temperature was $\vartheta_t = 120\,°C$ each time. $\Delta\vartheta_1$ errors are negligibly small when measuring metal temperatures ($\lambda_b > 40$ W/m K) at insertion depth $l > 3$ cm. According to Hunsinger (1966) it is advisable to ensure $l/d_T > 5$ for metals and $l/d_T > 15$ for non-metals.
2. To reduce $\Delta\vartheta_2$, the shape of the end of the sensor should match the shape of the bottom of the hole with the sensor pressed fully to the bottom.
3. Reduction of $\Delta\vartheta_3$ can be achieved by choosing the appropriate sensor design.

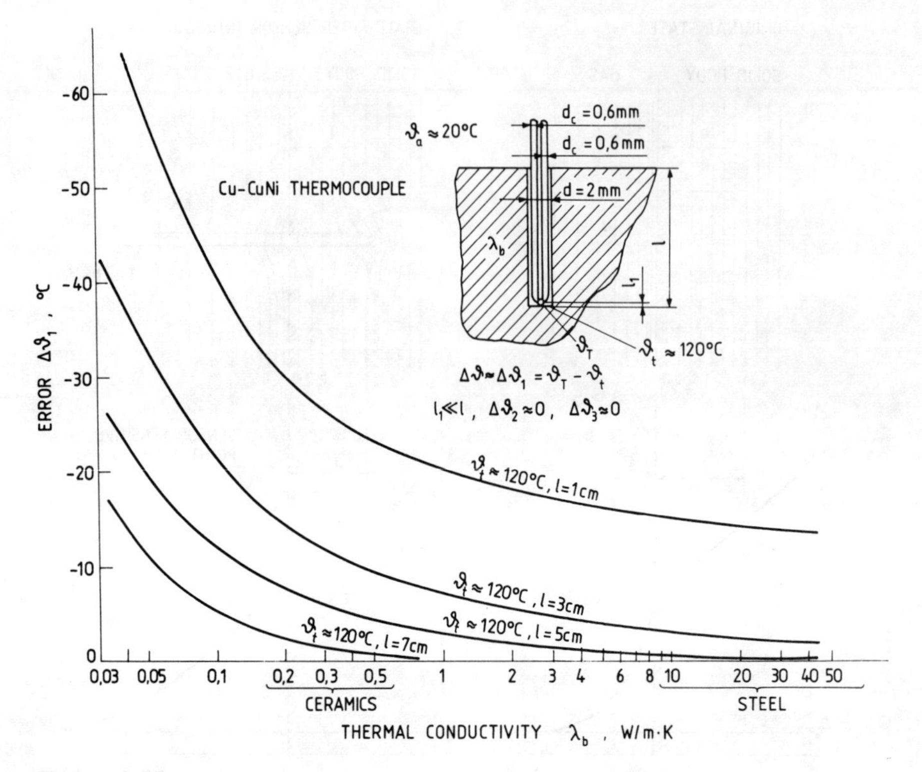

Figure 10.38
Internal temperature measurement in a solid body by a bare Cu–CuNi sensor versus thermal conductivity of the body λ_b.

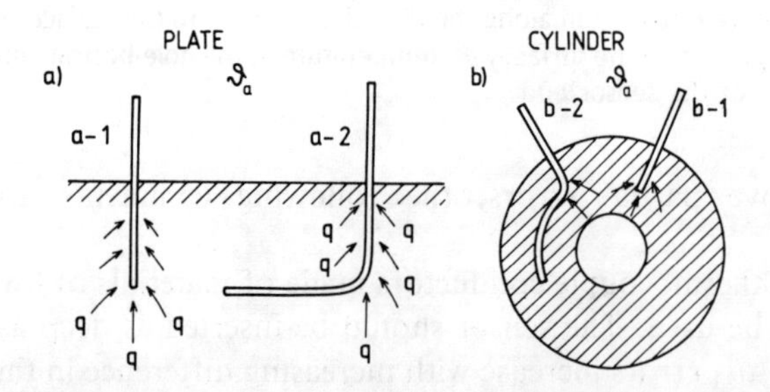

Figure 10.39
Internal temperature measurement in a solid body. a – 1, b – 1, sensor normal to surface; a – 2, b – 2, sensor along isotherms.

Although all three partial errors can be avoided by inserting the sensor along isotherms, as shown in Figure 10.39, it is not always easy to realize this for technical reasons. However, skewed insertion of the sensor frequently helps to reduce the errors. Only thin MI thermocouples and preferably, especially in non-metals, those with non-insulated measuring junction are used at present, for internal temperature measurements. More details concerning the measurement of the internal temperature of solids can be found in Baker *et al.* (1953) and Chyu and Bergles (1989).

Determination of the internal temperature of solids, without introducing a sensor inside the body is possible by using ultrasonic thermometers described in §15.4 or by models.

REFERENCES

Baker, H. D., Ryder, B. A. and Baker, N. H. (1953) *Temperature Measurement in Engineering*, Vol. 1, John Wiley and Sons, New York.

Benson, J. M. and Horne, S. (1962) Surface temperature of thin sheets and filaments. *Instrum. Control Systems* **35**(10), 115–117.

Boyer, M. W. and Buss, J. (1962) Measurement of surface temperature. A portable device compensated for heat losses. *Industr. Engng. Chem.* **8**(7), 728–729.

Bransier, J. (1975) Les temperature superficielles. Quelques méthodes de mésure par contact direct. *Mésures*, **40**(10), 54–61.

Browning, W. E. and Hemphill, R. L. (1962) Thermocouples for measurement of the surface temperature of nuclear fuel elements. *Temperature: Its Measurement and Control in Science and Industry*, Vol. 3, Part 2, Reinhold Publ. Co., New York, pp. 723–733.

Burton, E. J. (1953) Recent advances in radiation and immersion pyrometry. *Instruments*, **26**(10), 1524–1525.

Chyu, M. C. and Bergles, A. E. (1989) Locating method for temperature sensing elements in solid bodies. *Experimental Thermal and Fluid Science*, **2**(2), 247–249.

Drury, M. D., Perry, K. P. and Land, T. (1951) Pyrometers for surface measurements. *J. Iron Steel Inst.*, No. 11, pp. 245–250.

Emelyanenko, W. O. (1960) Surface temperature measurement of solids by heated thermocouples. *Inz.-Fiz. Journal*, **3**(10), 54–56 (in Russian).

Gatowski, J. A., Smith, M. K. and Alkidas, A. C. (1989) An experimental investigation of surface thermometry and heat flux. *Experimental Thermal and Fluid Science*, **2**(3), 280–292.

Hollander, M. (1959) An experimental measurement of the temperature distribution in the workpiece during metal-cutting. Ph.D. thesis, Columbia University, New York.

Hunsinger, W. (1966) *Temperaturmessung. Handbuch der Physik*. Springer-Verlag, Berlin.

Kalliomaki, K. and Wallin, P. (1971) Measurement of surface temperature with a thermally compensated probe. *J. Phys. E: Sci. Instrum.*, **4**(7), 771–774.

Kinzie, P. A. (1973) *Thermocouple Temperature Measurement*. John Wiley and Sons, New York.

Kulakov, M. W. and Makarov, B. I. (1969) *Surface Temperature Measurement of Solid Bodies*. Energiya, Moscow (in Russian).

Mackiewicz, E. (1976a) Dynamic compensating method of surface temperature measurement of solids. *Archiwum Budowy Maszyn*, **23**(3), 443–451 (in Polish).

Mackiewicz, E. (1976b) Static compensating method of surface temperature measurement of solids. *Archiwum Budowy Maszyn*, **23**(5), 549–557 (in Polish).

Michalski, L. (1978) Experimental determination of thermal contact resistance of metallic surfaces. *Archiwum Termodynamiki i Spalania*, **9**(1), 109–122.

Michalski, L. and Borowik, L. (1985) Continuous determination of the temperature at the inaccessible parts of electroheat plants. *Elektrowärme International*, **43**, B4, 189–193.

Moeller, C. E. (1963a) Special surface thermocouples. *Instr. Contr. Syst.*, **36**(5), 97–98.

Moeller, C. E. (1963b) Thermocouples for the measurement of transient surface temperatures. *Temperature: Its Measurement and Control in Science and Industry*, Vol. 3, Part 2, Reinhold Publ. Co., New York, pp. 617–623.

Moeller, C. E. (1972) Gold film thermometers for surface temperature measurements. *Temperature: Its Measurement and Control in Science and Industry*, Vol. 4, Part 2, Instrument Society of America, Pittsburgh, pp. 1049–1056.

Roeser, W. F. and Mueller, E. F. (1930) Measurement of surface temperatures. *J. Res. Nat. Bur. Stand.*, **9**(4), 793–802.

Sasaki, N. and Kamada, A. (1952) A recording device for surface temperature measurement. *Rev. Sci. Instrum.*, **23**(6), 261–263.

Stamper, J. A. (1963) Differential sensing controlled thermocouple. *Rev. Sci. Instrum.*, **34**(4), 444–445.

Stoer, J. (1972) *Einführung in die numerische Mathematik.*, Springer-Verlag, Berlin, Heidelberg.

Thomas, R. T. (1975) Extrapolation errors in thermal contact resistance measurements. *J. Heat Transfer*, **97**(5), 305–307.

Tye, R. P. (1969) *Thermal Conductivity*, Academic Press, London.

Wilcox, S. J. and Rohsenow, W. M. (1970) Film condensation of potassium using copper block for precise wall temperature measurement. *J. Heat Transfer*, **92**(8), 359–371.

Yarishev, N. A. and Minin, O. W. (1969) Extrapolation method of measurement of temperature and heat-flux. *Priborostroyenye*, **8**(1), 19–24 (in Russian).

Znichenko, W. M. (1963) Dynamic characteristics of thermocouples in surface temperature measurement. *Priborostroyenye*, **2**(9), 112–115 (in Russian).

11 Temperature Measurement of Transparent Solid Bodies

11.1 PYROMETRIC, CONTACTLESS METHOD

Discussions in §7.2 concerning the surface temperature measurement of solid bodies, have been limited to bodies having a transmission factor $\tau = 0$. These were non-transparent bodies in the whole range of wavelengths, used in optical pyrometry. All solid bodies which partially or totally transmit thermal radiation in the wavelength of visible and infrared radiation, are classified as transparent bodies.

Generally, in the background, behind the transparent body, 1, whose temperature is to be measured (Figure 11.1), there may be another body, 2, of temperature ϑ_2. Simultaneously, the investigated surface may be irradiated by another body, 3, having temperature ϑ_3. Thus, the radiation incident upon the pyrometer, is composed of the investigated surface radiation, the background radiation transmitted through the investigated body and the radiation from body, 3, reflected from the investigated surface. Consequently, the pyrometer reading will not be correct.

Errors caused by reflected radiation can be eliminated by either applying a shield shadowing the body, 3 (Figure 11.1), or by applying a pyrometer of zero-sensitivity in the wavelength range emitted by the body, 3, and reflected by 1. The error caused by background radiation can be eliminated by using a pyrometer of properly selected operating wavelength range, in which the spectral transmissivity of the body, 1, is as low as possible.

The mechanism of radiation of a semi-transparent body is as follows. Let us consider a homogeneous body of uniform temperature having an optically smooth surface and a thickness l (Figure 11.2). Neglect the background radiation and assume that the body has a transmissivity τ or the logarithmic absorption coefficient $k = \ln(1/\tau)$. Any layer of unity thickness at a depth x may send energy expressed by the heat-flux Φ'_x towards the left-hand surface. On its leftward path, of this total energy Φ'_x, the part, Φ'_{xa}, which is absorbed, is given by:

$$\Phi'_{xa} = \Phi'_x(1 - e^{-kx}) \qquad (11.1)$$

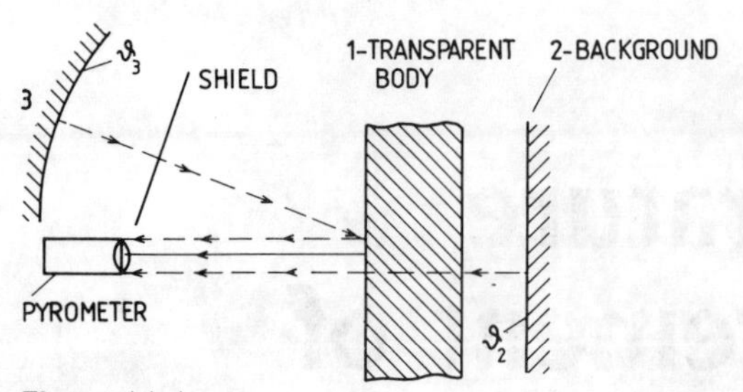

Figure 11.1
Pyrometric temperature measurement of a transparent body 1, against background 2, irradiated by body 3.

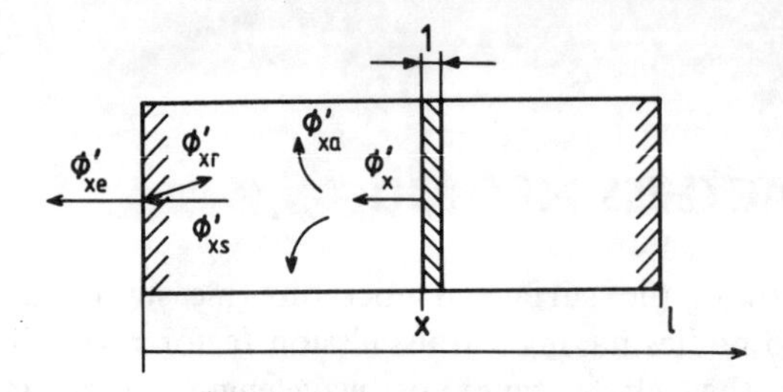

Figure 11.2
Radiation of a transparent body of thickness *l*.

Eventually the energy from one layer of unity thickness at a distance x, arriving at the left-hand surface is given by:

$$\Phi'_{xs} = \Phi'_x - \Phi'_{xa} = \Phi'_x \, e^{-kx} \tag{11.2}$$

Part of this energy:

$$\Phi'_{xr} = \Phi'_{xs}(1-\rho) \tag{11.3}$$

where ρ is the surface reflectivity, is reflected from the surface. Finally the energy coming from one layer of unity thickness at the depth x that is emitted by the left-hand surface is given by:

$$\Phi'_{xe} = \Phi'_{xs} - \Phi'_{xr} = \Phi'_x \, e^{-kx}(1-\rho) \tag{11.4}$$

The total energy, coming from all of the layers of the body of thickness l and emitted from its left-hand surface is given by the relation:

$$\Phi_{le} = \sum_{x=1}^{l} \Phi'_{xe} = \Phi'_x(1-\rho) \sum_{x=1}^{l} e^{-kx}$$
$$= \Phi'_x(1-\rho)[1 + e^{-k} + e^{-2k} + \ldots e^{-(l-1)k}] \tag{11.5}$$

It has been shown by Harrison (1960) that relation (11.5) is equivalent to:

$$\Phi_{le} = \Phi'_x(1-\rho)(1 - e^{-kl}) \tag{11.6}$$

Since a layer of unity thickness absorbs $(1-e^{-k})$ part of the incident energy which a black body would absorb, the equivalent emissivity ϵ_{e1} of this layer is:

$$\epsilon_{e1}=(1-e^{-k}) \tag{11.7}$$

Consequently when the reflection is also considered, the equivalent emissivity of the investigated body of thickness l is given by:

$$\epsilon_e=(1-e^{-kl})(1-\rho) \tag{11.8}$$

where k is a function of wavelength. In a similar way the equivalent spectral emissivity $\epsilon_{e\lambda}$ can be defined.

At sufficiently big kl-values, corresponding to sufficiently large thickness l, the factor $e^{-kl}\rightarrow 0$. Thus, the equivalent emissivity of a transparent thick body depends mainly on its reflectivity and consequently also on the surface state of the body. Energy, emitted at these wavelengths at which k is larger, is absorbed in a shorter distance x. Thus, at larger k the emitted energy originates from layers nearer to the surface. This phenomenon does not influence the pyrometer readings, if no temperature differences occur inside the body. When the temperature distribution inside the body is non-uniform, the energy emitted from each layer has to be analysed separately for each wavelength. Pyrometer readings then depend on the superposition of the radiation emitted from particular layers. MacGraw and Mathias (1962) have shown that those layers situated nearer to the surface always exert a predominant influence.

In industrial practice it is often necessary to measure the temperature of glass and plastics. Figure 11.3 presents relative spectral transmission and relative spectral reflectance of natrium-silicon glass and Figure 11.4 the spectral transmission of some plastics (Ircon Inc., USA). The thickness of transparent bodies exerts a vital influence on their equivalent, spectral emissivity $\epsilon_{e\lambda}$. This influence, in the case of some plastics, is shown in Figure 11.5. Effective wavelengths, λ_e, which are most convenient while measuring surface temperatures of some plastics, are listed in Table XXIV.

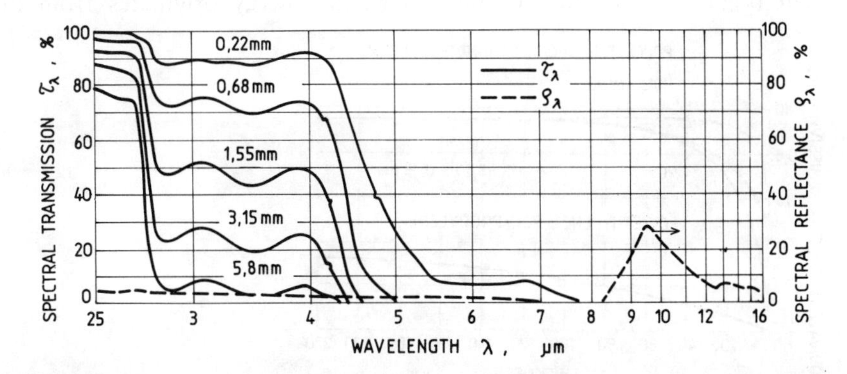

Figure 11.3
Relative spectral transmission τ_λ and reflectance ρ_λ of natrium-silicon glass versus wavelength λ. (Courtesy of Ircon Inc., USA.)

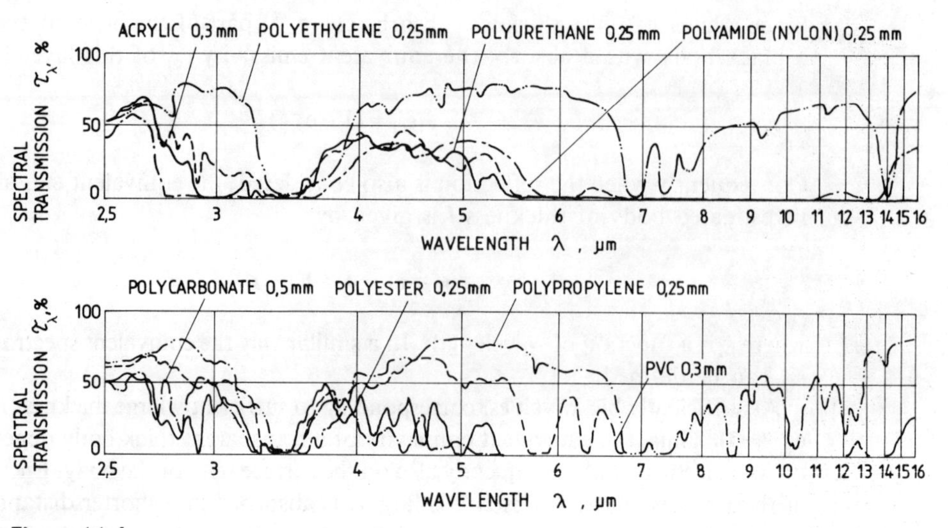

Figure 11.4
Relative spectral transmission τ_λ of some plastics versus wavelength λ. (Courtesy of Ircon Inc., USA.)

The spectral transmission of glass depends upon the wavelength λ_e. At longer wavelengths, glass becomes less transparent and thus more absorbing. When measuring the temperature of a glass surface it is advisable to use pyrometers operating in the wavelength range of 4.8 to 5.5 μm. As the spectral transmissivity, τ_λ, of glass (Figure 11.3) at these wavelengths is very low then simultaneously, the glass emissivity, ϵ_λ, approaches unity and the emitted radiation originates from a thin surface layer. Applications of such pyrometers are typically concerned with temperature measurements on electric lamps or during bottle production.

In the case of very thin glass layers, the wavelength range of 7.8 to 8.2 μm is used. When it is necessary to determine the glass temperature at a depth over 1 mm from the surface, Tenney (1986) indicates that pyrometers operating in the wavelength range of 3 to 4 μm are used. In this range the spectral transmissivity, τ_λ, of glass is relatively high (Figure 11.3) and the emitted radiant energy originates from the deeper

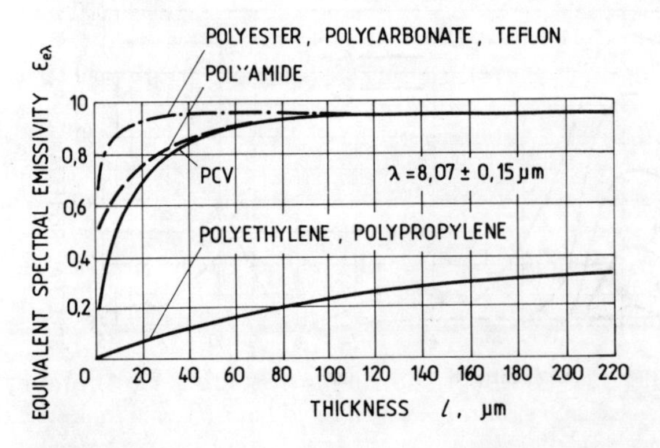

Figure 11.5
Equivalent spectral emissivity ϵ_λ versus thickness l of some plastic films.

glass layers. Any influence of surface radiation on the pyrometer readings is negligibly small. Temperature measurement of deeper glass layers is essential in judging the degree to which the glass has been through-heated while the temperature measurement of the glass surface is relevant in glass cooling processes. Simultaneous temperature measurement of glass surface and of internal glass layers permits the determination of the temperature gradients in the glass, a deciding factor in glass hardening processes. Two different pyrometers which are used simultaneously enable the formation of a differential signal. The wavelength ranges which are used should coincide with the so-called atmospheric windows (§7.4.1). This also enables measurements in the presence of gases and flames. Usually spectral or band photoelectric pyrometers are used. In some other cases only pyrometers with thermoelectric detectors, equipped with special cut-off filters, are utilized. Most producers offer pyrometers, which are especially intended for the temperature measurement of glass surfaces. Problems of pyromeric temperature measurement of sapphire and rubin are discussed by Lingart and Petrov (1981). To determine the temperature fields of surfaces, thermovision systems (§15.1.4), equipped with special correcting filters, are also used. Surface temperature measurement of plastics, commonly in the temperature range of 50 to 300 °C, is more difficult for two reasons. In the first place the wavelength ranges of low transmissivity are very narrow and secondly, thin plastic films enhance the errors due to background radiation.

11.2 CONTACT METHOD

Besides pyrometric methods, contact methods are also applied for surface temperature measurement. In some cases, the sensors get a certain amount of energy, not only by conduction from the surface itself, but also by radiation from the interior of the body. This phenomenon, which can cause some measuring errors, typically occurs in measuring the surface temperatures of thick material layers exhibiting large internal temperature differences.

REFERENCES

Garfunkel, J. H. (1972) Temperature control using infrared pyrometers with solid state detectors. *Temperature: Its Measurement and Control in Science and Industry*, Vol. 4, Part 3, Instrument Society of America, Pittsburgh, pp. 1299–1310.

Harrison, T. R. (1960) *Radiation Pyrometry and its Underlying Principles of Radiant Heat Transfer*, John Wiley and Sons, New York.

Ircon Inc., USA (undated) *Plastic Film Measurement*, Technical Note.

Lingart, J. K. and Petrov, A. V. (1981) Measurement of surface temperature of optical semi-transparent materials. *Temperature Measurement in Industry and Science*, First Symposium of IMEKO TC12 Committee, Czechoslovak Scientific and Technical Society, Praha, pp. 105–118.

MacGraw, D. A. and Mathias, R. G. (1962) Radiation pyrometry in glass-forming process. *Temperature: Its Measurement and Control in Science and Industry*, Vol. 3, Part 2, Reinhold Publ. Co., New York, pp. 381–390.

Tenney, A. S. (1986) *An Introductory Review of Radiation Thermometry*, Product Information Bulletin No. 9, Leeds & Northrup Co., North Wales, PA, USA.

12 Temperature Measurement of Gas and Liquid

12.1 LOW VELOCITY GAS

12.1.1 Contact sensors

Let us assume that a gas, with a temperature T_g and a flow-velocity below about 20 m/s flows in a tube as shown in Figure 12.1(a). In the tube there is a sheathed sensor, whose sensitive part, at temperature T_T, is placed in the sheath bottom. Such a case corresponds to a thermocouple with the measuring junction welded into the sheath bottom. Also let the temperature of the internal surface of the tube wall be T_w while the length to diameter ratio of the sensor sheath is assumed to be so large that heat transfer through the sheath bottom can be neglected. It is further assumed that the gas temperature is higher than the temperature of the tube wall ($T_g > T_w$). Consider a cylindrical sheath element of length dx, placed at a distance x from the sheath end as represented in Figure 12.1(b). In the thermal steady-state it is apparent that the heat balance must take account of all conduction, convection and radiation effects.

The convection heat flux from the gas to an element of length dx is:

$$d\Phi_k = \alpha_k \pi D dx (T_g - T_x) \tag{12.1}$$

where α_k is the convective heat transfer coefficient, T_g is the gas temperature, and T_x is the sheath temperature at a distance x from its end.

Assuming that the gas transmissivity equals unity, the heat flux dissipated from this element by radiation to the tube wall is given by:

$$d\Phi_r = \epsilon_T C_0 \pi D \, dx \left[\left(\frac{T_x}{100} \right)^4 - \left(\frac{T_w}{100} \right)^4 \right] \tag{12.2}$$

where ϵ_T is the emissivity of the sheath surface, C_0 is the radiation constant of a black body and T_w is the temperature of the tube wall and the other symbols are as in Equation (12.1).

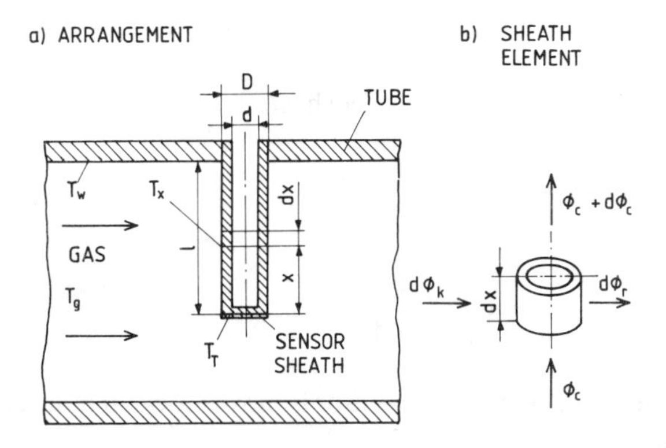

Figure 12.1
Gas temperature measurement in a tube by a sheathed sensor.

Relation (12.2) is valid under the assumption that the sheath surface is many times smaller than that of the tube (see Equation 7.20)), while relation (12.1) is valid for $\alpha_k = \text{constant}$.

The conductive heat flux along the sensor is described by the relation:

$$d\Phi_c = -\lambda A \frac{d^2 T_x}{dx^2} dx \tag{12.3}$$

where λ is the thermal conductivity of the sheath material,

$$A = \frac{\pi D^2}{4} - \frac{\pi d^2}{4}$$

and T_x is the temperature at a distance x from the sheath end.

A heat balance equation of element dx is then:

$$d\Phi_k = d\Phi_c + d\Phi_r \tag{12.4}$$

Substituting the corresponding values from (12.1), (12.2) and (12.3) into (12.4) yields:

$$\frac{d^2 T_x}{dx^2} \lambda A + \alpha_k \pi D (T_g - T_x) - \epsilon_T C_0 \pi D \left[\left(\frac{T_x}{100} \right)^4 - \left(\frac{T_w}{100} \right)^4 \right] = 0 \tag{12.5}$$

This nonlinear differential equation which cannot be solved by direct integration, can be solved in some cases by a graphical method or by introducing some application specific simplifications. In most cases, similarly to the convective heat transfer coefficient, α_k, an equivalent heat transfer coefficient by radiation, α_r, of constant value is introduced. Equation (12.5) then becomes:

$$\frac{d^2 T_x}{dx^2} \lambda A + \alpha_k \pi D (T_g - T_x) - \alpha_r \pi D (T_x - T_w) = 0 \tag{12.6}$$

This equation has the solution:

$$\frac{T_g - T_w}{T_x - T_w} = \frac{\alpha_k + \alpha_r}{\alpha_k}\left(\frac{\cosh nl}{\cosh nl - \cosh nx}\right) \tag{12.7}$$

where

$$n = \sqrt{\frac{(\alpha_k + \alpha_r)\pi D}{\lambda A}}$$

Considering that the sensitive point of the sensor is mostly located at its end ($x = 0$), relation (12.7) is simplified to:

$$\frac{T_g - T_w}{T_T - T_w} = \frac{\alpha_k + \alpha_r}{\alpha_k}\left(\frac{\cosh nl}{\cosh nl - 1}\right) \tag{12.8}$$

or after some transformations:

$$\Delta T_{c,r} = T_T - T_g = (T_w - T_g)\left(1 - \frac{\alpha_k(\cosh nl - 1)}{(\alpha_k + \alpha_r)\cosh nl}\right) \tag{12.9}$$

where $\Delta T_{c,r}$ is the overall error resulting from heat conduction along the sensor and from radiant heat exchange.

If radiant heat transfer can be neglected ($\alpha_T = 0$) relation (12.9) becomes:

$$\Delta T_c = T_T - T_g = \frac{T_w - T_g}{\cosh ml} \tag{12.10}$$

where ΔT_c is the error resulting from heat conduction along the sensor and

$$m = \sqrt{\frac{\alpha_k \pi D}{\lambda A}}$$

If the sensor is sufficiently long so that any error resulting from heat conduction along the sensor could be neglected, or if the product λA is sufficiently small and substituting $T_x = T_T$, Equation (12.5) is simplified to:

$$\alpha_k \pi D(T_g - T_T) - \epsilon_T C_0 \pi D\left[\left(\frac{T_T}{100}\right)^4 - \left(\frac{T_w}{100}\right)^4\right] = 0 \tag{12.11}$$

From Equation (12.11), the measuring error ΔT_r, resulting from radiant heat transfer, is expressed by:

$$\Delta T_r = T_T - T_g = \frac{\epsilon_T C_0}{\alpha_k}\left[\left(\frac{T_w}{100}\right)^4 - \left(\frac{T_T}{100}\right)^4\right] \tag{12.12}$$

The simplest way of solving this equation is a graphical one, as shown in Figure 12.2. Two curves are drawn there: $q_k = f_1(T_T)$ and $q_r = f_2(T_T)$, where $q_k = \alpha_k(T_g - T_T)$ is the convective heat flux density from the gas to the sensor surface, and

$$q_r = \epsilon_T C_0 \left[\left(\frac{T_T}{100}\right)^4 - \left(\frac{T_w}{100}\right)^4\right]$$

is the radiant heat flux density from the sensor surface to the tube wall.

In the thermal steady-state both these quantities are equal so that:

$$q_k = q_r \tag{12.13}$$

The point in Figure 12.2 where both curves intersect indicates the sensor temperature, T_T, and the radiant measuring error, ΔT_r.

Equations (12.9), (12.10) and (12.12) enable calculation of the measuring errors and the introduction of any necessary corrections in particular cases. This latter possibility is rarely used. Mostly based on these equations, conclusions may be drawn for the design of sensors, so that indication errors can be kept as small as possible.

Sometimes more complex models of a temperature sensor in a tube, through which gas flows, are also applied. In these models described by Blumröder (1981), Haas (1969) and Rudolphi (1969), the heat exchange between the part of the sensor sheath and sensor head protruding from the wall into the environment are considered.

Errors analysis of gas temperature measurement permits the practical establishment of the dependence of errors upon the sensor insertion depth into the medium or to establish the dependence of errors upon the values of the heat transfer coefficients between the sensor sheath and the gas.

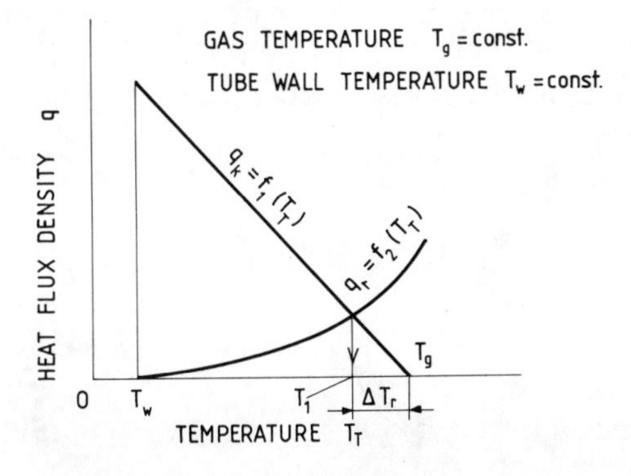

Figure 12.2
Graphical method of error ΔT_r determination (Equation (12.2)).

12.1.2 Methods of reducing errors in contact measurements

It follows from the preceding section, that reduction of the measuring errors requires:

increased heat flux Φ_k gained by convection,
decreased heat flux Φ_c lost by conduction,
decreased heat flux Φ_r lost by radiation.

Increase of heat flux Φ_k by convection can be achieved by:

(1) Increase of the convective heat transfer by using a finned sensor (Figure 12.3). The fins should be made of a material with a high thermal conductivity and low emissivity, to reduce the radiant heat losses.
(2) Increase of the convective heat transfer by applying high gas velocity in the tube, using a sensor sheath with as small a diameter as possible and by placing the sensor at an angle of about $\pi/2$ relative to the direction of the gas flow.
(3) Applying suction thermometers in which the gas velocity is increased only in the surrounding of the sensor. This method is especially appropriate in those cases in which an increase of the gas velocity in the tube is not possible.

The suction or aspiration thermometer was first proposed by R. Assmann (1892). Its operating principle, shown in Figure 12.4 (Wenzel and Schulze, 1926), has the end of a suction tube with a thermocouple inserted into a pipe-line or gas-filled enclosure, in which the gas temperature is to be measured. The compressed air produces suction of the gas in the nozzle. In this way, the gas, whose temperature is to be measured, streams past the thermocouple at high velocity. Thermocouple readings, which are a function of gas velocity, have the dependence on rate of gas flow shown in Figure 12.5 for a gas temperature of 100 °C. These thermometer readings are practically independent of the position of the measuring junction in the suction tube. Industrial suction thermometers are often equipped with one or a number of concentric radiation shields to reduce the radiant heat exchange. Suction thermometers which can be used up to about 1600 °C , are often water or air cooled, using MI thermocouples as temperature sensors. More detailed information can be found in the publications of Mullikin (1941), Mullikin and Osborn (1941) and Ribaud *et al.* (1959). It is also important to note that they can only be used when the quantity

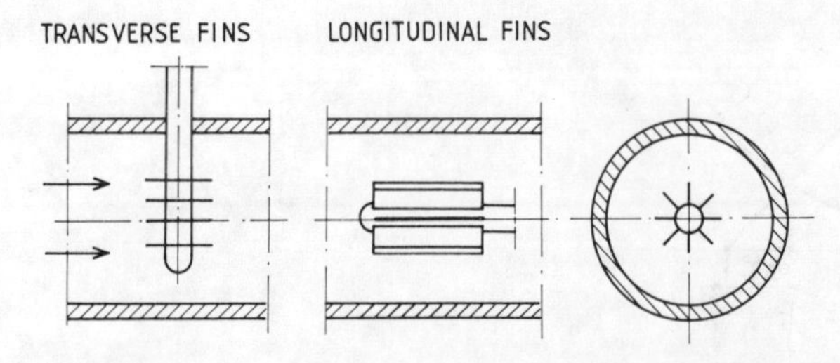

Figure 12.3
Finned sensors.

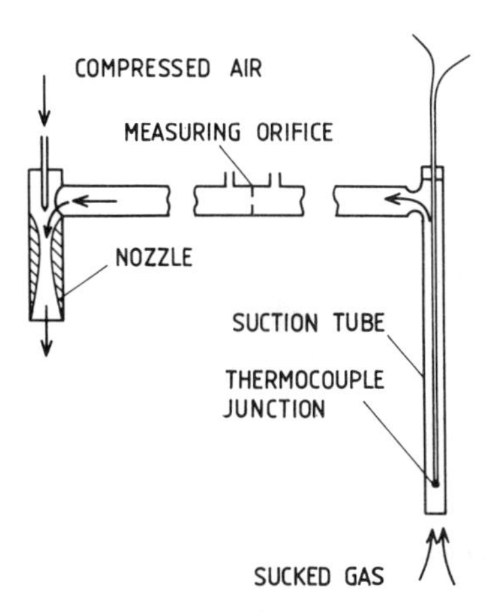

Figure 12.4
Suction thermometer by Wenzel and Schulze (1926).

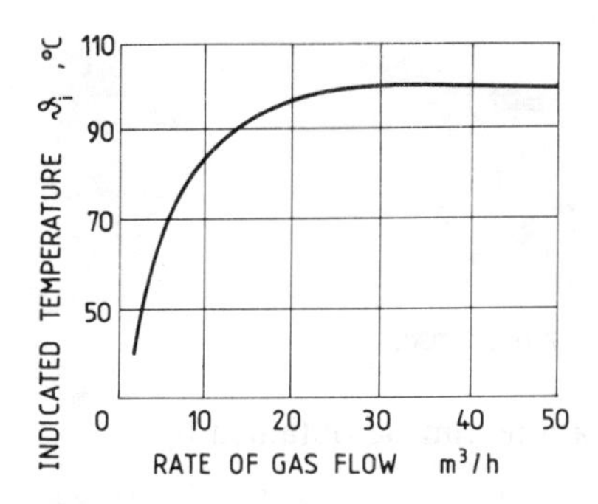

Figure 12.5
Thermocouple readings of the thermometer from Figure 12.4, versus rate of gas flow at $\vartheta_g = 100$°C.

of gas sucked by the thermometer is negligibly small compared with the rate of gas flow in the pipe-line itself.

Decrease of heat flux Φ_c lost by conduction can be obtained by:

1. Application of long sensors with small cross-sectional area, made from materials with low thermal conductivity, λ. Figure 12.6 presents some ways of installing rather long sensors in pipe-lines. It is advisable that the ratio of the sensor length, l, to its diameter, d, should be (a) $l/d \geqslant 6$ to 10 in flowing gas or (b) $l/d \geqslant 12$ to 15 in still gas.
2. Laying the sensor from the measuring point, along isotherms, as shown in Figure 12.6(c). Similar results can be obtained in arrangements like those in Figure 12.6(a) and 12.6(b).
3. Thermally insulating or heating of the sensor head to increase its temperature (Figure 12.7).

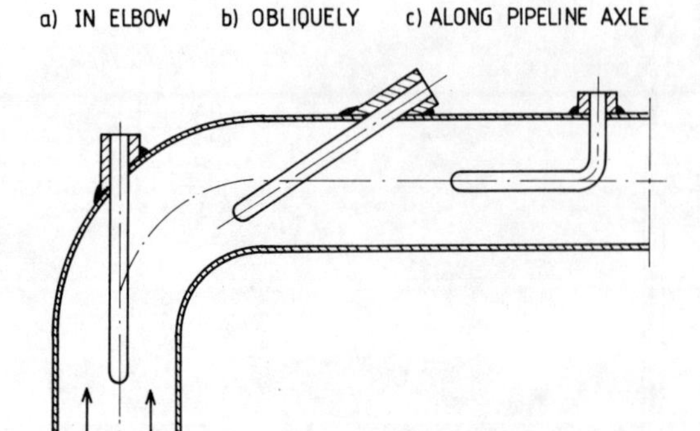

Figure 12.6
Methods of installing temperature sensors in a pipeline.

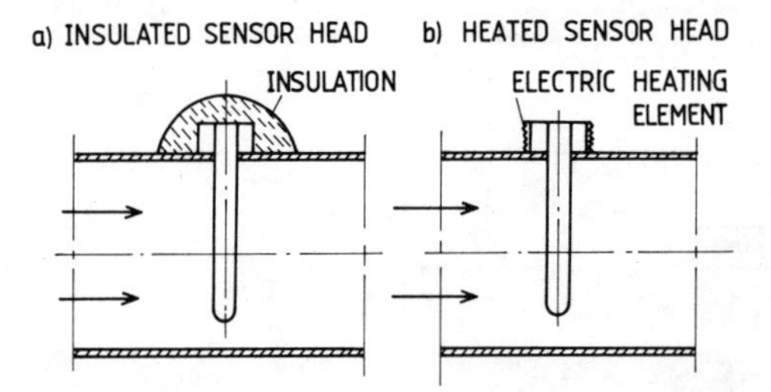

Figure 12.7
Methods of reducing heat-flux conducted along the sensor.

Decrease of heat flux Φ_r lost by radiation, can be obtained by:

1. *Covering the sensor* surface with materials such as gold, silver or platinum which have a low emissivity, ϵ_T.
2. *Application of radiation shields.* This is the most popular method of reducing radiant heat exchange between sensor and surrounding walls (Figure 12.8). A shield of low emissivity, ϵ_s, is placed between the sensor and tube wall. Assuming, that the shield is sufficiently long, to be able to neglect the influence of its open ends on the sensor heat balance, and that the shield inner surface is many times larger than that of the sensor, the measurement error due to radiant heat exchange can be calculated from Equation (12.12). In that relation, the shield temperature T_s, is substituted instead of the wall temperature, T_w. The relevant error $\Delta T_{r,s}$ is given by:

$$\Delta T_{r,s} = T_T - T_g = \frac{\epsilon_T C_0}{\alpha_k}\left[\left(\frac{T_s}{100}\right)^4 - \left(\frac{T_T}{100}\right)^4\right] \tag{12.14}$$

The shield temperature, T_s, is determined from the shield heat balance:

$$\Phi_{k,s} = \Phi_{r,s} \tag{12.15}$$

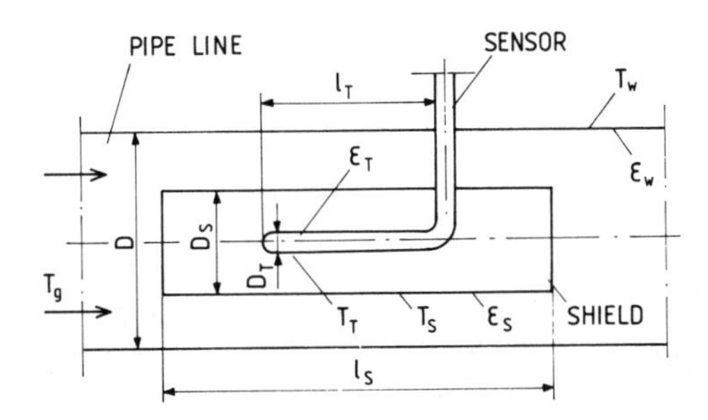

Figure 12.8
Sensor with a radiation shield.

where $\Phi_{k,s}$ is the heat flux gained by convection and $\Phi_{r,s}$ is the heat flux lost by radiation.

Heat flux $\Phi_{k,s}$ is given by

$$\Phi_{k,s} = 2\pi D_s l_s \alpha_{k,s}(T_g - T_s) \tag{12.16}$$

where $\alpha_{k,s}$ is the convective heat transfer of the shield and the other symbols are as in Figure 12.8.

In relation (12.16) double, internal and external shield surfaces, have been considered. Heat flux $\Phi_{r,s}$ is described by:

$$\Phi_{r,s} = \pi D_s l_s \epsilon_s C_0 \left[\left(\frac{T_s}{100}\right)^4 - \left(\frac{T_w}{100}\right)^4\right] \tag{12.17}$$

where ϵ_s is the shield surface emissivity, and the other symbols are as in Figure 12.8.

Relation (12.17), based on Equation (7.20b), is valid under the assumption that the external shield surface is many times smaller than that of the surrounding surface of the tube wall.

After some simplifications, the heat balance equation in steady-state, becomes:

$$2\alpha_{k,s}(T_g - T_s) = \epsilon_s C_0 \left[\left(\frac{T_s}{100}\right)^4 - \left(\frac{T_w}{100}\right)^4\right] \tag{12.18}$$

The value of shield temperature, T_s, can be found graphically as in Equation (12.11). Assuming that $T_g > T_w$ the shield temperature, T_s, is always higher than that of the wall, T_w. Comparing the relations (12.12) and (12.14) it is clearly seen, that error due to the radiant heat exchange between the sensor and its surroundings decreases due to the application of the shield.

Analogous reasoning can be made for two, three and more shields. Each consecutive shield decreases the error resulting from the radiant heat exchange with a progressively smaller and smaller influence. According to King (1943) this error, $\Delta T_{r,ns}$, for n shields is given by the approximate relation:

$$\Delta T_{r,ns} \approx \frac{\Delta T_r}{1+n} \tag{12.19}$$

A more detailed analysis of the influence of shields on the readings of a gas measuring thermometer is presented by Moffat (1952). The distances between the shields should be large enough to enable a free gas flow.

3. *Application of heated radiation shields* (Mullikin, 1941). In this method a thermocouple is placed on the axis of a shield which is heated by an additional low-power heating element. A second thermocouple, which measures the shield temperature, allows adjustment of the heating power until the readings of both thermocouples are the same. When this state is reached, the thermocouple which measures the gas temperature, does not exchange any energy with the surroundings by radiation so that its readings are correct. It is a rather time consuming method unless it is automated.

The simultaneous fulfilment of all three conditions for reducing measuring errors may be realized with the use of a bare thermocouple of very small wire diameter. The diameter of the thermocouple wire and its surface area are prime influences which determine radiation heat exchange. As a thin gas film always exists around a wire, the convective heat exchange depends upon the wire diameter plus double film thickness. With wire diameter approaching zero the convective heat exchange is mainly determined by the double film thickness while the radiant heat exchange disappears. Thus as only convective heat exchange remains, any radiant measuring errors ΔT_r disappear. At the same time, with wire diameter approaching zero any conductive measuring errors, ΔT_c also disappear. In practice it is advisable to use bare or MI thermocouples which are as thin as possible. The lower diameter limit is imposed by the mechanical strength and the corrosion resistance of the wires. This method is suitable for both laboratory and industrial applications.

Another very precise method of gas temperature measurement is the extrapolation method. Gas temperature is simultaneously measured by a number of bare or MI thermocouples of different diameters. The results, which are displayed graphically as a function of the diameters of the thermocouple wire, are extrapolated to the zero diameter. This value is then the true gas temperature. As an example, Figure 12.9 presents results obtained while using bare NiCr–NiAl thermocouples to measure the temperature of hot air with velocity 8 m/s flowing through a tube having a wall temperature of 15 °C. The extrapolated value was 175 °C. Before the measurements it is very important to check the identity of the thermoelectric characteristics of the thermocouples used. MI thermocouples (see §4.3.3), which are produced with a diameter as thin as 0.3 mm, are very convenient in the extrapolation method.

An important source of errors occurs in the temperature measurement of flowing gas when a non-uniform distribution of gas temperature occurs across the tube section. To get readings approaching the average gas temperature a number of sensors is used (Figure 12.10). If thermocouples are used, they have to be connected in series.

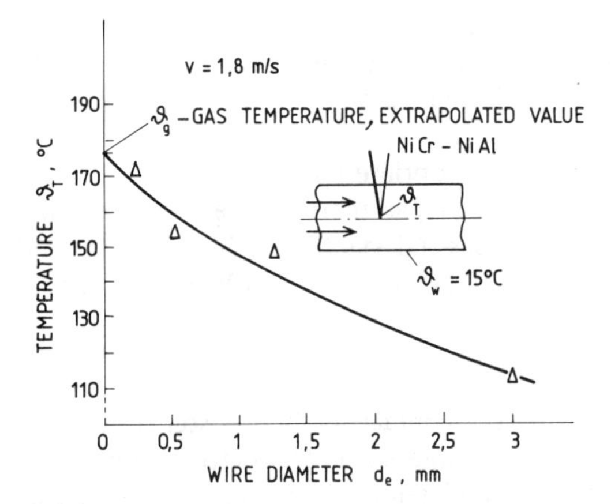

Figure 12.9
Temperature indicated ϑ_T by bare thermocouples in flowing air of velocity $v = 1.8$ m/s, versus wire diameter d.

The measured thermal e.m.f. must then be converted into a temperature value. A correct average temperature can be obtained directly using parallel connected thermocouples, provided all of the thermocouples used have precisely the same resistance. A merit of the parallel connection is that correct readings are obtained even in the case of a broken circuit in one of the measuring loops (see §15.5).

For gas temperature measurement in the temperature range of from about 1000 to 3000 °C, only thermocouples of metal group, Pt, Rh and Ir, whose properties have been described in §4.4, are used. However, besides the changes in their characteristics, described before, these thermocouples can also act as catalysts in oxidization and combustion processes. These phenomena result in additional heating of the thermocouples causing important additional errors. A detailed analysis of these phenomena, which can also occur in any other temperature sensors with protecting shields made of the previously mentioned metals, is presented in Ash and Grossmann (1972) and Thomas and Freeze (1972).

Many references and methods for the experimental detection of these phenomena are also given by these authors. As an example, in measuring the temperature of hot

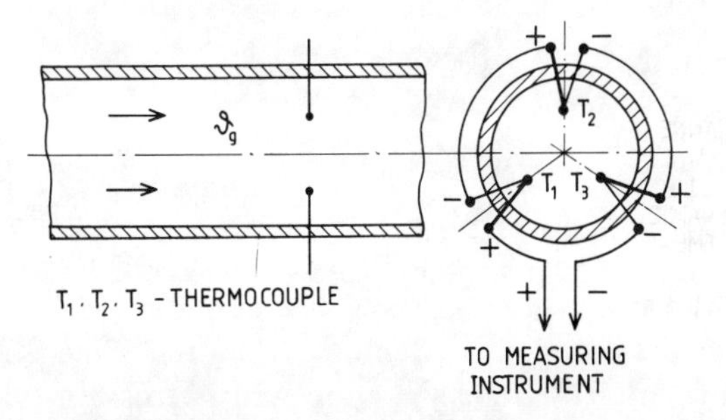

Figure 12.10
Series thermocouple connection for measuring the average temperature of a flowing gas.

completely burned exhaust gases of internal combustion engines or of gas turbines at about 1000 to 1500 °C, measuring errors can be as high as about 400 °C (Thomas and Freeze, 1972). To prevent additional catalytic heating, thermocouples or their protective sheaths as pointed out by Kinzie (1973), have to be covered by a layer of BeO, ZrO, Si_2O_3, Cr_2O_3 or other appropriate materials.

Detailed information on methods of gas temperature measurement and many references can be found in papers by Baas and Mai (1972), Benson and Brundrett (1962), Moffat (1952), Mullikin (1941), Mullikin and Osborn (1941) and Torkelsson (1980).

12.1.3 Indirect pyrometric measurements

Gas temperature measurement by the contact method is sometimes very difficult, especially when no measuring lag is acceptable, as occurs for example, in closed loop temperature control. In addition, at high temperatures and in corrosive atmosphere, when the sensors have to be equipped with heavy protective sheaths, indirect pyrometric measurements are used. A photoelectric or total radiation pyrometer could be directed at the inner ceramic lining of the channel, although the temperature drop near the surface or the thermal inertia of the lining could result in measuring errors. As this method is excluded when the walls are cooled, it is advisable, therefore, to place in the channel, a special ceramic element at which the pyrometer would be directed (Figure 12.11). This element should be hollow inside to increase the surface of convective heat transfer and to reduce the amount of heat conducted to the channel walls. Hence such an element should be made of a material with a high surface emissivity and a low thermal conductivity. The part of the element at which the pyrometer is directed must be placed in a zone of highest gas temperature and biggest gas velocity, which usually occurs as far away from the wall as possible. The application of a closed-end ceramic tube, placed in flowing gas, with pyrometer directed at the tube bottom from inside, is based on the same principle. Sufficiently long tubes can be regarded as black bodies.

12.1.4 Direct pyrometric measurements

Direct pyrometric measurement of gas temperature is possible, using a pyrometer of effective wavelength corresponding to the absorption wavelength range of the gas,

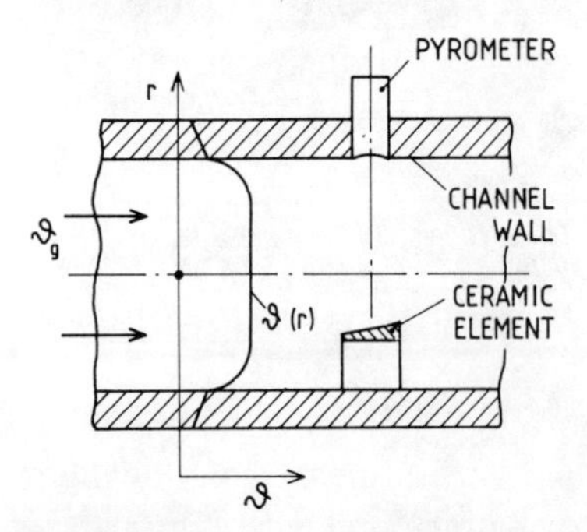

Figure 12.11
Indirect pyrometric measurement of a flowing gas with a radial temperature distribution $\vartheta(r)$.

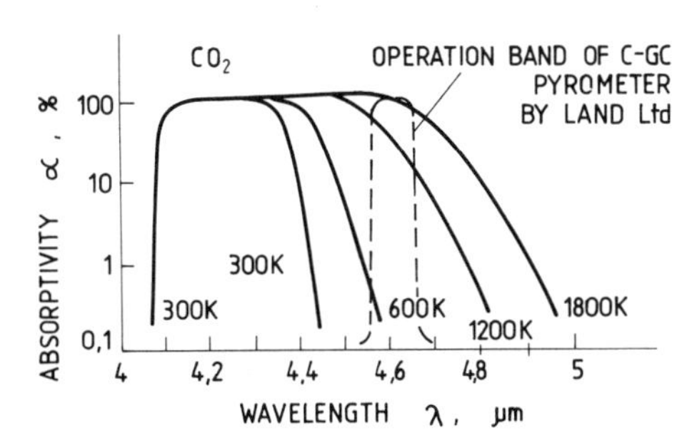

Figure 12.12
CO_2-absorptivity, α, versus wavelength, λ. (Courtesy of Land Infrared Ltd, UK.)

whose layer thickness has to be large enough to eliminate any background influence. As an interesting example of the necessary precision of choosing the effective wavelength, the C-GC pyrometer by Land Infrared Ltd can be mentioned. Due to dependence of the absorptivity of CO_2 upon temperature, this pyrometer can be used to measure the temperature of a hot CO_2 layer through a layer of cool CO_2. In the chosen wavelength range this absorptivity at 1800 K is about 1000 times larger than at 300 K (Figure 12.12). Similar problems also concern the temperature measurements of flames. As such measurement is rather complicated it is outside the scope of this book. Interested readers will find further details in Baker *et al.* (1953) and Green (1987).

12.2 HIGH VELOCITY GAS

New problems arise in measuring the temperature of gases flowing at velocities over 20 m/s. A sensor placed in a high-velocity gas stream causes the gas movement velocities to slow down resulting in the sensor being heated to a temperature ϑ_{tot} which is higher than the gas temperature ϑ_g. At high speed, the kinetic energy of the gas is high. During adiabatic expansion, when the gas velocity reduces to zero, this kinetic energy is transformed into heat. According to the first principles of thermodynamics, when the total kinetic energy of the gas is transformed into heat in ideal conditions, an energy balance gives,

$$\frac{mv^2}{2} = mc_p(\vartheta_{tot} - \vartheta_g) \tag{12.20}$$

or after some transformations:

$$\vartheta_{tot} - \vartheta_g = \frac{v^2}{2c_p} \tag{12.21}$$

where m is the mass of the gas, v is the velocity of the gas, c_p is the specific heat of the gas at constant pressure, ϑ_g is the temperature of gas flowing at a velocity v and ϑ_{tot} is the gas temperature at the adiabating point where the velocity is reduced to zero.

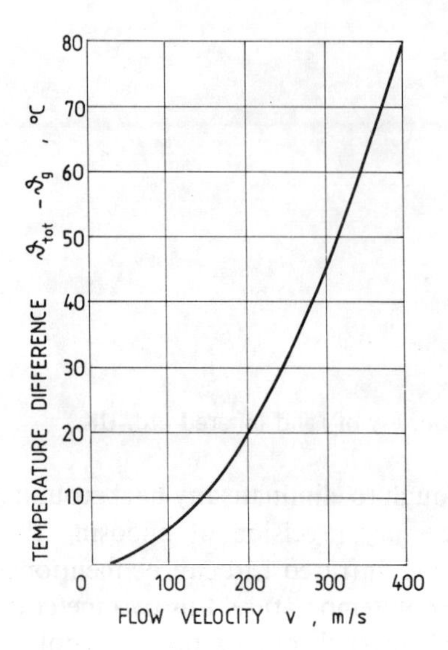

Figure 12.13
Difference of total, ϑ_{tot}, and gas, ϑ_g, temperatures, versus flow velocity, v.

Figure 12.13 presents the difference $(\vartheta_{tot} - \vartheta_g)$ versus flow velocity v in which ϑ_{tot} is called the total gas temperature and ϑ_g the true or static gas temperature. The temperature ϑ_g can also be defined as the gas temperature, which would be indicated by a minute, point-like thermometer moving along in the same direction as the gas flow and with the same velocity.

At the surface of a real sensor the gas velocity does not fall to zero and there are also heat losses. For these two reasons the sensor temperature ϑ_T is lower than ϑ_{tot} given by Equation (12.21). Using the symbols in Equation (12.21) the average sensor temperature ϑ_T indicated by a measuring instrument connected to it is given by:

$$\vartheta_T = \vartheta_g + r\frac{v^2}{2c_p} \tag{12.22}$$

The factor r, in (12.22), which is called the characterizing recovery factor of a given sensor, depends upon the sensor design and the gas parameters. If the indicated sensor temperature ϑ_T and factor r are known, it is possible to calculate the true gas temperature ϑ_g.

Gas temperature is often measured by thin, bare thermocouples. The r-values for different wire diameters and thermocouple designs, which are given in many publications such as Baker *et al.* (1961), Jacob (1957) and Hottel and Kalitinsky (1945), unfortunately differ from one author to another. Figures 12.14 and 12.15 present the experimental data by Hottel and Kalitinsky (1945), for bare thermocouples. Data for MI thermocouples can be found in Breitkopf *et al.* (1980). As there is uncertainty in determining the value of a recovery factor, r, due to its dependence on the measuring conditions, some special sensors having high values for r are constructed. The principal idea of their design is based on the application of a protecting shield, which serves

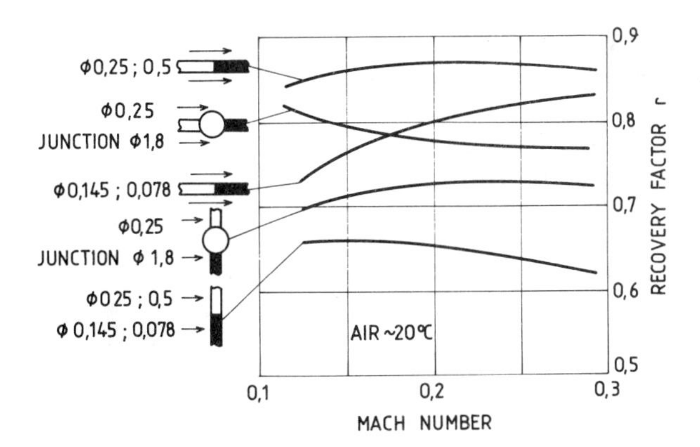

Figure 12.14
Recovery factor, *r*, of bare Fe–CuNi thermocouples versus Mach number in air, at 20 °C (Hottel and Kalitinsky, 1945).

to brake the gas and at the same time also as a shield to prevent any radiant heat exchange with the environment. It is not possible to achieve a value of $r = 1$, as this would correspond to a velocity of zero around the measuring junction and consequently also to a value of zero for convective heat transfer coefficient α_k on the junction surface. Practically, values of the recovery factor, *r*, lie mostly between 0.95 and 0.98.

Figure 12.16(a) shows a typical sensor construction for temperature measurement of high-velocity gases by Pratt-Whitney Aircraft (USA), described by Baker *et al.* (1961). Figure 12.16(b) presents values of the recovery factor, *r*, as a function of air velocity. Some more detailed information can be found in Baker *et al.* (1961), and Moffat (1952).

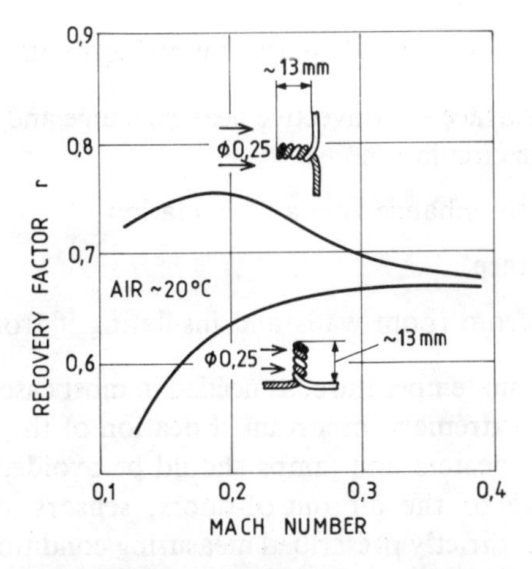

Figure 12.15
Recovery factor, *r*, of bare Fe–CuNi thermocouples versus Mach number in air, at 20 °C (Hottel and Kalitinsky, 1945).

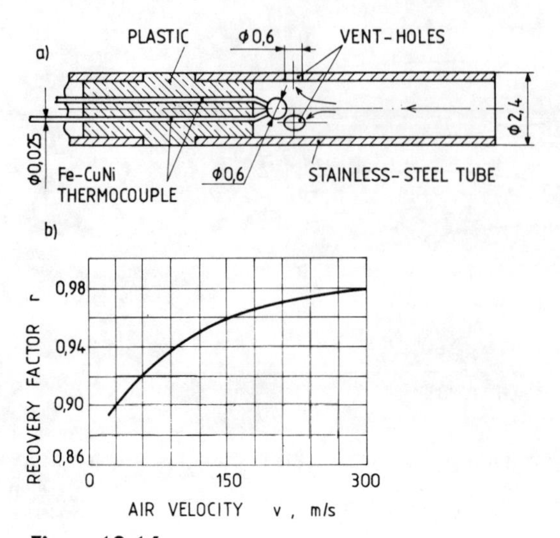

Figure 12.16
Sensor for the temperature measurement of a high-velocity gas (Baker *et al.* 1961).

12.3 STILL AIR

A very special case of the temperature measurement of slowly moving gas is that of gas moving under the influence of natural convection. This happens among others, while measuring the temperature of air in rooms. Considering only normal room temperature it is noted that the intensity of heat transfer between the sensor and air is very low. At the same time, as the sensor and its surrounding are at nearly the same temperature, errors owing to conductive and radiant heat exchange are very small. Increase in the sensor time constant resulting from this low intensity of heat transfer may cause dynamic errors.

Correct sensor design should be characterized by the following features:

- Possibly small sensor mass, large surface of convective heat exchange and if needed an arrangement to force local air circulation.
- Radiation shield of vertical axis to enhance free air circulation.
- Low emissivity of the sensor surface.
- Placing the sensor at a distance from room walls and insulating it from them.

In order to measure the average room temperature, as needed in most cases, correct placing of the sensor in the room is extremely important. Location of the sensor in close proximity to windows, doors, heaters and lamps should be avoided.

When measuring the temperature of the air out-of-doors, sensors should be protected from direct solar exposure. Strictly prescribed measuring conditions which are described by Jakob (1958), are defined for meteorological purposes. Often special suction thermometers are used. Liquid-in-glass thermometers, resistance temperature detectors and thermistors may all be used for still-air temperature measurements.

12.4 LIQUIDS

In this book, considerations of the temperature measurement of liquids are limited to problems related to either slowly flowing or still liquids which may be water, oil or other well known types. The temperature range is also limited below about 200 °C, where the conditions for only convective heat transfer occur so that Equation (12.10) is relevant. In liquids, the convective heat transfer coefficient is many times higher than in gases. As a result, for a given error, the necessary immersion depths, are much smaller.

An example of the dependence for a Pt-100 Ω sensor in a steel sheath, considered by Blumröder (1981), is presented in Figure 12.17. The permissible errors ΔT are below 0.1% or 2%. Typical ranges of convective heat transfer coefficients are also indicated.

12.5 HIGH TEMPERATURE GAS AND PLASMA

Gas and plasma temperatures of several thousands of °C, which far exceeds the application range of normal and high-temperature thermocouples, often need to be measured. A method, in which a pulsed thermocouple is periodically inserted into the high temperature medium (Figure 12.18), can be used for this purpose. Kretschmer *et al.* (1977) have described how its temperatures does not exceed its normal application range provided its immersion time periods are short enough. Such a sensor, which should be chemically inert and should have no catalytic properties, takes up the heat

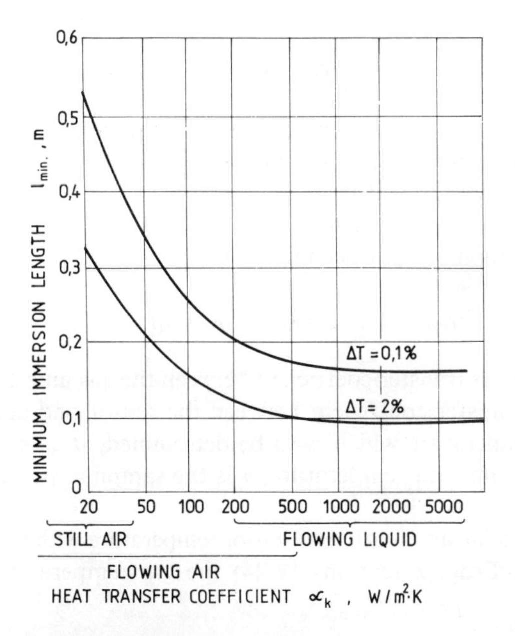

Figure 12.17
Minimum sensor immersion length, l_{min}, of a Pt-100Ω sensor in steel sheath for permissible errors, ΔT, versus convective heat transfer coefficient, α_k.

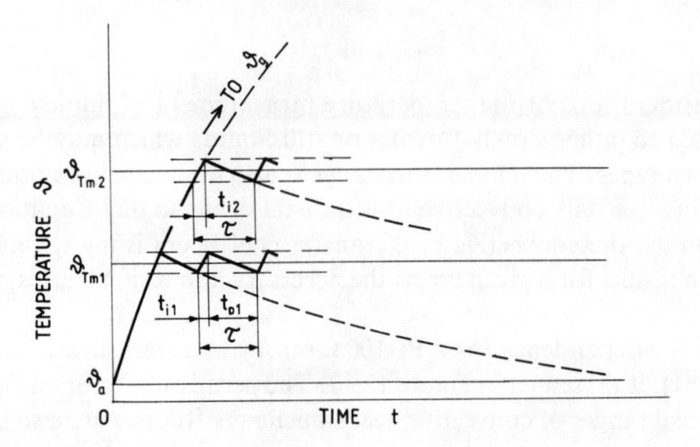

Figure 12.18
Pulsed thermocouple junction temperature, ϑ_T, versus time, t, at sampling time, τ, and two different immersion time, t_i, values.

by convection and radiation. If the actual sensor temperature is kept sufficiently low, there is no need to calculate any corrections for radiant heat exchange with the surrounding.

In those parts of the sampling time, τ, in which the sensor is withdrawn from the high-temperature medium, it is cooled down. The mean sensor temperature ϑ_{Tm}, corresponds to a thermal steady-state, in which the thermal energy (heat) taken up during the time, t_i, equals the thermal energy given up during the time t_0. The corresponding equation is:

$$\int_0^{t_i} \Phi_{g-T} \, dt = \int_{t_i}^{\tau} \Phi_{T-a} \, dt \tag{12.23}$$

where t_i is the time duration of the immersion, τ is the sampling period, Φ_{g-T} is the heat flux taken up from the medium during the time, t_i, Φ_{T-a} is the heat flux, given up to the ambient during the withdrawal time, t_0.

Assuming that both heat fluxes Φ_{g-T} and Φ_{T-a} are linear functions at the relevant temperature difference, Equation (12.23) becomes:

$$k_{g-T}(\vartheta_g - \vartheta_{Tm})t_i = k_{T-a}(\vartheta_{Tm} - \vartheta_a)(\tau - t_i) \tag{12.24}$$

where k_{g-T} is the equivalent heat transfer coefficient between the gas and the sensor, k_{T-a} is the equivalent heat transfer coefficient between the sensor and ambient of withdrawal, ϑ_g is the gas temperature which is to be determined, ϑ_{Tm} is the mean sensor temperature, ϑ_a is the ambient temperature, τ is the sampling period, and t_i is the immersion time.

Assume, for simplicity that a linear variation of sensor temperature with time occurs during heating and cooling. From Equation (12.24) the gas temperature can be expressed as:

$$\vartheta_g = \frac{k_{T-a}}{k_{g-T}}(\vartheta_{Tm} - \vartheta_a)\frac{\tau - t_{i1}}{t_{i1}} + \vartheta_{Tm1} \tag{12.25}$$

For simplicity, substitute the relations:

$$k' = \frac{k_{T-a}}{k_{g-T}} \quad \text{and} \quad t_i' = \frac{\pi - t_i}{t_i} \tag{12.26}$$

Holding the sampling time, τ, the gas temperature, ϑ_g, and the ambient temperature, ϑ_a, all constant, the experiments are conducted for two different immersion times $t_i = t_{i1}$ and $t_i = t_{i2}$ with the corresponding mean temperature readings, ϑ_{Tm1} and ϑ_{Tm2}. From relations (12.25) and (12.26) two different formulae for gas temperature are derived, as follows:

$$\vartheta_g = k' t_{i1}'(\vartheta_{Tm1} - \vartheta_a) + \vartheta_{Tm1} \tag{12.27}$$

$$\vartheta_g = k' t_{i2}'(\vartheta_{Tm2} - \vartheta_a) + \vartheta_{Tm2} \tag{12.28}$$

The gas temperature, which can be determined by eliminating k' from relations (12.27) and (12.28), is given by

$$\vartheta_g = \left(\vartheta_{Tm1} - \vartheta_{Tm2} \frac{t_{i1}'(\vartheta_{Tm1} - \vartheta_a)}{t_{i2}'(\vartheta_{Tm2} - \vartheta_a)}\right)\left(1 - \frac{t_{i1}'(\vartheta_{Tm1} - \vartheta_a)}{t_{i2}'(\vartheta_{Tm2} - \vartheta_a)}\right)^{-1} \tag{12.29}$$

or after some transformations:

$$\vartheta_g = \frac{\vartheta_{Tm1} t_{i2}'(\vartheta_{Tm2} - \vartheta_a) - \vartheta_{Tm2} t_{i1}'(\vartheta_{Tm1} - \vartheta_a)}{t_{i2}'(\vartheta_{Tm2} - \vartheta_a) - t_{i1}'(\vartheta_{Tm1} - \vartheta_a)} \tag{12.30}$$

Thus, the gas temperature, ϑ_g, may be determined from relation (12.30) from the two mean sensor temperatures, ϑ_{Tm1} and ϑ_{Tm2}, corresponding to the two respective values of relative time, t_{i1} and t_{i2}. To ensure that the amplitudes of the sensor temperature variations are small compared with the mean sensor temperature, ϑ_{Tm}, the sampling time, τ, has to be short enough. When the sampling frequency is too low the amplitudes of the sensor temperature are so high that they cause a decrease in the precision of the measurement of the mean sensor temperature, ϑ_{Tm}, and hence of the gas temperature, ϑ_g. On the other hand, if the immersion frequency is too high the resulting mechanical accelerations required may cause some structural damage to the sensor. According to Roeser and Olsen (1962), the sampling frequency should be about 0.5 to 5 Hz.

Thermocouples with bare measuring junction have a water-jacket (Figure 12.19) out of which the unsheathed end should protrude by a length of about 6 to 10 wire diameters. As thermocouples containing platinum are too brittle for the mechanical stresses they have to stand, it is advisable to use NiCr–NiAl types. The junction movement should be very rapid, so that the time needed for transfer from the gas to the water-jacket is negligibly small compared with the immersion time, t_i. One of the proven solutions is the application of an electromagnetic drive by two coils and a ferromagnetic core (Figure 12.19).

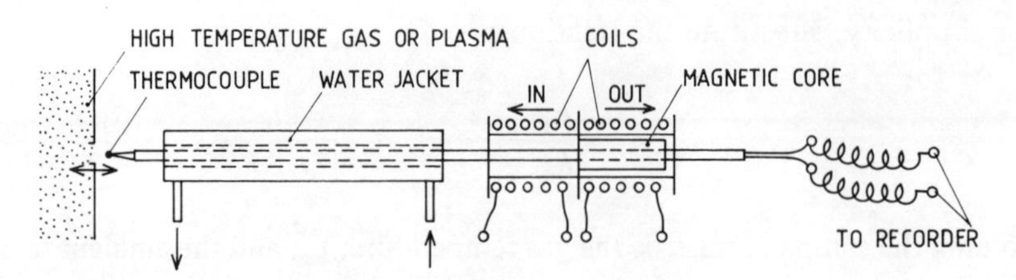

Figure 12.19
Principle of a pulsed thermocouple.

Choice of the ratio $(\tau - t_i)/t_i$ should be so that the mean sensor temperature, ϑ_{Tm}, does not exceed about 700 °C. This corresponds to the value from which the cooling-down curve of the sensor may still be approximated by a straight line.

REFERENCES

Ash, R. L. and Grossmann, G. R. (1972) Catalytic considerations in temperature measurement. *Temperature: Its Measurement and Control in Science and Industry*, Vol. 4, Part 3, Instrument Society of America, Pittsburgh, pp. 1663–1670.

Assmann, R. (1892) Das Aspirationsthermometer. *Abhandl. kgl. preuss. Meteorolog. Inst.*, **5**, 117.

Baas, P. B. and Mai, K. (1972) Trends of design in gas turbine temperature sensing equipment. *Temperature: Its Measurement and Control in Science and Industry*, Vol. 4, Part 3, Instrument Society of America, Pittsburgh, pp. 1811–1822.

Baker, H. D., Ryder, E. A. and Baker, N. H. (1953) *Temperature Measurement in Engineering*, Vol. 1, John Wiley and Sons, New York.

Baker, H. D., Ryder, E. A. and Baker, N. H. (1961) *Temperature Measurement in Engineering*, Vol. 2, John Wiley and Sons, New York.

Benson, R. S. and Brundrett, G. W. (1962) Development of resistance wire thermometer for measuring transient temperatures in exhaust system of internal combustion engines. *Temperature: Its Measurement and Control in Science and Industry*, Vol. 3, Part 2, Reinhold Publ. Co., New York, pp. 631–654.

Blumröder, G. (1981) *Beitrag zur Beschreibung und Ermittlung des statischen thermischen Fehlerverhaltens industrieller Berührungsthermometer*, PhD thesis, T. H. Ilmenau.

Breitkopf, G., Witting, S. and Kim, S. (1980) Recovery factor des frontal angeströmten zylindrischen Mantelthermoelementes mit ebener Stirnfläche. *Wärme und Stoffübertragung*, **13**(4), 287.

Green, S. F. (1987) Temperature in flames and gases. *Measurement and Control* **20**(6), 19–22.

Haas, A. (1969) Einfluss des Thermometerkopfes auf die statischen und dynamischen Eigenschaften industrieller Thermometern, *m.s.r*, **12**(12), 141–142.

Hottel, H. C. and Kalitinsky, A. (1945) Temperature measurement in high velocity air streams. *J. Appl. Mechanics*, **12**(3), A25–A32.

Jakob, M. (1957) *Heat Transfer*, Vol. 2, John Wiley and Sons, New York.

Jakob, M. (1958) *Heat Transfer*, Vol. 1, John Wiley and Sons, New York.

King, W. J. (1943) Measurement of high temperatures in high velocity gas streams. *Trans. ASME*, **65**, pp. 421.

Kinzie, P. A. (1973) *Thermocouple Temperature Measurement*, John Wiley and Sons, New York.

Kretschmer, D., Odgers, J. and Schlader, A. F. (1977) The pulsed thermocouple for gas turbine applications. *Trans. ASME*, **99**.

Moffat, E. M. (1952) Multiple shielded high temperature probes, comparison of experimental and calculated errors. *SAE*, T-13, No. I.

Mullikin, H. F. (1941) Gas temperature measurement and the high velocity thermocouple. *Temperature: Its Measurement and Control in Science and Industry*, Reinhold Publ. Co., New York, pp. 775–805.

Mullikin, H. F. and Osborn, W. J. (1941) Accuracy tests of the high velocity thermocouple. *Temperature: Its Measurement and Control in Science and Industry*, Reinhold Publ. Co., New York, pp. 805–830.

Ribaud, G. *et al.* (1959) *Etudes de pyrométrie pratique*, Editions Eyrolles, Paris.

Roeser, S. D. and Olsen, H. L. (1962) The intermittent thermometer—a new technique for the measurement of extreme temperatures. *Temperature: Its Measurement and Control in Science and Industry*, Vol. 3, Part 2, Reinhold Publ. Co., New York, pp. 901–906.

Rudolphi H. (1969) Einfluss der Wärmeübergangsbedingungen auf die Messgenauigkeit eines Temperaturfühlers in Luftkanälen. *ATM*, Blatt V2165–5/6.

Tauras, J. A. (1972) Some designs using sheathed thermocouple wire for jet engine applications. *Temperature: Its Measurement and Control in Science and Industry*, Vol. 4, Part 3, Instrument Society of America, Pittsburgh, pp. 1805–1810.

Thomas, D. B. and Freeze, P. D. (1972) The effects of catalysis in measuring the temperature of incompletely-burned gases with noble-metal thermocouples. *Temperature: Its Measurement and Control in Science and Industry*, Vol. 4, Part 3, Instrument Society of America, Pittsburgh, pp. 1671–1676.

Torkelsson, S. A. (1980) A new type of resistance thermometer for accurate temperature measurement in high gradient thermal boundary layers. *J Sci. Instrum.*, **57**(5), 549–552.

Wenzel, M. and Schulze E. (1926) Versuche mit Durchflusspyrometern. *Mitteilungen der Wärmestelle No. 92 des Vereins Deutscher Eisenhüttenleute.*

West, W. E. and Westwater, J. W. (1953) Radiation-conduction correction for temperature measurements in hot gases. *Industrial and Eng. Chemistry*, **45**(10), 2152–2156.

13 Temperature Measurement of Moving Bodies

13.1 INTRODUCTION

The method of measuring the temperature of moving bodies depends upon the character of the movement. For instance in the flight of rockets, meteorological probes or satellites the only possible way is by wireless signal transmission.

Another group of techniques, covering the temperature measurement of bodies in oscillatory or rotational movement (Hanus, 1967; Stanforth, 1962), will be discussed in detail in this chapter. As the simplest way of temperature measurement of rotating bodies uses pyrometric techniques this possibility should be the starting point of all analytical considerations. Pyrometric methods, irrespective of all the limitations resulting from the shape and state of the investigated surfaces, are only applicable to surfaces which are clearly visible from outside with no obstruction in the line of sight of the pyrometer. Measuring the temperature at any particular point of a rotating body, which is not visible in this way, requires the fastening of a temperature sensor to the moving body. Its output signal is then transmitted to a stationary external measuring instrument.

The majority of problems to be discussed in this chapter, concerns the temperature measurement of machine parts, motors and other moving objects in a range from 20 to several hundred degrees Celsius. In this temperature range, temperature indicators, already described in §3.5, are also very useful.

13.2 PYROMETRIC CONTACTLESS METHODS

Pyrometric measurement principles are described in detail in Chapter 7. In the temperature range mentioned above, low-temperature total radiation pyrometers and spectral and band photoelectric pyrometers are applicable. One of the main difficulties, which is due to the low emissivities of metallic surfaces, may be remedied in some cases by covering the investigated surface with a thin layer of black varnish or teflon. Another source of errors may be any radiation from other sources, reflected by the surface whose temperature is to be measured. Here again a remedy may be either the application of radiation shields or the use of a pyrometer of properly chosen spectral sensitivity. Depending on the thermal inertia or lag of the pyrometer and on the rotational speed of the investigated body, an applied pyrometer, can indicate either the average value of the temperature or it can follow the variations in the temperature round the body's periphery.

In most photoelectric pyrometers, the inertia of the associated instrument is much higher than that of the pyrometer itself. Analogue instruments or potentiometers indicate average values while the digits of digital instruments change so quickly that they cannot be read. In this last case it is advisable to use a peak-picking device (§7.7).

Special purpose dedicated pyrometers are also produced. Examples are the pyrometer TBP from Land Infrared Ltd, designed for the temperature measurement of gas-turbine blades, and the Ardomot pyrometer from Siemens AG (Germany), used in the temperature measurement of slip-rings in asynchronous motors.

13.3 SIGNAL TRANSMISSION FROM ROTATING SENSORS TO STATIONARY METERS

13.3.1 Sliding contact circuits

In the transmission of small currents and voltages by sliding contacts, slip-ring/brush systems are the most popular. In their utilization the problems which can arise, are:

- variations of contact resistance of an irregular, impulsive character,
- high contact resistance,
- parasitic e.m.fs across the slip-ring/brush contact,
- accelerated brush wear.

Contact resistance across slip-ring/brush contact is caused by a layer of semiconductive oxides. This resistance is especially high with low voltage signals. At higher voltages break-through of the oxides layer, causes transitional, punctual, metallic connections, resulting in a decrease of the contact resistance. As an example Figure 13.1 shows the dependence of the contact resistance, between two graphite brushes of 60 mm^2 cross-sectional area, and a brass slip-ring at a velocity of 4 m/s as a function of applied voltage (Thomas and Horvat, 1954). For instance, in Baker *et al.* (1961) it is shown that the contact resistance of a silver brush against a copper ring is about 0.0002 Ω at a force of about 5 N, increasing to about 0.01 Ω at 0.1 N. These authors also demonstrate that good results can be obtained with silver, which besides oxidizing very slowly, is reduced to pure silver again under the influence of local overheating. As metal–metal sets rapidly wear down, even when lubricated with vaseline or graphite powder to reduce the friction, they can only be used for short-time measurements. In general, it can be concluded that contact resistances are a function of the materials used, the contact forces and the peripheral speed. However, the data provided by different authors differ considerably.

Parasitic e.m.fs across the contact resistance of sliding contacts are usually within the range from a few microvolts to some hundreds of microvolts. As said before,

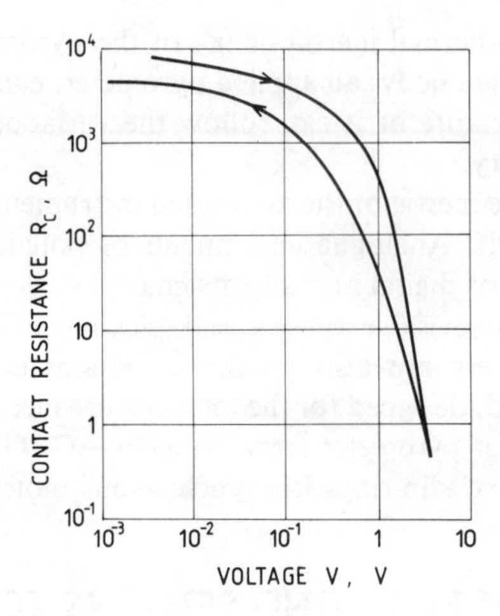

Figure 13.1
Contact resistance between two graphite brushes and a brass slip-ring at $v = 4$ m/s versus applied voltage.

and reported by Amey and Menge (1977) and Thomas and Horvat (1954), the contact resistance itself depends upon materials, contact forces and peripheral speed. So far, although their origin is not quite clear they are mainly due to thermal e.m.fs resulting from local overheating when slip-rings and brushes are made of different materials. According to Weiss (1961), the contact between a brush and a slip-ring can be regarded as an arrangement of two series connected thermocouples due to the brush-to-oxides layer and oxides-layer to slip-ring. However, some parasitic e.m.fs can be observed, even if both brushes and slip-rings are made of the same material. Weiss (1961) suggests, that under the influence of air humidity, local voltaic cell effects may occur at the contact of graphite-brush and graphite slip-ring. For brushes and slip-rings of the same metal, Baker *et al.* (1961) suggest that parasitic e.m.fs are caused by high local overheating, resulting in levels of electrical noise. Table 13.1 gives the approximate values of parasitic e.m.fs. Chavernoz (1966) advises the use of silver wire of 0.6 mm diameter for brushes around a silver ring. Up to a rotational speed of 4000 rpm parasitic e.m.fs are below 10 μV. A similar arrangement of copper wire around a copper ring is described by Alimov (1960). Parasitic e.m.fs always increase with increasing circumferential speed (Thomas and Horvat, 1954).

Table 13.1
Parasitic voltage, *E*, across two 6 × 6 mm brushes and a slip-ring at peripheral speed 0.35 m/s (Baker *et al.*, 1961).

Slip-ring	Brush	Contact force (*N*)	*E* (μV)
Silver	Graphite	50	0.3
Silver	Graphite-silver	50	0.3–2.5
Rhodium	Graphite	30	0.8
Rhodium	Graphite-silver	60	0.5
Gold	Graphite	40	0.6
Copper oxidized	Graphite	50	1–11

Good performance is obtained using mercury contacts. A two-terminal mercury sealed slip-ring assembly for thermocouples, by Omega Engineering Inc. (USA), is characterized by the following data:

- dimensions ϕ 16×33 mm,
- sealed ball-bearing alignment,
- speed, 0–2000 rpm in either direction,
- operation in any position,
- transmitted voltage, 1 μV to 120 V,
- transmitted current, 1 μA to 1 A,
- frequency, d.c. to 10 MHz,
- ambient temperature, −25 to 70 °C.

The properties of sliding contacts, and especially their high contact resistances which are discussed above, makes the application of resistance thermometer detectors in moving objects rather difficult. As reported by Mouly (1959), thermistor thermometers, having a high resistance, are more convenient when used in deflection type bridge circuits, with graphite-copper sliding contacts since they give overall measuring errors below 1 °C. Nevertheless, thermocouples are the most popular. A simplified diagram of a typical thermoelectric arrangement is shown in Figure 13.2 in which the conductors of a thermocouple, rotating together with the body under measurement, are connected to slip-rings fastened on the cylinder axle. Connection from the brushes to the point of stabilized reference temperature, ϑ_r, is made either by thermocouple wires or by appropriate compensating leads. Varying contact resistance imposes the necessity of using the potentiometric measuring system. The equivalent circuit of the arrangement is presented in Figure 13.3. Assuming that both brushes and both rings are respectively at the same temperatures, the equivalent e.m.f of the circuit according to the method described in §4.1, is given by:

$$E = e_{AB}(\vartheta_1) + e_{BC}(\vartheta_2) + e_{CD}(\vartheta_3) + e_{DB}(\vartheta_4) + e_{BCu}(\vartheta_r) + e_{CuA}(\vartheta_r) + e_{AD}(\vartheta_4) + e_{DC}(\vartheta_3) + e_{CA}(\vartheta_2) \qquad (13.1)$$

Valid relations for the thermoelectric circuit of Figure 13.3 are:

$$e_{BC}(\vartheta_2) + e_{CA}(\vartheta_2) = -e_{AB}(\vartheta_2)$$

$$e_{AD}(\vartheta_4) + e_{DB}(\vartheta_4) = e_{AB}(\vartheta_4)$$

$$e_{BCu}(\vartheta_r) + e_{CuA}(\vartheta_r) = -e_{AB}(\vartheta_r)$$

Inserting these into Equation (13.1) gives

$$E = e_{AB}(\vartheta_1) - e_{AB}(\vartheta_r) + e_{AB}(\vartheta_4) - e_{AB}(\vartheta_2) \qquad (13.2)$$

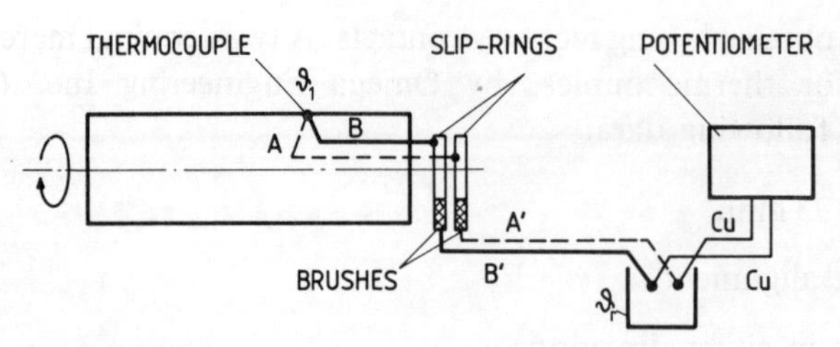

Figure 13.2
Thermocouple temperature measurement of a rotating body. Slip-ring signal transmission.

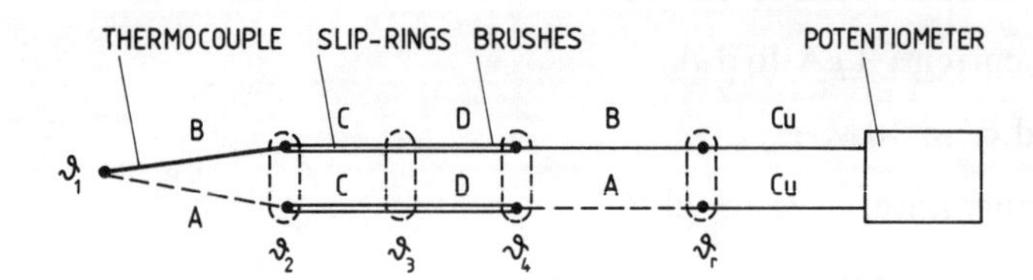

Figure 13.3
Equivalent circuit of the arrangement in Figure 13.2.

or, after further transformations:

$$E = E_{AB}(\vartheta_1, \vartheta_r) + E_{AB}(\vartheta_4, \vartheta_2) \tag{13.3}$$

Considering, that the correct value of the overall e.m.f. at the measuring temperature, ϑ_1, and at the reference temperature, ϑ_r, should be $E_{AB}(\vartheta_1, \vartheta_r)$, thc measuring error would be $E_{AB}(\vartheta_4, \vartheta_2)$. This error is zero in the case when $\vartheta_2 = \vartheta_4$ occurring when the ring temperature, ϑ_2, at the junction with the thermocouple wires equals the temperature, ϑ_4, at the junction of the compensating leads and brushes. The condition $\vartheta_2 = \vartheta_4$ is not easy to achieve because of the heat generated by friction of the brushes. Also, to fulfil the condition of the temperature equality of both the brushes and the rings discussed before, it is necessary to ensure that the contact forces of both brushes are absolutely identical. Although the circuit shown in Figure 13.2 is used sometimes, its precision is not high.

Lower errors in signal transmission are obtained with the circuit of Figure 13.4. In this circuit both rings are inserted into conductor A which is cut in two and connected with conductor B in a thermally isolated chamber. The junction of conductors A and B freely rotates in the chamber where the compensating stationary thermocouple is also placed. Figure 13.5 shows the corresponding electric circuit. Assuming that both brushes and both rings are at the same temperatures, the equivalent e.m.f. is given by the relation:

$$E = e_{AB}(\vartheta_1) + e_{BA}(\vartheta_2) + e_{AC}(\vartheta_3) + e_{CD}(\vartheta_4) + e_{DCu}(\vartheta_5) +$$
$$e_{CuA}(\vartheta_6) + e_{AB}(\vartheta_2) + e_{BA}(\vartheta_r) + e_{ACu}(\vartheta_6) + e_{CuD}(\vartheta_5) + e_{DC}(\vartheta_4) + e_{CA}(\vartheta_3) \tag{13.4}$$

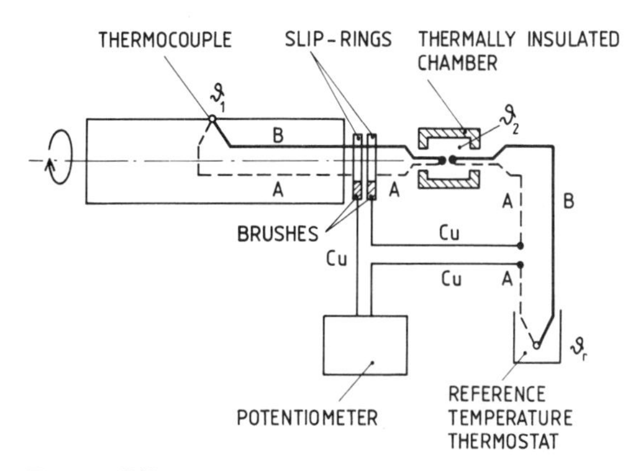

Figure 13.4
Thermocouple temperature measurement of a rotating body with elimination of parasitic e.m.fs. Slip-ring signal transmission.

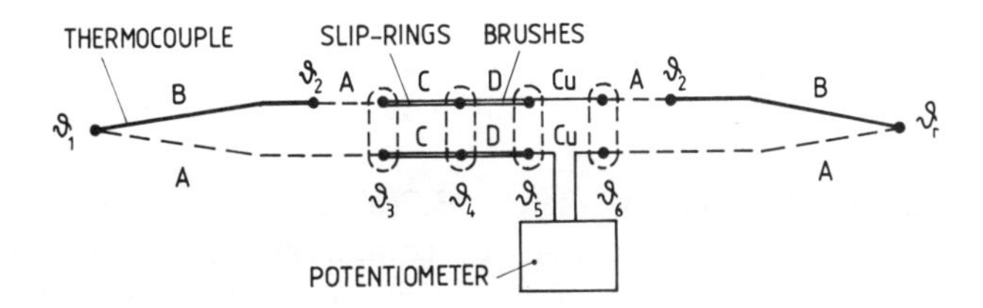

Figure 13.5
Equivalent circuit of the arrangement in figure 13.4.

Considering the general rule $e_{xy}(\vartheta_n) = -e_{yx}(\vartheta_n)$ Equation (13.4) becomes

$$E = e_{AB}(\vartheta_1) - e_{AB}(\vartheta_r) = E_{AB}(\vartheta_1, \vartheta_r) \tag{13.5}$$

It is evident from Equation (13.5) that no parasitic e.m.fs occur.

13.3.2 Inductive circuits

Inductive circuits can be used to avoid any problems arising from the application of sliding contacts, especially at higher rotational speed. Signal transmission from a thermocouple, rotating together with the body under measurement, can occur by an axial (Figure 13.6(a)) or radial (Figure 13.6(b)) magnetic field. In the axial field arrangement the current in the axial coil of Figure 13.6(a) which is proportional to measured temperature, generates a constant magnetic field, whose strength may be measured by a Hall generator. The radial field, applied by the rotating coil in Figure 13.6(b), induces an alternating voltage in the stationary receiver coil, whose value is proportional to the measured temperature. Both of these methods are not precise, because the transmitted signal depends upon the air gap dimension, the core saturation and, in the radial arrangement, upon the rotational speed of the shaft.

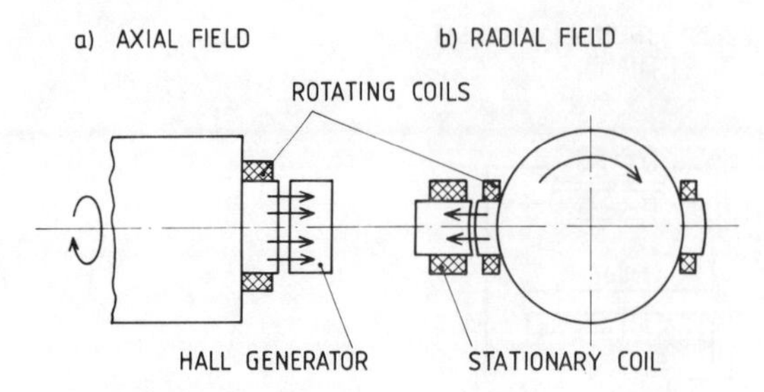

Figure 13.6
Inductive signal transmission from rotating bodies.

Inductive compensating circuits, shown in Figure 13.7, may be used with the above methods. In the axial field arrangement of Figure 13.7(a) at the moment of measurement, the magnetic field, excited by the rotating sender coil, is compensated by a contrary directed magnetic field induced by a stationary coil. The state of full compensation, which corresponds to zero magnetic field strength in the air gap, is detected by a special probe, most commonly a Hall generator in the axial field arrangement.

In the radial field arrangement of Figure 13.7(b) a flat coil, connected to an oscilloscope, is used as a zero detector. At full compensation, which occurs when both the rotating sender and the stationary compensation coils coincide, the voltage induced in the flat detector coil is zero. At partial compensation, corresponding to over or under compensation, a differential signal of changing phase is observed on the oscilloscope screen. In both the axial and radial arrangements, the measured temperature is determined from the readings of an ammeter measuring the compensating current. Both of these arrangements, which have to be adjusted manually, can only be used for spot measurements.

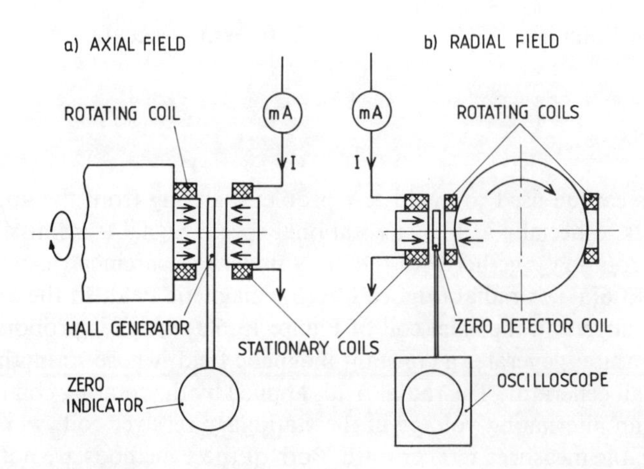

Figure 13.7
Inductive compensating circuit for signal transmission from rotating bodies.

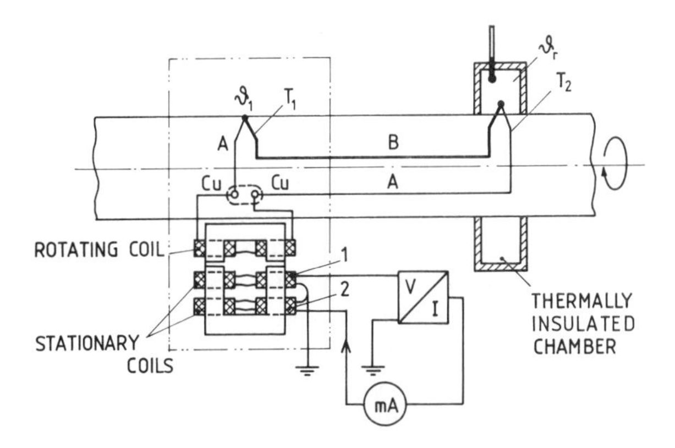

Figure 13.8
Automatic inductive compensating circuit for signal transmission from rotating bodies.

A similar but automatic system, given in Figure 13.8, was described by Weiss (1961). The temperature, ϑ_1, is measured by a thermocouple, T_1, rotating together with the body being measured. An auxiliary rotating thermocouple, T_2, whose measuring junction is inside a stationary thermally insulated chamber of known temperature, ϑ_r, was assumed as the reference temperature. It allows any necessary corrections to the thermocouple readings to be calculated. Under the influence of the temperature difference $(\vartheta_1 - \vartheta_r)$ a current flows in the rotating sender coil with a core. In every revolution this rotating core passes a stationary core, with measuring, 1, and compensating, 2, coils. The voltage induced in the measuring stationary coil is transformed into a current in the compensating coil using an electronic current. This current, which is automatically set by the electronic circuit so that the voltage of the compensating stationary measuring coil is zero, is thus proportional to the measured temperature ϑ_1, indicated by the miliammeter, graduated in degrees of temperature. Within the

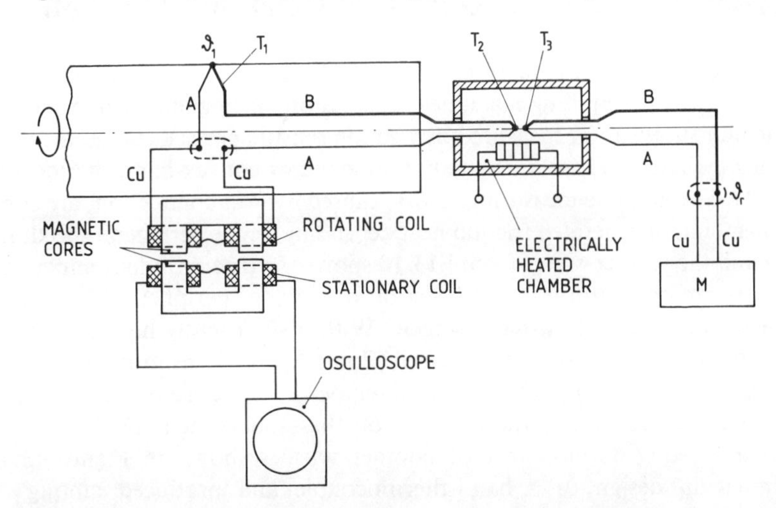

Figure 13.9
Inductive signal transmission from rotating bodies. Circuit with thermal compensation.

measuring range of 40 to 450 °C the temperature dependence of the compensating current is stable for the speed range of 190 to 18 000 rpm. The overall measuring error is kept below about ±2%.

Figure 13.9 shows an inductive circuit with thermal compensation. Rotating together with the investigated body the measuring thermocouple, T_1, is connected with reverse polarity, in series with a compensating thermocouple, T_2, whose measuring junction is placed in a stationary, electrically heated chamber. Also in series with both thermocouples is a rotating coil magnetically coupled with the stationary one. The difference between the e.m.fs of the thermocouples, T_1 and T_2, results in a current flow in the rotating coil inducing an e.m.f. in the stationary coil, which e.m.f. is read on an oscilloscope screen. Setting of the chamber temperature is such that there is no current in the rotating coil, which is checked on the oscilloscope screen. Under these conditions the temperatures of T_1 and T_2 are equal, while the chamber temperature, measured by the thermocouple, T_3, and the measuring instrument, M, equals the desired measured temperature. With an air gap of 0.2 mm and rotational speed up to 6000 rpm, the sensitivity threshold of the arrangement is about 5 °C. For multipoint measurement an additional commutator, rotating with the body and operated by a shifting ring, is used. The circuit has then only one compensating thermocouple.

CMR, France, manufactures a system for the inductive transmission of signals in the temperature measurement of the connecting rod of the internal combustion engine. A thermistor detector, moving with the connecting rod, and a coil are used. During each revolution this coil passes between stationary sender and receiver coils. Damping of the signal transmitted between these two coils is a function of thermistor resistance and thus of the measured temperature. In the temperature range from 75 to 250 °C the measuring error is 2 °C at speeds from 600 to 1200 rpm. The sender-coil is fed from a transistor generator.

13.4 FRICTION SENSORS AND 'QUASI-CONTACTLESS' METHOD

Surface temperature measurement of smooth metallic cylinders, which is typically necessary in paper-making machines, plastics processing and also in the rubber and textile industries can be measured by friction sensors pressed to the surface of slowly rotating cylinders. These are usually bow-band or convex-band thermocouples (see §10.3.1). The negative measuring errors, caused by heat conduction, are compensated by an amount of generated friction heat depending on the peripheral speed, the applied force and the surface state. Figure 13.10 shows the true cylinder temperature ϑ_t and the sensor temperature ϑ_T, as a function of cylinder peripheral speed for different contact forces, for a bow-band sensor. With a sufficiently high contact force it is seen that, at one given peripheral speed (Point A), compensation of both influences, happens at one and only one measured temperature. Therefore, it is rather difficult to base any correction of the readings on these phenomena (Krüger, 1959).

An interesting development of contact sensors shown in Figure 13.11, which is similar in design to a band-thermocouple and produced among others by Keller GmbH (Germany), is the so called emissivity converter. An elastic teflon band, when pressed to the rotating cylinder surface, adopts the surface temperature.

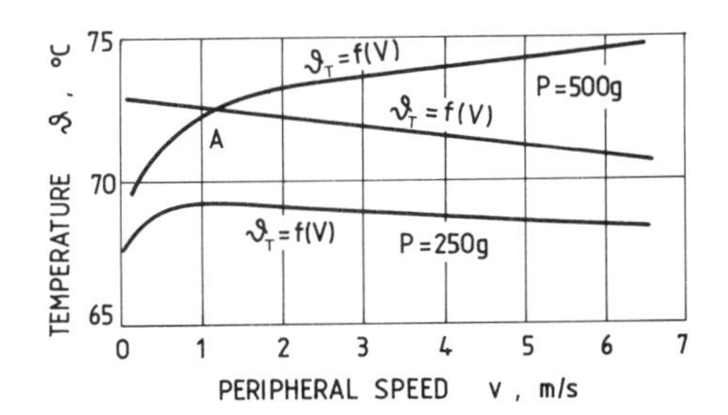

Figure 13.10
True cylinder temperature, ϑ_t, and temperature, ϑ_T, indicated by a bow-band friction sensor versus peripheral cylinder speed. A—point of error compensation.

The PF-03 pyrometer, which is directed at the other side of the band measures its temperature at a constant and well defined emissivity ($\epsilon = 0.95$), set at the indicating instrument. Application of the device is permissible up to about 250 °C and up to peripheral speeds of 10 m/s. The low friction coefficient of teflon eliminates the generation of friction heat.

The quasi-contactless method (§10.4) which allows more precise measurements, is typically applied in the temperature measurement of the surfaces of rotating cylinders.

Figure 13.11
Emissivity converter for the temperature measurement of rotating cylinders. (Courtesy of Keller GmbH Germany.)

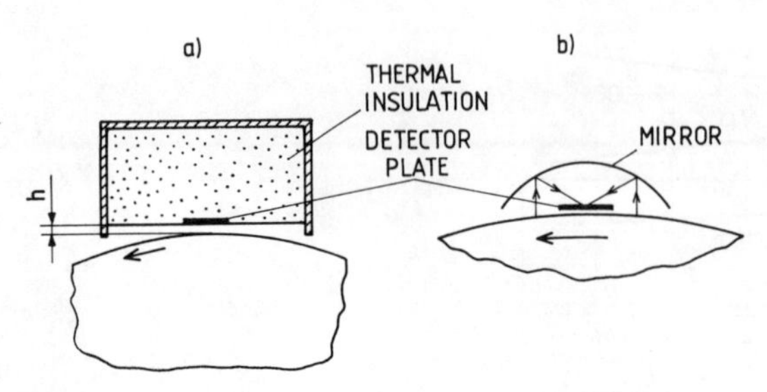

Figure 13.12
Quasi-contactless method for the temperature measurement of rotating cylinders.

A thermocouple, or a resistance thermometer detector encapsulated in a thin plate placed near the investigated surface, is heated by radiation, convection and conduction. To reduce its heat losses to the environment, either thermal insulation as in Figure 13.12(a) or concentration mirror as in Figure 13.12(b), are applied. The measuring errors, $\Delta\vartheta$, depend upon the distance, h, from the surface (Figure 13.12(a)) and the peripheral speed and temperature of the surface. This dependence, at $\vartheta_t = 120\,°C$, is shown in Figure 13.13. The errors, which increase with increasing air-gap as a result of the temperature drop across the surface air film, also increase with increasing peripheral speed, as then more cool air is sucked into the air-gap. Similar dependence of errors on air-gap and peripheral speed are also exhibited by those sensors with concentrating mirror depicted in Figure 13.12(b).

For the temperature measurement of moving cylindrical and flat surfaces, Weichert *et al.* (1976), describe the use of thermovibulators, which intensify the convective heat transfer between the surface and the sensor using a blower. Two other modifications are similarly based upon the convective method. In the first one, described by Fothergill (1975), an air stream of temperature, ϑ_a, is forced along the body's surface with the air temperature along the stream direction measured by two sensors. From the measured temperature difference, the body temperature, ϑ_b, at

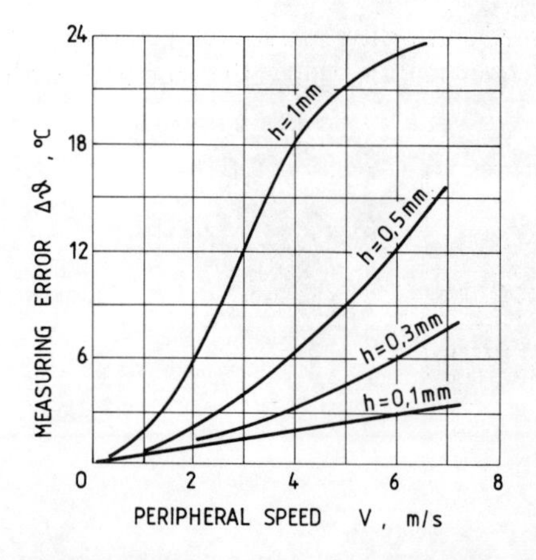

Figure 13.13
Measuring errors of the system in Figure 13.12(a), versus cylinder peripheral speed.

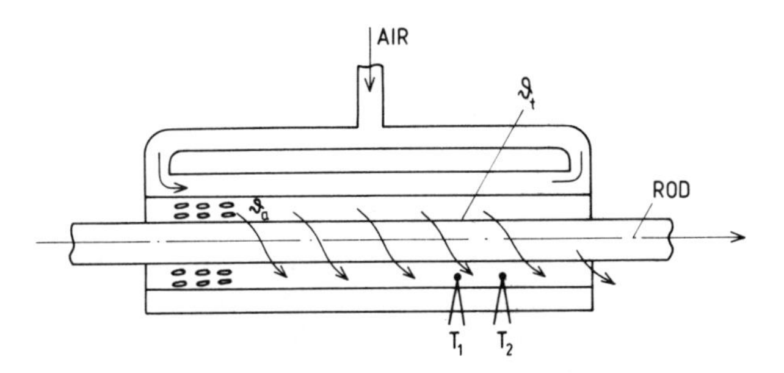

Figure 13.14
Convective temperature measurement of a moving Al rod.

$\vartheta_b > \vartheta_a$ is calculated. In the second modification the air-stream along the body is first heated. As long as both sensors read the same temperature, the air temperature is then equal to the temperature of the body.

Both modifications are applicable for flat as well as cylindrical bodies. As an example, in the temperature measurement of a moving aluminium band of $\epsilon \sim 0.1$ and $\vartheta_b \sim 300\,°C$, the errors were about $\pm 1\%$. The cooling of the band caused by the air-stream was below 0.5% of ϑ_b. A band displacement of ± 5 cm transverse to the direction of movement, did not cause any significant measurement errors. Fothergill (1975) describes the temperature measurement of a moving aluminium rod using the arrangement shown in Figure 13.14. Compressed air of temperature, ϑ_a, is forced through a set of holes, to stream round the rod in a circular movement. With the air temperature measured by thermocouples T_1 and T_2 the measured rod temperature can be determined from the measured temperature difference $(\vartheta_{T_1} - \vartheta_{T_2})$.

The second modification of the method, based on preheated air, can also be used to obtain higher precision of measurements if the thermocouples T_1 and T_2 can be replaced by thermopiles of four thermocouples around the rod. In this solution the readings, which are independent of any radial displacement of the rod, have errors well below $\pm 2\%$.

Similar to the thermally compensated sensor for surface temperature measurement (§10.3.3), Hornbaker (1972) has described a convective method with thermal

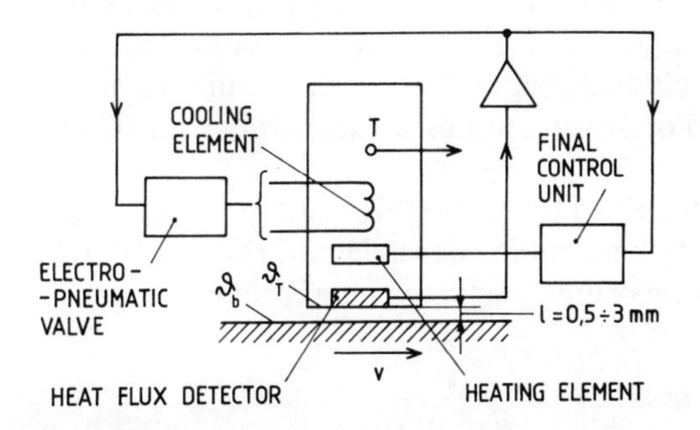

Figure 13.15
Convective method with thermal compensation for the temperature measurement of moving bodies.

compensation, as shown in Figure 13.15. A thermally compensated sensor is placed at a distance of 0.5 to 3 mm from the moving surface, exchanging heat with it through convection. When the sensor temperature, ϑ_T, is below that of the body, ϑ_b ($\vartheta_b > \vartheta_T$), the signal from the heat-flux detector which is positive, then controls the power of the heating element through a phase-sensitive amplifier and final control element. If $\vartheta_T > \vartheta_b$, the negative signal from the heat-flux detector causes cooling of the element by activating a pneumatic valve. At thermal equilibrium ($\vartheta_T = \vartheta_b$), the temperature readings of the sensor are equal to the measured temperature value. The same device can be operated at a nearly steady thermal state where the detector output signal, s, is not zero, but is given by the equation:

$$s = f(\vartheta_b, l, v) \tag{13.6}$$

where l is the distance between the sensor and the investigated surface, and v is the velocity of the moving surface.

In this modification the device can be operated at a distance between the sensor and the surface of 0.5 to 3 mm in the temperature range 0 to 500 °C. Variations in the velocity of the moving surface from about 0.5 to 6 m/s, at $l = 0.75$ mm, did not cause any errors above $\pm 5\%$.

Using another type of sensor on a similar principle, Hornbaker (1972) also describes a system for the temperature measurement of moving fibres and wires, of diameter 0.2 to 3 mm and velocity from 0 to 50 m/s. Measuring errors in the temperature range up to 200 °C were about ± 2.5 °C.

Typical non-restrictive examples of the application of the systems described above occur in the paper making and textile industries, as well as in the production of metal wires and bands.

13.5 WIRELESS SYSTEMS

In the temperature measurement of rotating or moving bodies transmission of the measuring signals is frequently only possible with wireless systems. Temperature sensor, transmitter, batteries and antenna move along with the body while the signal receiver and the recorder are stationary. The main problems in the design of this type of rotating transmitting arrangements are their necessarily small dimensions and their robustness against the very big acceleration forces, which can even amount to 300 N. Thermocouple and thermistor sensors are used exclusively with IC amplifiers and transmitters.

A typical arrangement, described by Adler (1971) and shown in Figure 13.16, uses a thermistor sensor, whose temperature dependent variations of resistance, R_T, change the frequency of a relaxation oscillator in accordance with

$$f = \frac{1}{R_T C + K} \tag{13.7}$$

where R_T is the thermistor resistance, C is the capacitance, and K is a constant.

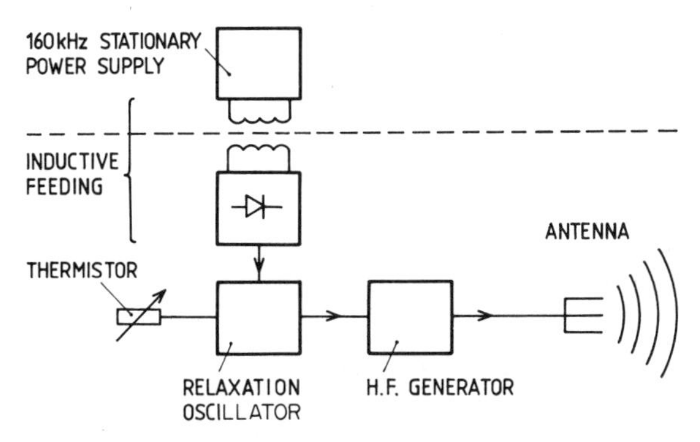

Figure 13.16
Wireless system or signal transmission from moving bodies with a thermistor sensor.

The output signal of the oscillator which modulates the 90 MHz carrier frequency of a HF generator coupled to the antenna, is linearized in the receiver. In the temperature range from 0 to 150 °C, the measuring error of the system which is about ±0.5 °C, is insensitive to supply voltage variations of about ±25%. The batteries, which can operate at ambient temperatures up to 150 °C last for about 50 to 200 h of continuous operation. They contribute to a total volume of transmitter and batteries of some few cubic centimetres. As the battery lifetime is limited, it is advisable to power the system using the inductive feeding scheme shown in Figure 13.16 at higher acceleration. A stationary sender coil generates an alternating magnetic field of 160 kHz which induces a high frequency signal in the receiver-coil mounted with the rotating transmitter. Subsequently the induced high frequency signal is rectified and stabilized. One sender-coil can consecutively supply several rotating measuring arrangements with the distance between both coils even up to 25 mm. If thermocouple sensors are applied, transistorized FET-converters are most commonly used. These operate at a frequency of 2 to 4 kHz, with output signals, which supply an integrated high frequency amplifier, of frequency about 100 MHz. The system operates on the basis of amplitude modulation and its output stage on frequency modulation. Adler (1971) points out that such a system gives a lower noise level than systems with double frequency modulation. Some disturbances and additional frequency modulation may occur owing to variations in the distance between the rotating transmitter and stationary metal masses, which can be caused by changing parasitic capacitances.

Transmitter shielding and the application of an additional amplifier separating the antenna from the environment, are suitable precautions to reduce these effects.

13.6 OTHER METHODS

Using temperature indicators is an easy and simple method for determining the temperature of rotating bodies. Observing or photographing them, allows a judgement of whether the permissible temperature value has been reached. It is advisable to use stroboscopic lighting if taking colour pictures.

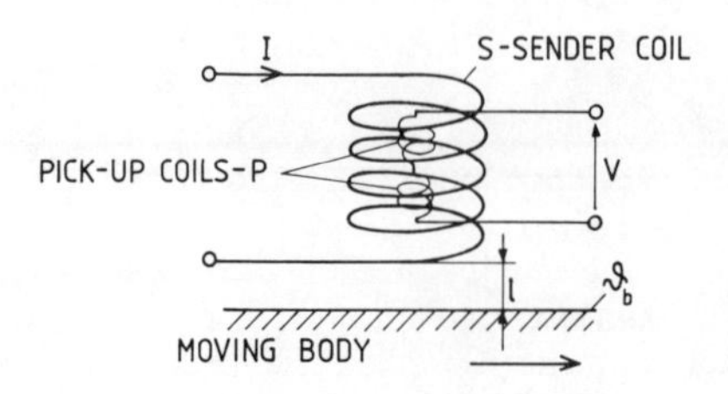

Figure 13.17
Inductive method of temperature measurement of moving bodies.

Another very popular method is the use of thermography, described in §15.1, or its modification the thermostroboscope method (AGA, undated). This last method permits stationary pictures of rotating car tyres while driving, of ball-bearings, of brake drums and so on, to be taken. Black and white or colour displays with marked isotherms are possible. The method is based on replacing normal scanning by vertical scanning, which is synchronized with the rotations. This synchronization is mostly realized by photoelectric signals, reflected from a small mirror rotating together with the body. An additional electronic phase shifting circuit permits aiming of the stroboscopic device at any desired point on the circumference of the body. Evidently, this phase shift is independent of rotational speed.

Fluoroptic thermometers, described in §15.3, are also used for the temperature measurement of rotating bodies. The measuring signal can be transmitted across an air or water gap, due to the existing possibility of cutting the light guide of the thermometer in two. At the moment when both the rotating and stationary ends of the light guide coincide with each other, once per rotation, the measurement is taken. In another method the fluorescent element, which is placed alone on the rotating body, is approached by the stationary end of the light guide.

More than fifty years ago Keinath (1934) described an inductive method for the temperature measurement of rotating cylinders based upon the temperature dependence of the eddy currents, which are induced in a rotating steel cylinder. A modification of this method, which was proposed recently by Keller (1980), applies one sender-coil, as in Figure 13.17, and two pick-up coils, P, all axially arranged. An alternating current in the sender coil, s induces eddy-currents in the rotating body. These eddy-currents themselves again induce a magnetic field. The voltage, V, in the pick-up coils has the same frequency as the exciting current but with an angle φ relative to the current I.

Both voltage, V, and phase-angle, φ, which are functions of body temperature, also depend upon the distance, l. The phase angle, φ, depending only insignificantly upon the distance, l, is chosen as a measuring signal and transformed into frequency. This allows both precise measurements and undisturbed data transmission. Calibration of the system for a given material is required. Although it also permits the measurement of the temperature of metallic cylinders through a non-conducting layer of paper, plastic and so on, its application is restricted to ferromagnetic materials only.

REFERENCES

Adler, A. (1971) Transmission des signaux électriques des jauges de contrainte et thermocouples par radio-télémésure. *Mésures et Control Industr.*, **36**, (1/2), 72–77.

AGA (undated) Thermostrobe adapter, Publ. No. 558017.

Alimov, R. Z. (1960) Device for checking the quality of sliding contacts for temperature measurement of rotating bodies, *Ismeritelnaya Technika*, No. 4, pp. 31–32 (in Russian).

Amey, W. and Menge K. (1977) Ardomot, ein neues Gerät zur Temperaturüberwachung von Läufern grosser Käfig-Läufermotoren. *Siemens Zeitschr.*, **51**, 117–119.

Anon. (1961) Das Übertragen kleiner elektrischer Messwerte von schnell rotierenden Maschinen auf ruhende Anzeigegeräte. *ETZ*, **13**(2), 29–34.

Baker, H. D., Ryder, E. A. and Baker, N. H. (1961) *Temperature Measurement in Engineering*, Vol. 2, John Wiley and Sons, New York.

Braun, F. (1981) Messung von Oberflachentemperaturen mit Farbindikatoren an bewegten Körpern, *Messen and Prüfen*, **17**(9), 574–577.

Chavernoz, R. (1966) Transmetteur de signaux électriques issues de pieces en rotation. *Mésures, Regulation, Automatisme*, **31**(1), 73–76.

Fothergill, I. R. (1975) Non-contact temperature measurement using forced air convection. *Temperature Measurement, Conference Series No. 26*, Institute of Physics, London, 1975, pp. 409–414.

Hanus, G. (1967) Télémésures dans les machines tournantes. *Mésures, Regulation, Automatisme*, **32**(7/8), 37–42.

Hornbaker, D. R. (1972) The convective null-heat-balance concept for non-contact temperature measurement of sheets, rolls, fibers and wire. *Temperature: Its Measurement and Control in Science and Industry*, Vol. 4, Part 1, Instrument Society of America, Pittsburgh, pp. 737–748.

Keinath, G. (1934) Induktive Temperaturemessung. *ATM*, No. 1, V215-2.

Keller, P. (1980) Berührungslose Temperaturmessung an rotierenden Walzen mit einem neuen Verfahren. *Fachberichte Hüttenpraxis Metallverarbeitung*, **18**(2), 105–108.

Krüger, H. (1959) Messung und Regelung der Oberflächentemperatur umlaufender Walzen. *VDI-Z*, **101**(9), 343–346.

Mouly, R. J. (1959) An impedance bridge for surface temperature measurement. *Trans. AIEE*, **78**, 388–393.

Stanforth, C. M. (1962) Temperature measurement of rotating parts. *Temperature: Its Measurement and Control in Science and Industry*, Vol. 3, Part 1, Reinhold Publ. Co., New York.

Takami, K. *et al.* (1972) Capacitance thermometer applied to a railway traction motor. *Temperature: Its Measurement and Control in Science and Industry*, Vol. 4, Part 2, Instrument Society of America, Pittsburgh, pp. 1311–1320.

Thomas, W. A. and Horvat, R. J. (1954) Measurement of induction-motor rotor temperatures. *Instruments*, **27**(4), 410–414.

Weichert, L. *et al.* (1976) *Temperaturmessung in der Technik-Grundlagen und Praxis*, Lexika-Verlag, Essen.

Weiss, H. (1961) Ein Messgerät für die Temperaturmessung mit Thermoelementen auf sehr schnell umlaufenden Maschinen. *ETZ*, **13**(13), 353–357.

14 Temperature Measurement in Industrial Heating Appliances

14.1 CHAMBER FURNACES

14.1.1 Internal furnace temperature

Electric or gas chamber furnaces which usually operate at temperatures from 600 to 1200 °C, are typically applied in the heating of metal charges for plastic working, in the heat treatment of metals, in stress relieving heating, in the firing of ceramic or porcelain wares, and so on. The charge temperature is the most relevant variable in thermal technological processes. In most cases, in the thermal steady state, it can be assumed to be equal to the internal furnace temperature as measured by thermocouples or resistance thermometer detectors. These sensors, supplied as integral elements of the furnace by the manufacturer, are placed in a position where they do not interfere with furnace charging (Figure 14.1).

Assuming that the heat transfer inside the furnace chamber is predominantly by radiation, it can be shown (Michalski, 1966) that the indicated sensor temperature does not depend upon the position of the sensor in the chamber, provided the sensor is no nearer to the heating element than about 5 to 7 cm. This holds true for all sensors of very small junction dimensions, such as bare or MI thermocouples. Sheathed sensors, which have to be placed a little further away from heating elements, must also be inserted sufficiently deeply into the chamber to eliminate any errors caused by heat conduction along the sheath as described in Chapter 12.

In the thermal steady state or when the furnace temperature changes slowly, Michalski (1966) has shown that the temperature of a minute, point-like sensor junction is expressed by the relation:

$$T_{\mathrm{T}} \approx \sqrt[4]{\left\{ \epsilon_{\mathrm{w}} \sum_{n=\mathrm{I}}^{\mathrm{VI}} T_{\mathrm{n}}^{4} \varphi_{T \to n} + \epsilon_{\mathrm{w}}(1-\epsilon_{\mathrm{w}}) \sum_{n=\mathrm{I}}^{\mathrm{VI}} \left(\sum_{\substack{k=\mathrm{I} \\ k \neq n}}^{\mathrm{VI}} T_{k}^{4} \varphi_{n \to k} \right) \varphi_{T \to n} \right\}} \tag{14.1}$$

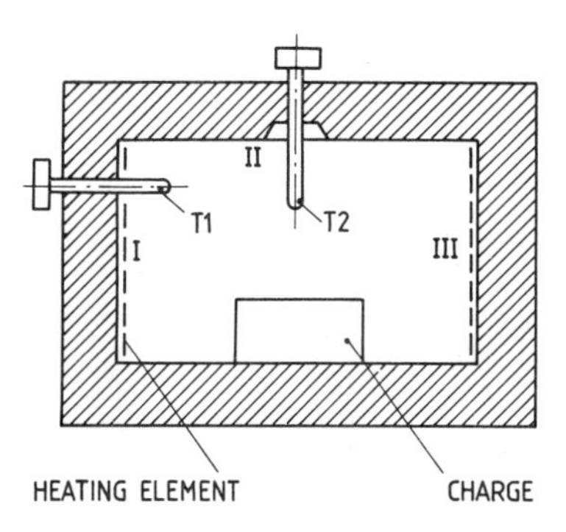

Figure 14.1
Typical arrangement of temperature sensors in a chamber furnace.

where ϵ_w is the total emissivity of the chamber walls, n is the consecutive number of the six chamber walls, k is the consecutive number of the five chamber walls, sixth wall excluded, T_n, T_k is the temperature of a particular one of the six chamber walls, $\varphi_{T\to n}$ is the angle factor of the sensor T, relative to wall n (Jakob, 1958), and $\varphi_{n\to k}$ is the angle factor of wall n, relative to wall k (Jakob, 1958).

In relation (14.1), it is assumed that all of the chamber walls have the same emissivity ϵ_w, which most frequently has a value $\epsilon_w \simeq 0.9$. For approximate calculations, when it can be assumed that the emissivity of all of the walls is $\epsilon_w = 1$ and that the chamber can be regarded as a cube for the relative dimensions of the chamber walls confined from 1:1:1 to 1:2:4, then relation (14.1) is modified to:

$$T_T \approx \sqrt[4]{\frac{1}{6}\sum_{n=I}^{VI} T_n^4} \tag{14.2}$$

Considering a furnace with a charge in its chamber, the temperature measuring device indicates a certain mean value of temperature between that of the walls and the charge. As the charge temperature approaches that of the walls ($T_c \approx T_w$) the indicated value of the temperature is nearly equal to the charge temperature.

14.1.2 Charge temperature

The temperature of the charge inside the furnace chamber can be measured either by contact or pyrometric methods. As a sensor has to be placed on the charge surface or inside it and its conductors led outside the furnace chamber, the contact method is rarely used. When this method is appropriate, thin, bare thermocouples of wire diameter about 0.5 to 2 mm in flexible insulation are usually used. They are fed through holes in the walls or in the door, to the outside. The method is used for short-time measurements as for example in the determination of charge through-heating time. Once this time is determined it is then repeated in batch production.

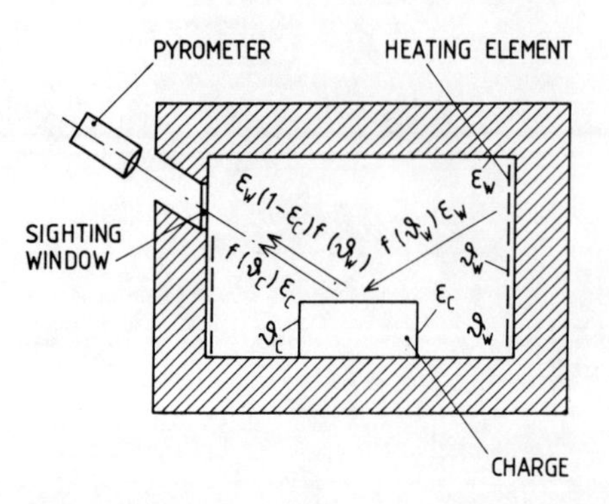

Figure 14.2
Pyrometric charge temperature measurement in a furnace chamber. ϑ_w, ϑ_c—walls and charge temperature, ϵ_w, ϵ_c—walls and charge emissivity.

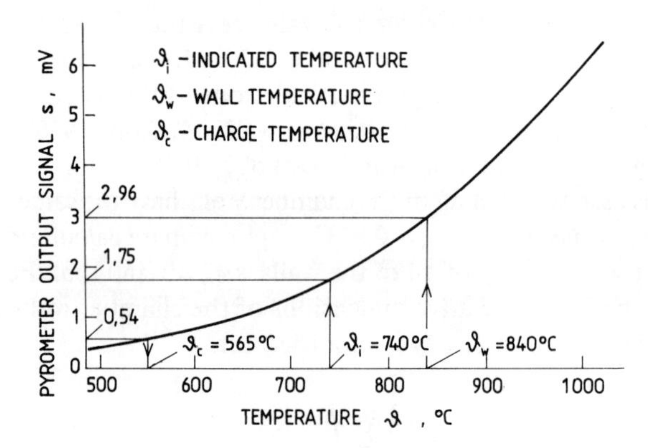

Figure 14.3
Method of determining the true charge temperature, ϑ_c, using the characteristic of the Ardometer 20 total radiation pyrometer.

Continuous temperature measurement of a charge inside a furnace chamber is possible using a pyrometer, by aiming it at the charge through a sighting window as shown in Figure 14.2. Assuming a non-transparent charge, the pyrometer reading depends upon a signal emitted by the charge surface, $\epsilon_c f(\vartheta_c)$, and upon a signal, $(1-\epsilon_c)f(\vartheta_w)\epsilon_w$, emitted by the walls of temperature ϑ_w and emissivity ϵ_w, reflected from the surface of the charge, having emissivity ϵ_c. The overall signal, s, determining the pyrometer readings is therefore given by:

$$s = \epsilon_c f(\vartheta_c) + (1-\epsilon_c) f(\vartheta_w)\epsilon_w \tag{14.3}$$

Any nonlinear dependence of the pyrometer readings upon the measured temperature accounted for by the function, $f(\vartheta_c)$, and $f(\vartheta_w)$ depends on the type of pyrometer applied. Similarly the emissivity of walls, ϵ_w, and of charge, ϵ_c, which can be either total, band or spectral emissivity, also depend upon the pyrometer type.

When $\vartheta_w < \vartheta_c$, measurement errors resulting from reflected wall radiation are usually negligibly small, especially when the charge emissivity is correctly set on the emissivity corrector scale of the pyrometer. It is then advisable to apply a pyrometer with a short effective wavelength, λ_e. No errors are observed when $\vartheta_w = \vartheta_c$.

If $\vartheta_w > \vartheta_c$ the pyrometer readings are too high, giving errors which increase with increasing wall temperature, ϑ_w, and decreasing charge emissivity, ϵ_c. As the correct setting of the emissivity corrector does not prevent these errors, it is advisable to apply pyrometers with a long effective wavelength, λ_e. A numerical example illustrates the calculation of the true charge temperature.

Numerical example

The temperature of a cast-iron charge was measured by a Siemens AG Ardometer 20 total radiation pyrometer, having the measuring temperature range: 500 to 1000 °C. A charge of total emissivity $\epsilon_c = 0.5$ was placed in a chamber of wall temperature $\vartheta_w = 840$ °C and of wall emissivity $\epsilon_w \simeq 1$. If the temperature indicated by the pyrometer was $\vartheta_i = 740$ °C, find the true charge temperature ϑ_c.

Solution: Based on the pyrometer characteristic, shown in Figure 14.3, the pyrometer output signal corresponding to an indicated temperature of 740 °C was 1.75 mV, whereas the signal corresponding to 840 °C was 2.96 mV. Inserting these data into relation (14.3) gives.

$$1.75 = 0.5f(\vartheta_c) + (1 - 0.5) \times 2.96$$

or

$$f(\vartheta_c) = \frac{1}{0.5}(1.75 - 0.5 \times 2.96) = 0.54 \text{ mV}.$$

From the pyrometer characteristic in Figure 14.3, the corresponding true charge temperature is:

$$\vartheta_c = 565 \text{ °C}.$$

In some cases, total radiation pyrometers are advantageous for the temperature measurement of charges inside a furnace. For example, when measuring the temperature of a steel charge of emissivity $\epsilon_c = 0.8$, placed in a furnace having a wall temperature about 100 °C above that of the charge, the readings of a total radiation pyrometer (Ardometer) were only 22 °C above the true charge temperature of 1000 °C (Figure 14.4). Applying a two-colour pyrometer is of no use.

To reduce the influence of radiation from the walls, the following methods can be used:

1. A water, air or nitrogen cooled sighting tube, protects the sighted area from wall radiation (Figure 14.5) (Lieneweg, 1964). Water-cooled tubes can be even 2 to 5 m long (Land Infrared Ltd).

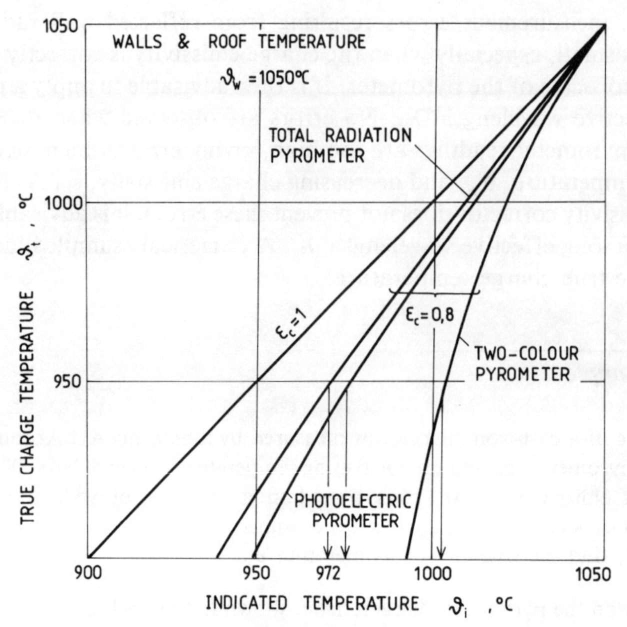

Figure 14.4
Pyrometric measurements of the charge temperature of ϵ_c=0.8, in a chamber furnace. Wall temperature about 100 °C higher than the charges. (Courtesy of Siemens AG, Germany.)

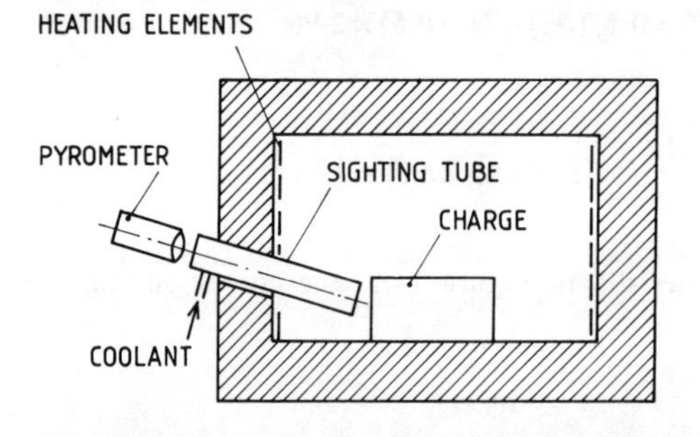

Figure 14.5
Pyrometric charge temperature measurement in a chamber furnace.

2. A two-pyrometer method, described by Beynon (1981) and Roney (1982), is based on the application of the difference-signal of two pyrometers. The first one, which is aimed at the charge, has an output signal given by the relation:

$$S_1 = \epsilon_c f(\vartheta_c) + (1 - \epsilon_c)\epsilon_w f(\vartheta_w) \tag{14.4}$$

In this relation, the first term is the signal depending on the temperature ϑ_c of the charge with emissivity, ϵ_c, and the second term is the signal depending on radiation from the wall, at a temperature, ϑ_w, and emissivity, e_w, reflected from the charge. The function f describes the scale defining equation of the pyrometer.

The second pyrometer, with an identical scale defining equation, is aimed at part of the wall whose temperature, ϑ_w, equals the average temperature of all of the walls, with emissivity, ϵ_w. The signal of the second pyrometer is given by the relation:

$$s_2 = \epsilon_w f(\vartheta_w) \tag{14.5}$$

Correcting the signal, s_2, in the function of the charge emissivity ϵ_c, the resulting differential signal determines the readings free from any influence of the wall radiation. A microprocessor conditions the signal. Indication errors are due to uncertainty in estimating the charge emissivity ϵ_c and to absorption of the signal by the furnace atmosphere.

3. In two-pyrometer methods, with an additional, cooled, reference radiation source, the first pyrometer is aimed at the charge. The second one is aimed at a water-cooled reference radiation source, placed in the furnace chamber and only emitting the radiation reflected from the walls. It is advisable, that this reference source has the same emissivity ϵ_r as the charge emissivity, ϵ_c ($\epsilon_r = \epsilon_c$). The difference signal of both pyrometers is:

$$s = \epsilon_c f(\vartheta_c) + (1 - \epsilon_c) f(\vartheta_w)\epsilon_w - (1 - \epsilon_r) f(\vartheta_w)\epsilon_w \tag{14.6}$$

When $\epsilon_r = \epsilon_c$ this difference signal becomes:

$$s = \epsilon_c f(\vartheta_c) \tag{14.7}$$

It is clear that the readings are free from any influence due to radiation from the wall.

4. A pyrometer method with additional thermocouple, described by Beynon (1981), which measures the average chamber temperature, is a cheaper modification of the two-pyrometer method. A special microprocessor system, which can handle the signals from two sources of different characteristics, has to be used to form a difference signal.
5. Ettwig (1986) applied a ratio-pyrometer with a polarizer for measuring the temperature in a slab-heating furnace. A ratio pyrometer, equipped with a polarizer, heat-protecting glass and quartz window, proved to be very useful. At visible light wavelengths with $\lambda < 0.6\,\mu m$, the optical properties of steel covered with scale in parallel polarized light at an angle of about 70° approaches $\epsilon \simeq 1$. Hence any measuring errors caused by reflected environmental radiation were eliminated. The application of a ratio-pyrometer eliminated all the errors caused by flames, gases or dirt-covered optics. Band-radiation pyrometers with polarizers, operating at visible light wavelengths, at temperatures over about 800 °C, also proved to be successful.
6. Pyrolaser pyrometer, produced by the Pyrometer Instrument Co., Inc. (Northvale, NJ, USA), presents quite a new method for coping with charge temperature measurement in a chamber whose walls are hotter than the charge. The emissivity of the charge is obtained using a laser beam directed to and reflected from the

charge, at the same location, temperature and wavelength as the infrared temperature measurement itself. Based on knowing the charge emissivity, and hence its reflectivity, and measuring the wall temperature with the same pyrometer, the true charge temperature can be determined. Of course, the necessary correction of charge emissivity is also considered. Measurements can be triggered manually, remotely or self-initiated. Pyrolaser, which is microprocessor based, can display and store up to 700 sets of data such as time, emissivity, maximum, minimum and average temperature.

Sighting window. In vacuum furnaces or in hermetic furnaces with protective atmospheres, all pyrometric measurements which must be made through glass or quartz windows, need to use pyrometers calibrated together with the window. Any dirt film on the window surface may cause additional errors.

Flames: In gas furnaces, when measurement of the charge temperature occurs through a flame containing CO, CO_2 and H_2O, the applied pyrometers should operate at wavelengths outside the absorption bands of these gases (§7.8).

Measurement of the internal charge temperatures of massive charges is very important in the estimation of the time necessary for through-heating. When the charge is hollow a temperature sensor, mostly a thermocouple, is placed inside it. In most cases, however, the internal temperature can only be determined indirectly, in the following ways:

1. Charges with simple shapes like cylinders, spheres, or cubes which can be easily simulated, allow the internal temperatures to be calculated as described by Gröber *et al.* (1957) or determined by using a model (Endress, 1986; Michalski, 1986). It is necessary to linearize the material properties and to assume that their mean values are constant. Knowledge of the heat-transfer conditions between the walls of the furnace chamber and the charge is also required.
2. Determination of internal temperature fields, taking account of the temperature dependence of material properties and of the boundary conditions, is only possible using digital computers.
3. The sole feasible, non-invasive method of internal temperature measurement is by ultrasonic thermometers (§15.4).
4. A rough estimation of the time to heat through charges with higher thermal capacity can be done by measuring the mean heating power versus time, starting at the moment when the furnace is charged. Stabilization of the value of supplied power means that the through-heating process is finished.

14.1.3 External surface temperature

The temperature of the external surfaces of a furnace is measured to check the state and condition of its thermal insulation. This temperature which is below 150 °C in most cases, is measured by portable or stationary contact sensors at characteristic points of the surface. Low temperature pyrometers and thermovision (§15.1.4) are also used. The thermovision method is especially well adapted for the purpose as it immediately gives the overall picture of the surface temperature field. Its precision is about ± 0.1 °C.

14.2 CONTINUOUS FURNACES

The following methods are used for continuous temperature measurement of moving charges.

1. Drawn thermocouples are insulated, elastic and of about 0.5 to 2 mm diameter. As the measuring junction is fastened to the charge, the thermocouple, connected by elastic conductors to a stationary recorder, is drawn through the furnace by the moving charge. After passing along the whole length of the furnace the thermocouple, most conveniently a NiCr–NiAl type, is cut off. Unfortunately thermocouples of this type also break sometimes.
2. A thermally insulated storing device, moving along the furnace together with the charge, stores the measured temperature values, at preset time intervals. After passing the whole length of the furnace, the digitally stored data are read out and recorded. For example the device STOR from Ultrakust GmbH (Germany) (Figure 14.6), of diameter 100 mm and length 310 mm, which may have 1 to 6 measuring channels, also stores up to 256 data points. In the standard version STOR can operate with thermocouple or RTD sensors at ambient temperatures

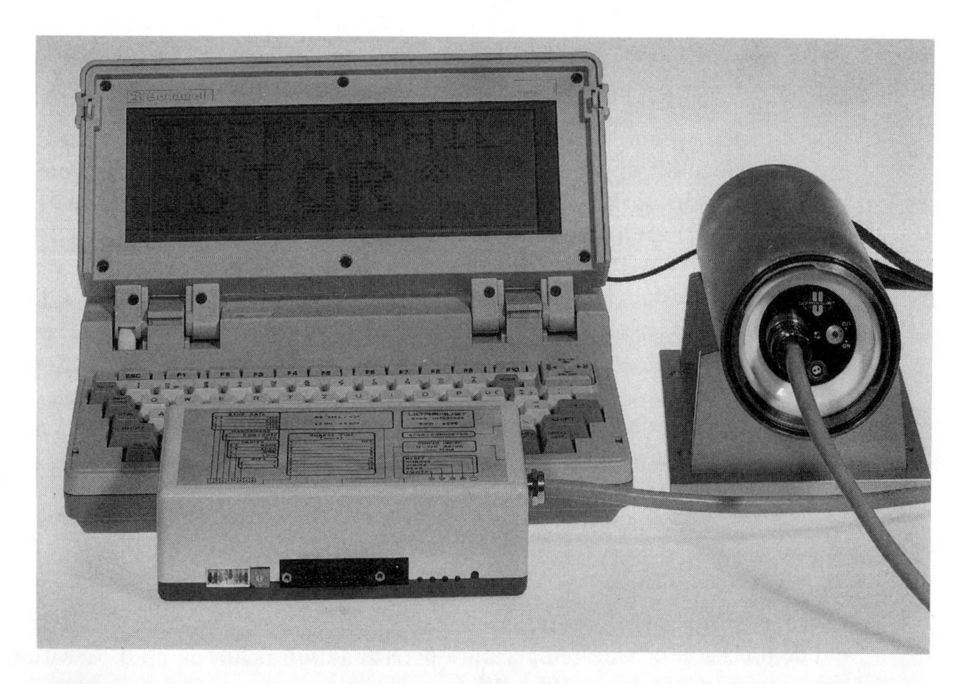

Figure 14.6
Storing device STOR with a data read-out computer. (Courtesy of Ultrakust Gerätebau GmbH, Germany.)

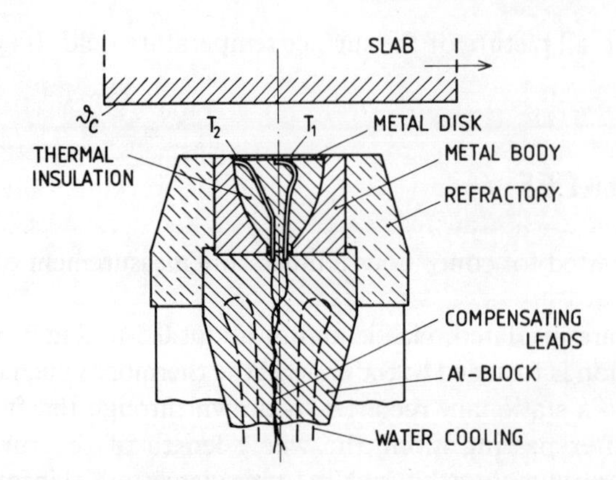

Figure 14.7
Heat-flux sensor for the temperature measurement of moving slabs.

up to 500 °C. It can remain up to 4 h at temperatures as high as 1350 °C, when operated inside an additional insulating housing of diameter 210 to 500 mm and length of 430 to 600 mm. A built-in NiCd battery ensures correct operation for times up to 10 h. Typically, the device is used in temperature recording in continuous furnaces in the metallurgical and glass industries, in drying processes and so on. Similar devices have been described by Tamura *et al.* (1982) and by Waters (1980).

For measuring the temperature of a moving charge at one given point along the length of a continuous furnace, there are two main methods. Pyrometers may be used, provided all the problems described in §14.1.2 are taken into consideration. Heat flux sensors of the type described by Ashcroft and Norris (1975) and Hornbaker and Rall (1972) are placed in the chamber, underneath moving slabs as shown in Figure 14.7. Two MI thermocouples are welded into a thin metal disk, one T_1 at its centre and the other, T_2, at its radius. The whole assembly is then welded around the circumference of the disk on to the end of a cylindrical metal body. Thermocouple compensating leads are led through a water-cooled aluminium block which is confined in a refractory outer cap. Experimentally calibrating the instrument, permits a determination of the charge temperature ϑ_c as a function of the temperatures, ϑ_1 and ϑ_2 respectively measured by T_1 and T_2 using the relation:

$$\vartheta_c = a\vartheta_1 + b(\vartheta_1 - \vartheta_2) + c\frac{d\vartheta_1}{dt} \tag{14.8}$$

where a, b and c are constants.

The precision of slab temperature determination is about ± 20 °C with a sensor time-constant of about 10 s. When each consecutive slab is loaded, the sensor is vibrated by a special device to clean its surface from scale, oxides and other impurities.

14.3 ELECTRIC SALT-BATH FURNACES

Salt-bath furnaces, which operate at temperatures up to 1300 °C, are used mostly for the heating of steel tools in hardening processes. Some special alloy steels require very precise temperature control of about ±2 °C. When the working temperatures are very high and also when the heat transfer between the molten slab and steel charge is very intense, sufficient precision may be achieved by only measuring the salt temperature. Continuous measurement is necessary.

Methods employed depend upon the specific conditions. For example, immersion thermocouples, of type K are used up to 1000 °C. Of type S and B, which are used up to 1300 °C, type B (Pt30Ph–Pt6Rh) is the most stable one. When there is a highly corrosive salt influence with high operating temperatures, deciding factors are a correct choice of the sheath material and sheath design, as well as proper application and use. Heat resisting alloy steels and mullit are the most commonly used sheath materials. Non-porous ceramic protection tubes are necessary to prevent contamination of the thermocouple which could lead to changes in the e.m.f. versus temperature characteristics. Metal sheaths should either be cast or drilled from full cylinders. Welded tubes are unsuitable because welds are not corrosion resistant. Before immersion they should be heated up to nearly normal operating temperature and then very slowly immersed, to prevent any cracking. If all of these precautions are taken, for example a drilled CrNi-steel sheath of diameter 27 mm could stand continuous operation at 1250 °C in a special hardening salt (91% BaCl, 2.5% MgF_2, Borax) for up to 7 to 10 days. To minimize the highly corrosive influence of salt vapours, all electrical equipment should be well protected or placed in another room.

Salt temperature can also be measured by total radiation pyrometers or photoelectric pyrometers. Although the equivalent emissivity of molten salt equals unity, most errors are caused by the slag layer on its surface which is always cooler than the salt itself. Immersing and withdrawing of charges also temporarily obscures the field of view of the pyrometer. Best results are obtained by applying pyrometers with peak-pickers.

In indirect measurements by pyrometer the instrument is directed inside a ceramic tube with a closed end, immersed in the molten salt. At $l/d \geq 6$ (§7.2) the interior of the tube can be regarded as a black body. Application of such a sighting tube, which limits the disposable bath volume to some extent, requires a certain air-pressure to be kept inside the ceramic tubes to prevent salt from diffusing into it. Sighting tubes equipped with a prism, which permits the pyrometer to be mounted horizontally, are also used.

14.4 GLASS TANK FURNACES

Characteristic points for temperature measurements in a glass tank, which are shown in Figure 14.8, are described in an application sheet of Land Pyrometer Ltd (undated). The temperatures of these points are either relevant for the technological process or for the worktime of the tank lining. Pyrometers and thermocouples may both be applied. Thermocouples, which are used for the temperature measurement of molten glass, at temperature up to 1300 °C, may be either Pt30Rh–Pt6Rh or more rarely

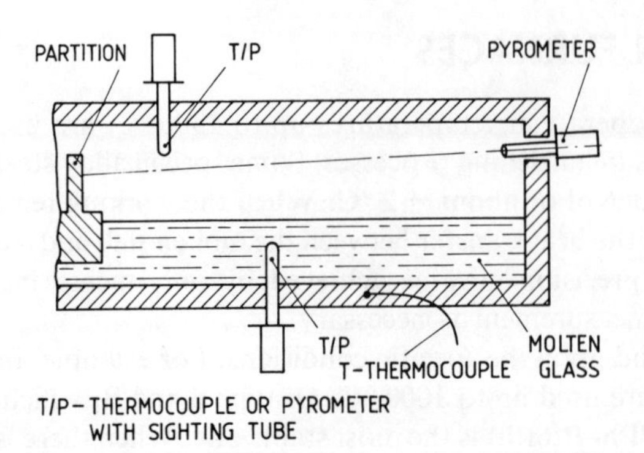

Figure 14.8
Temperature measurement in a glass tank. (Courtesy of Land Infrared Ltd, UK.)

Pt10Rh–Pt types. Protecting tubes which should resist corrosion and have sufficient mechanical strength, should neither colour nor pollute the molten glass in any way. Greenberg (1975) asserts that molybdenum or platinum are most commonly used. The thermocouples are mounted through holes in the bottom. MI thermocouples, when also used, are placed in permanently mounted Al_2O_3 protecting tubes, permitting easy exchange of the thermocouple from the outside (Greenberg, 1975). The application of photoelectric pyrometers, aiming at the inside of Al_2O_3 sighting tubes, mounted from below through the bottom and protected outside by the molybdenum tubes, is a competitive solution. Water-cooling of the pyrometer housing is usual.

The temperature of the furnace roof can be measured by sheathed thermocouples inserted from above. As in the given operating conditions they rapidly change their thermoelectric characteristics, measuring errors can amount to about 70 °C after only one month of operation. A certain extension of their lifetime can be achieved by maintaining nitrogen at high-pressure inside the sheath. Total radiation or photoelectric pyrometers give better performance. They are directed inside a closed end ceramic sighting tube, inserted through the roof for a length of about three times diameter. Gas diffusion into the tube must be prevented by maintaining an air-overpressure inside it. The best method of measuring the roof temperature uses a pyrometer which is directed at the roof through a hole in one of the walls. A proper choice of the spectral response of the pyrometer can only be made after the presence of flames in the furnace has been considered.

Early detection of places, where the wall lining may fail, requires periodic checking of the temperature of the outside wall surface. Portable contact thermometers, pyrometers, thermovision and sometimes thermocouples placed in the wall-lining during bricklaying (Figure 14.8) might be used for this purpose.

For measurement of the temperature of furnace partitions, photoelectric pyrometers of high distance ratio are used. They have to operate at effective wavelength outside the absorption bands of CO_2 and air vapour.

For spot measurements inside the furnace the disappearing filament pyrometer is used.

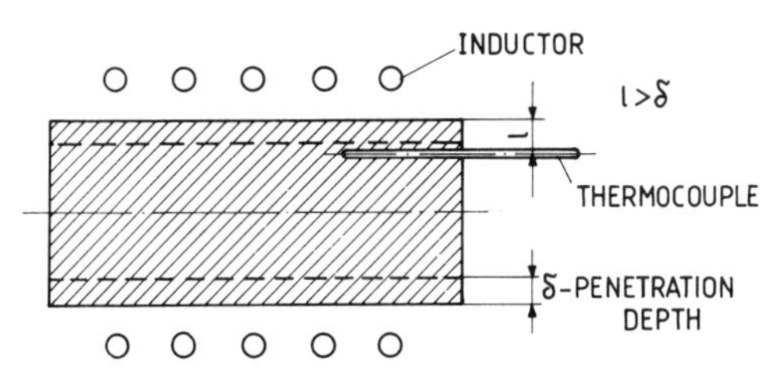

Figure 14.9
Thermocouple sensor in an induction heated charge.

14.5 INDUCTION HEATED CHARGES

In induction heating, where metallic or semiconductor charges are placed in an alternating magnetic field with a frequency ranging from 50 Hz to about 1 MHz, the induced a.c. currents cause charge heating. Electromagnetic waves penetrating the charge are damped, with heating confined mainly to a depth δ, called the penetration depth. This depth, which depends on frequency and material properties, is given by:

$$\delta = 5050\sqrt{\frac{\rho}{\mu_r f}} \quad \text{cm} \tag{14.9}$$

where ρ is the charge resistivity, in Ω cm, μ_r is the relative magnetic permeability of the charge, and f is the frequency in Hz.

In contact methods only thermocouples are used. The presence of alternating magnetic fields causes some additional problems due to induction heating of the thermocouples themselves, especially at medium (500 to 10 000 Hz) or high (60 kHz to 1 MHz) frequencies. Michalski and Eckersdorf (1981) assert that the use of thin thermocouple wires can prevent this effect. The specific power density in a thermocouple wire, when considered as a cylinder, is:

$$P_s = H^2\frac{\rho}{\delta}F_r \quad \text{W/cm}^2 \tag{14.10}$$

where H is the magnetic field strength in A/cm, ρ is the resistivity of thermocouple wires in Ω cm, δ is the penetration depth in cm given by Equation (14.9), and F_r is the shape factor of a cylinder as described in Anon. (1974).

When the wire diameter $d<0.5\delta$ and for $F_r \to 0$, parasitic thermocouple heating can be neglected. As platinum has the lowest specific resistivity of all the thermocouple materials, it also has the smallest penetration depth at a given frequency. Consequently platinum wires should have the smallest of all diameters. For example at 1000 °C the penetration depth of platinum is 3.6 cm at 8 kHz and 0.5 cm at 300 kHz. Evidently the condition $d<0.5\delta$ is easily met for all thermocouple materials, especially as $\mu_r = 1$ in most cases. Another way of avoiding parasitic thermocouple heating is to place them in the charge at a depth λ larger than the penetration depth δ as shown in Figure 14.9 (Michalski and Eckersdorf, 1981).

Parasitic e.m.fs, induced in a thermoelectric circuit, may cause both measuring errors and damage to the indicating instrument. To prevent these the following precautions can be taken:

- as mentioned above, place the thermocouple deeper than the penetration depth,
- shielding of the thermocouple, for example by using a MI thermocouple with isolated measuring junction,
- applying grounded shields along the compensating and connecting leads,
- applying transposed connecting leads,
- arranging thermocouples along the equipotential lines of the magnetic field,
- applying filters across the terminals of the measuring instrument to remove the a.c. component (RC-filter at 50 Hz and RL-filters at high frequency). Filters cause a slight increase in the inertia of the indications.

The charge, the shield and the measuring instrument essentially require a single point ground only at the shield of the thermocouple, irrespective of whether the charge is to be grounded or not.

Only pyrometers can measure the temperature of moving, induction heated charges. The main difficulty is the non-homogeneity of the charge surface which may be partially covered by scale or oxides. A good solution here, offered by Land Infrared Ltd (undated), uses scanning pyrometers with a peak-picker (§7.4.6). Sometimes, light-guide pyrometers are used, where the charge is sighted between the inductor windings.

14.6 CONTINUOUS, MOVING CHARGES

Wires and metal bands can be continuously heated using either direct resistance heating or induction heating. In direct resistance heating the electrical current flows along the charge between two rotating or sliding electrodes. Heat is developed in induction heating, by the alternating magnetic field in the charge itself.

The transverse movements and vibrations of wires of extremely small diameters require the application of special pyrometers with a rectangular field of view, where the charge covers a constant percentage of the field of view (Figure 14.10) irrespective

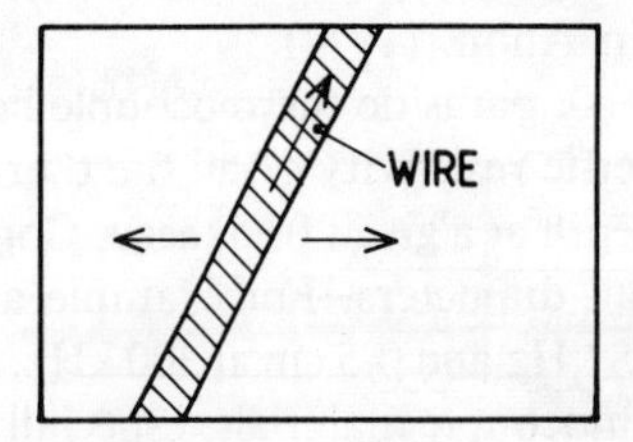

Figure 14.10
Rectangular field of view of a pyrometer for the temperature measurement of moving wires, exhibiting simultaneous transverse movements.

of its movements. Pyrometer calibration must take account of only partial coverage of the field of view. In any continuous production process the actual pyrometer readings can be a measure of the repeatability of production, even though they display emissivity errors. Some two-colour pyrometers give correct readings even at 5% coverage of the normal field of view (Ircon Inc., USA). No special shape of field of view is then needed.

Another method for measuring the temperature of moving wires, bands or fibres, described by Weichert *et al.* (1976), is based on the application of two identical pyrometers. In the background of the moving wire or band a flat heating element is heated up to a temperature equal to the measured value. One pyrometer measures the temperature of the wire or band and background together, while the second one only measures the background temperature. The measured temperature is determined when the readings of the pyrometers are identical. The heating element should have the same emissivity as the target with its temperature automatically controlled until both pyrometers exhibit the same readings. A similar method is also advised for the temperature measurement of moving fibres, even as thin as 0.05 mm (Weichert *et al.*, 1976).

Dr G. Maurer GmbH, FRG, produces a special scanning photoelectric pyrometer equipped with an oscillating optical system with a scanning frequency of 12 Hz and a peak-picker. Its typical applications are in the temperature measurement of moving and swinging wires and bands. The smallest target size has a diameter of 0.5 mm within the measuring range of 70 to 2000 °C.

14.7 DIELECTRIC HEATED CHARGES

Although it is easy to measure the surface temperature of dielectric heated charges using a pyrometer, it is often the internal temperature of the charge which is required. Internal charge temperatures, measured by the contact method, provide a true reference value for implementing or optimizing a technological process. The main difficulty with the insertion of a contact sensor is how to prevent any breakdown between the sensor and condenser plate (Figure 14.11). Thin MI thermocouples with isolated measuring junction (§4.3) are best suited for this purpose. The thermocouple sheath has to be grounded, with the measuring instrument protected by a filter against the high frequency voltage. A Fluoroptic thermometer (§15.3) in which no breakdown phenomena occur is also suitable for the temperature measurement of dielectric heated charges.

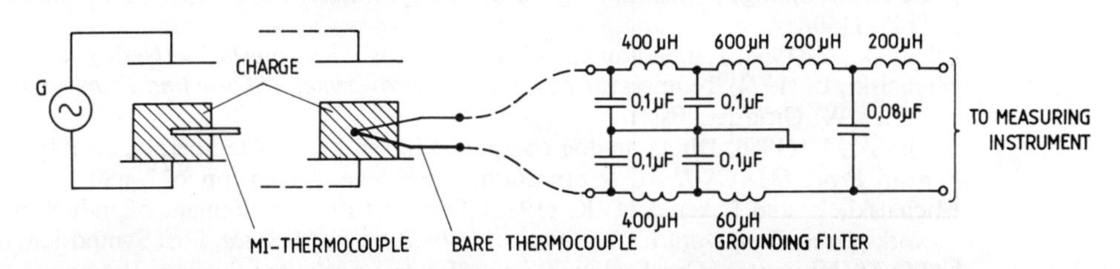

Figure 14.11
Temperature measurement of dielectric heated charge.

Very thin, bare thermocouple wires of 0.02 to 0.05 mm diameter are used for temperature measurement in medicine and biology. In the case of big differences of wire resistances, as occur in Cu–CuNi thermocouples ($\rho_{Cu} = 1.7 \times 10^{-8}$ Ωm, $\rho_{CuNi} = 49 \times 10^{-8}$ Ωm) capacitive leakage currents flow predominantly along the lower resistance wire. At the contact point of this thermocouple lead and the Cu conductor, any observed local overheating results in large measuring errors. A grounding filter, described by Chakraborty and Brezovich (1980) as shown in Figure 14.11, should be used.

REFERENCES

Anon. (1974) *Elektrowärme, Theorie und Praxis*, Verlag, W. Girardet, Essen.

Ashcroft, I. R. and Norris, P. A. (1975) Heat flux pyrometer. *Temperature Measurement: Conference Series, No. 26*, The Institute of Physics, London, 1975, pp. 321–328.

Beynon, T. G. R. (1981) Pyrometric measurement of slab temperature in a reheat furnace: the simplest two-thermometer approach. *Temperature Measurement in Industry and Science*, First Symposium of IMEKO TC-12 Committee, Czechoslovak Scientific and Technical Society, Prague.

Chakraborty, D. P. and Brezovich, I. A. (1980) A source of thermocouple error in radio-frequency electric fields. *Electronics Letters*, **16**(22), 853–854.

Dixon, J. (1987) Industrial radiation pyrometry. *Measurement and Control*, **20**(7), 11–16.

Emschermann, H. H., Fuhrman, B. and Huhnke, D. (1980) Anwendung der EFH-Messwertspeicherung zur Temperaturmessung in Durchlauföfen. *Draht*, **66**(6), 411–413.

Endress, H. (1986) Ungekühlte Elektroden in elektrischen Glasschmelzwannen. *Elektrowärme International*, **44**(B3), 123–131.

Ettwig, H. H. (1986) *Optische Messung der Realtemperatur im Hubherdofen*. Mannesmann Forschungsinstitut GmbH, Untersuchungsbericht 31/1986.

Greenberg, H. J. (1975) Measuring molten glass temperatures. *Instr. and Control Systems*, **48**(7), 19–24.

Gröber, H., Erk, S. and Grigull, U. (1957) *Die Grundgesätze der Wärmeübertragung*, Springer-Verlag, Berlin.

Hornbaker, D. R. and Rall, D. (1972) The convective null-heat-balance concept for non-contact temperature measurements of sheets, rolls, fibers and wire. *Temperature: Its Measurement and Control in Science and Industry*, Vol. 4, Part 1, Instrument Society of America, Pittsburgh, pp. 737–748.

Jakob, M. (1953) *Heat Transfer*, John Wiley and Sons, New York.

Land Pyrometer Ltd (undated) *Application Data Sheet 11, Temperature Measurement at the Glass Tank*.

Lieneweg, F. (1964) Fehler und Einflüsse bei der optischen Temperaturmessung mit Gesamtstrahlungs-, Teilstrahlungs- und Farbpyrometern. *Arch. f Eisenhüttenwesen*, **35**(12), 1145–1150.

Michalski, L. (1966) Temperatur eines Kammerofens. *Elektrotechniek* (Neth.), **44**(20), 466–471.

Michalski, L. (1974) Temperaturmessung. *Elektrowärme, Theorie und Praxis*, Chapt. III. 4, Verlag W. Girardet, Essen.

Michalski, L. (1986) Direct analog computers in synthesis of temperature control in electro-heat. *Proc. IMACS-IFAC Symposium*, Villeneuve d'Ascq, pp. 677–680.

Michalski, L. and Eckersdorf, K. (1981) Temperature measurement of induction hardened work-pieces. *Temperature Measurement in Industry and Science*, First Symposium of IMEKO TC-12 Committee, Czechoslovak Scientific and Technical Society, Prague.

Raudszus, G. and Zimermann, F. Z. (1986) Exakte Temperaturmessung die Voraussetzung für die Qualitatssicherung. *Elektrowärme International*, **44**(B6), 279–283.

Roney, J. E. (1982) Steel surface temperature measurement in industrial furnaces by compensation of reflected radiation errors. *Temperature: Its Measurement and Control in Science and Industry*, 6th International Symposium, Instrument Society of America, Washington, pp. 485–490.

Rosspeinter, M., Kolbe, E. and Ehrhardt, W. (1972) Temperaturmessung an induktiv erwärmten Werkstücken, *m.s.r*, **15**(2), 62–66.

Tamura, Y. *et al.* (1982) Temperature measurement of steel in the furnace. *Temperature: Its Measurement and Control in Science and Industry*, 6th International Symposium, Instrument Society of America, Washington, pp. 505–513.

Waters, W. A. (1980) Insulated portable instrument container. *RCA Technical Notes*, June, pp. 1–13.

Weichert, L. *et al.* (1976) *Temperaturmessung in der Technik-Grundlagen und Praxis*, Lexica-Verlag, Grafenau.

15 Chosen Methods and Problems

15.1 IMAGING OF TEMPERATURE FIELDS

At temperatures above about 650 °C, which corresponds to the lower limit of visible radiation, temperature values and temperature distribution can be estimated by the human eye or by taking colour photographs. It is also possible to apply those pyrometers already described. Another possibility, especially below 650 °C is the application of the methods presented below. Agerskans (1975) reviews thermal imaging.

15.1.1 *Photography and optoelectronic converters*

In these systems, the whole image of the temperature field appears simultaneously. Photography on infrared sensitive films applies special films, which are sensitive to the radiation of wavelengths up to 1.3 μm. The sensitivity of these films, which have to be stored at low temperatures, is far below that of the human eye with an upper application range of about 550 °C (Weichert *et al.*, 1976).

Optoelectronic image converters operate upon the principle of optoelectronic amplification and imaging electronics (Figure 15.1). The target image is formed by a lens on the photocathode, which emits a number of free electrons depending on radiation intensity. These are accelerated in an electric field, creating a visible image of the target on a luminescent screen. Modern image converters enable the handling of temperature fields up to a wavelength of 1.3 μm, corresponding to temperatures above 400 °C.

15.1.2 *Vidicon systems*

A vidicon system is a modification of a television camera in which the tube screen is covered by a photoconducting layer (Figure 15.2). This layer which is periodically scanned by an electron-beam, has a lens to concentrate the infrared radiation emitted by the target whose image is formed. The coil C produces the scanning movement of the electron beam. At those points irradiated by infrared radiation, the photoconducting layer starts conducting with the adoption of a potential which is positive relative to the cathode. At consecutive scanning movements, these points are

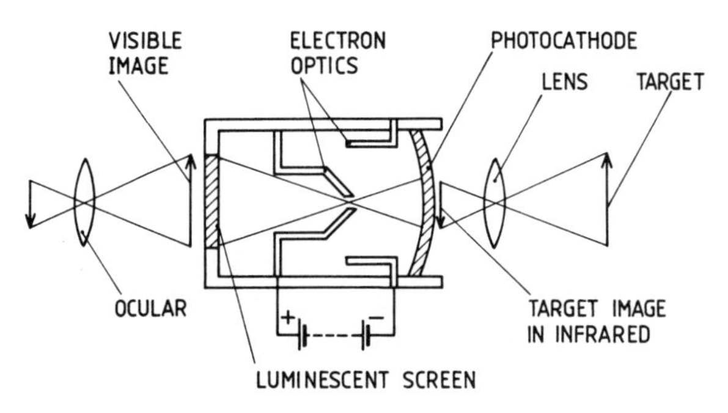

Figure 15.1
Optoelectronic image converter.

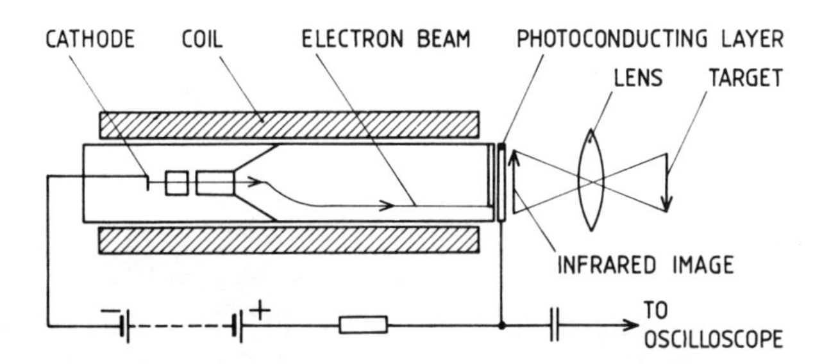

Figure 15.2
Vidicon system.

suddenly discharged. On the other side of the layer, impulses are generated which are transmitted to a video amplifier. In vidicon systems the charges are conveniently preserved during the time intervals between consecutive scannings, because the photoconducting layer is continuously irradiated by infrared. This phenomenon gives a very high system sensitivity. As their high spectral sensitivity enables temperature fields up to a wavelength of about 2 μm, corresponding to temperatures as low as 250 °C to be handled, materials, such as Pb_2O_3 or PbSO, are used as photoconductors. The investigated temperature field can be observed on an oscilloscope screen.

15.1.3 *Pyroelectric systems*

A pyroelectric system, in which the photoconducting layer is replaced by a pyroelectric crystal layer, is a specific modification of vidicon systems. The applied triglycine-sulphate crystals (TGS), allowing a spectral sensitivity range even up to 40 μm, are characterized by a temperature dependent polarization. Temperature variations, being a function of infrared irradiation, produce local surface charges. As for the vidicon system, the surface is scanned by an electron beam with the generated impulses transmitted to an amplifier. Although the pyroelectric layer, which is only sensitive to temperature variations, cannot cope with stationary temperature fields, modulation of incoming infrared radiation can eliminate this drawback. The system ensures a temperature resolution of about ± 0.5 °C.

15.1.4 Thermovision

In infrared cameras, which have opto-mechanical scanning systems, based on rotating or oscillating mirrors or prisms, all of the target surface points are scanned sequentially. The resulting series of signals is transformed in a detector into electrical signals. After amplification they are displayed on a monitor screen as a visible image of the temperature field.

A technical example of thermovision systems, which will now be presented, is the AGEMA Thermovision 870/880 Systems whose camera is illustrated in Figure 15.3.

Optical system. To avoid errors caused by absorption of the infrared by the atmosphere and by water vapour, the following wavelength bands are applied (Figure 7.21):

short waveband: 2 to 5.6 μm,
long waveband: 8 to 12 μm.

The lenses used have a viewing angle from 2.5 ° to 40 ° depending on target size and distance. Corresponding target surfaces are from 0.6×0.6 m to 6.4×6.4 m at a distance of 10 m.

Figure 15.3
Thermovision camera. (Courtesy of AGEMA Infrared Systems, UK.)

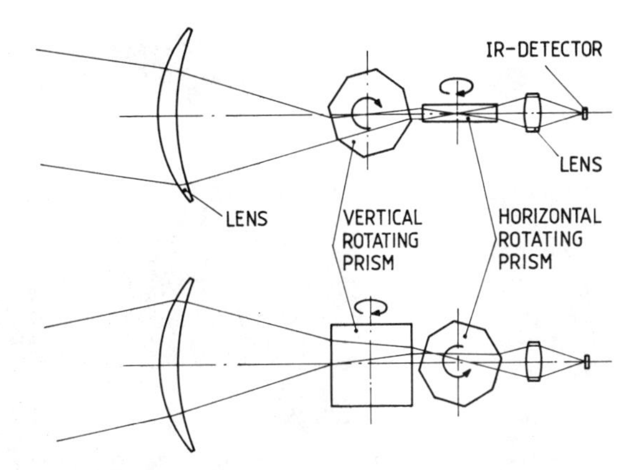

Figure 15.4
Thermovision scanning system. (Courtesy of AGEMA Infrared Systems, UK.)

A rotating mirror drum and an oscillating plane mirror are used in a scanning system (Figure 15.4). At a scanning frequency of 25 Hz a spatial resolution of 100 to 175 measuring elements per line is achievable. Detectors of InSb or $(HgTe)_x$ $(CdTe)_{1-x}$ which are cooled either thermoelectrically or by liquid nitrogen, ensure a temperature resolution of 0.1 °C in a measuring range of 30 °C. The measurement ranges are between − 20 and + 1500 °C having a sensitivity of 0.07 to 0.1 °C and accuracy of ±2% or ±2 °C. To achieve maximum measurement accuracy the Thermovision 870/880 instruments include a microprocessor controlled system in which two miniature reference sources are built into the scanner. By scanning these reference sources at every horizontal scan, 2500 times per second, the gain and level of the system can be precisely controlled.

The temperature field may be displayed on the screen, in one of the following ways:

- black, grey and white picture,
- black, grey and white picture with white marked isotherms adjustable within a selected thermal range,
- colour picture with up to 16 colour grades (Figure 15.5 illustrates a typical example),
- colour picture with black isotherm,
- black and white picture with colour isotherms.

A microprocessor based TIC-8000 system vastly extends the imaging possibilities for the evaluation of real-time measurements and of frozen images. Possible examples are: spot measuring, profiles, areas of special interest, statistics and histograms, digital temperature display, time–temperature plots, capturing of transient events, dissection of thermal events into microseconds, subtraction of images to detect changes over time or deviation from an expected ‘good’ thermal picture, digital magnification and so on. The software EQUAL enables calculation of the correct emissivity for each point of the image for objects with varying emissivities and to achieve a thermogram compensated for emissivity differences. A typical application images the temperature

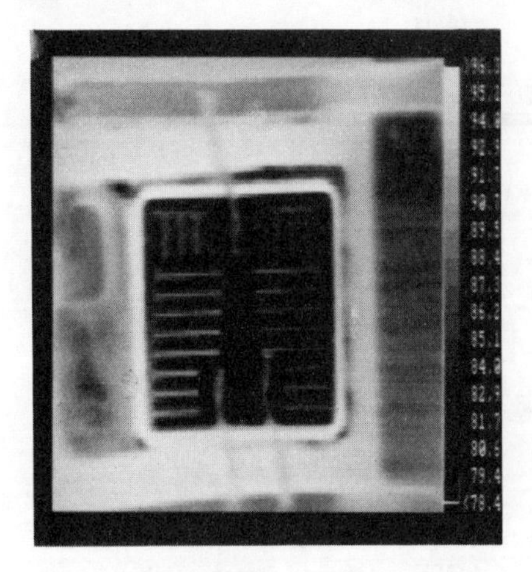

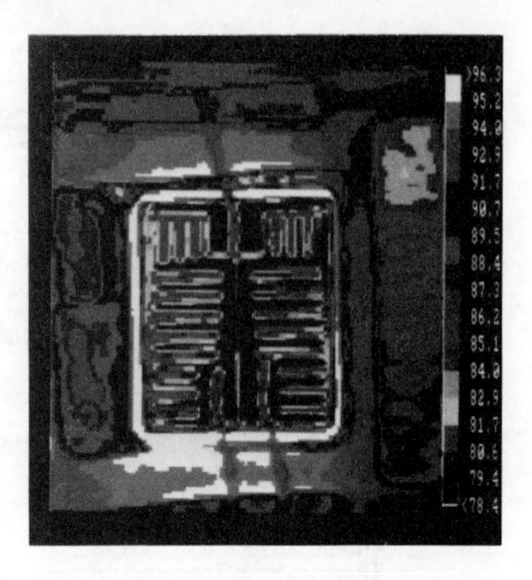

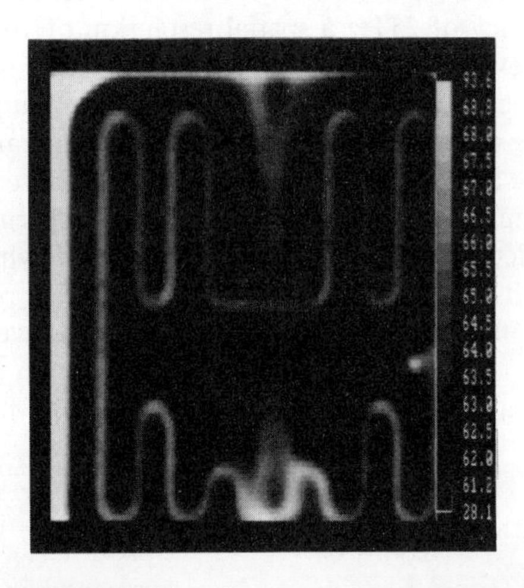

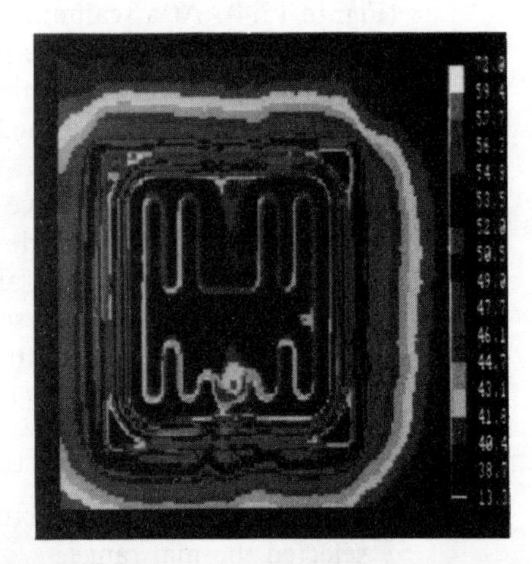

Figure 15.5
A typical thermograph from the Thermovision 870/880 system. (Courtesy of AGEMA Infrared Systems, UK.)

distribution over a printed circuit board. For example, the thermograms of a circuit board powered for up to 10 min and of the same board externally heated, form the basis for calculating and displaying a true thermogram with emissivity correction.

This system which is one of the most versatile types displaying temperature fields, covers the following typical applications:

- condition monitoring by identifying heat losses and tracing leaks in heating systems, buildings, furnaces and cooling systems,
- chemical installations, high voltage power systems and so on,

- quality assurance and process control, in automated manufacturing processes, which also require a system for rejection of any products exhibiting departure from an ideal reference, such as printed circuit boards in particular,
- research and development of acquisition and evaluation of any thermal data,
- medical diagnosis,
- monitoring in aviation and missile exploration,
- geology and meteorology,
- monitoring the temperature of rotating bodies.

A complete Thermovision 880 and TIC-8000 system is shown in Figure 15.6.

In the future, as observed by Weichert *et al.* (1976), the elimination of opto-mechanical scanning can be expected. Single detectors will be replaced by linear or matrixed arrays, which will make thermovision simpler and cheaper. Consequently Yamada *et al.* (1982) state that it will surely contribute to still larger development of the application range of the system. Thermovision systems undoubtedly represent a significant advance in temperature measurement.

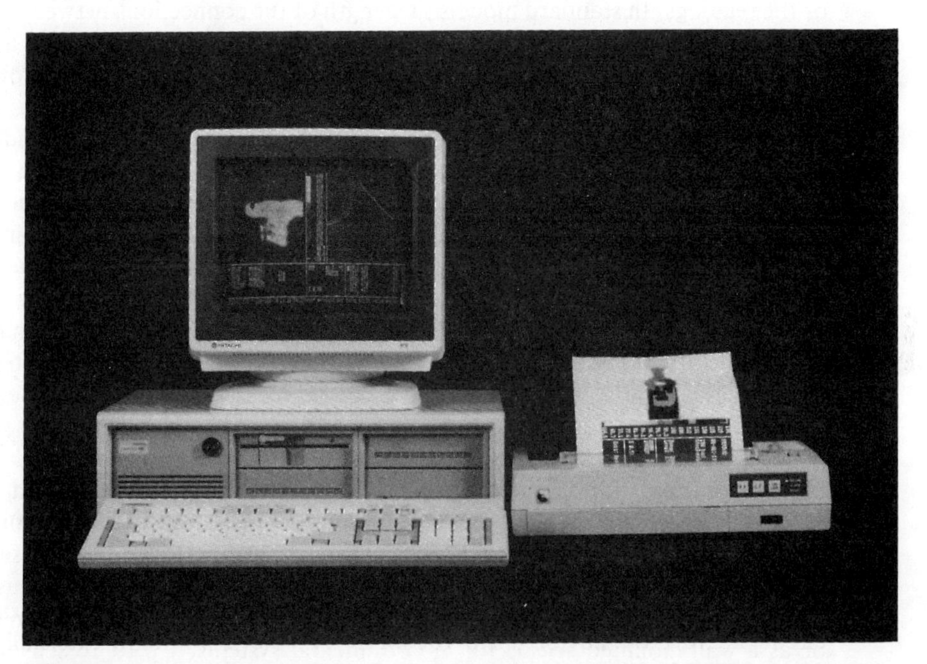

Figure 15.6
Complete system configuration of Thermovision 880 and TIC-8000. (Courtesy of AGEMA Infrared Systems, UK.)

15.2 QUARTZ THERMOMETERS

All of the temperature sensors described so far in this book give analogue output signals, which can be converted when necessary. The only temperature sensor directly delivering a frequency output signal, is the quartz thermometer (Schöltzel, 1970; Ziegler, 1987). This system is very promising for future applications according to Benjaminson and Rowland (1972) and Berger and Balko (1972). The resonant frequency of a quartz oscillator as a function of measured temperature is given by the relation

$$f(\vartheta) = f_0(1 + \alpha\vartheta + \beta\vartheta^2 + \gamma\vartheta^3) \tag{15.1}$$

where f_0 is the frequency at a reference temperature, in most cases 0 °C, and α, β and γ are coefficients.

By proper choice of the cutting plane of the crystal, the coefficients β and γ in Equation (15.1) can be made zero, so that the resonant frequency is a linear function of temperature. A quartz thermometer by Hewlett Packard (USA) (shown in Figure 15.7) is based on this principle. Inside the temperature sensor there is a small, polished gold-plated quartz disk of about 6.4 nm diameter. The resonant frequency of the plates which is about 28 MHz at the reference temperature, varies with temperature with a sensitivity of about 1000 Hz/K. This gives an overall resolution of measurement as high as 0.0001 °C, with excellent long-term stability of the readings. In standard models the length of the connections between the sensor and the indicating instrument is 3.7 m increasing to 60 m by applying a separate oscillator or up to 1300 m if additional signal amplifiers are used. A microprocessor based digital indicator ensures interchangeability of the sensors, using a PROM corrector circuit. The measuring instrument, which is designed for two independent sensors, also permitting the measurement of temperature difference, with a resolution set at 0.01, 0.001 or 0.0001 °C. Connection of the instrument to an analogue or digital recorder or a computer is accompanied by the inclusion of a self-checking device, which warns the operator if any error occurs. Periodical calibration of the thermometer is made at the ice-point, by setting the indicated value to zero. Measuring errors are ±0.075 °C in a temperature range from −80 to +250 °C and 0.045 °C in the range −50 to +150 °C. The achievable resolution, which is well above that of a Pt-resistance thermometer, can be improved to 0.000 001 °C by extending the impulse counting period to 100 s.

Typical applications of the thermometer are in the calibration and standardization of thermometers, in precision calorimetry, in distant measurement of sea- or river-temperatures, continuous long-lasting temperature measurements and so on. Quartz thermometers are free from any problems owing to lead resistance and noise pick-up.

A complete quartz measuring system QuaT is offered by Heraeus GmbH (Germany). Temperature sensors are delivered in a steel sheath of 6 mm diameter or as a wall-mounted sensor for environmental systems. Up to 16 sensors can be connected and multiplexed within 1 s. Commutation is performed by opto-couplers. As there is also the possibility of direct connection to a computer or printer, the sensor signals can be computer conditioned. Measuring errors are about ±0.1 °C in a temperature range of −20 to +130 °C and ±0.2 °C within −40 to +300 °C, with a distance between the sensor and stationary instrument up to 1000 m. There are also portable, battery-operated units with one or two sensors, connected by a 200 m long cable (Figure 15.8). Indications of all sensors can also be read by a wireless system. Each QuaT sensor has LED indicators continuously displaying temperature values,

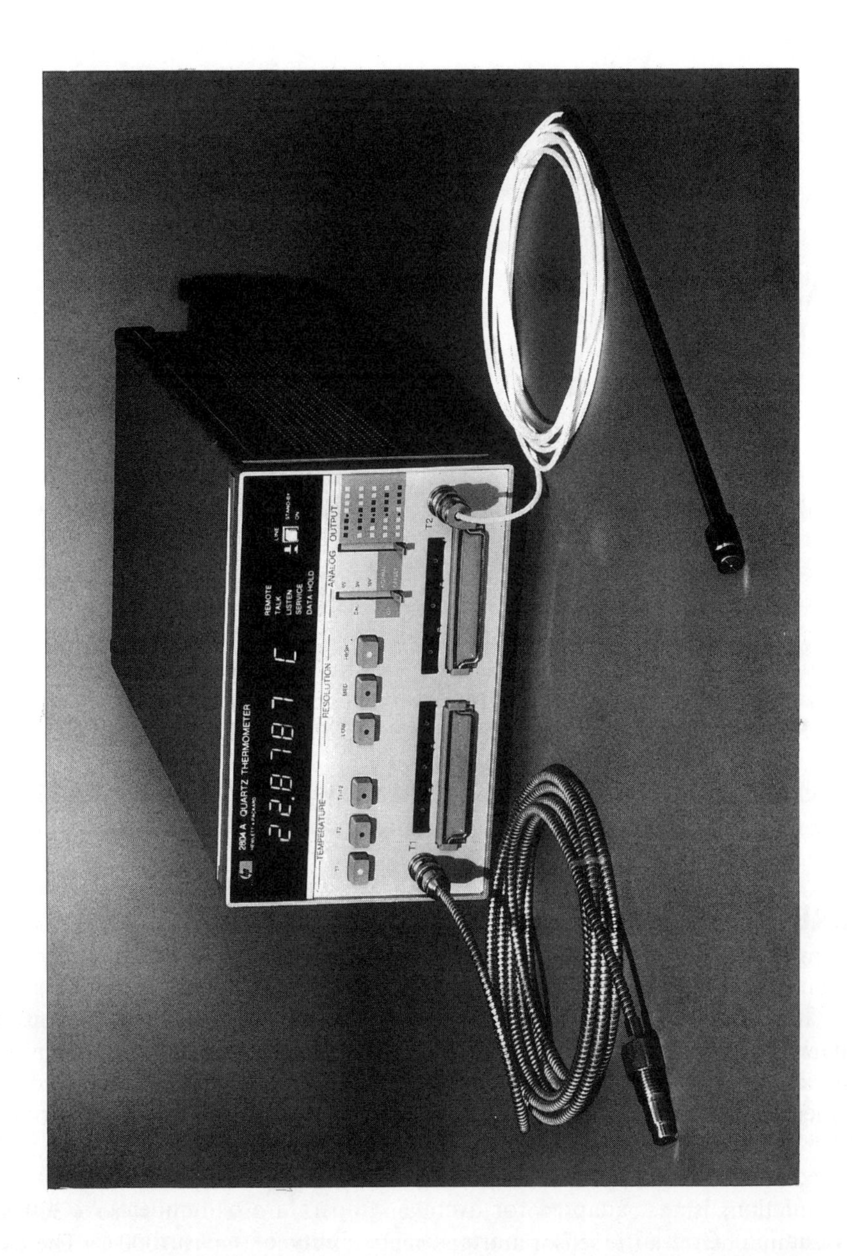

Figure 15.7
Quartz Thermometer with two probes. (Courtesy of Hewlett-Packard, USA.)

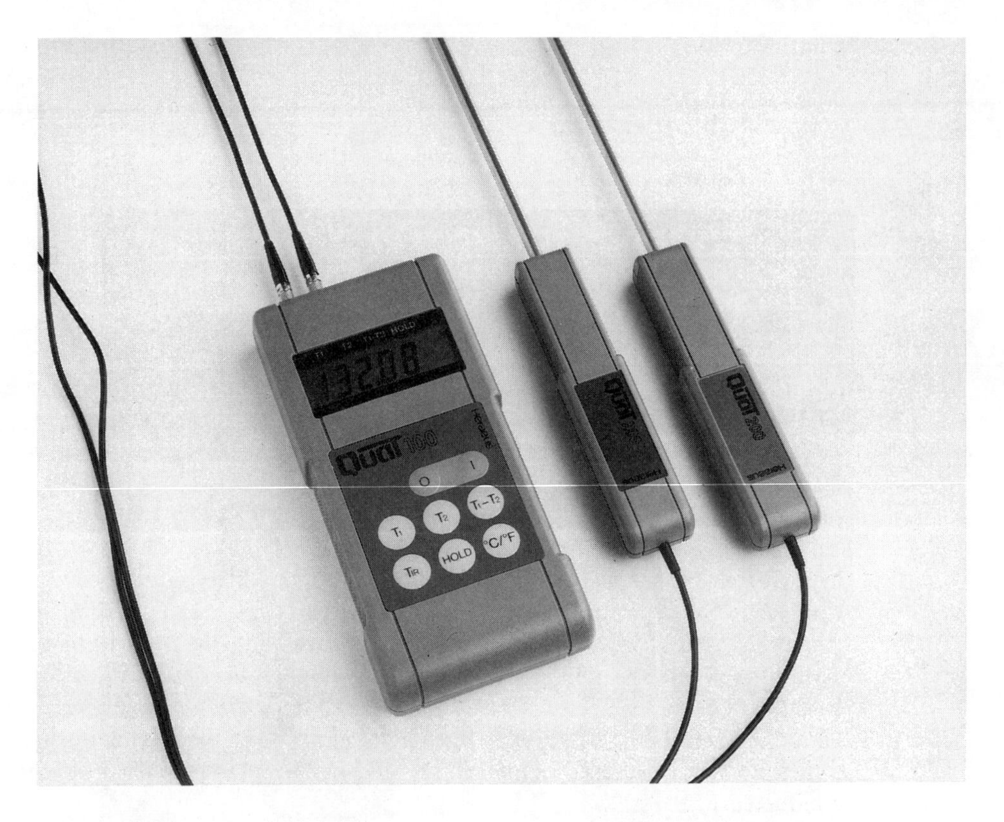

Figure 15.8
Quartz measuring system QuaT with two handheld probes. (Courtesy of Heraeus GmbH, Germany.)

which can be read, from a distance of 30 cm, by a QuaT 100 indicating instrument. The QuaT system is intended for use in an industrial environment.

15.3 FIBRE OPTIC THERMOMETERS

The rapid development of different applications of optical fibre techniques, discussed by Grattan (1987), also includes new applications in temperature measurement. Besides the well-known application as light-guides in pyrometry described in Chapter 7, the number of new thermometers based on fibre optic applications, may be divided as either temperature sensors, making direct use of optical fibre properties, or as optical fibre signal conductors from or to the sensor. *Direct use* of optical fibres, which is based on the Raman effect of quartz fibres, observed as temperature dependent light modulation, is best adapted for average temperature or temperature distribution evaluation. Grattan (1987) reports the possibility of measuring the temperature distribution at 200 sites along a distance of 1 km. Special Eu^{3+} and Nd^{3+} doped silica fibres based upon the differential spectral absorption characteristics, are also used. In Nd^{3+} doped fibres an absorption band centred at 0.86 μm decreases in

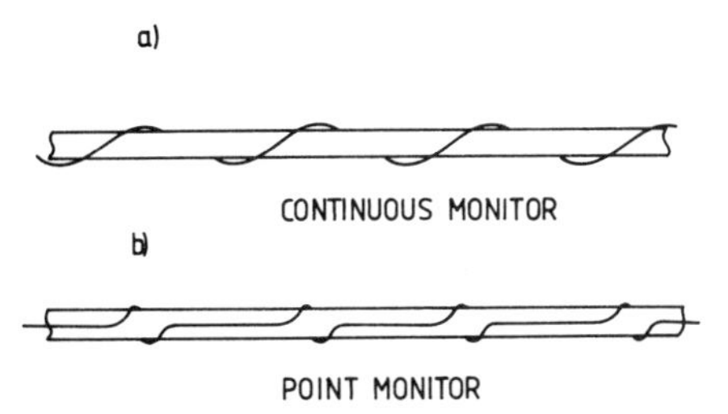

Figure 15.9
Fibre optic temperature monitoring.

absorption intensity, while a band centred at 0.84 μ increases in absorption intensity. The ratio of both absorption intensities is a function of temperature.

There are also other methods of direct use of optical fibres as temperature sensors. For example, Gottlieb and Brandt (1979) describe the temperature induced change of refractive index of optical fibre for measuring the average temperature along its length. Temperature dependent optical losses at bends in the fibre can guarantee an average temperature value along a uniformly wound fibre (Figure 15.9(a)). The fibre can also be bent lightly over short regions connected by straight sections (Figure 15.9(b)). Since the sensitive points are connected by low loss straight sections, a very long length can be used. Another possibility, also described by Gottlieb and Brandt (1979), is to use fibres with a bend radius changing with temperature by using various thermomechanical effects, such as a bimetal or others. Marcuse (1971) outlines a method which uses the temperature dependent optical coupling between parallel conducted light guides.

Typical uses of fibre optics in direct sensors cover monitoring, or measurement of average temperature of large surfaces or long objects such as pipelines (Sandberg and Haile, 1987), measurement in remote locations, in the presence of strong electromagnetic fields, at high voltage or in hazardous environments. Fibre optics act here as intrinsic, distributed parameter temperature sensors.

Indirect use of fibre optics technique covers all cases in which optical fibres carry light to and from the detector element, placed at the fibre end. This is sometimes referred to as extrinsic sensing. The detector itself can use thermochromic materials such as cobalt chloride solution in water/alcohol (Brenci *et al.*, 1984). This substance, which shows intensive colour change between 25 °C and 75 °C, has a spectral absorption with temperature given in Figure 15.10. In the design principle of the system, shown in Figure 15.11, light from the halogen lamp is chopped and transmitted by a silica optical fibre to the small probe containing the thermochromic solution. Incoming light is reflected from the mirrored probe bottom to another optical fibre bifurcated in two filtered light beams at $\lambda_1 = 0.655\ \mu m$ and $\lambda_2 = 0.800\ \mu m$. The light at $\lambda = 0.655\ \mu m$ is amplitude modulated by the temperature while the light beam at $\lambda = 0.800\ \mu m$, which is practically temperature independent (Figure 15.10), serves as a reference value. Both signals are detected by photodiodes, amplified and a.c./d.c. converted. Subsequently their ratio, which is formed in a microprocessor, gives a measure of temperature. The readings are independent of source fluctuations and of transmission losses because both signals are transmitted along one common optical

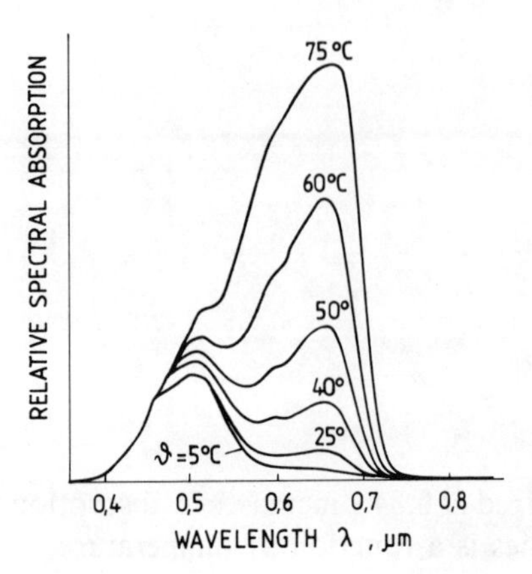

Figure 15.10
Spectral-absorption of cobalt chloride solution in water/alcohol at different temperatures ϑ (Brenci *et al.*, 1984).

path. The diameter of the probe is 1.5 mm and its length 10 mm with a measurement accuracy of about ± 0.2 to 0.4 °C.

Another indirect fibre optic instrument is the fluorescent thermometer like the Fluoroptic by Luxtron (USA) (Grattan, 1987; Luxtron, undated). The thermometer is based on measuring the ratio of emitted fluorescent radiation of a rare earth phosphor sample at two wavelengths. This visible radiation is enhanced by ultra-violet radiation. The measured ratio of emitted visible light is an explicit and repeatable function of the temperature of the phosphor sample which is placed at the end of an elastic light guide of diameter 0.7 mm. This light guide, 2 to 15 m long, is covered by a protecting layer. Phosphor ultra-violet excitation occurs in the same light guide through which the emitted visible light is received. Although typical measurement ranges are 20 to 200 °C or 20 to 240 °C a special range, with a low limit of -50 °C, is also offered. The accuracy is about 1 to 2 °C. Measured temperatures which are read on a digital instrument, are updated every 0.25, 1, or 4 s, with an analogue output of sensitivity about 10 mV/°C.

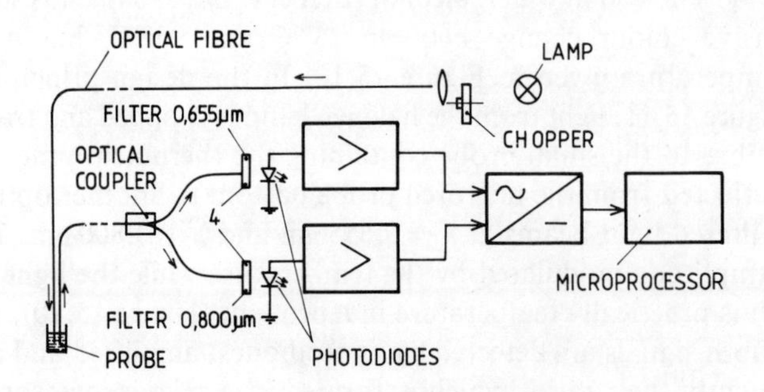

Figure 15.11
Principle of the thermochromic, optical fibre thermometer.

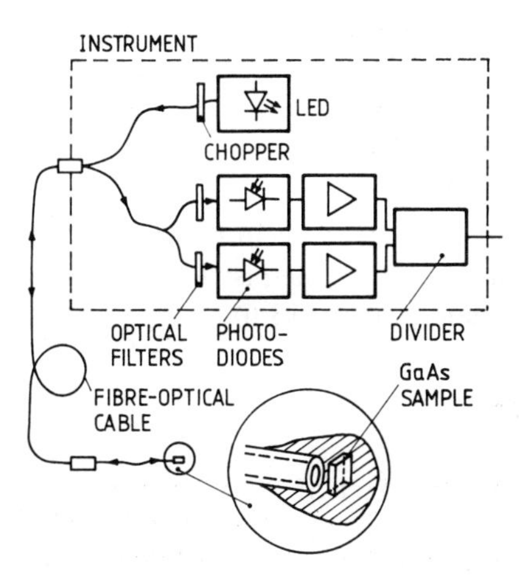

Figure 15.12
Optoelectronic arrangement of GaAs, fibre optic thermometer.

A similar thermometer, offered by ABB (Sweden), uses GaAs as a fluorescent material. Its operating principle, which is shown in Figure 15.12, irradiates the GaAs sample by the modulated light from an LED. The measured temperature value is obtained from the ratio of the emitted visible light at two chosen wavelengths. Grattan (1987) states that the readings are independent of the radiation intensity.

There are also fluorescent thermometers based on the decay-time concept, of a periodically excited probe. Different fluorescent materials used are ZnCdS and ZnSe among others. Figure 15.13 shows the use of either the half-value or one-fourth-value times. In the Fluoroptic 750 thermometer from Luxtron (USA) magnesium fluorogermanate, irradiated by ultra-violet light is used. This thermometer, which is used in the temperature range from -30 to $+300$ °C, has a decay time of some milliseconds. A similar thermometer, which is built by Degussa GmbH (Germany), is described by Fehrenbach (1989). Compared with other electric thermometers,

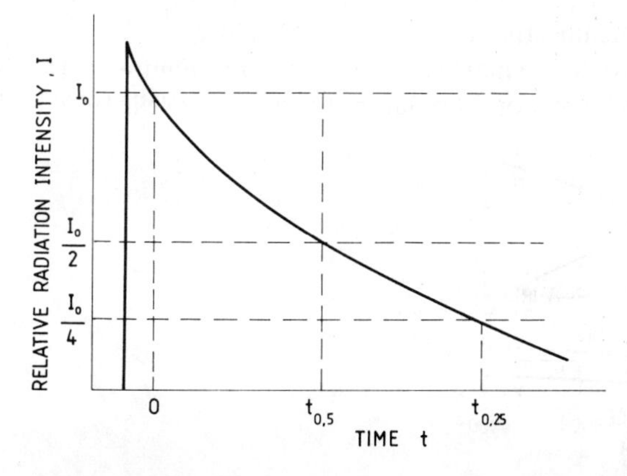

Figure 15.13
Decay times (e.g. $t_{0.5}$ or $t_{0.25}$) of periodically excited fluorescent probe.

fluorescent thermometers have similar advantages to fibre optic thermometers. The most important of these are its applicability in the presence of strong electromagnetic fields, such as in induction or dielectric heating, with a sensor, which is immune to any break-downs or short-circuits. The sensor also has extremely small dimensions, low heat conductivity and low thermal inertia.

Typical applications of fluorescent thermometers occur in medicine, biology, high voltage systems, dielectric or microwave heating, chemistry and physics and in the temperature measurement of rotating bodies (see §13.6).

15.4 ULTRASONIC THERMOMETERS

Ultrasonic thermometers are based on the effect of the temperature of a medium upon the velocity of sound waves in the medium, as formulated by Newton in 1687. Realistic development of ultrasonic thermometry, has only taken place in the last twenty years (Bell, 1972; Dadd, 1983). Lynnworth and Carnevale (1972) give the following formulae.

In gases:

$$v = \sqrt{cRT/M} \tag{15.2}$$

In liquids:

$$v = \sqrt{\beta(T)/\rho(T)} \tag{15.3}$$

In solids:

$$v = \sqrt{M/\rho(T)} \tag{15.4}$$

where c is the ratio of specific heats c_p/c_v, R is the gas constant, T is the temperature in kelvins, M is the molecular weight, β is the volumetric modulus of elasticity, and ρ is the specific density.

As $\beta(T)$ and $\rho(T)$ are not precisely known, ultrasonic thermometers have to be calibrated in gases approaching the ideal one, which has $c = 5/3$.

The sound velocity, v, for some materials is presented in Figure 15.14. Use of the formulae given above, which are most popular in the resonant (Bell, 1972) and impulse

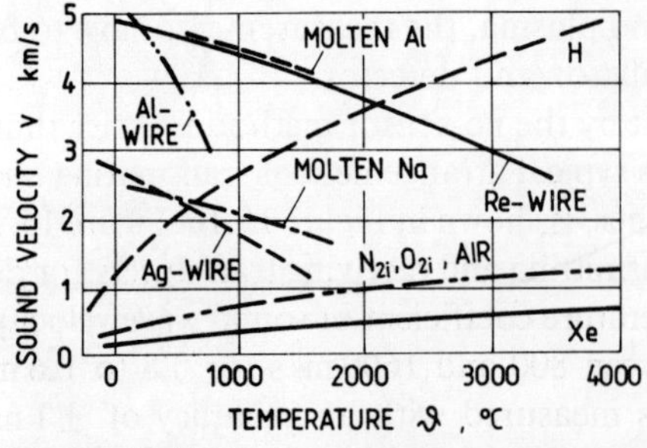

Figure 15.14
Sound velocity in some materials versus temperature.

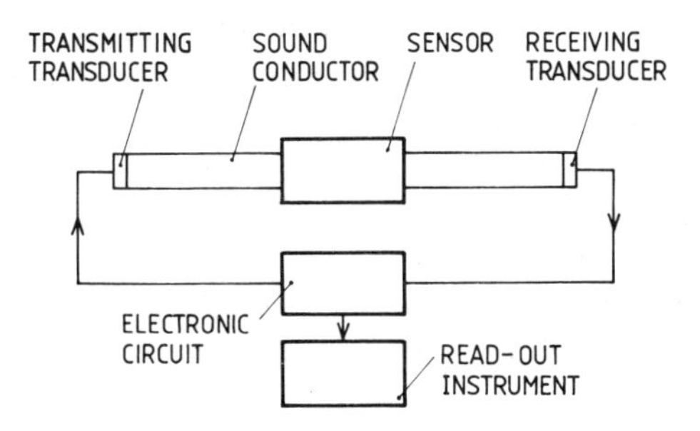

Figure 15.15
Ultrasonic thermometer.

(Green, 1986) systems described below, apply in the basic diagram of an ultrasonic thermometer shown in Figure 15.15. Oscillations of 0.1 to 3 MHz, which are generated in an electronic circuit, are transformed into acoustic waves in a piezoelectric or magnetostrictive transmitting transducer before transmission to the sensors by a sound conductor. This sensor is placed in the medium whose temperature is to be measured. The sonic signals, which can also be transmitted directly to the investigated medium, arrive at the receiving transducer with a certain medium temperature dependent lag. Measurement of this lag is performed by an electronic circuit and displayed on a read-out instrument. Single signals, or a series of signals in the case of high noise level are applied.

A wire or rod immersed in the investigated medium or even, what is very important, the investigated medium itself can also be applied as a sensor. Rods, inserted from one side which are connected on the other side to the transmitting transducer, are also used. These transmitting transducers also serve as receiving transducers which receive signals reflected from the end of the rod. The time difference between the two signals is used to determine the temperature. For a given sensor, the indicating instrument can be calibrated directly in temperature degrees.

Gas temperature measurement, by the no-sensor method is, to some degree, hindered by signal attenuation in gases and also by gas turbulence. Nevertheless, the very high upper application limit of the method in the range 300 to 17 000 K enables measurement of the temperature of ionized gases and plasma, which is not feasible by any other method. Green (1986) asserts that, for the determination of the temperature distribution in gases and plasma, the measurements have to be repeated for different arrangements of oscillator and detector.

Liquid temperature measurement by the no-sensor method is easier than in gases, owing to lower signal attenuation. A typical arrangement of transmitting and receiving transducers outside the liquid container, is shown in Figure 15.16. Figure 15.17 presents some different arrangements of transmitting and receiving transducers for determining temperature distribution. The temperature coefficient of sound wave velocity in liquids is about 4 m/s K at velocities between 800 and 1600 m/s (or 0.8 to 1.6 mm/μs). If the transit time of sound waves is measured with an accuracy of ± 1 ns, along a distance of 1 m, the thermometer resolution is about 1 mK. Over a distance of 10 mm the thermometer resolution would be about 0.1 K.

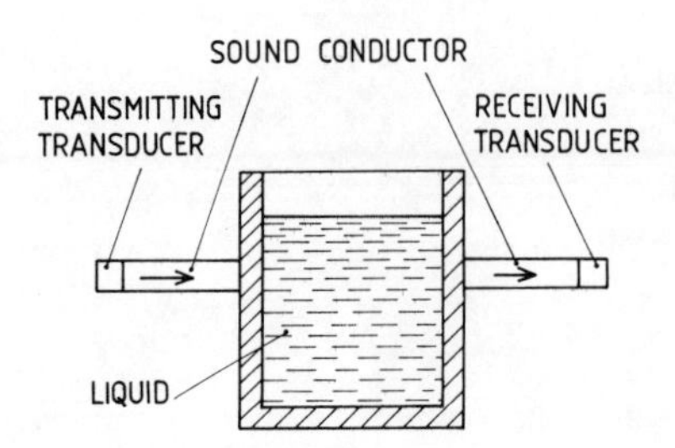

Figure 15.16
Ultrasonic temperature measurement in a liquid container, by the no-sensor method.

The velocity of sound waves in molten metals is about 5000 m/s with a corresponding temperature coefficient of about -0.25 m/s K. This coefficient for plastics is about -3 m/s K. Detailed data of the sound velocity in different materials, at different temperatures, are given by Lynnworth and Carnevale (1972).

Additional measuring errors can be caused by gas bubbles, impurities, non-regular container geometries and the existence of deposits on the walls of containers and on stirrers.

Solid body temperature measurement is done either by the no-sensor method or using wire, rod, band or tube sensors, in thermal equilibrium with the investigated body. The ultrasonic method allows new possibilities for measuring the average temperature of bulky solid bodies, using the body itself as the temperature sensor. A solid body can be regarded as bulky when its characteristic dimension is greater than 5λ, where λ is the wavelength of the applied ultrasonic oscillations. Readings can also be taken during short ($t < 0.1$ s) contact periods with the body under measurement. The ultrasonic method is the sole non-aggressive technique for average temperature measurement of bulky solid bodies. Such measurements are needed to check the degree of through-heating of solid bodies. Temperature distribution inside bulky solid bodies can be determined using a wire-sensor which has a number of equally spaced indents along its length (Figure 15.18). For example, along a 1-m long

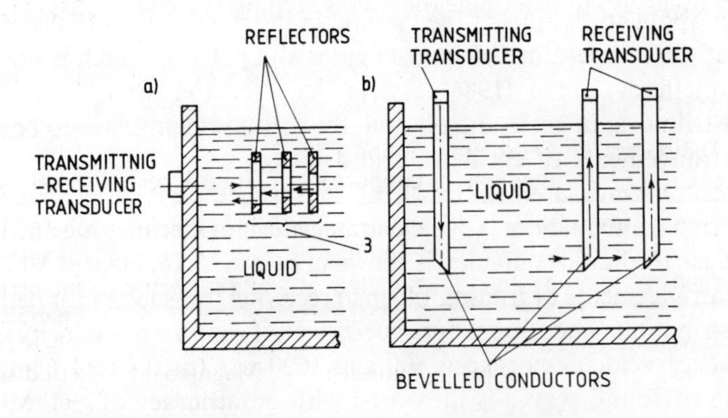

Figure 15.17
Ultrasonic determination of the temperature distribution in a liquid container.

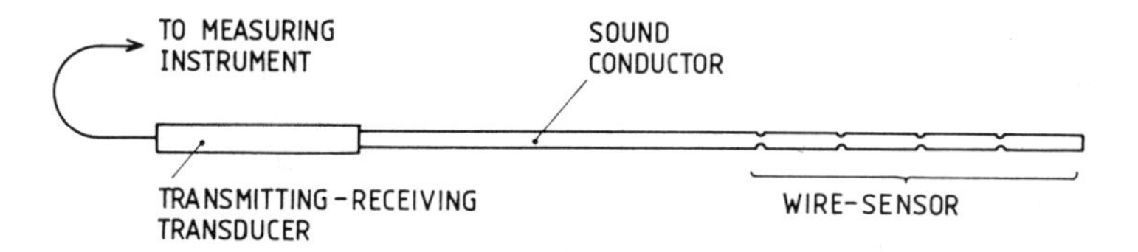

Figure 15.18
Wire sensor of an ultrasonic thermometer.

wire sensor, there are indents every 100 mm, which locally diminish the wire cross-section. Measuring the time lags of the sound wave reflected from these consecutive indents allows the temperature distribution along the wire to be measured. The main advantage of wire-sensors and similar ones is the wide choice of possible sensor materials such as aluminium up to 600 °C, stainless steel to 1100 °C, sapphire to 1700 °C, molybdenum to 2100 °C and tungsten rhenium alloys to 2700 °C. No high-temperature electrical insulators are needed. Instead of wires, monolithic ribbons and tubes can also be used.

Table 15.1 (Lynnworth and Carnevale, 1972) presents some chosen examples of the implementation of ultrasonic thermometers. Typical applications of ultrasonic thermometers are concerned with temperature measurements:

- over 2000 K,
- in nuclear reactors,
- in tanks or inside solid bodies,
- in furnaces, in which no sensors can be introduced,
- in the presence of high frequency electromagnetic fields using non-metallic sensors.

15.5 AVERAGE VALUE MEASUREMENT

In the case of non-uniform temperature distribution on a surface or inside a solid body it is often necessary to determine the average temperature value. The most direct,

Table 15.1
Ultrasonic thermometers.

Measured object	Temperature (°C)	Sensor or medium	Sound conductor	Transducer	Signal
Gas	1500–15 000	N_2	Quartz	Piezoelectric	One period 1 MHz
Liquid	1000	Molten Na	Stainless steel	Piezoelectric	One period or period series 3–10 MHz
Solids with holes	3000	Re-wire	Tungsten	Magneto-strictive	Period series 0.1 MHz
Solids with-out holes	1500	Steel wire	Steel	Piezoelectric	Period series 1 MHz

but not the easiest way, would be to calculate an average value of n particular measurements. There are also methods to measure an average value by direct means (§§15.3, 15.4).

Non-electric dilatation thermometers (§3.2) directly indicate an average value of the temperature along the sensor length which is mostly from 20 to 30 cm. Flat resistance thermometer detectors act in a similar way. Total radiation pyrometers and photoelectric pyrometers aimed at a surface of non-uniform temperature distribution also indicate a certain weighted mean value. This results from the fact that the radiant flux, emitted by particular surface points, is a nonlinear function of their relevant temperatures. Even if it is assumed that all surface points have the same emissivity (total, band or spectral emissivities, depending on applied pyrometer type), the parts of higher temperature will influence the readings in a way which is greater than in direct proportion (§7.8).

When n thermocouples are used for measuring average temperature values, they should be series connected with the same reference temperature ϑ_r (Figure 15.19). From this measured thermal e.m.f., ΣE_k, the average value, E_m, is calculated as:

$$E_m = \frac{\sum_{k=1}^{n} E_k}{n} \tag{15.5}$$

where E_k is the thermal e.m.f. of thermocouple no. k. The average temperature value can then be found from the thermocouple characteristics. If thermocouples of non-linear e.m.f. versus temperature characteristic are used, the calculated average temperature could be subject to important errors, especially if the particular temperature values differ substantially from one another.

It is also possible to measure an average temperature using n thermocouples connected in parallel (Figure 15.20). The average value of e.m.f. E_m then has the value:

$$E_m = \frac{\sum_{k=1}^{n} \frac{E_k}{R_k}}{\sum_{k=1}^{n} \frac{1}{R_k}} \tag{15.6}$$

where E_k is the thermal e.m.f. of thermocouple no. k and R_k is the resistance of measuring loop no. k including the thermocouple and its compensating and connecting leads.

Readings are correct, only if all measuring loops have identical resistances. If the temperature differences between the particular measuring points are not too large, the measured e.m.f., E_m, can be directly converted into a corresponding average temperature value. Alternatively this last value can be directly read from an indicating instrument calibrated in temperature degrees. When the particular loop resistances are different, those of lower resistance will influence the readings to a higher extent. For this reason the parallel connection is rarely used. All thermocouples should also have the same reference temperature in this method. Thermocouples with insulated measuring junctions, such as MI thermocouples, should be used for measuring the average temperature of metallic bodies, to avoid short-circuits.

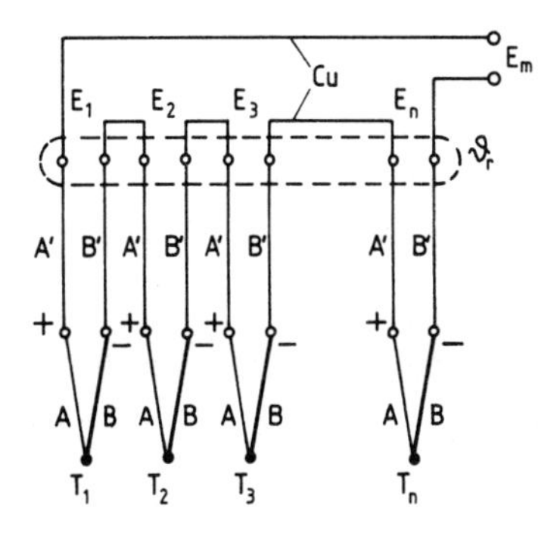

Figure 15.19
Average temperature measurement by series connected thermocouples.

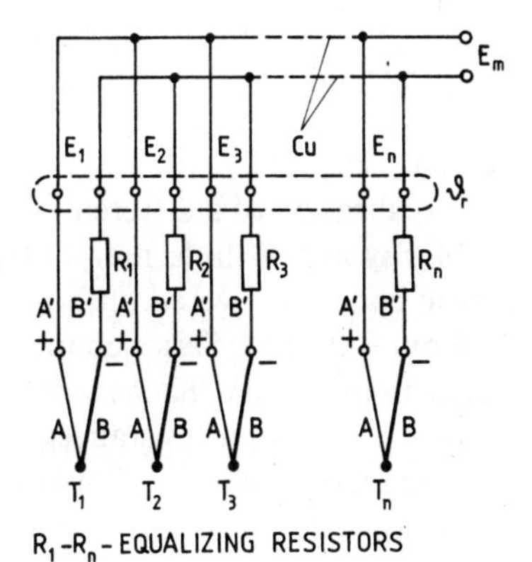

Figure 15.20
Average temperature measurement by parallel connected thermocouples.

Series connected resistance thermometer detectors, when applied for average temperature measurement, should have the same resistances of their connecting leads. However, this method is rarely used.

Applications of average temperature measurement are typical in outer surface temperature measurement of furnaces, kilns or buildings to estimate their heat losses.

15.6 MEASUREMENT OF SPATIAL TEMPERATURE DIFFERENCES

The simplest way of determining a temperature difference at two given points is to make two measurements and calculate their difference. If both readings are not taken at the same moment, errors may result when both temperatures change in time. This method is not very convenient as a continuous measurement of temperature differences is needed in most cases (Eggers, 1975). A thermocouple thermometer, which by its very nature measures a temperature difference between a measuring and a reference junction,

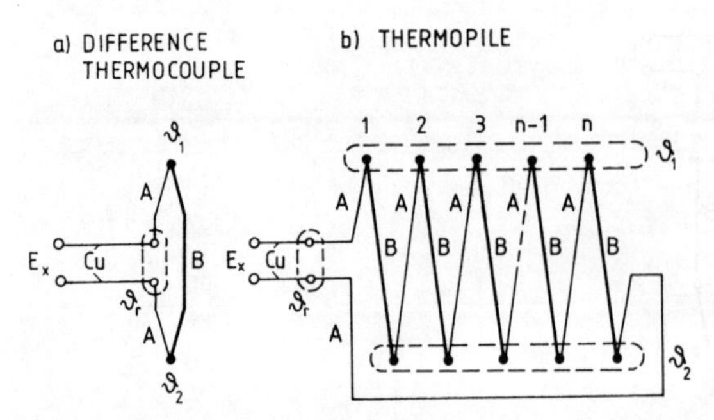

Figure 15.21
Measurement of temperature difference, $\Delta\vartheta = \vartheta_2 - \vartheta_1$, by thermocouples.

is best suited here. The simplest thermocouple circuit, whose indicating instrument is calibrated directly in difference values, $\Delta\vartheta = \vartheta_2 - \vartheta_1$, is shown in Figure 15.21(a). Its readings, $\Delta\vartheta$, are independent of the measured temperature level, only in the case when both thermocouples are identical and have linear e.m.f. versus temperature characteristics. To increase the measurement sensitivity thermopiles are used (Figure 15.21(b)). An example of the technical realization of a differential thermopile of very low thermal inertia is described by Muray and Wunderman (1972). A thin film thermopile, deposited on a Mylnar plastic band, is covered on both sides with a protecting thin Mylnar layer of thickness about 0.05 mm. Series connected Ni–Cu thermocouples, 34 in number, ensure a sensitivity of 0.5 mV/K. As some measuring errors may occur, caused by the noise level in the very high resistances of thin film thermopiles, Muray and Wunderman (1972) pointed out that they should be checked. This type of thermopile is also used as a heat flux detector.

Resistance thermometer detectors can also be used for difference temperature measurements (Clark, 1968). In early forms deflection type bridge circuits were exclusively used (Figure 15.22(a)). As their off-balance voltage depends upon the difference of resistances R_{T2} and R_{T4} it also depends upon the difference of both temperatures. To increase the sensitivity, double resistance bridges are used (Figure 15.22(b)) with two resistances placed at each temperature ϑ_1 and ϑ_2. All deflection type bridges are subject to errors depending on the average value of both temperatures. These errors can be compensated in a special compensating bridge circuit (Figure 15.23) which uses double resistance detectors, R_{T2} and R_{T4}. The resistances, R'_{T2} and R'_{T4}, correspond to the resistances, R_{T2} and R_{T4}, of simple difference bridges (Figure 15.22(a)) while the resistances, R''_{T2}, and R''_{T4}, in series measure the average temperature and influence the bridge current. The higher the average temperature value, the higher is the sum $(R''_{T2} + R''_{T4} + R_5)$. At constant supply voltage, V, the ratio of currents I_1 and I_2 changes and I_1 increases, compensating for a higher average temperature value. A similar solution has been proposed by Lieneweg (1975), while a special double bridge circuit for the simultaneous measurement of temperature difference and average temperature is described by Hammeke (1966). This bridge is used in thermal energy counters of running hot water.

Although the application of thermistors in temperature difference measurements is quite possible, Hajiev *et al.* (1972) state that care should be taken to ensure that both thermistors have identical characteristics.

a) SIMPLE BRIDGE

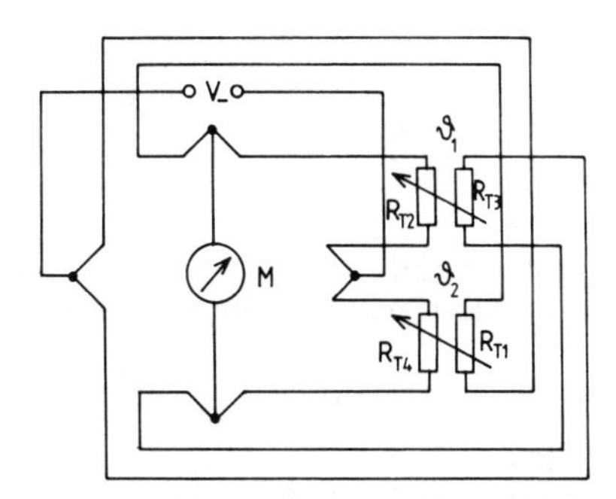

Figure 15.22
Measurement of temperature difference, $\Delta\vartheta = \vartheta_2 - \vartheta_1$, by deflection type resistance bridge.

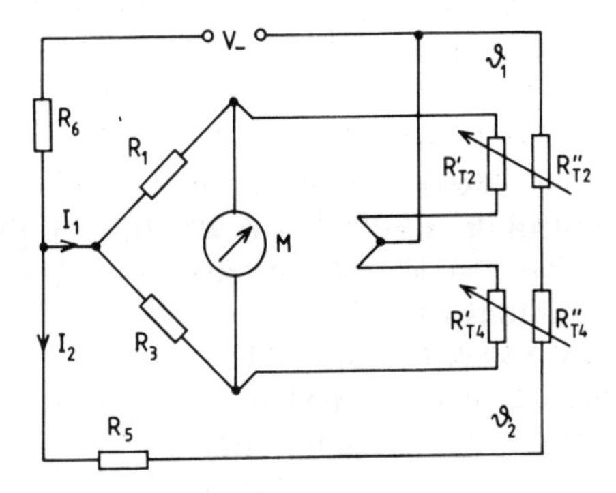

R_1, R_3-RATIO RESISTORS
R_5, R_6-AUXILIARY RESISTORS

Figure 15.23
Measurement of temperature difference, $\Delta\vartheta = \vartheta_2 - \vartheta_1$, by a compensating resistance bridge.

Typical applications of temperature difference measurements occur in gas analysers, heat flux measurement (van der Graaf, 1987a), calorimetry (White and Downes, 1988), thermal energy counters (van der Graaf, 1987b), thermal difference analysis, medicine, biology (Reed, 1972), heat treatment of metallic charges to avoid too high temperature differences and thus too high thermal stresses (Crawford, 1963) and numerous other scientific applications.

15.7 CONTINUOUS SENSORS

Some continuous sensors or distributed parameter sensors make direct use of fibre optic sensors, as described in §15.3. Degussa GmbH (Germany) is offering a new continuous FTW semiconductor system, not exactly conceived for temperature measurement but for hot-spot detecting. The sensor is built as a 2.5-mm diameter, corrosion resistant tube in lengths up to 30 m, with an axially placed central conductor, filled inside with a temperature sensitive NTC semiconductor mass (Figure 15.24).

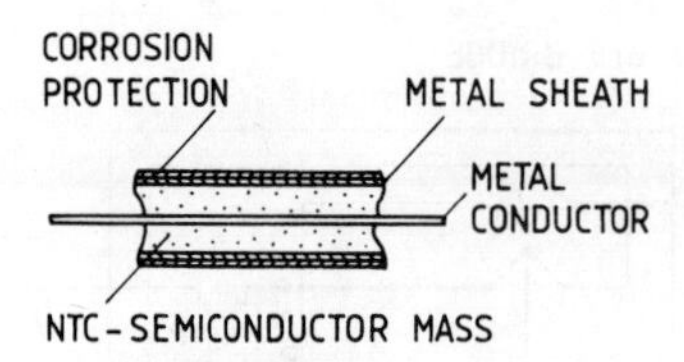

Figure 15.24
Continuous semiconductor temperature sensor.

It can be used for supervising the temperatures of long elements or placed, as meanders or spirals, on larger surfaces. The whole sensor length which is temperature sensitive has a resistance between the tube and central conductor decreasing with increasing temperature. Owing to the non-linear resistance versus temperature characteristic, overheating of even a small part, results in a marked decrease of the resistance of the section, signalling a hot spot. For correct operation the difference between the normal working temperature and its alarm level has to be from 50 to 200 °C.

An electronic instrument, supplied along with the sensors, can supervise or record the temperature of 6, 12 or 24 channels. For example, consider a 15-m long sensor operating at a normal working temperature of 200 °C. A hot spot of about 0.5 m diameter at 400 °C, results in a 50% deflection of the recorder measuring range, whereas under normal working conditions its deflection is only 5%. By using a 4 to 20 mA converter an output signal which is immune to external disturbances is available. Typical applications of this system cover the hot-spot detection in chemical reactors, furnaces, power cables in their trenches or trunking, fire protection and storage of different kinds of goods.

A continuous thermocouple described by Smith (1988) is composed of two thermocouple conductors embedded in special ceramic insulation of NTC characteristic. This whole assembly is placed in a stainless steel protective sheath similar to that in MI thermocouples. These two thermocouple wires, either K or E type are not welded together nor even touch each other. Heating of the continuous thermocouple at any given point results in a localized decrease of the insulation resistance between the two conductors so forming a 'temporary measuring junction'. A measure of the actual temperature at that point is provided by the thermoelectric force developed there. The connection resistance at the hottest point of the cable which is much lower than at any other part of the cable, determines the overall output signal. Within the application range of -29 to 900 °C an accuracy about ± 1 °C is possible. However, in some heating patterns of the cable, some loss in output signal can be observed.

Continuous thermocouples are mostly used to prevent any abnormal operation of industrial equipment such as:

- temperature changes above normal ambient,
- temperature exceeding pre-set absolute values,
- abnormal rate of temperature increase.

They are typically applied for controlling the temperature in computer cabinets, power stations, industrial process installations, warehouses, reactor covers and so on as well as for sometimes activating fire alarms. One of the main advantages of continuous thermocouples is that they do not need any power supply. Continuous thermocouples are supplied in some 10 to 20-m long sections.

REFERENCES

Agerskans, J. (1975) Thermal imaging, a technical review. *Temperature Measurement Conference Series, No. 26*, The Institute of Physics, London, pp. 375–388.

Bell, J. F. W. (1972) Ultrasonic thermometry using resonance techniques. *Temperature: Its Measurement and Control in Science and Industry*, Vol. 4, Part 1, Instrument Society of America, Pittsburgh, pp. 709–713.

Benjaminson, A. and Rowland, F. (1972) The development of the quartz resonator as a digital thermometer with a precision of 1.10^{-4}. *Temperature: Its Measurement and Control in Science and Industry*, Vol. 4, Part 1, Instrument Society of America, Pittsburgh, pp. 701–708.

Berger, R. L. and Balko, B. (1972) Thermal sensor coatings suitable for rapid response biomedical applications. *Temperature: Its Measurement and Control in Science and Industry*, Vol. 4, Part 3, Instrument Society of America, Pittsburgh, pp. 2169–2192.

Brenci, M. *et al.* (1984) Thermochromic transducer optical fibre temperature sensor, *Proc. 2nd Optical Fibre Sensors Conference*, Stuttgart, pp. 155–160.

Clark, R. C. (1968) A simplified equation for calculation of small temperature difference from resistance thermometer measurements. *J Sci. Instr.*, **33**, Series 2.

Crawford, R. B. (1962) Industrial applications of temperature difference measurements, *Temperature: Its Measurement and Control in Science and Industry*, Vol. 3, Part 2, Reinhold Publ. Co., New York, pp. 913–925.

Dadd, M. W. (1983) Acoustic thermometry in gases using pulse techniques. *TEMCON Conference*, London.

Eggers, H. R. (1975) Messung kleiner Temperaturdifferenzen und Temperatur-Anderungen. *ATM Messtechnische Praxis*, No. 11, pp. 197–200.

Fehrenbach, G. W. (1989) Faseroptisches Temperaturmessystem auf der Basis der Lumineszenzabklingzeit. *Techn. Messen*, **56**(2), 85–88.

Gottlieb, M. and Brandt, G. B. (1979) Measurement of temperature with optical fibers. *Fibre Optic Conference*, Chicago, pp. 236–242.

Graaf, F., van der (1987a) Thermopiles applied in heat flux transducers and in sensitive T sensors. *Proc. 3rd Int. Conf. TEMPMEKO 87*, Sheffield.

Graaf, F., van der (1987b) A temperature-difference sensor to measure the energy change in flowing air in an air circuit. *Proc. 3rd Int. Conf. TEMPMEKO 87*, Sheffield.

Grattan, K. T. V. (1987) The use of fibre optic techniques for temperature measurement. *Measurement and Control*, **20**(6), 32–39.

Green, S. F. (1986) Acoustic temperature and velocity measurement in combustion gases. *8th International Heat Transfer Conference*, San Francisco.

Hajiev, S. N., Agarunov, M. J. and Nurullaev, H. G. (1972) Precision measurement of temperature differences with thermistors by a simple technique. *Temperature: Its Measurement and Control in Science and Industry*, Vol. 4, Part 2, Instrument Society of America, Pittsburgh, pp. 1065–1070.

Hammeke, K. (1966) Messung von Differenz und Mittelwert zweier Temperaturen mit Widerstandsthermometern. *ATM*, pp. 227–230.

Kamper, R. A. (1972) Survey of noise thermometry. *Temperature: Its Measurement and Control in Science and Industry*, Vol. 4, Part 1, Reinhold Publ. Co., New York, pp. 349–354.

Lawless, W. N. (1972) A low temperature glass-ceramic capacitance thermometer. *Temperature: Its Measurement and Control in Science and Industry*, Vol. 4, Part 1, Instrument Society of America, Pittsburgh, pp. 1143–1152.

Lieneweg, F. (1975) *Handbuch, Technische Temperaturmessung*, F. Vieweg, Braunschweig.

Luxtron, U.S.A., Katalog Model 1000, Fluoroptic Thermometer.

Lynnworth, L. C. and Carnevale, E. H. (1972) Ultrasonic thermometry using pulse techniques. *Temperature: Its Measurement and Control in Science and Industry*, Vol. 4, Part 1, Instrument Society of America, Pittsburgh, pp. 715–732.

Marcuse, D. (1971) The coupling of degenerate modes in two parallel dielectric waveguides. *Bell System Technical Journal*, July.

Mouly, R. J. (1962) Temperature measurement with eddy currents. *Temperature: Its Measurement and Control in Science and Industry*, Vol. 3, Part 2, Reinhold Publ. Co., New York, pp. 1009–1013.

Muray, J. J. and Wunderman, I. (1972) Thermopile detectors for biomedical temperature measurements. *Temperature: Its Measurement and Control in Science and Industry*, Vol. 4, Part 3, Instrument Society of America, Pittsburgh, pp. 2151–2158.

Reed, R. P. (1972) High resolution thermometry in the biological context—some problems and techniques. *Temperature: Its Measurement and Control in Science and Industry*, Vol. 4, Part 3, Instrument Society of America, Pittsburgh, pp. 2137–2149.

Sandberg, C. and Haile, L. (1987) Fiberoptic Application in Pipes and Pipelines. *IEEE Transactions on Industry Applications*, **IA-23**(6), 1061.

Schöltzel, P. (1970) Temperaturmessung mit Quarzsensoren. *VDI-Z*, **112**(3), 14–18.

Smith, C. (1988) Loss prevention through process instrumentation—the continuous thermocouple. *Measurement and Control*, **21**(Dec./Jan. 1988/89), 297–301.

Walther, L. and Gerber, D. (1981) *Infrarotmesstechnik*, VEB Verlag Technik, Berlin.

Weichert, L. *et al.* (1976). *Temperaturmessung in der Technik—Grundlagen und Praxis*, Lexica-Verlag, Grafenau.

White, D. R. and Downes, C. J. (1988) Differential thermometer for high temperature flow calorimetry. *J Physics E: Sci. Instr.*, **22**(2), 79–81.

Yamada, T., Harada, N. and Koyanegi, M. (1982) Temperature distribution measurement with silicon photodiode array. *Temperature: Its Measurement and Control in Science and Industry*, Proceedings of the Sixth International Symposium, Instrument Society of America, Washington, pp. 395–401.

Ziegler, H. (1987) Temperaturmessung mit Schwing-Quarzen. *Techn. Messen*, **54**(4), 124–129.

16 Calibration and Testing of Temperature Measuring Instruments

16.1 INTRODUCTION

The following main terms are used in the calibration and testing of temperature measuring instruments, necessary for the maintenance and dissemination of ITS-90.

- Calibration of a thermometer is the sum of activities concerned with the determination of its thermometric characteristics. These characteristics define the function correlating the chosen property of the thermometer with the temperature. If a thermometer directly indicates the measured temperature, its calibration depends on correlating certain numerical values with the scale graduation. For example, this concerns liquid-in-glass thermometers.
- Testing a thermometer is the sum of activities concerned with verifying that the thermometer complies with the relevant regulations.
- Primary standards are thermometers used for reproduction of ITS-90, as well as for international comparisons.
- Transfer standards are thermometers used for the transfer of temperature units to other thermometers, which thus have lower accuracy than these standards. They comprise secondary, tertiary and other standards, which occupy important transfer levels in what is called the chain of traceability of standards.
- Working standards are thermometers destined for the calibration of other working standards, situated lower in the traceability hierarchy. They are also used in the calibration of industrial thermometers.
- Industrial thermometers are thermometers used in the day-to-day practice of temperature measurement.
- Laboratory thermometers are thermometers, used in laboratories.

Calibration and testing procedures for thermometers comprise a general scheme which defines the hierarchy of thermometers. They also determine the methods for and accuracy of, transferring the temperature unit from primary standards to industrial thermometers (Richardson, 1963; Gray and Chandon, 1972; Gray and Finch, 1972).

Most industrialized countries have established national standard laboratories which are equipped for reproducing ITS-90 through the calibration and testing of standard thermometers.

Laboratories which are among the more well known, are the National Physical Laboratory (NPL), UK; National Institute for Standards and Technology (NIST) (formerly NBS), USA; WNIIM, USSR; Physikalisch-Technische Bundesanstalt (PTB), Germany; National Standards Laboratory (NSL), Australia; Institut National de Metrologie, France.

Industrial thermometers are tested in regional and industrial laboratories.

Instrumentation and methods currently used for reproducing IPTS-68 supplemented by those used for reproducing ITS-90 (Preston-Thomas, 1990) are described below. Further alterations may occur when the supplementary information for ITS-90 appears.

16.2 INSTRUMENTATION FOR REPRODUCTION OF INTERNATIONAL TEMPERATURE SCALES

16.2.1 Fixed points

Some of the more important and commonly used defining fixed points of Table 1.1 and secondary reference points of Table 1.2 applied in the temperature range from 13.81 K to 83.798 K are not dealt with, as this book does not cover extremely low temperatures.

The boiling point of oxygen, 90.188 K (– 182.962 °C) is realized following the design of NPL, shown in Figure 16.1 (Hall, 1956). In a copper block, there is a small container, where pure liquid oxygen is in equilibrium with oxygen vapour. Resistance thermometer detectors are placed in copper-nickel tubes for calibration. A Dewar vessel filled with liquid oxygen contains the whole assembly, with the pressure in the small container C measured by a mercury manometer.

To measure the vapour pressure in the container its temperature has to be the lowest in the system. Therefore the container connecting tube has to be protected by a vacuum jacket to prevent undercooling. The measured values of pressure assist in the determination of the temperature of oxygen vapour. Thus the achievable accuracy in the temperature is better than 0.002 K (Hall and Barber, 1964). A precise description of the boiling point of oxygen at NIST (previously NBS) is given by Stimson (1956).

The triple point of water: 0.01 °C (273.16 K), shown in Figure 16.2, is one of the defining fixed points of ITS-90 which is of primary importance. Its glass cell is filled with distilled water under vacuum. After cooling the water down to about 0 °C a layer of some millimetres of ice is formed around the inside of the tube by means of the powdered solid CO_2. With the solid CO_2 removed, the inner tube is filled with water at about + 20 °C for a short time, until a thin layer of ice is melted and replaced by a thin layer of water. The inner tube is then filled by water at 0 °C to enhance the heat transfer to the calibrated thermometer. Keeping the cell in the ice-water mixture maintains the triple point temperature with an accuracy better than 0.0001 K to 0.0003 K for at least 24 h. A more detailed description is given by Stimson (1956) and Hall and Barber (1964).

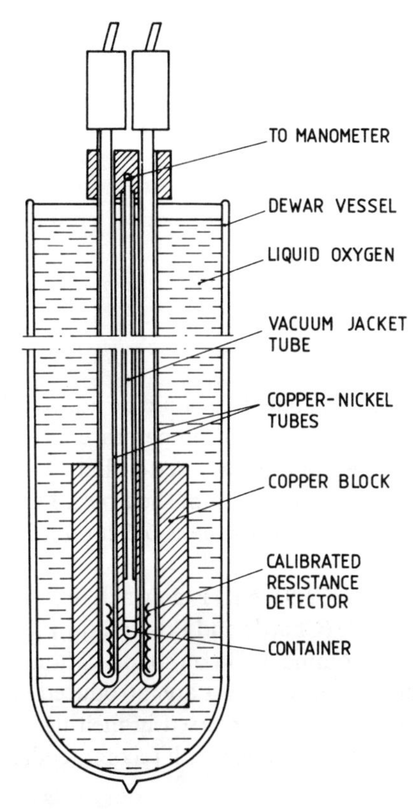

Figure 16.1
Boiling point of oxygen.

The ice point: 0 °C (273.15 K) is no more a defining fixed point of ITS-90; nevertheless it is commonly used for calibration of thermometers when an accuracy not exceeding ± 0.001 °C is required (Figure 16.3). Thermometer or sensors to be calibrated are immersed in a mixture of finely shaved pure ice and distilled water contained in a Dewar vessel. The water which should be air-saturated, should also be well shaken

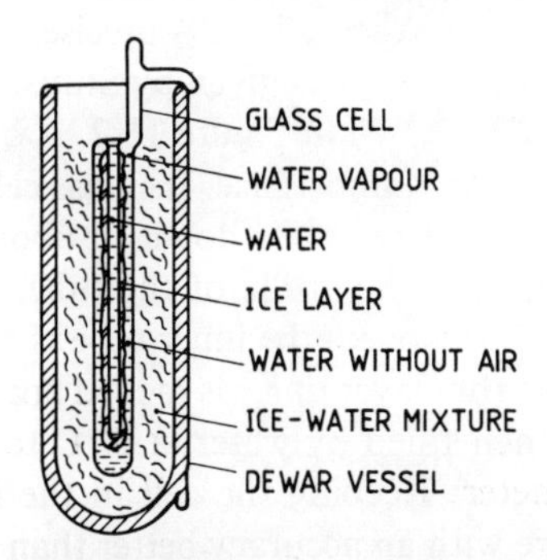

Figure 16.2
Triple point of water.

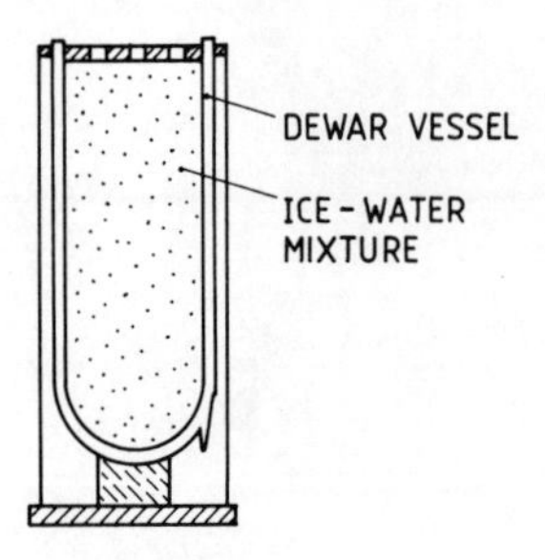

Figure 16.3
Ice-point.

with air at a temperature near 0 °C. If needed, some ice should be added and surplus water removed (Hall and Barber, 1964). The constant temperature ice bath Model 911A, produced by Rosemount Inc., ensures an accuracy better than ±0.002 °C. In this apparatus good air saturation of the bath is accomplished by a high agitation rate, which draws air into the bath. The calibration zone is 6.5 cm in diameter and about 20 cm deep.

The melting point of gallium: 29.7646 °C (302.9146 K) is now a defining fixed point of ITS-90. A relevant apparatus built in NPL (Chattle and Pokhodun, 1987) is shown in Figure 16.4. Owing to an increase in the volume of Ga on solidification, it is placed in an elastic PTFE container. Standardized thermometers are introduced into a nylon tube with a lining of A1. The sealed cell which is immersed in melting Ga, is filled with an atmosphere of pure argon. Each cell is supplied with recommendations from

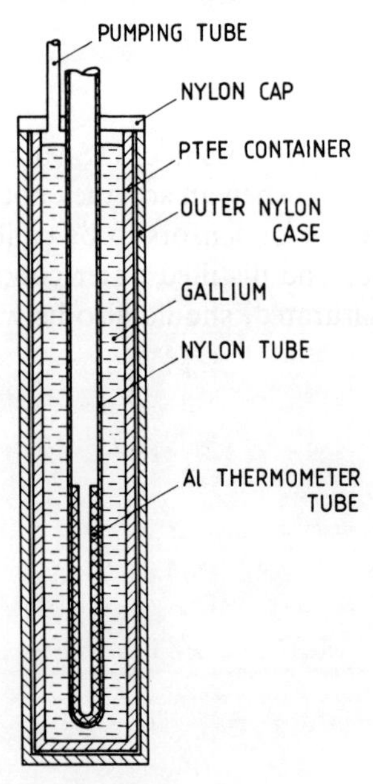

Figure 16.4
Melting point of gallium.

the manufacturer specifying the necessary immersion depth in the bath. An accuracy in the reproduction of the melting point of gallium of about ±0.4 mK can be attained (Chattle and Pokhodun, 1987). A commercially made melting point apparatus of gallium is produced by Yellow Springs Instrument Co., Inc (USA). The cell which can be removed for freezing and then returned to the apparatus, holds the melting temperature for many hours, ensuring a precision of about ±0.002 °C.

The boiling point of water: 100 °C (373.15 K), although no longer a defining fixed point of ITS-90, is still very popular and simple in operation. It can be built as an apparatus which is open to the atmosphere or which is enclosed. In the latter the pressure is stabilized (Hall and Barber, 1964). In both cases the following precautions are necessary to avoid systematic errors:

- sufficient immersion depth of the sensor, to avoid conduction losses along the sensor,
- the sensor should be well shielded to avoid radiant heat exchange with the surrounding walls,
- no parts of the apparatus should be allowed to dry, so preventing superheating of the vapour.

An open system which is used, when an accuracy below ±0.005 °C is sufficient (Hall and Barber, 1964), is suitable for testing several thermometers simultaneously (Figure 16.5). Water in the boiler is kept boiling by an electric heater. Saturated vapour flowing along the radiation shield, washes the calibrated sensors before eventually condensing in the water-cooled condenser and returning to the boiler. A water manometer indicates the internal pressure.

In a closed system (Figure 16.6), the accuracy is about ±0.001 °C (Hall and Barber, 1964). A heating element heats a small amount of water. The silver wires projecting from the water, limit a too high intensity of vapour generation. Tubular sheaths containing calibrated sensors, are surrounded by radiation shields. A special tubular connection links the apparatus with a pressure controlling

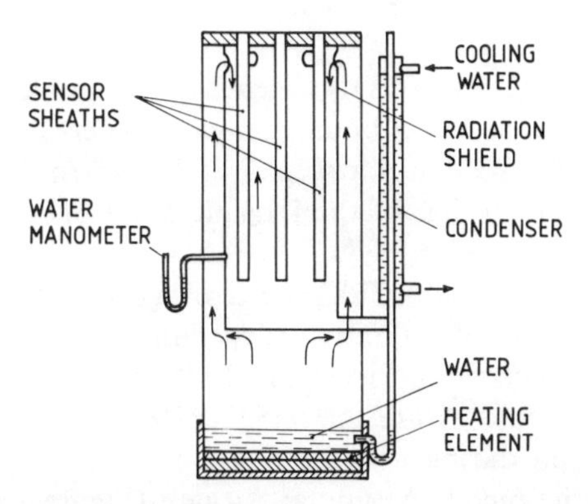

Figure 16.5
Boiling point of water in an open system.

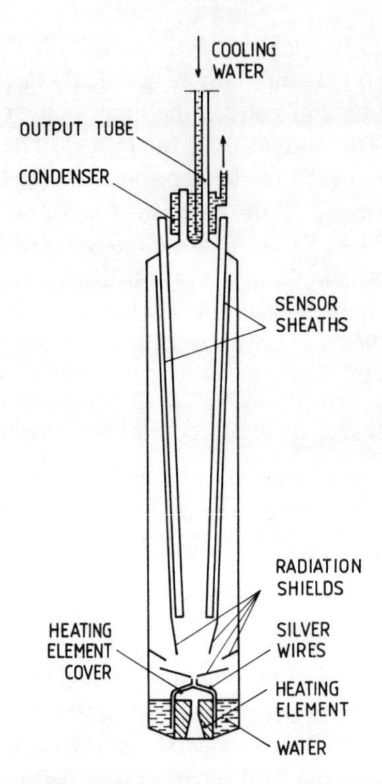

Figure 16.6
Boiling point of water in a closed system.

device and a manometer. A more detailed description of the apparatus is given by Stimson (1956).

The freezing point of indium: 156.5985 °C (429.7485 K) has been introduced as one of the defining fixed points of ITS-90, replacing the boiling point of water. A description of a similarly constructed melting point of indium, conceived as a small transportable cell, is given by Hanafy *et al.* (1982). The cell is built in teflon, operating in an argon atmosphere.

The freezing point of tin: 231.928 °C (505.078 K) which is one of the defining fixed points of ITS-90, is preferred to the use of the boiling point of water. Its achievable reproducibility is well within ±0.001 °C (Furukawa *et al.*, 1981) when the tin purity is 99.9999%. A construction similar to that of the freezing point of zinc, given in Figure 16.7, is used.

The freezing point of zinc: 419.527 °C (692.677 K) (Furukawa *et al.*, 1981; Hall and Barber, 1964) ensures a reproducibility of ±0.001 °C depending upon the purity of the zinc. For more precise measurements this purity should be 99.999%. The apparatus used to produce the fixed point should assure the necessary zinc purity as well as a uniform temperature distribution during its solidification, with equality of the temperatures of the sensor and the metal. A tubular furnace (Figure 16.7) with a copper block can ensure this uniformity of temperature. Inside the block there is a graphite crucible with a lid, through which a sheath for carrying calibrated sensors

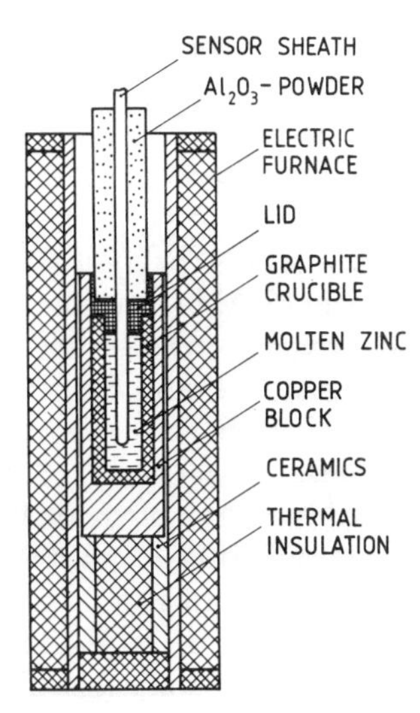

Figure 16.7
Freezing point of zinc.

is inserted. At the beginning of metal solidification undercooling by some hundredth of a degree which can be observed, may be eliminated by withdrawing the calibrated sensor for a short time. After reaching ambient temperature the sensor is re-inserted. The temperature which then quickly rises to the freezing point, stays constant at the freezing point temperature for a long period of time. Zinc purity should be periodically tested. More detailed information can be found in Preston-Thomas (1956).

The freezing point of antimony: 630.74 °C (908.89 K), which was one of the secondary reference points of IPTS-68, is measured by a standard Pt-resistance thermometer. It is then applied for calibrating standard thermocouples. The applied apparatus is quite similar to that for the freezing point of zinc. To avoid oxidation of the metal a slow stream of nitrogen is passed through the apparatus. A tendency of the antimony to undercool on freezing may be reduced by precisely controlling the heating power. It is advisable to stir with a graphite rod. In this manner, any undercooling can be reduced from about 30 °C to a fraction of degree. A constancy of the freezing temperature can be maintained to within 0.01 °C for a period as long as half an hour (Hall and Barber, 1964).

The freezing point of aluminium: 660.323 °C (933.473 K) is built in a similar way as for zinc. Graphite fibre insulation and a graphite crucible should be used. An atmosphere of argon gas is needed to prevent oxidation. This freezing point is now a defining fixed point of ITS-90. Compared with the nearby antimony point (630.74 °C) it has many merits, being cheaper, less toxic and less prone to undercooling (Furukawa *et al.*, 1975).

The freezing point of silver: 961.78 °C (1234.93 K) is realized in a similar apparatus as for other metals. Even though silver does not oxidize easily, it should still be protected from air contact, since it absorbs oxygen in the molten condition, resulting in depression of the freezing point. As oxygen absorption starts at a temperature of about 30 °C above the melting point, any unnecessary overheating of the metal should be avoided (Hall and Barber, 1964). The same ways of protection against oxidation as for antimony are used.

The freezing point of gold: 1064.18 °C (1337.33 K) was used in IPTS-68 for calibration of standard thermocouples and as a defining fixed point for pyrometers. It remains a defining fixed point of ITS-90.

Realization of the freezing point of gold for thermocouple calibration takes place, as described before, in a graphite, steatite, porcelain or Al_2O_3 crucible. All of these materials can be used as they are sufficiently temperature resistant and do not contaminate gold. No oxidation of gold takes place (Hall and Barber, 1964).

For pyrometer calibration a model of a black body must be constructed (Kostkowski and Lee, 1962). Figure 16.8 presents such a furnace, as built by NIST (Lee, 1966). Gold is contained in high-purity graphite. It has high specific emissivity, high thermal conductivity and is easily machined. An effective emissivity of $\epsilon \simeq 0.999\,99$ of the sighted opening is achieved with a sufficiently high ratio of length to diameter of the opening. A furnace has three independent heating elements placed in graphite muffles, which ensure uniform temperature distribution along the opening. Two extreme heating elements are used for automatic temperature control at a level of 1063 ± 5 °C. The middle heating element, used for controlling the process of gold melting and freezing, assists in attaining an error of realization of the gold point, in cooperation with a spectropyrometer, of well below 0.02 °C.

16.2.2 *Primary standards*

Primary standards are instruments specified in the text of ITS-90 for interpolation between the fixed points.

Standard resistance thermometers are used for interpolation from −259.3467 °C to 961.78 °C which is the freezing point of silver. Following ITS-90 the thermometer resistor must be strain-free, annealed pure platinum, wound from 0.05 to 0.5-mm Pt wire. It is advisable to use resistors of 25 Ω at 0 °C. In the upper temperature range,

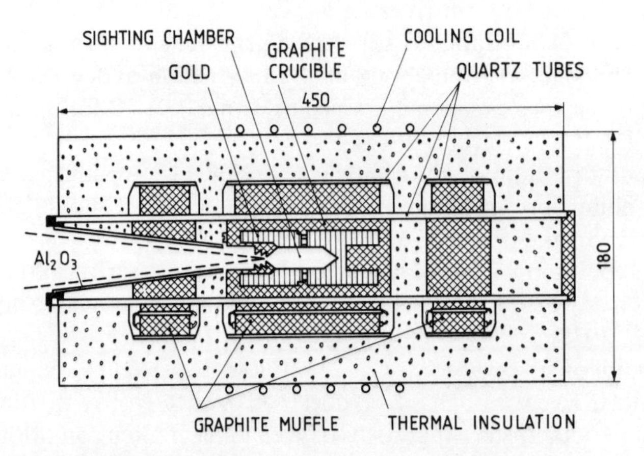

Figure 16.8
Freezing point of gold.

0.1 to 2.5 Ω resistors are recommended. The resistor should be enclosed in a hermetic sheath filled by dry, neutral gas with an addition of oxygen. At the lower temperature range, up to 13 K, they should be helium filled. The resistor should be annealed at a temperature higher than the highest expected working temperature, but in any case never below 450 °C (except for cryosensors). Quality of a sensor, its design and annealing is verified during calibration, determining the constants from interpolation equations and checking the stability of the resistance (Curtis, 1972; Foster, 1972).

Depending on working temperature range, there are three types of resistance temperature sensors:

- −260 °C to 0 °C —low temperature sensor (encapsulated type),
- −180 °C to 650 °C —normal sensors (long-stem sensor),
- 650 °C to 1065 °C—high temperature sensors.

Low temperature sensors, which are beyond the scope of this book, are described by Curtis (1972) and Sparks and Powell (1972).

Resistance sensors used as interpolation standards of ITS-90 from −180 °C to 650 °C (long-stem type) have undergone many modifications to increase their accuracy and stability, to reduce their size and to intensify the heat transfer between the resistor and sheath and between the sheath and environment (Baker *et al.*, 1953).

One of the contemporary designs is presented in Figure 16.9. Platinum wire, wound in a spiral of about 1 mm diameter is placed in a thin-walled Pyrex tube matching the spiral diameter and shaped as shown in Figure 16.9. Platinum terminals which are soldered to both spiral ends are sealed in glass in such a way that the spiral is totally strain-free. It is also important for the spiral to remain strain free during its further working life. These terminals are extended by low resistance gold wires in ceramic insulation. The whole assembly, which is encapsulated in a glass sheath, is hermetically sealed after careful drying at about 400 °C. Resistance of the sensor is about 25 Ω at 0 °C. As an example, the standard Pt resistance sensor produced by Rosemount Inc. (USA) (Berry, 1982), is hermetically sealed in a metal sheath containing a helium–oxygen atmosphere. Its stability within a specified temperature range of −200 °C to +650 °C is better than 0.01 °C per year. The self-heating rise in temperature is less than 0.002 °C with an insulation resistance from the resistor to the outside sheath greater than 5000 MΩ at 100 V d.c. while its nominal resistance is about 25 Ω at 0 °C.

High temperature resistance sensors operating from +650 °C to +1100 °C replace the Pt10Rh–Pt thermocouple in the temperature range up to +961.78 °C (freezing point of silver). The Pt10Rh–Pt thermocouple was a primary standard of IPTS-68.

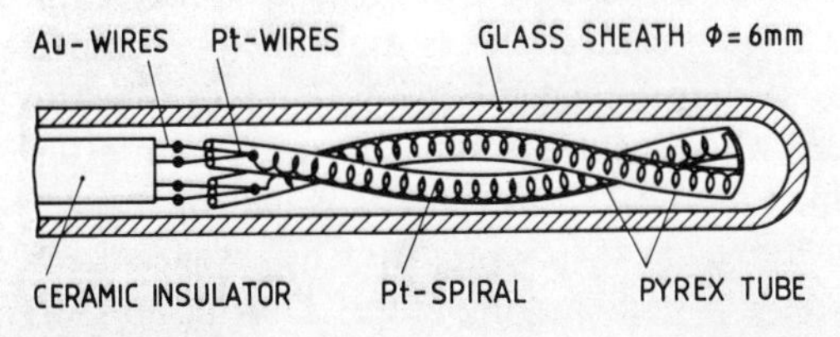

Figure 16.9
Standard resistance sensor.

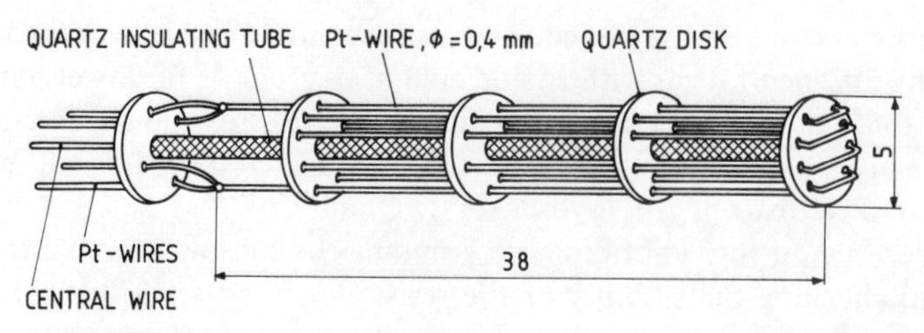

Figure 16.10
High temperature standard resistance sensor.

Design of high temperature resistance sensors is the subject of many publications (Anderson, 1972; Evans and Burns, 1962; Chattle, 1972; Curtis, 1972; Evans, 1972; Furukawa *et al.*, 1981). One of the designs proposed by Evans and Burns (1962) which is shown in Figure 16.10, has a resistor made from eight platinum wires of diameter 0.4 mm, connected in series and fixed in isolating quartz disks. Four platinum terminals are used for measurement of the sensor resistance with a fifth, central one, serving the measurement of the sensor insulation. The sensor is encapsulated in a hermetic quartz sheath, not shown in Figure 16.10, filled by dry air or by an argon–oxygen mixture. Resistance of the sensor which is 0.25 Ω at 0 °C, is much less than the maximum allowable value of 2.5 Ω at 0 °C. High temperature sensors which must be made specifically for the purpose, using selected and prepared components, must undergo stringent quality control (NPL-ITS-90). For practical details and appropriate heat treatment see the supplementary information for ITS-90.

A standard photoelectric spectropyrometer is used for extrapolation above 961.78 °C (freezing point of silver). In one of the pyrometers used which is similar to the disappearing filament pyrometer, the radiant intensities of the target and of the filament are not compared in a subjective way by the observer. The comparison is quite objective, using a photomultiplier. Applications of spectropyrometers for equalizing the temperatures of IPTS were first reported in USSR and USA and later between 1956 to 1960 in other countries (Kandyba and Kowalewskij, 1956; Lee *et al.*, 1972; Nutter, 1972; Ruffino *et al.*, 1972).

Lee (1966) reports that the accuracy of a photoelectric spectropyrometer exceeds that of the disappearing filament pyrometers formerly applied in IPTS by about five times.

The optical layout of a spectropyrometer developed in NIST (Lee *et al.*, 1972) is shown in Figure 16.11. Three identical vacuum lamps are used so that any changes in the characteristics of any lamp may be detected. The lamp filaments are made of tungsten strips of such a length that the temperature of its central part is not influenced by changes of ambient temperature. This central part is specially marked. A set of band filters, comprising a red filter and intereference filters, determines a narrow (12.5 nm) working band of the pyrometer, whose centre is at 654.5 nm. The interference filters have polarizing properties which are not advantageous when the target itself emits partially polarized radiation as tungsten strip lamps may do. Hence the need for the pyrometer optical system to have a depolarizing filter. During measurements the pyrometer lamp is periodically moved vertically by about ±0.2 mm. This causes in the mirror split an image of the lamp to alternately appear with the target image in

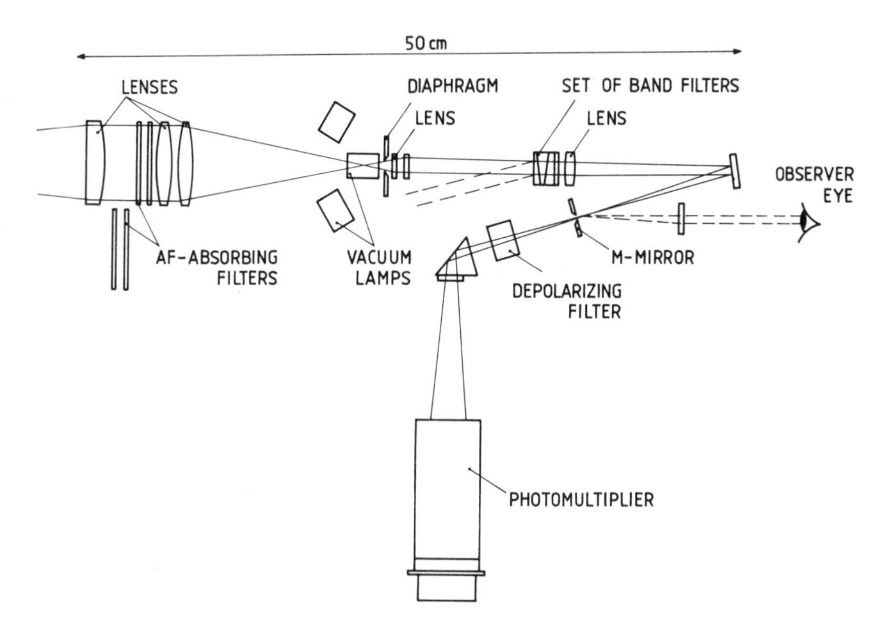

Figure 16.11
Standard photoelectric spectropyrometer.

the mirror. Both images are transmitted to a photomultiplier, at whose output two electrical signals are generated. The pyrometer lamp current is set so that both photomultiplier signals are equal. This means that the radiances of the lamp filament and of the target are also equal. In this system a recorder is used to detect the equality of both photomultiplier signals. The target image, reflected from the mirror M, is used for precise aiming of the pryometer at the target. Absorbing filters AF are applied at temperatures above 1529 K. The pyrometer is calibrated (Lee, 1966; Lee *et al.*, 1972) at the freezing points of silver (1234.93 K), gold (1337.33 K) or copper (1357.77 K). Between 1337.33 K and 1529 K the pyrometer is calibrated against a device for radiance multiplying. In the range from 1529 K to 8843 K calibration is performed using standard tungsten strip lamps and absorbing filters.

For practical details and current good practice for optical pyrometry see the 'Supplementary Information for the ITS-90' published by the International Committee of Weights and Measures.

Standard tungsten strip lamps are used for interpolation in the temperature range from 1337 to 2600 K. Vacuum lamps can be used up to about 2000 K, whereas above this temperature the use of gas-filled lamps (Figure 16.12) is advised. The strip length must be big enough to prevent any substantial influence of ambient temperature on the strip temperature. A 'place' is also marked on the strip where the measurements should be made. The sighting angle of the pyrometer is given by two points. One point is on the sighting window and another is on the strip. These two points should coincide during measurements. To prevent any reflection of the radiation which might be a source of errors, both lamp windows (Figure 16.11) are situated at an angle of 5° to the lamp axis.

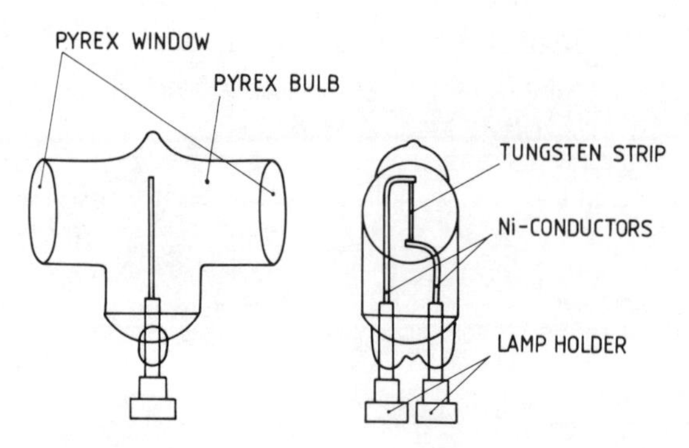

Figure 16.12
Standard tungsten strip lamp.

The dependence of the temperature of the lamp's strip upon the lamp current is called the thermometric characteristic of the lamp. To achieve high stability of this characteristic the lamps are degassed many times during the production process before being finally glued and annealed (Quinn and Lee, 1972).

Calibration of tungsten strip lamps is made by a comparison method based on the readings of a photoelectric spectropyrometer and simultaneous measurement of the lamp current at different strip temperatures.

Tungsten strip lamps must be fed by a direct current, maintaining the correct polarity during the measurements, because the temperature distribution along the strip depends upon the current direction through the Thomson effect.

16.3 WORKING STANDARDS

The primary standards described above are used for the realization of ITS and for international comparisons. Working standards are used for calibration and testing of other thermometers.

Resistance sensors are used in the temperature range from -180 to 1100 °C for the calibration and testing of resistance sensors of lower accuracy. Below 0 °C they are used for the calibration and testing of Cu–CuNi thermocouples and semiconductor thermometers. The nominal value of the reference resistance R_0 (at 0 °C) of these sensors is about 10 Ω, 25 Ω or 100 Ω. These values do not apply to first order sensors (transfer standards) from 0 to 650 °C, for which only 10 and 25 Ω are permitted. This is also true for high-temperature sensors of lower resistance. General outlines of the design of working standard resistors are the same as for standard resistance thermometers. Calibration of resistance sensors, can be made at fixed points. In many laboratories the ice-point is used instead of the triple point of water. The comparison method in testing baths is also used, comparing the readings of the thermometer to be calibrated with those of a thermometer of higher accuracy. Thermostats which are applied are described in §16.5.2. To avoid errors arising from conduction losses along the sheath and leads, the recommended minimum depth of immersion of sensors is about 20 cm. All sensors should be annealed prior to calibration at 500 °C for about

30 min. Self-heating errors are usually less than 0.004 °C, provided the sensor resistance for which R_0 (at 0 °C) is about 25 Ω are generally measured at a current of 1 mA d.c.

Transfer standards such as mercury-in-glass and mercury-thalium-in-glass thermometers are used for calibration and testing of other glass thermometers, as well as that of manometric, resistance and thermocouple thermometers. Different liquid-in-glass thermometers are used. Their permissible errors depending on the temperature range, are ±0.01 °C to ±3 °C. To increase the accuracy of measurement by mercury-in-glass thermometers errors due to zero changes, external pressure variations and variations in temperature of the mercury column should be eliminated. Detailed information on the design of precision mercury-in-glass thermometers, including consideration of the glass and quartz used, as well as measuring technique and errors is given by Hall and Barber (1964).

Calibration of glass thermometers is done at fixed points or by the comparison method in baths.

Thermocouple sensors used as working standards are Pt10Rh–Pt, Pt30Rh–Pt6Rh and some other thermocouples of non-rare metals, calibrated against a resistance thermometer and at higher temperature against standard optical pyrometers. Working standard Pt10Rh–Pt thermocouples are used in the temperature range from 300 to 1450 °C for the calibration and testing of Pt10Rh–Pt thermocouples of lower accuracy as well as other thermocouples. They have to be annealed carefully, in most cases by direct current flow. They are mounted in twin-hole insulation of pure Al_2O_3. The thermocouple conductors are of diameter from 0.35 mm to 0.65 mm. For example, the Standard Rosemount thermocouple Model 163A has an accuracy of ±0.75 °C in the range 0 to 1100 °C and ±2 °C in the range from 1100 to 1450 °C. Working standard thermocouples Pt30Rh–Pt6Rh can be used in the temperature range from 650 °C to 1770 °C. Working standard thermocouples which are calibrated at fixed points or by the comparison method have their reference junction usually kept at 0 °C, the ice-point (Roberts, 1980).

Tungsten strip lamps are used in the temperature range from 1100 to 3100 K for the calibration and testing of similar lamps of lower accuracy as well as for calibration and testing of pyrometers. These working standard lamps which are constructed in a similar way to standard tungsten strip lamps, are calibrated by photoelectrical spectropyrometry.

Disappearing filament pyrometers are used in the temperature range from 1100 to 2800 K for the calibration of working standard tungsten strip lamps and for testing of similar pyrometers of lower accuracy. Grey filters are used for expanding their measuring range. These pyrometers are calibrated against standard tungsten strip lamps, based on their thermometric characteristics.

16.4 TESTING OF INDUSTRIAL THERMOMETERS

16.4.1 Introduction

During normal operation industrial thermometers are subject to various factors, such as high temperatures, mechanical and chemical influence and so on. Under the impact of such factors, their thermometric characteristics can significantly vary. Thus they have to be checked periodically and recalibrated or repaired if needed. The frequency of such periodical checking which depends upon the working conditions, should be determined by the maintenance service.

16.4.2 *Testing at fixed points*

The following fixed points are usually used for the testing of industrial thermometers:

1. Ice-point (0 °C) is used for testing of liquid-in-glass thermometers, resistance thermometers and for ensuring constant reference temperature in thermoelectric thermometers.
2. Boiling point of water (100 °C) is used mainly for the testing of resistance thermometers.
3. Freezing points of different metals such as tin (232 °C), lead (327 P°C), zinc (419 °C), aluminium (660 °C), copper (1085 °C) and of some non-metals such as sodium chloride (801 °C) are applied for the testing of thermoelectric thermometers.
4. Melting points of some metals, such as gold (1064 °C), palladium (1554 °C) or platinum (1772 °C), are used for testing of thermocouples following the wire-method (Figure 16.14).

The ice-point and the boiling point of water have been discussed in §16.2.1. For the testing of industrial thermometers the boiling point of water in an open system is used (Figure 16.5). Designs of secondary reference points are described in §16.5.1.

Testing at fixed points and especially at the freezing points of metals is precise. However, as it is time consuming and rather expensive, it is rarely used for industrial thermometers.

16.4.3 *Comparison method*

This method, commonly used for the testing of industrial thermometers, consists in comparing the readings of the thermometer under test with those of standard thermometers. Location of the sensitive points of both thermometers at the same temperature is essential. The thermometers under test and the standard thermometer are placed in a testing bath or furnace, construction as described in §§16.5.2 and 16.5.3, ensuring a uniform temperature distribution. The testing accuracy depends upon the accuracy of the standard thermometer, on uniformity and constancy of temperature and to a great extent on the ability and experience of the operator (see §§16.4.4 to 16.4.6).

In the testing of industrial pyrometers by the comparison method, both standard and tested pyrometers have to be directed at the same radiation source (see §§16.4.8 and 16.4.9).

16.4.4 Liquid-in-glass thermometers

Testing of liquid-in-glass thermometers comprises visual examination and testing of thermometer accuracy.

Visual examination covers control of the scale and of its position relative to the capillary. In mercury-in-glass thermometers the mercury column should neither be broken nor contain observable impurities. Any thermometer in which dirt is observed or moisture traces exist on the capillary walls should be rejected. Before testing, broken mercury must be rejoined.

Thermometer accuracy is tested by the comparison method using a stirred-liquid bath and two standard liquid-in-glass thermometers (§16.5.2). Readings are taken while the temperature is slowly rising, completing two measuring cycles. During each cycle the routine of measurements should be: standard thermometer no. 1, tested thermometers, standard thermometer no. 2, standard thermometer no. 2 again, tested thermometers in reverse succession and standard thermometer no. 1. Calculated mean values of the four readings for each tested thermometer correspond to the mean temperature values determined by the standard thermometers.

Thermometers which must be tested completely immersed, should normally be tested at seven but not less than five evenly-spaced temperatures covering 80% of the scale range (NPL, 1982).

16.4.5 Manometric thermometers

Testing of manometric thermometers consists of visual examination, testing of the thermometer accuracy, testing of the hysteresis of the indications and testing of the variation of the indications.

Visual examination covers verifying the correctness of the markings. The general state of the thermometer, including that of the sensors, capillary, pointer, scale and elastic element, should also be assessed.

Testing of accuracy and hysteresis is by the comparison method in a stirred-liquid bath. Because of large time lags in the indications of manometric thermometers during their testing, it is important to keep the sensor in the bath for as long as necessary until the readings are stabilized. Meanwhile the bath temperature should be kept constant. Measurements are taken at several temperatures, starting at the lowest one, going up to the highest and then reversing the procedure on the way back down. For accuracy tests, the mean values of readings at each temperature are considered. The difference in the indications between increasing and decreasing temperature, is a measure of hysteresis.

Indication variations are determined by measuring the same temperature under constant measuring conditions several times and observing differences in the readings. The accuracy and variations of the indications should be evaluated from sensor and read-out instrument, obtained at the same level.

16.4.6 Resistance thermometer detectors

Testing of resistance thermometer detectors requires visual examination, checking of the resistance stability, testing of thermometric characteristic of the resistor, testing both the break-down strength of the sensor insulation and the insulation resistance of the sensor.

During visual examination of the sensor, its general state is checked and defects observed. Correct marking of the working range, type of resistor, sheath material and so on are to be verified.

Resistance stability checking which is performed if the resistor is used over 300 °C is made by measuring the resistance R_0 at 0 °C before and after 4 h of heating at a temperature 5% higher than the highest working temperature. The changes of resistance R_0 should not exceed one third of the permissible resistance tolerance, as defined by relevant standards.

Testing of the thermometric characteristic has to verify that it conforms to relevant standards in a particular country. For conformity to ITS-90 the following measurements are made:

- measuring of the sensor resistance at the triple point of water $R_{0.01}$ and at the gallium melting point (29.7646 °C) R_{Ga},
- comparison with a standard resistor in a thermostat at some chosen temperature, within the sensor operating range.

The measured $R_{0.01}$ and R_{Ga} values, as well as their ratio, have to be compared with relevant standardized values (Equation (1.17)). Usually these readings are taken simultaneously for a number of resistors.

Nowadays bridge or voltage comparison systems equipped with digital read-out instruments as well as computerized data acquisition and processing systems, as described in Chapter 8, are increasingly popular.

Testing of break-down strength and of resistance of the sensor has to be done to ensure conformity with relevant national standards.

16.4.7 Thermocouples

Testing of thermocouples (Roeser and Wensel, 1941) covers:

- visual examination,
- testing of e.m.f. versus temperature characteristics,
- testing of sensor insulation resistance.

Visual examination is carried out after removing the thermocouple from its sheath and removing the ceramic wire insulators. Thermocouples exhibiting stains and scale are rejected.

Before testing the e.m.f. versus temperature characteristic, rare metal thermocouples are chemically cleaned by 50% hydrochloric acid, carefully washed in distilled water and annealed by direct current flow. Thermocouples of Pt10Rh–Pt and Pt13Rh–Pt are annealed at 1150 ± 50 °C, which corresponds to a current of about 11 A for a diameter of 0.3 mm. Thermocouples of Pt30Rh–Pt6Rh are annealed at 1400 ± 50 °C (about 13.5 A by ϕ 0.5 mm).

The e.m.f. versus temperature characteristic has to be compared with the relevant standards of particular countries. Testing methods used are the comparison method, the differential comparison method, the measurement of e.m.f. at fixed points and the measurement of e.m.f. at fixed points by the wire method.

The comparison method is used for all of the following standardized industrial thermocouples in the manner described below.

Thermocouples Pt10Rh–Pt and Pt30Rh–Pt3Rh in ceramic insulators are placed together with the standard thermocouple in a tubular electrical furnace, having a normal working temperature of at least 1200 °C. The measuring junction of the thermocouples to be tested, should be placed simultaneously in the middle of the furnace length (see §16.5.3) and should not exceed five in number. To ensure that the temperature of the measuring junctions of standard and tested thermocouples is equal, they should be bound by platinum wire or welded together.

Thermocouples NiCr–NiAl are put together with the standard NiCr–NiAl thermocouple in ceramic insulators and placed in a metal block in a tubular furnace (§16.5.3), so that the measuring junctions are in direct contact with the metal block. The working temperature of the furnace should be at least 1000 °C. If a Pt10Rh–Pt thermocouple is used as a standard thermocouple, it is also placed in one of the block holes. In that case the thermocouple is not bare but is placed in a gas-tight glass or ceramic sheath.

To ensure better equalization of the temperature of tested and standard thermocouples, in more precise measurements, the measuring junction of the standard thermocouple is placed in a hole, drilled in the junction of the tested thermocouple which is always larger. Contamination is prevented by protecting the wires of the standard thermocouple with refractory cement. Inhomogeneity in the junction itself does not influence the results as long as no temperature differences occur in it. When testing a number of thermocouples, they can be welded together to form one common measuring junction.

Thermocouples Fe–CuNi and Cu–CuNi are tested in baths, up to 300 °C (§16.5.2), using a mercury-in-glass thermometer as the standard. From 300 °C to 700 °C, Fe–CuNi thermocouples are tested in the same way as are NiCr–NiAl. The number of simultaneously tested thermocouples should not exceed six.

In all thermocouple testing, their reference junction temperature should be at the ice-point.

Usually e.m.f. measurements are made with slowly rising temperature, not exceeding about 0.5 °C/min at equal time intervals in the order of standard thermocouple, tested thermocouples No: 1, 2, 3, 4, 4, 3, 2, 1 and standard thermocouple again. As a result an average value of two measurements is taken for each thermocouple. Thus the found mean e.m.f. values of particular thermocouples correspond to the average temperature, measured by the standard thermocouple. The measurements should be taken by a potentiometer or other instrument with an accuracy of at least 0.01%.

More and more frequently, thermocouple testing by the comparison method is conducted using electric furnaces with programmed temperature control (Jones and

Egan, 1975; Sergejev, 1979) and a totally computerized data conditioning system (Kirby, 1982).

The differential comparison method is only used for rare-metal thermocouples with the same type of thermocouple used both as the standard and as the tested thermocouple. Conductors with the same polarity of both the standard and tested thermocouples, up to four at a time, are bound together by a platinum wire as near their measuring junctions as possible to ensure as good a thermal and electrical contact as possible. The same type of tubular furnace is used as in the comparison method. Measurements of the differential e.m.f. values between conductors of the same polarity of the standard and the tested thermocouples are taken from the first to the last thermocouple and then in reverse succession. At the beginning and at the end of each cycle, the true-furnace temperature is measured by the standard thermocouple. The relevant electric circuit for the Pt10Rh–Pt thermocouples is shown in Figure 16.13.

Following the American National Standard ASTM E220-80, the differential comparison method has the following advantages, relative to the comparison method:

- Measured differential e.m.fs are small relative to the relevant thermocouple e.m.f. at the given temperature and thus do not need to be measured very precisely.
- During testing much higher rates of temperature increase can be applied, because the differences of the thermoelectric characteristics of the tested and the standard thermocouples, insignificantly vary as a function of temperature. Also the furnace temperature does not need to be precisely measured.

E.m.f. measurement at fixed points is very popular, especially in USA (Richardson, 1962; Trabold, 1962). The tested thermocouple is immersed consecutively in crucibles with metals and salts of different freezing temperatures. Convenient for this purpose are the freezing points of tin (232 °C), lead (327 °C), zinc (419 °C) and sodium chloride (801 °C), as described in Section 16.5.1. The tested thermocouples should be immersed to an adequate depth to prevent heat flow from the measuring junction along the thermocouple conductors. Any small changes in immersion depth should not affect the measured e.m.f. values provided this depth is adequate. The e.m.f. values, corresponding to relevant fixed point temperatures, are recorded as a function of time with the horizontal part of the e.m.f. versus temperature curve determining the sought e.m.f. value.

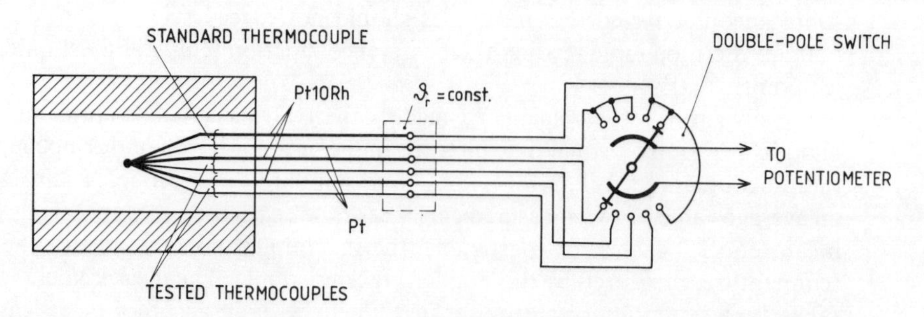

Figure 16.13
Testing of Pt10Rh–Pt thermocouples by differential comparison method.

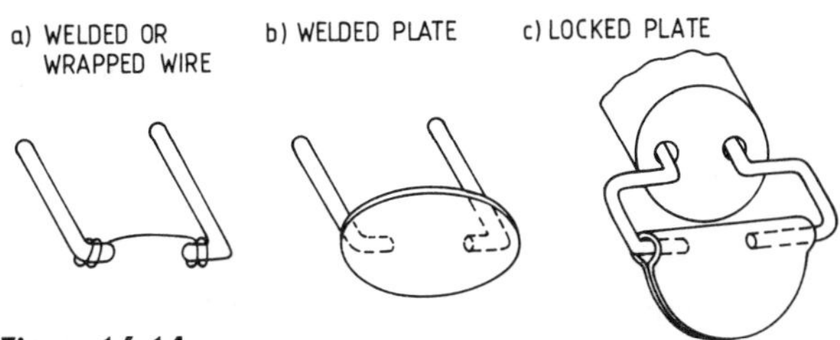

Figure 16.14
Testing of thermocouples by the wire method.

E.m.f. measurement at fixed points by the wire method (Hall and Barber, 1964; Trabold, 1962) is applied mainly for rare-metal thermocouples. The measuring junction of the tested thermocouple is cut in two with a short pure metal wire then soldered to both thermocouple conductors (Figure 16.14(a)). Thus prepared, the thermocouple is slowly heated-up in a tubular furnace, till the inserted wire melts. The recorded e.m.f. versus time value clearly indicates the constant temperature part of the curve, till the circuit is broken. Instead of wire an Au, Pd or Pt plate can also be used (Figure 16.14(b)(c).

In situ testing of thermocouples is becoming increasingly more important. Thermocouples after a longer working life or after being contaminated should not be tested in a laboratory environment.

The e.m.f. of a non-homogeneous thermocouple depends upon the temperature distribution along the thermocouple and thus the e.m.f. values measured by the laboratory method do not correspond to *in-situ* readings. Testing is made by the comparison method in which the standard thermocouple is placed alongside the tested one so that their measuring junctions are at the same temperature. At a constant measured temperature, both thermocouples can be placed alternatively at the same place. If many thermocouples are used in one installation, an additional empty sheath is introduced for placing of the standard thermocouple during testing. Such an empty ceramic sheath is sometimes placed together with two other sheaths in one common outer ceramic tube. This third sheath which is kept empty is only to house the standard thermocouple temporarily during testing (Kortvelyessy, 1987).

Testing should be carried out at some temperatures distributed over the working range. Although the method described above is not as precise as the laboratory method, it is very useful in many cases. If a temperature sensor is always connected to the same measuring instrument then the whole installation is tested instead of testing the thermocouple alone. Readings of the whole installation are compared with those of a standard thermometer.

16.4.8 Disappearing filament pyrometers

Testing of industrial disappearing filament pyrometers involves both visual examination, and testing of the precision of readings.

Visual examination concerns the optical system and filter as well as smoothness of movements of the objective.

Precision of readings of industrial pyrometers is tested against a tungsten-strip lamp (§16.2.2) or a set of lamps, in the temperature range from 700 to 2000 °C. The comparison method in which the readings are compared against the standard pyrometer, using the strip lamp as a transfer instrument, is also used. After setting the lamp current corresponding to the desired check-point temperature, two series of readings are taken. Each series covers four measurements of the luminance temperature (§7.5.3) of the lamp strip, alternatively for rising and falling strip luminance. During each measuring series the lamp current has to be kept constant. The readings are taken at all marked scale points, at rising and then at falling strip temperatures. After each change of lamp current the readings can be taken when a new thermal steady-state of the strip is reached (in practice about 5 min). To calculate the pyrometer error at a given scale point an average value of both series of readings is taken. This is then compared with the temperature found from the lamp characteristic with the scatter of the readings calculated as the maximum difference of the readings in a given series of measurements.

Error in the measurement of lamp current which is most commonly measured by a potentiometric method, measuring the voltage drop across a standard resistor in series with the lamp, should not exceed 0.1% of the measured value.

Testing of a pyrometer by the comparison method, using a standard pyrometer, is explained in Figure 16.15. Both the tested and standard pyrometers are aimed through identical optical systems at the radiation source, which is a tungsten-strip lamp. Its temperature is set by means of a regulating transformer. The applied standard pyrometer has a known thermometric characteristic, given as the measured temperature versus voltage drop across the pyrometer lamp. When the brightness of the tungsten strip lamp and of both pyrometer filaments are made equal, the voltage drop across the standard pyrometer lamp is measured by the potentiometric method. Simultaneously the temperature indicated by the tested pyrometer is measured.

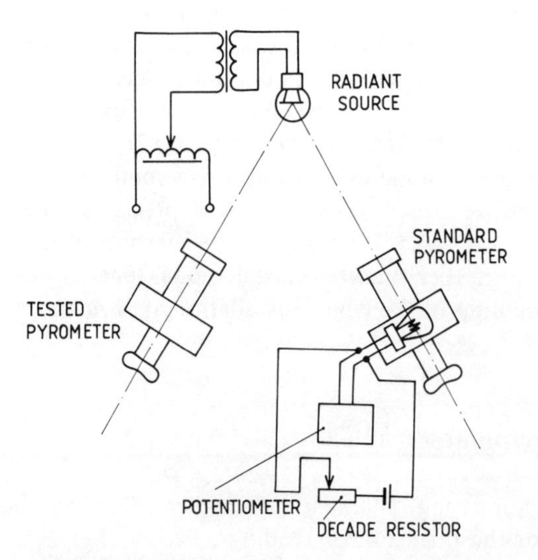

Figure 16.15
Testing of a disappearing pyrometer by the comparison method.

16.4.9 *Total radiation, photoelectric and two-colour pyrometers*

The above mentioned pyrometers are tested by the comparison method using a radiation source, which approaches a black body as closely as possible. A disappearing filament pyrometer, a pyrometer of the same type as the one tested or a thermocouple may be used as a standard instrument (NPL, 1982). A high-temperature tubular furnace with PtRh heating element is used as the radiation source. When the tube length of such a furnace is sufficiently larger than its diameter (Figure 7.2(d)), the furnace can be regarded as a black body source (Bedford, 1972). Recently heat-pipe black body sources have also been used (see §16.5) (Aa and Gelok, 1987; Neuer and Brost, 1975).

When the standard pyrometer has the same optical system as the tested one, the radiation source does not need to present a black body model. Nickel plates, heated by current flow, or tungsten-strip lamps, equipped with an optical magnifying system to get a sufficiently large radiating surface, are convenient.

16.5 AUXILIARY EQUIPMENT

16.5.1 *Secondary reference points*

The most commonly used secondary reference points, which are the freezing points of metals, and salts listed in §16.4.2, were standardized under the regulations of IPTS-68. They are built in a similar way to the fixed point of tin, zinc, antimony, silver and gold which were described in §16.2.1. Depending on the required accuracy different purity of metals is needed. In testing of industrial thermocouples a purity of 99.9% is sufficient. Easily oxidizing metals are melted in graphite crucibles while the other kinds are melted in steel crucibles. Crucibles are covered by graphite or steel lids. Tin and zinc are sufficiently protected against oxidation by a layer of graphite powder on top of the metal. The tested thermocouple is placed in a protecting sheath, immersed at a depth sufficient to prevent any temperature gradients in the neighbourhood of the measuring junction of the tested thermocouple. Placing the crucible in a vertical tubular furnace which can be used for many different crucibles, ensures a uniform temperature distribution in them. A typical temperature versus time display of a freezing process which is shown in Figure 16.16 illustrates the metal dependent under-cooling phenomenon observed at the beginning of the freezing process. The measurements start at the moment, the metal temperature is stabilized with the measured e.m.f. values then corresponding to the relevant fixed point temperature ϑ_f.

16.5.2 *Thermometer testing baths*

Thermometer testing baths are used for thermometer calibration and testing by the comparison method. A uniform temperature distribution around both standard and tested thermometers should be ensured.

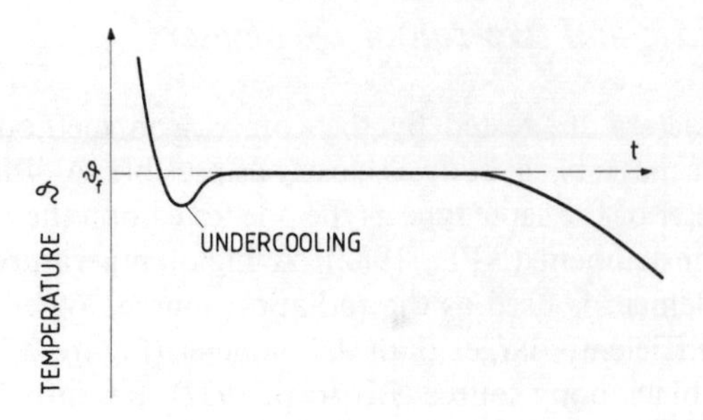

Figure 16.16
Temperature versus time of a freezing process with ϑ_f the fixed point temperature.

In liquid testing baths this uniformity is ensured by efficient stirring and liquid circulation. Although liquid heating is done by electric heating elements, previously cooled liquid, or liquid presently cooled by the addition of solid CO_2, or by a cooling coil can be used, for temperatures below ambient.

Liquid media which are used in testing baths should be characterized by good thermal conductivity, high specific heat and low viscosity. Best suited is water, but its application range is only 0 to 100 °C. Other liquids, applicable from −175 °C up to 630 °C, are listed in Table 16.1 (Hall and Barber, 1964).

Construction of an acetone bath, applicable in the temperature range from −80 to 0 °C, is shown in Figure 16.17. Acetone in a Dewar vessel which is then placed in a metal container, is stirred by air blown through a tube having a number of holes. Cooling is done by the addition of small amounts of solid CO_2 and by varying the rate of air blowing. More intensive air blowing which promotes acetone evaporation thus causes more intensive cooling. However, no steady rate of cooling is possible. A slow, steady rise of temperature is achievable, ensuring sufficient precision for the comparison testing of liquid-in-glass thermometers for example (Hall and Barber, 1964).

A water-bath which is used from 0 to 100 °C (Figure 16.18) consists of two cylindrical containers, connected at top and bottom by horizontal channels. The tested

Table 16.1
Media applied in testing baths and fluidized bed thermostats.

Thermostat	Medium	Temperature range (°C)
Liquid bath	Freon	−175 to −40
	Alcohol	−100 to 0
	Acetone	−80 to +20
	Water	0 to 100
	Mineral* and silicon oil	50 to 250
	Salt mixture 52% KNO_3 + 44.8% $NaNO_3$	180 to 630
Fluidized bed	Al_2O_3 powder	70 to 1100
	Zr powder	

*Upper application limit should be about 50 °C below the inflammation point.

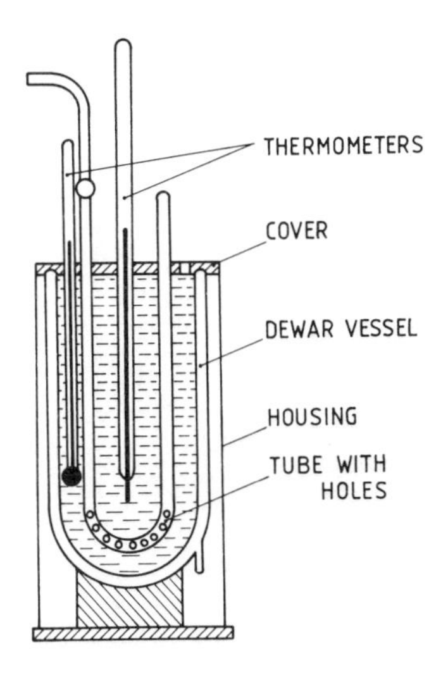

Figure 16.17
Acetone bath.

and standard thermometers are placed in the larger of two containers while the smaller one contains the heating elements and a propeller stirrer. Water is continuously fed from above. The whole assembly is thermally insulated.

Liquid-baths for temperatures over 100 °C, are designed as in Figures 16.18 or 16.19. In the bath of Figure 16.19, liquid circulation is achieved by pumping. The bath is composed of three containers, a larger one of tested and standard thermometers and a smaller one for the air supply.

The air-inlet tube is periodically connected to an air pump. In one half of the cycle the compressed air lifts the valve 1 and allows the liquid to flow from B to A with simultaneous overflow of the excess liquid into the outer container C. The liquid level in C then becomes higher than in B. In the second half of the cycle the air-inlet tube

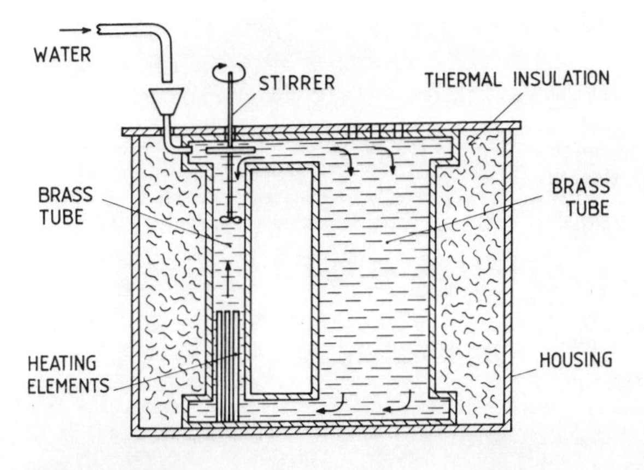

Figure 16.18
Water bath.

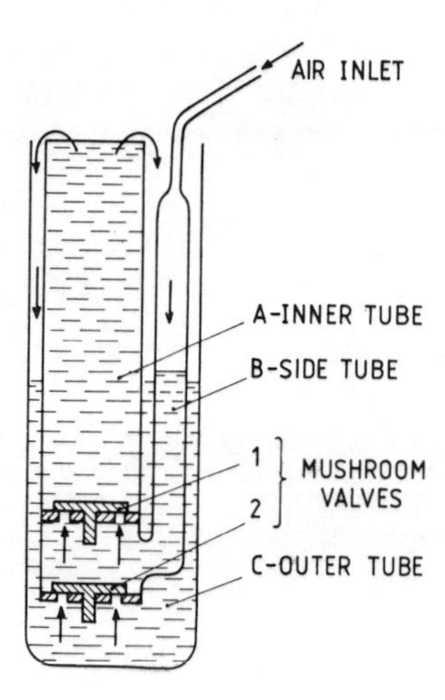

Figure 16.19
Liquid bath for temperatures over 100 °C.

is disconnected from the pump while the atmospheric pressure is restored in the tube B. The valve 2 opens and the liquid levels in B and C become equal. Another design of a liquid bath, applicable up to 2000 °C, is given by Marcarino *et al.* (1981).

In fluidized bed thermostats temperature uniformity is reached by fluidization of powder particles (Staffin and Rim, 1972). They can be applied in the temperature range from about 70 to 1100 °C. A schematic diagram of a fluidized bed, which is shown in Figure 16.20, illustrates how the standard and tested thermometers are placed in a container with Al_2O_3 or Zr powder. Air is let in from below, through a mesh which enforces a uniform distribution of the air-stream. The air, whose flow rate is stabilized, is heated by electric heating elements to give a temperature uniformity in the bed of about ± 0.1 to ± 0.5 °C.

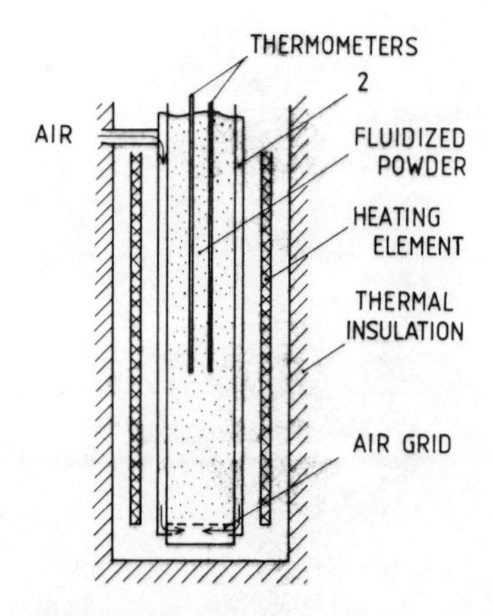

Figure 16.20
Fluidized bed thermostat.

Between the fluidized powder and the heated thermometers, the heat transfer coefficient is about 300 W/m^2 K. Although fluidized bed thermostats exhibit lower accuracy than liquid baths, their main advantages are that one powder sort covers a very large temperature range, no danger of liquid inflammation exists and no corrosion occurs.

16.5.3 Thermometer testing furnaces

At temperatures above the application range of liquid-baths, the electric tubular furnaces which are used do not have such high temperature uniformity as liquid-baths. The heating elements, wound on ceramic tubes, are chrome-nickel alloys up to 1100 °C, Kanthal wires up to 1300 °C and Pt, or PtRh, wires up to about 1600 °C. To ensure a long and possibly uniform temperature zone, the following should be considered.

- the ratio of tube length to diameter has to be as high as possible (>20:1),
- to avoid temperature drops at the ends of the furnace the linear power density has to be higher at the end than in the middle, three-zone furnaces with separate controllers for each zone are most popular,
- a metal block of high thermal conductivity should be inserted at the middle of the furnace.

When combined, all these methods can secure a uniformity to within $\pm$0.05 °C over a length of 12 cm at 630 °C (Hall and Barber, 1964).

A tubular electric furnace with metal block given in Figure 16.21 shows the standard and the tested thermocouples placed in holes in the block. At lower temperatures up to 500 °C Al and Cu blocks are used, up to 750 °C bronze blocks are used and up to 900 °C nickel blocks are used. The internal exchangeable ceramic tube is there to improve the longitudinal temperature uniformity as well as to protect the outer ceramic tube from contamination.

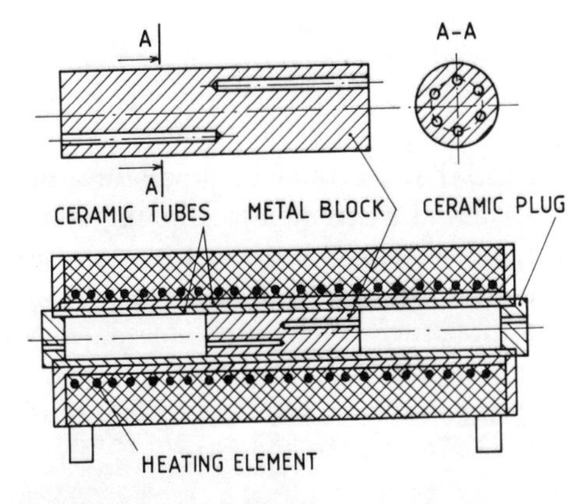

Figure 16.21
Tubular electric furnace with a metal block.

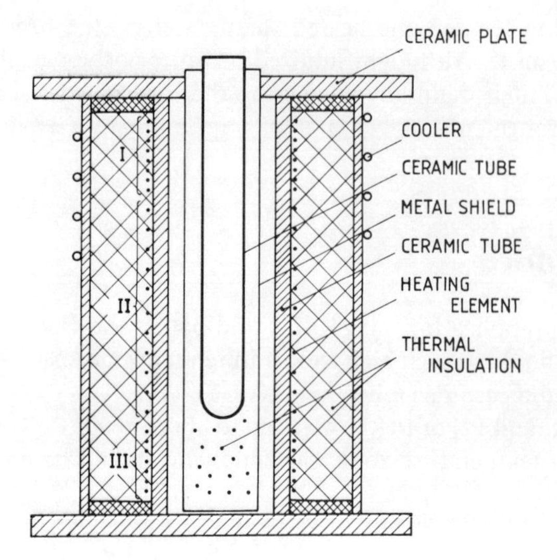

Figure 16.22
Three-zone tubular furnace for thermocouple testing.

Inside the furnace which should only be used for testing, utmost cleanliness must be preserved to protect the very sensitive Pt10Rh–Pt standard thermocouples. Rare-metal thermocouples, which have to be tested in different furnaces from ordinary thermocouples, are tested without metal blocks. The testing furnaces are equipped with automatic temperature control (Jones and Egan, 1975; Sergejev, 1979).

For automatic testing, tubular furnaces with programmed temperature control are used (Jones and Egan, 1975) (Figure 16.22). To achieve a uniform temperature distribution the heating elements are divided into three zones. Testing of the thermocouples is done during a continuous increase of temperature with a computer system for temperature control and data recording (Figure 16.23).

While a rare-metal standard thermocouple is used for testing rare-metal thermocouples, they are joined in one common measuring junction. In the case of common metal thermocouples, they are all joined together in a common measuring junction with a hole pierced to take the rare-metal standard thermocouple.

Testing or calibration is made in the temperature range from 400 to 1000 °C at about 25 °C intervals first for rising and then for falling temperature. Programmed temperature control and sampling of tested thermocouples is fully automatic.

The particular points of the e.m.f. versus temperature characteristic are recorded after being computed by the least squares method. By these means the total testing time is significantly shortened. In addition, measuring errors are smaller than $\pm 3\,\mu V$ which is of the same order as for non-automatic testing.

Many recent furnaces are based on the heat-pipe principle of Figure 16.24 (Busse *et al.*, 1975; Coville and Laurencier, 1975; Neuer and Brost, 1975), which gives excellent temperature uniformity.

The heat-pipe systems (Figure 16.24) with sheath tubes containing the standard and tested thermocouples are confined in the chamber of an electric furnace. A layer

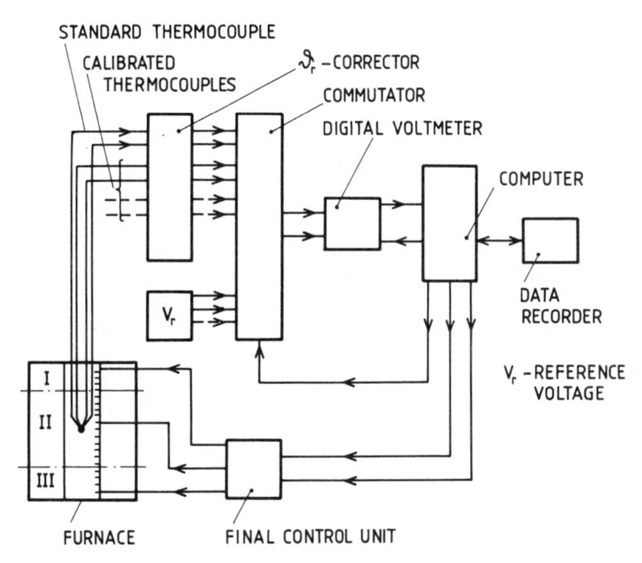

Figure 16.23
Block diagram of automatic thermocouple testing.

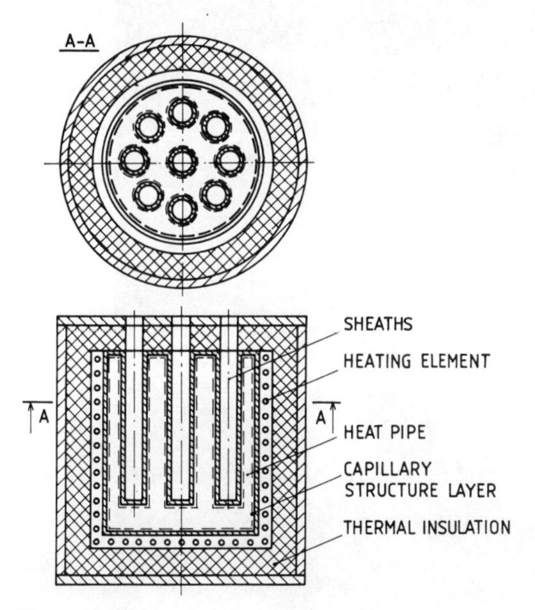

Figure 16.24
Basic diagram of a heat-pipe furnace.

of capillary structures covers the internal wall of the heat-pipe. The constructional materials and liquids used, which are compiled in Table 16.2, are chosen on the basis of the necessary temperature range. Liquid, heated by heating elements, evaporates to condense in cooler parts of the system which is connected with heat dissipation. The condensed vapour returns to the warmer heat-pipe parts, due to the capillary effect.

Table 16.2
Liquid and wall materials in the construction of heat-pipes.

Liquid	Wall material	Temperature range (°C)
Freon	Copper, aluminium	120–300
Ammonia	Stainless steel, nickel, aluminium	230–330
Acetone	Copper, stainless steel	230–420
Water	Copper, nickel, titanium	300–550
Organic liquids	Stainless steel, steel	400–600
Mercury with admixtures	Stainless steel	450–800
Potassium	Stainless steel, nickel	700–1000
Sodium	Stainless steel, nickel, inconel	800–1350

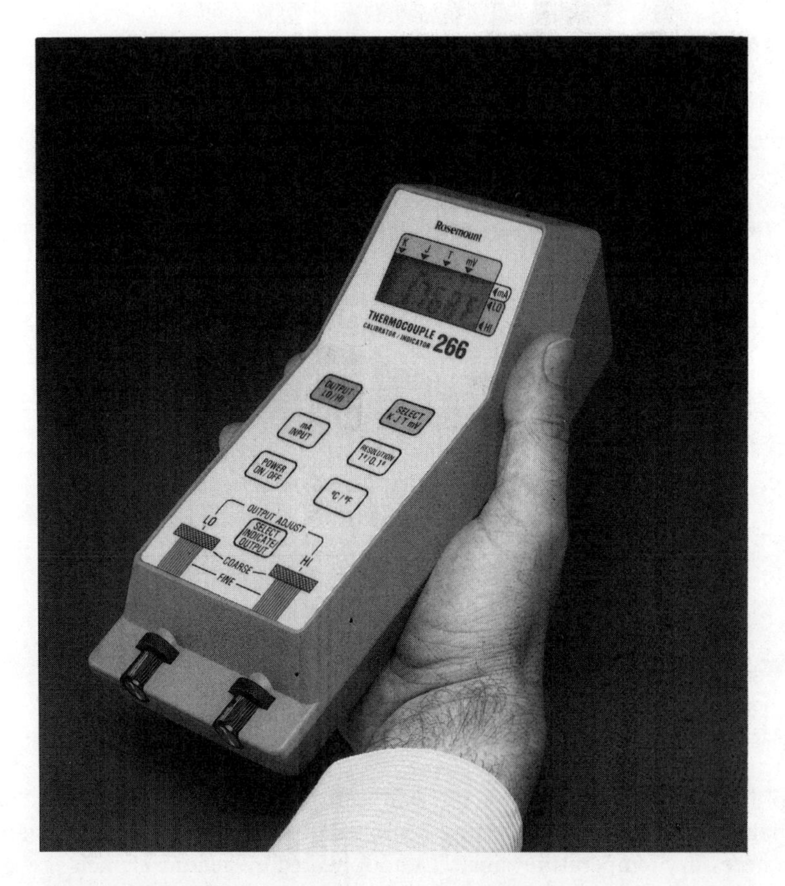

Figure 16.25
Hand-held thermocouple calibrator. (Courtesy of Rosemount Inc., USA.)

This highly efficient heat transfer process, which proceeds continuously, contributed to a very high temperature uniformity.

Heat-pipes are also used in the construction of black body models, used for calibration of radiation pyrometers (Neuer and Brost, 1975; Aa and Gelok, 1987). A heat-pipe based black body, constructed by Institut für Kerntechnik of Stuttgart University, has an equivalent emissivity of 0.998 with temperature differences at the opening bottom, below 1 °C. This apparatus operates in the temperature range from 340 to 1400 °C.

16.5.4 Calibrators

Hand-held, rugged calibrators have been developed for daily maintenance work and periodical testing of temperature, measuring equipment in laboratories and for field use. As good examples of modern calibrators two calibrators by Rosemount Inc. (USA) will be presented. The Model 266 calibration indicator (Figure 16.25), built for thermocouples, can measure mV signals, output signals from a thermocouple type K, I or T with ensured cold junction compensation and also 4 to 20 mA d.c. signals from transmitters. In the calibrator mode it functions as a signal simulator by providing an adjustable type I, K or T thermocouple output for testing temperature measuring instruments, transmitters or recorders. Both input and output values are indicated on a liquid crystal display. Setting of operation mode, thermocouple type, range, resolution and temperature e.m.fs is easily done by push-button switches in the keyboard. This microprocessor based battery-operated calibrator, which is intrinsically safe for operation in hazardous environments, has an accuracy of 0.1% of reading ±0.5 °C or 0.1% of reading ±1 °C, depending on chosen resolution value.

A similar calibrator, Model 267, which is offered for RTD sensors, can measure the output signals from Pt100 Ω and Cu 10 Ω resistors. It can also measure ohm sources and simulate sensors while testing measuring instruments, recorders or transmitters. The accuracy of this type is the same as described above.

REFERENCES

Aa van der, H. H. M. and Gelok, E. J. (1987) Design and use of a small, fixed temperature heat pipe black-body source. *3rd International Conference TEMPMEKO 87*, Sheffield, The Institute of Measurement and Control, London, pp. 87–98.

Anderson, R. L. (1972) The high temperature stability of platinum resistance thermometers. *Temperature: Its Measurement and Control in Science and Industry*, Vol. 4, Part 2, Instrument Society of America, Pittsburgh, pp. 927–934.

Baker, H. D., Ryder, E. A. and Baker, N. H. (1953) *Temperature Measurement in Engineering*, Vol. 1, John Wiley and Sons, New York.

Bedford, R. E. (1972) Effective emissivities of black-body cavities—a review. *Temperature: Its Measurement and Control in Science and Industry*, Vol. 4, Part 1, Instrument Society of America, Pittsburgh, pp. 425–434.

Berry, R. I. (1982) Oxidation, stability and insulation characteristics of Rosemount standard platinum resistance thermometers. *Temperature: Its Measurement and Control in Science and Industry*, Vol. 6, Instrument Society of America, Washington, pp. 753–762.

Busse, C. A. *et al.* (1975) The gas-controlled heat pipe—a temperature-pressure transducer. *Temperature Measurement*, Conference Series, No. 26, The Institute of Physics, London, 1975, pp. 428–438.

Chattle, M. V. (1972) Platinum resistance thermometry up to the gold point. *Temperature: Its Measurement and Control in Science and Industry*, Vol. 4, Part 2, Instrument Society of America, Pittsburgh, pp. 907–918.

Chattle, M. V. and Pokhodun, A. I. (1987) An intercomparison between fixed-point cells made at VNIIM (USSR) and NPL (UK) for the realisation of the melting and triple points of gallium and the solidification points of indium and cadmium, *Proc. 3rd International Conference TEMPMEKO 87, Sheffield*, The Institute of Measurement and Control, London, pp. 41–50.

Coville, P. and Laurencier, A. (1975) Inter-calibration of temperature transducer with a heat pipe furnace. *Temperature Measurement*, Conference Series, No. 26, The Institute of Physics, London, 1975, pp. 439–445.

Curtis, D. I. (1972) Platinum resistance interpolation standards. *Temperature: Its Measurement and Control in Science and Industry*, Vol. 4, Part 2, Instrument Society of America, Pittsburgh, pp. 951–961.

Evans, J. P. (1972) High temperature platinum resistance thermometry, *Temperature: Its Measurement and Control in Science and Industry*, Vol. 4, Part 2, Instrument Society of America, Pittsburgh, pp. 899–906.

Evans, I. P. and Burns, G. W. (1962) A study of stability of high temperature platinum resistance thermometers. *Temperature: Its Measurement and Control in Science and Industry*, Vol. 4, Part 1, Instrument Society of America, Pittsburgh, pp. 899–906.

Foster, F. B. (1972) A fixed point calibration procedure for precision platinum resistance thermometers. *Temperature: Its Measurement and Control in Science and Industry*, Vol. 4, Part 2, Instrument Society of America, Pittsburgh, pp. 1403–1414.

Furukawa, G. T. *et al.* (1975) The freezing point of aluminium as a temperature standard. *Temperature Measurement*, Conference Series, No. 26, The Institute of Physics, London, 1975, pp. 389–397.

Furukawa, G. T. *et al.* (1981) Temperature research above 0 °C at the National Bureau of Standards. *Temperature Measurement in Industry and Science*, Second Symposium of IMEKO TC-12 Committee, Czechoslovak Scientific and Technical Society, Praha, pp. 39–47.

Gray, G. N. and Chandon, H. C. (1972) Development of a comparison calibration capability. *Temperature: Its Measurement and Control in Science and Industry*, Vol. 4, Part 2, Instrument Society of America, Pittsburgh, pp. 1369–1378.

Gray, W. T. and Finch, D. J. (1972) Accuracy of temperature measurement. *Temperature: Its Measurement and Control in Science and Industry*, Vol. 4, Part 2, Instrument Society of America, Pittsburgh, pp. 1381–1392.

Hall, J. A. (1956) The international temperature scale. *Temperature: Its Measurement and Control in Science and Industry*, Vol. 2, Reinhold, New York, pp. 115–139.

Hall, J. A. and Barber, C. R. (1964) Calibration of temperature measuring instruments. *Notes on Applied Science*, No. 12, National Physical Laboratory, London.

Hanafy, M., Moussa, M. R. and Omar, H. J. (1982) A small transportable indium cell for use as temperature reference. *Temperature: Its Measurement and Control in Science and Industry*, Vol. 6, Instrument Society of America, Pittsburgh, pp. 347–350.

Jones, T. P. and Egan, T. M. (1975) The automatic calibration of thermocouples in the range 0–1100 °C. *Temperature Measurement*, Conference Series, No. 26, The Institute of Physics, London, 1975, pp. 211–218.

Kandyba, V. V. and Kowalewskij, V. A. (1956) A phototelectric spectropyrometer of high precision. *Doklady Akad. Nauk. SSSR*, **108**, 633–670 (in Russian).

Kirby, C. G. M. (1982) Automation of a thermometer calibration facility. *Temperature: Its Measurement and Control in Science and Industry*, Vol. 6, Instrument Society of America, Pittsburgh, pp. 1293–1298.

Kortvelyessy, L. (1987). *Thermoelement Praxis*, 2nd edition, Vulkan-Verlag, Essen.

Kostkowski, H. I. and Lee, R. D. (1962) Theory and methods of optical pyrometry. *Temperature: Its Measurement and Control in Science and Industry*, Vol. 3, Part 1, Reinhold, New York, pp. 449–481.

Lee, R. D. (1966) The NBS photoelectric pyrometer and its use in realizing the International Practical Temperature Scale above 1063 °C. *Metrologia*, **2**(4), 150–162.

Lee, R. D. *et al.* (1972) Intercomparison of the IPTS-68 above 1064 °C by four National Laboratories. *Temperature: Its Measurement and Control in Science and Industry*, Vol. 4, Part 1, Instrument Society of America, Pittsburgh, pp. 377–393.

Marcarino, P. *et al.* (1981) An oil bath for high accuracy thermometer comparison. *Temperature Measurement in Industry and Science*, Second Symposium of IMEKO, TC-12 Committee, Czechoslovak Scientific and Technical Society, Praha, pp. 56–61.

Neuer, G. and Brost, O. (1975) Heat pipes for the realization of isothermal conditions at temperature reference sources. *Temperature Measurement*, Conference Series, No. 26, The Institute of Physics, London, pp. 446–452.

NPL-1982, National Physical Laboratory, Measurement Services, Temperature, Teddington.

NPL-ITS-90 (1990) *Adoption of the International Temperature Scale of 1990*. National Physical Laboratory, Teddington.

Nutter, G. D. (1972) A high precision automatic optical pyrometer. *Temperature: Its Measurement and Control in Science and Industry*, Vol. 4, Part 1, Instrument Society of America, Pittsburgh, pp. 519–530.

Preston-Thomas, H. (1956) The zinc point as a thermometric fixed point. *Temperature: Its Measurement and Control in Science and Industry*, Vol. 2, Reinhold Publ. Co., New York, pp. 169–177.

Preston-Thomas, H. (1990) The International Temperature Scale of 1990 (ITS-90). *Metrologia*, **27**(1), 4–10.

Quinn, T. I. (1975) Temperature standards. *Temperature Measurement*, Conference Series, No. 26, The Institute of Physics, London, pp. 1–16.

Quinn, T. I. and Lee, R. D. (1972) Vacuum tungsten strip lamps with improved stability as radiance temperature standards. *Temperature: Its Measurement and Control in Science and Industry*, Vol. 4, Part 1, Instrument Society of America, Pittsburgh, pp. 395–411.

Richardson, S. C. (1962) Temperature standards and practices in a large industrial company. *Temperature: Its Measurement and Control in Science and Industry*, Vol. 3, Part 2, Reinhold Publ. Co., New York, pp. 39–44.

Roberts, P. I. (1980) The importance of thermocouple calibrations. *Measurement and Control*, **13**(6), 213–217.

Roeser, W. F. and Wensel, H. T. (1941) Methods of testing thermocouples and thermocouple materials. *Temperature: Its Measurement and Control in Science and Industry*, Reinhold Publ. Co., New York, pp. 284–314.

Ruffino, C., Righini, F. and Rosso, A. (1972) A photoelectric pyrometer with effective wavelength in the near infrared. *Temperature: Its Measurement and Control in Science and Industry*, Vol. 4, Part 1, Instrument Society of America, Pittsburgh, pp. 531–537.

Sergeyev, C. A. (1979) Instruments for temperature and thermophysical measurements. *Ismer. Technika*, **4**, 16–19 (in Russian).

Sparks, L. L. and Powell, R. L. (1972) Calibration of capsule platinum resistance thermometers at the triple point of water. *Temperature: Its Measurement and Control in Science and Industry*, Vol. 4, Part 2, Instrument Society of America, Pittsburgh, pp. 1415–1421.

Staffin, H. K. and Rim, C. H. (1972) Calibration of temperature sensors between 538 °C (1000 °F) and 1092 °C (2000 °F) in air fluidized solids. *Temperature: Its Measurement and*

Control in Science and Industry, Vol. 4, Part 2, Instrument Society of America, Pittsburgh, pp. 1359–1368.

Stimson, H. F. (1956) Precision resistance thermometry and fixed points. *Temperature: Its Measurement and Control in Science and Industry*, Vol. 2, Reinhold Publ. Co., New York, pp. 141–168

Trabold, W. G. (1962) An industrial thermocouple calibration facility. *Temperature: Its Measurement and Control in Science and Industry*, Vol. 3, Part 2, Instrument Society of America, Pittsburgh, pp. 45–50.

Appendix
Measurement Theory and Temperature Scales

A1.1 INTRODUCING CONTEMPORARY MEASUREMENT THEORY

The tortuous path in the development of a scale of temperature is similar to that in the development of the contemporary theory of measurement. Galileo Galilei programmatically told his associates 'what is not measurable make measurable'. Subsequently, Newton's theory of arithmetic magnitudes was implicitly supported when Lord Kelvin drew attention to the cardinal importance of attaching numbers to the description of physical phenomena. The work of Helmholtz (1887) provided the British physicist N. R. Campbell with a thorough theoretical foundation for his development of what is now commonly known as extensive measurement (Campbell, 1957). All of these pioneering efforts, which led to the development of the contemporary theory of measurement, are important as they give a good contextual understanding of it. Measurement is a fundamental enabling function of information technology. A thorough and well developed theory is of cardinal importance in measurement technology as its seminal activity is instrumental in the gathering of information which may lead to the acquisition and subsequent distribution of knowledge. It is easy to underestimate the importance and essential nature of such a theory as well as to dismiss its relevance.

The contemporary formal theory of measurement is essentially representational. This description arises from the fact that numbers are used in an empirical, but nevertheless objective, way to represent some properties of either objects or events in the real world. As temperature describes such a quality, any general representational theory of measurement may be applied to it. Because of the formalism, which now exists regarding this theory, no text on temperature measurement can be complete without taking account of it.

A representational theory of measurement has four parts (Finkelstein, 1982). The first part acknowledges the existence of some empirical relational system corresponding to the qualitative property which is to be represented. A representation condition, which is embodied in a homomorphic mapping, defines the manner in which the empirical relational system may be mapped into a representative relational system in the field of numbers. This numerical relational system, which constitutes the third element of the theory, must also be uniquely representative. Uniqueness is essential to the representation to ensure that different manifestations of the same quality in the same or different objects are represented by different numbers. Of course an essential corollary to this requirement is that similar manifestations are assigned the same number.

When the above four conditions are fulfilled, it is apparent that, in principle, the quality manifestation may be measured. The next essential part of the process of measurement is the proposal of a suitably meaningful scale of measurement. Scales, which may be either direct or indirect, are further sub-classified in Figure A1. As temperature is an intensive quality manifestation it is only possible to propose an associative scale for its measurement.

Formally, the foregoing theory can be interpreted in the context of temperature measurement in the following manner.

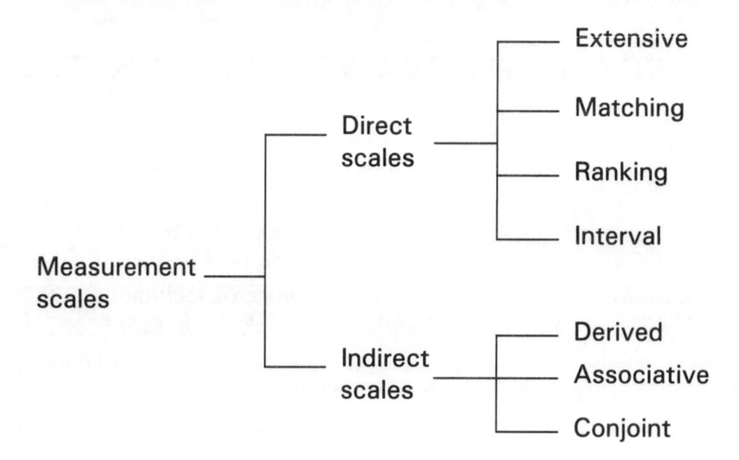

Figure A.1.
Classification of measurement scales.

A1.2 A REPRESENTATIONAL THEORY FOR TEMPERATURE MEASUREMENT

A1.2.1 Hotness as an empirical relational system

Consider the class of all objects, which manifest the quality of hotness or coldness, defined as the set:

$$\Omega=\{w_1, w_2, w_3, \ldots, w_i, \ldots\} \tag{A1.1}$$

Suppose that each member, w_i, of this class, Ω, may exhibit the quality of hotness, H, defined by the set of manifestations:

$$H=\{h_1, h_2, h_3, \ldots, h_i, \ldots\} \tag{A1.2}$$

Some bodies may possess a similar degree of hotness. Alternatively some bodies may be hotter (or cooler) than these. Consequently, a set of relations must exist on, H. The first of these relations describes the idea of empirical similarity, also called indifference, which is normally denoted by the symbol $\sim$. The second relation describes empirical transitiveness, symbolized by $\succ$. An empirical set of relations, $\mathscr{R}_H$, which consists of similarity and transitiveness, thus exists on the hotness. This set of empirical relations may be written:

$$\mathscr{R}_H = \langle \sim, \succ \rangle \tag{A1.3}$$

The hotness quality, H, upon which the empirical relations exist, may thus be described by the empirical relational system:

$$\mathscr{H} = \langle H, \mathscr{R}_H \rangle \tag{A1.4}$$

A1.2.2 *Temperature as a number relational system representing hotness*

Let a class of numbers called temperature, T, exist. It is further noted that certain values of T are equal, symbolized by $=$, while still others may be greater than (i.e. $>$) or less than (i.e. $<$) these. It is evident that the relations which exist upon the numbers called temperature are described as:

$$\mathscr{R}_T = \langle =, > \rangle \tag{A1.5}$$

Temperature is thus the number relational system:

$$\mathscr{T} = \langle \mathrm{T}, \mathscr{R}_T \rangle \tag{A1.6}$$

In order that the hotness quality, H, may be represented by the numbers, temperature, T, it is essential that the relations, $\mathscr{R}_H$, on H imply and are implied by the relations, $\mathscr{R}_T$, on temperature, T. Formally temperature measurement is defined as the objective empirical operation:

$$M_T : H \rightarrow T \tag{A1.7}$$

in such a way that the hotness quality, $\mathscr{H} = \langle \mathrm{H}, \mathscr{R}_H \rangle$, is mapped homomorphically on to $\mathscr{T} = \langle \mathrm{T}, \mathscr{R}_T \rangle$ by M_T and F_T. The operation F_T, which is a one-to-one mapping with domain $\mathscr{R}_H$ and range $\mathscr{R}_T$, may be written:

$$F_T : \mathscr{R}_H \rightarrow \mathscr{R}_T \tag{A1.8}$$

so that the relations $=$ or $>$ in $\mathscr{R}_T$ respectively imply and are respectively implied by the relations $\sim$ and $\succ$ in $\mathscr{R}_H$.

Temperature measurement is homomorphic because the operation M_T maps separate but indistinguishable manifestations of hotness to the same temperature T. Thus, it is evident that the quadruplet:

$$\mathscr{S}_T = \langle \mathscr{H}, \mathscr{T}, M_T, F_T \rangle \qquad \text{(A1.9)}$$

constitutes a scale of temperature for $t_i = M_T(h_i)$. The elemental temperature, t_i, of the set of temperatures, T, which is the image of the elemental hotness, h_i, of hotness qualities, H, under the operation M_T is called the measure of hotness on the temperature scale $\mathscr{S}_T$.

A1.2.3 Associative scales for temperature measurement

As hotness is an intensive, empirical quality manifestation there is no operation of combining hotness values to obtain a new hotness value by an operation analogous to concatenation or addition of these two values. For this reason it is not possible to set up an extensive scale for temperature measurement. It becomes necessary to propose a suitable temperature scale by associating the measurement of temperature, representing hotness, with some other, more easily measurable qualities.

The behaviour of many materials depends upon their hotness. Among others example effects are:

(a) the number relational coefficients of linear and cubic thermal expansion of materials, respectively given as α_{elT} and β_{ecT},
(b) the number relational thermoelectric e.m.f. in the junction of dissimilar metals, e_{eT},
(c) the number relational temperature dependence of the resistance of conductors, $\mathscr{R}_{RT}$,
(d) the number relational temperature dependence of carrier concentration in semiconductors, $\mathscr{P}_{pT}$
(e) the number relational general temperature dependent material property $\mathscr{G}_{mpT}$.

The existence of a set of measures of the above qualities, associated with temperature, provides the basis of an associative scale of temperature.

To establish meaningful associative scales of temperature three conditions must be fulfilled. The first of these requires that the relations between the hotness manifestations in a physical object or event must be isomorphically represented, that is on a one-to-one basis, by the relations between the general material parameter, $\mathscr{G}_{mpT}$, possessed by the identical object or event. Of equal importance are the existence of both a scale of measurement for this material parameter itself and some function defining the manner in which temperature is associated with it. This function, say $G = f(T)$, may be empirically derived but will generally require to have a monotonic increasing character. When these are complied with it is apparent that an associative scale of temperature may be proposed using the associated general temperature dependent parameter of the thermometric material.

Consider the class of all objects defined by Equation (A1.1), which may exhibit the quality of hotness in such manner as to enable the definition of the scale of temperature $\mathscr{S}_T$ given by Equation (A1.9). Suppose now that each member, w_i, of this class, Ω, also exhibits the general material parameter, G_{mpT}. Let this be defined in a similar manner as the hotness quality, H, of Equation (A1.4). Thus the general material parameter is described by the empirical and number relations and transforms:

$$\mathscr{G}=\langle G_{\mathrm{mpT}}, \mathscr{R}_{\mathrm{G}}\rangle, \mathscr{N}=\langle N, \mathscr{R}_{\mathrm{N}}\rangle; F_{\mathrm{G}}:\mathscr{R}_{\mathrm{G}}\rightarrow\mathscr{R}_{\mathrm{N}}; M_{\mathrm{G}}:G_{\mathrm{mpT}}\rightarrow N \qquad \text{(A1.10)}$$

while its scale of measurement is the quadruplet:

$$\mathscr{S}_{\mathrm{G}}=\langle\mathscr{G}, \mathscr{N}, M_{\mathrm{G}}, F_{\mathrm{G}}\rangle \qquad \text{(A1.11)}$$

Since the scale of temperature, $\mathscr{S}_{\mathrm{T}}$, exists as Equation (A1.9), the scale of material parameter, $\mathscr{S}_{\mathrm{G}}$, exists as Equation (A1.11) it follows that the representation condition giving G as a measure of T, which is a measure of H, is (Finkelstein, 1975; Leaning and Finkelstein, 1979):

$$h_1 \succ_{\sim \mathrm{H}} h_2 \rightleftharpoons t_1 \geqslant_{\mathrm{T}} t_2 \rightleftharpoons f[M_{\mathrm{T}}(h_1)] \geqslant_{\mathrm{G}} f[M_{\mathrm{T}}(h_2)] \qquad \text{(A1.12)}$$

where $\rightleftharpoons$ means 'implies and is implied by' and f is some known function.

Other relevant considerations about the implications of associative scales for temperature measurement are given in §1.2.

A1.2.4 *Uncertainty in temperature measurement*

The above theory of temperature measurement is totally deterministic. It is therefore a model or paradigm of actual, practical measurement which must always contain an element of uncertainty. A probabilistic theory, which purports to deal with the uncertainty of measurement, has been proposed by Leaning and Finkelstein (1979).It uses an extension of the theory of probabilistic relational systems.

The important conceptual ingredient of this probabilistic theory, as it applies to temperature measurement, is the replacement of the deterministic representation theorem of Equation (A1.8) by a suitable probabilistic representation theorem which includes a probabilistic homomorphism. It may be concluded that there are two contributory factors to uncertainty in indirect associative temperature measurement. The first indicates that any uncertainty in the empirical relations themselves must give rise to errors in the measurement. The second source of error is due to uncertainties associated with the doubts regarding the veracity of the mathematical law. There are two factors which may give rise to this. The first concerns the structural validity of the mathematical law, $f(T)$, which may be called the inter-model relations, while the second is due to inaccuracies in measuring the associated parameter on the scale represented by Equation (A1.11). This probabilistic theory of temperature measurement will not be considered further.

REFERENCES

Campbell, N. R. (1920) *Physics: The Elements*. Cambridge University Press, Cambridge (Reprinted 1957 under the title *Foundations of Science*, Dover, New York).

Campbell, N. R. (1957) *Foundations of Science*, Dover, New York (Reprinted from (1920) *Physics: The Elements*, Cambridge University Press, Cambridge).

Finkelstein, L. (1975) Fundamental concepts of measurement: definitions and scales. *Measurement and Control,* **8**, 105–111.

Finkelstein, L. (1982) Theory and philosophy of measurement, in Sydenham, P. H. (Ed.) *Handbook of Measurement Science*. Vol. I *Theoretical Fundamentals*, John Wiley and Sons, Chichester.

Helmholtz, H. V. (1887) Zahlen und messen Erkenntnis–theoretische Betracht, Philosophische Aufsaetze Eduard Zeller gewidmet, Leipzig. (Translated by C. L. Brian (1930) under the title *Counting and Measuring*, Van Nostrand, New York.)

Leaning, M. and Finkelstein, L. (1979) A probabilistic treatment of measurement uncertainty in the formal theory of measurement, *Acta IMEKO VIII*.

Auxiliary Tables

Table I
Conversion of Fahrenheit to Celsius scale of temperature.

°F	°C	°F	°C	°F	°C	°F	°C	°F	°C	°F	°C	°F	°C	°F	°C	F°	°C
-459.4	-273.1	0	-17.8	50	10.0	100	38	500	260	1000	538	1500	816	2000	1093	2500	1371
-450	-268	1	-17.2	51	10.6	110	43	510	266	1010	543	1510	821	2010	1099	2510	1377
-440	-262	2	-16.7	52	11.1	120	49	520	271	1020	549	1520	827	2020	1104	2520	1382
-430	-257	3	-16.1	53	11.7	130	54	530	277	1030	554	1530	832	2030	1110	2530	1388
-420	-251	4	-15.6	54	12.2	140	60	540	282	1040	560	1540	838	2040	1116	2540	1393
-410	-246	5	-15.0	55	12.8	150	66	550	288	1050	566	1550	843	2050	1121	2550	1399
-400	-240	6	-14.4	56	13.3	160	71	560	293	1060	571	1560	849	2060	1127	2560	1404
-390	-234	7	-13.9	57	13.9	170	77	570	299	1070	577	1570	854	2070	1132	2570	1410
-380	-229	8	-13.3	58	14.4	180	82	580	304	1080	582	1580	860	2080	1138	2580	1416
-370	-223	9	-12.8	59	15.0	190	88	590	310	1090	588	1590	866	2090	1143	2590	1421
-360	-218	10	-12.2	60	15.6	200	93	600	316	1100	593	1600	871	2100	1149	2600	1427
-350	-212	11	-11.7	61	16.1	210	99	610	321	1110	599	1610	877	2110	1154	2610	1432
-340	-207	12	-11.1	62	16.7	212	100	620	327	1120	604	1620	882	2120	1160	2620	1438
-330	-201	13	-10.6	63	17.2	220	104	630	332	1130	610	1630	888	2130	1166	2630	1443
-320	-196	14	-10.0	64	17.8	230	110	640	338	1140	616	1640	893	2140	1171	2640	1449
-310	-190	15	-9.44	65	18.3	240	116	650	343	1150	621	1650	899	2150	1177	2650	1454
-300	-184	16	-8.89	66	18.9	250	121	660	349	1160	627	1660	904	2160	1182	2660	1460
-290	-179	17	-8.33	67	19.4	260	127	670	354	1170	632	1670	910	2170	1188	2670	1466
-280	-173	18	-7.78	68	20.0	270	132	680	360	1180	638	1680	916	2180	1193	2680	1471
-273	-169	19	-7.22	69	20.6	280	138	690	366	1190	643	1690	921	2190	1199	2690	1477
-270	-168	20	-6.67	70	21.1	290	143	700	371	1200	649	1700	927	2200	1204	2700	1482
-260	-162	21	-6.11	71	21.7	300	149	710	377	1210	654	1710	932	2210	1210	2710	1488
-250	-157	22	-5.56	72	22.2	310	154	720	382	1220	660	1720	938	2220	1216	2720	1493
-240	-152	23	-5.00	73	22.8	320	160	730	388	1230	666	1730	943	2230	1221	2730	1499
-230	-146	24	-4.44	74	23.3	330	166	740	393	1240	671	1740	949	2240	1227	2740	1504
-220	-140	25	-3.89	75	23.9	340	171	750	399	1250	677	1750	954	2250	1232	2750	1510
-210	-134	26	-3.33	76	24.4	350	177	760	404	1260	682	1760	960	2260	1238	2760	1516
-200	-129	27	-2.78	77	25.0	360	182	770	410	1270	688	1770	966	2270	1243	2770	1521
-190	-123	28	-2.22	78	25.6	370	188	780	416	1280	693	1780	971	2280	1249	2780	1527
-180	-118	29	-1.67	79	26.1	380	193	790	421	1290	699	1790	977	2290	1254	2790	1532
-170	-112	30	-1.11	80	26.7	390	199	800	427	1300	704	1800	982	2300	1260	2800	1538
-160	-107	31	-0.56	81	27.2	400	204	810	432	1310	710	1810	988	2310	1266	2810	1543
-150	-101	32	0	82	27.8	410	210	820	438	1320	716	1820	993	2320	1271	2820	1549
-140	-95.6	33	0.56	83	28.3	420	216	830	443	1330	721	1830	999	2330	1277	2830	1554
-130	-90.0	34	1.11	84	28.9	430	221	840	449	1340	727	1840	1004	2340	1282	2840	1560
-120	-84.4	35	1.67	85	29.4	440	227	850	454	1350	732	1850	1010	2350	1288	2850	1566
-110	-78.9	36	2.22	86	30.0	450	232	860	460	1360	738	1860	1016	2360	1293	2860	1571
-100	-73.3	37	2.78	87	30.6	460	238	870	466	1370	743	1870	1021	2370	1299	2870	1577
-90	-67.8	38	3.33	88	31.1	470	243	880	471	1380	749	1880	1027	2380	1304	2880	1582
-80	-62.2	39	3.89	89	31.7	480	249	890	477	1390	754	1890	1032	2390	1310	2890	1588
-70	-56.7	40	4.44	90	32.2	490	254	900	482	1400	760	1900	1038	2400	1316	2900	1593
-60	-51.1	41	5.00	91	32.8			910	488	1410	766	1910	1043	2410	1321	2910	1599
-50	-45.6	42	5.56	92	33.3			920	493	1420	771	1920	1049	2420	1327	2920	1604
-40	-40.0	43	6.11	93	33.9			930	499	1430	777	1930	1054	2430	1332	2930	1610
-30	-34.4	44	6.67	94	34.4			940	504	1440	782	1940	1060	2440	1338	2940	1616
-20	-28.9	45	7.22	95	35.0			950	510	1450	788	1950	1066	2450	1343	2950	1621
-10	-23.3	46	7.78	96	35.6			960	516	1460	793	1960	1071	2460	1349	2960	1627
0	-17.8	47	8.33	97	36.1			970	521	1470	799	1970	1077	2470	1354	2970	1632
		48	8.89	98	36.7			980	527	1480	804	1980	1082	2480	1360	2980	1633
		49	9.44	99	37.2			990	532	1490	810	1990	1088	2490	1366	2990	1643
				100	37.8											3000	1649

Data for interpolation

°F	1	2	3	4	5	6	7	8	9	10
°C	0.56	1.11	1.67	2.22	2.78	3.33	3.89	4.44	5.00	5.56

Example: 1622 °F = 822 + 1.11 = 883.11 °C

Table II
Reference calibration table for thermocouple copper/constantan, code T (Cu–Cu,Ni), conforms to IEC 584. Reference temperature 0°C; e.m.f. in mV.

°C	0	−10	−20	−30	−40	−50	−60	−70	−80	−90	−100
−200	−5.603	−5.753	5.889	−6.007	−6.105	−6.181	−6.232	−6.258	—	—	—
−100	−3.378	−3.656	−3.923	−4.177	−4.419	−4.648	−4.865	−5.069	−5.261	−5.439	−5.603
0	0	−0.383	−0.757	−1.121	−1.475	−1.819	−2.152	−2.475	−2.788	−3.089	−3.378
°C	**0**	**10**	**20**	**30**	**40**	**50**	**60**	**70**	**80**	**90**	**100**
0	0	0.391	0.789	1.196	1.611	2.035	2.467	2.908	3.357	3.813	4.277
100	4.277	4.749	5.227	5.712	6.204	6.702	7.207	7.718	8.235	8.757	9.286
200	9.286	9.820	10.360	10.905	11.456	12.011	12.572	13.137	13.707	14.281	14.860
300	14.860	15.443	16.030	16.621	17.217	17.816	18.420	19.027	19.638	20.252	20.869
400	20.869	—	—	—	—	—	—	—	—	—	—

Table III
Reference calibration table for thermocouple copper/constantan, code U (Cu–CuNi), conforms to DIN 43710. Reference temperature 0°C; e.m.f. in mV.

°C	0	−10	−20	−30	−40	−50	−60	−70	−80	−90	−100
−200	−5.70	—	—	—	—	—	—	—	—	—	—
−100	−3.40	−3.68	−3.95	−4.21	−4.46	−4.69	−4.91	−5.12	−5.32	−5.51	−5.70
0	0	−0.39	−0.77	−1.14	−1.50	−1.85	−2.18	−2.50	−2.81	−3.11	−3.40
°C	**0**	**+10**	**+20**	**+30**	**+40**	**+50**	**+60**	**+70**	**+80**	**+90**	**+100**
0	0	0.40	0.80	1.21	1.63	2.05	2.48	2.91	3.35	3.80	4.25
100	4.25	4.71	5.18	5.65	6.13	6.62	7.12	7.63	8.15	8.67	9.20
200	9.20	9.74	10.29	10.85	11.41	11.98	12.55	13.13	13.71	14.30	14.90
300	14.90	15.50	16.10	16.70	17.31	17.92	18.53	19.14	19.76	20.38	21.00
400	21.00	21.62	22.25	22.88	23.51	24.15	24.79	25.44	26.09	26.75	27.41
500	27.41	28.08	28.75	29.43	30.11	30.80	31.49	32.19	32.89	33.60	34.31
600	34.31	—	—	—	—	—	—	—	—	—	—

Table IV

Reference calibration table for thermocouple iron/constantan, code J (Fe–Cu,Ni), conforms to IEC 584. Reference temperature 0°C; e.m.f. in mV.

°C	0	−10	−20	−30	−40	−50	−60	−70	−80	−90	−100
−200	−7.890	−8.096	—	—	—	—	—	—	—	—	—
−100	−4.632	−5.036	−5.426	−5.801	−6.159	−6.499	−6.821	−7.122	7.402	−7.659	−7.890
0	0	−0.501	−0.995	−1.481	−1.950	−2.437	−2.892	−3.344	−3.785	−4.215	−4.632

°C	0	10	20	30	40	50	60	70	80	90	100
0	0	0.507	1.019	1.536	2.058	2.585	3.115	3.649	4.186	4.725	5.268
100	5.268	5.812	6.359	6.907	7.457	8.008	8.560	9.113	9.667	10.222	10.777
200	10.777	11.332	11.887	12.442	12.998	13.553	14.108	14.663	15.217	15.771	16.325
300	16.325	16.879	17.432	17.984	18.537	19.089	19.640	20.192	20.743	21.295	21.846
400	21.846	22.397	22.949	23.501	24.054	24.607	25.161	25.716	26.272	26.829	27.388
500	27.388	27.949	28.511	29.075	29.642	30.210	30.782	31.356	31.933	32.513	33.096
600	33.096	33.683	34.273	34.867	35.464	36.066	36.671	37.280	37.893	38.510	39.130
700	39.130	39.754	40.382	41.013	41.647	42.283	42.922	43.563	44.207	44.852	45.498
800	45.498	46.144	46.790	47.434	48.076	48.716	49.354	49.989	50.621	51.249	51.875
900	51.875	52.496	53.115	53.729	54.341	54.948	55.553	56.155	56.753	57.349	57.942
1000	57.942	58.533	59.121	59.708	60.293	60.876	61.459	62.039	62.619	63.199	63.777
1100	63.777	64.355	64.933	65.510	66.087	66.664	67.240	67.815	68.390	68.964	69.536
1200	69.536	—	—	—	—	—	—	—	—	—	—

Table V

Reference calibration table for thermocouple iron/constantan, code L (Fe–CuNi), conforms to DIN 43710. Reference temperature 0°C; e.m.f. in mV.

°C	0	−10	−20	−30	−40	−50	−60	−70	−80	−90	−100
−200	−8.15	—	—	—	—	—	—	—	—	—	—
−100	−4.75	−5.15	−5.53	−5.90	−6.26	−6.60	−6.99	−7.25	−7.56	−7.86	−8.15
0	0	−0.51	−1.02	−1.53	−2.03	−2.51	−2.98	−3.44	−3.89	−4.33	−4.75

°C	0	+10	+20	+30	+40	+50	+60	+70	+80	+90	+100
0	0	0.52	1.05	1.58	2.11	2.65	3.19	3.73	4.27	4.82	5.37
100	5.37	5.92	6.47	7.03	7.59	8.15	8.71	9.27	9.83	10.39	10.95
200	10.95	11.51	12.07	12.63	13.19	13.75	14.31	14.88	15.44	16.00	16.56
300	16.56	17.12	17.68	18.24	18.80	19.36	19.92	20.48	21.04	21.60	22.16
400	22.16	22.72	23.29	23.86	24.43	25.00	25.57	26.14	26.71	27.28	27.85
500	27.85	28.43	29.01	29.59	30.17	30.75	31.33	31.91	32.49	33.08	33.67
600	33.67	34.26	34.85	35.44	36.04	36.64	37.25	37.85	38.47	39.09	39.72
700	39.72	40.35	40.98	41.62	42.27	42.92	43.57	44.23	44.89	45.55	46.22
800	46.22	46.89	47.57	48.25	48.94	49.63	50.32	51.02	51.72	52.43	53.14
900	53.14	—	—	—	—	—	—	—	—	—	—

Table VI

Reference calibration table for thermocouple nickel-chromium/nickel-aluminium, code K (NiCr–NiAl), conforms to IEC 584. Reference temperature 0°C; e.m.f. in mV.

°C	0	−10	−20	−30	−40	−50	−60	−70	−80	−90	−100
−200	−5.891	−6.035	−6.138	−6.262	−6.344	−6.404	−6.441	−6.458	—	—	—
−100	−3.553	−3.852	−4.138	−4.410	−4.669	−4.912	−5.141	−5.354	5.550	−5.730	−5.891
0	0	−0.392	−0.777	−1.156	−1.527	−1.889	−2.243	−2.586	−2.920	−3.242	−3.553

°C	0	10	20	30	40	50	60	70	80	90	100
0	0	0.397	0.798	1.203	1.611	2.022	2.436	2.850	3.266	3.681	4.095
100	4.095	4.508	4.919	5.327	5.733	6.137	6.539	6.939	7.338	7.737	8.137
200	8.137	8.537	8.938	9.341	9.745	10.151	10.560	10.969	11.381	11.793	12.207
300	12.207	12.623	13.039	13.456	13.874	14.292	14.712	15.132	15.552	15.974	16.395
400	16.395	16.818	17.241	17.664	18.088	18.513	18.938	19.363	19.788	20.214	20.640
500	20.640	21.066	21.493	21.919	22.346	22.772	23.198	23.624	24.050	24.476	24.902
600	24.902	25.327	25.751	26.176	26.599	27.022	27.445	27.867	28.288	28.709	29.128
700	29.128	29.547	29.965	30.383	30.799	31.214	31.629	32.042	32.455	32.866	33.277
800	33.277	33.686	34.095	34.502	34.909	35.314	35.718	36.121	36.524	36.925	37.325
900	37.325	37.724	38.122	38.519	38.915	39.310	39.703	40.096	40.488	40.879	41.269
1000	41.269	41.657	42.045	42.432	42.817	43.202	43.585	43.968	44.349	44.729	45.108
1100	45.108	45.486	45.863	46.238	46.612	46.985	47.356	47.726	48.095	48.462	48.828
1200	48.828	49.192	49.555	49.916	50.276	50.633	50.990	51.344	51.697	52.049	52.398
1300	52.398	52.747	53.093	53.439	53.782	54.125	54.466	54.807	—	—	—

Table VII

Reference calibration table for thermocouple nickel-chromium/constantan, code E (NiCr–CuNi), conforms to IEC 584. Reference temperature 0°C; e.m.f. in mV.

°C	0	−10	−20	−30	−40	−50	−60	−70	−80	−90	−100
−200	−8.824	−9.063	−9.274	−9.455	−9.604	−9.719	−9.797	−9.835	—	—	—
−100	−5.237	−5.680	−6.107	−6.516	−6.907	−7.279	−7.631	−7.963	−8.273	−8.561	−8.824
0	0	−0.581	−1.151	−1.709	−2.254	−2.787	−3.306	−3.811	−4.301	−4.777	−5.237

°C	0	+10	+20	+30	+40	+50	+60	+70	+80	+90	+100
0	0	0.591	1.192	1.801	2.419	3.047	3.683	4.329	4.983	5.646	6.317
100	6.317	6.996	7.683	8.377	9.078	9.787	10.501	11.222	11.949	12.681	13.419
200	13.419	14.161	14.909	15.661	16.417	17.178	17.942	18.710	19.481	20.256	21.033
300	21.033	21.814	22.597	23.383	24.171	24.961	25.754	26.549	27.345	28.143	28.943
400	28.943	29.744	30.546	31.350	32.155	32.960	33.767	34.574	35.382	36.190	36.999
500	36.999	37.808	38.617	39.826	40.236	41.045	41.853	42,662	43,470	44,278	45,085
600	45.085	45.891	46.697	47.502	48.306	49.109	49.911	50.713	51.513	52.312	53.110
700	53.110	53.907	54.703	55.498	56.291	57.083	57.873	58.663	59.451	60.237	61.022
800	61.022	61.806	62.588	63.368	64.147	64.924	65.700	66.473	67.245	68.015	68.783
900	68.783	69.549	70.313	71.075	71.835	72.593	73.350	74.104	74.857	75.608	76.358
1000	76.358	—	—	—	—	—	—	—	—	—	—

Table VIII

Reference calibration table for thermocouple Nicrosil-Nisil, code N (NiCrSi-NiSi), conforms to IEC 65 B (Secr.). Reference temperature 0°C; e.m.f. in mV.

°C	0	+10	+20	+30	+40	+50	+60	+70	+80	+90	+100
0	0	0.261	0.525	0.793	1.064	1.340	1.619	1.902	2.188	2.479	2.774
100	2.774	3.072	3.374	3.679	3.988	4.301	4.617	4.936	5.258	5.584	5.912
200	5.912	6.243	6.577	6.914	7.254	7.596	7.940	8.287	8.636	8.987	9.340
300	9.340	9.695	10.053	10.412	10.773	11.135	11.499	11.865	12.233	12.602	12.972
400	12.972	13.344	13.717	14.092	14.467	14.844	15.222	15.601	15.981	16.362	16.744
500	16.744	17.127	17.511	17.896	18.282	18.668	19.055	19.443	19.831	20.220	20.609
600	20.609	20.999	21.390	21.781	22.172	22.564	22.956	23.348	23.740	24.133	24.526
700	24.526	24.919	25.312	25.705	26.098	26.491	26.885	27.278	27.671	28.063	28.456
800	28.456	28.849	29.241	29.633	30.025	30.417	30.808	31.199	31.590	31.980	32.370
900	32.370	32.760	33.149	33.538	33.927	34.315	34.702	35.089	35.476	35.862	36.248
1000	36.248	36.633	37.018	37.403	37.786	38.169	38.552	38.934	39.316	39.696	40.076
1100	40.076	40.456	40.835	41.213	41.590	41.966	42.342	42.717	43.091	43.464	43.836
1200	43.836	44.207	44.578	44.947	45.315	45.682	46.048	46.413	46.777	47.140	47.502
1300	47.502	—	—	—	—	—	—	—	—	—	—

Table IX

Reference calibration table for thermocouple platinum-rhodium/platinum, code S (Pt10Rh-Pt), conforms to IEC 584. Reference temperature 0°C; e.m.f. in mV.

°C	0	−10	−20	−30	−40	−50	−60	−70	−80	−90	−100
0	0	−0.053	−0.103	−0.150	−0.194	−0.236	—	—	—	—	—
°C	**0**	**10**	**20**	**30**	**40**	**50**	**60**	**70**	**80**	**90**	**100**
0	0	0.055	0.113	0.173	0.235	0.299	0.365	0.432	0.502	0.573	0.645
100	0.645	0.719	0.795	0.872	0.950	1.029	1.109	1.190	1.273	1.356	1.440
200	1.440	1.525	1.611	1.698	1.785	1.873	1.962	2.051	2.141	2.232	2.323
300	2.323	2.414	2.506	2.599	2.692	2.786	2.880	2.974	3.069	3.164	3.260
400	3.260	3.356	3.452	3.549	3.643	3.743	3.840	3.938	4.036	4.135	4.234
500	4.234	4.333	4.432	4.532	4.632	4.732	4.832	4.933	5.034	5.136	5.237
600	5.237	5.339	5.442	5.544	5.648	5.751	5.855	5.960	6.064	6.169	6.274
700	6.274	6.380	6.486	6.592	6.699	6.805	6.913	7.020	7.128	7.236	7.345
800	7.345	7.454	7.563	7.672	7.782	7.892	8.003	8.114	8.225	8.336	8.448
900	8.448	8.560	8.673	8.786	8.899	9.012	9.126	9.246	9.355	9.470	9.585
1000	9.585	9.700	9.816	9.932	10.048	10.165	10.282	10.400	10.517	10.635	10.754
1100	10.754	10.872	10.991	11.110	11.229	11.348	11.467	11.587	11.707	11.827	11.947
1200	11.947	12.067	12.188	12.308	12.429	12.550	12.671	12.792	12.913	13.034	13.155
1300	13.155	13.276	13.397	13.519	13.640	13.761	13.883	14.004	14.125	14.247	14.368
1400	14.368	14.489	14.610	14.731	14.852	14.973	15.094	15.215	15.336	15.456	15.576
1500	15.576	15.697	15.817	15.937	16.057	16.176	16.296	16.415	16.534	16.653	16.771
1600	16.771	16.890	17.008	17.125	17.243	17.360	17.477	17.594	17.711	17.826	17.942
1700	17.942	18.056	18.170	18.282	18.594	18.304	18.612	18.708	—	—	—

Table X

Reference calibration table for thermocouple platinum-rhodium/platinum, code R (Pt13Rh–Pt), conforms to IEC 584. Reference temperature 0°C; e.m.f. in mV.

°C	0	−10	−20	−30	−40	−50	−60	−70	−80	−90	−100
0	0	−0.051	−0.100	−0.145	−0.188	−0.226	—	—	—	—	—
°C	**0**	**10**	**20**	**30**	**40**	**50**	**60**	**70**	**80**	**90**	**100**
0	0	0.054	0.111	0.171	0.232	0.296	0.363	0.431	0.501	0.573	0.647
100	0.647	0.723	0.800	0.879	0.959	1.041	1.124	1.208	1.294	1.380	1.486
200	1.486	1.557	1.647	1.738	1.830	1.923	2.017	2.111	2.207	2.303	2.400
300	2.400	2.498	2.596	2.695	2.795	2.896	2.997	3.099	3.201	3.304	3.407
400	3.407	3.511	3.616	3.721	3.826	3.933	4.039	4.146	4.254	4.362	4.471
500	4.471	4.580	4.689	4.799	4.910	5.021	5.132	5.244	5.356	5.469	5.582
600	5.582	5.696	5.810	5.925	6.040	6.155	6.272	6.388	6.505	6.623	6.741
700	6.741	6.860	6.979	7.098	7.218	7.339	7.460	7.582	7.703	7.826	7.949
800	7.949	8.072	8.196	8.320	8.445	8.570	8.696	8.822	8.949	9.076	9.203
900	9.203	9.331	9.460	9.585	9.718	9.848	9.978	10.109	10.240	10.371	10.503
1000	10.503	10.636	10.768	10.902	11.035	11.170	11.304	11.439	11.574	11.710	11.846
1100	11.846	11.983	12.119	12.257	12.394	12.532	12.669	12.808	12.946	13.085	13.224
1200	13.224	13.363	13.502	13.642	13.782	13.922	14.062	14.202	14.343	14.483	14.624
1300	14.624	14.765	14.906	15.047	15.188	15.329	15.470	15.611	15.752	15.893	16.035
1400	16.035	16.176	16.317	16.458	16.599	16.741	16.882	17.022	17.163	17.304	17.445
1500	17.445	17.585	17.726	17.866	18.006	18.146	18.286	18.425	18.564	18.703	18.842
1600	18.842	18.981	19.119	19.257	19.395	19.533	19.670	19.807	19.944	20.080	20.215
1700	20.215	20.350	20.483	20.616	20.746	20.878	21.006	—	—	—	—

Table XI

Reference calibration table for thermocouple platinum-rhodium/platinum-rhodium, code B (Pt30Rh–Pt6Rh), conforms to IEC 584. Reference temperature 0°C; e.m.f. in mV.

°C	0	10	20	30	40	50	60	70	80	90	100
0	0	−0.002	−0.003	−0.002	0	0.002	0.006	0.011	0.017	0.025	0.033
100	0.033	0.043	0.053	0.065	0.078	0.092	0.107	0.123	0.140	0.159	0.178
200	0.178	0.199	0.220	0.243	0.266	0.291	0.317	0.344	0.372	0.401	0.431
300	0.431	0.462	0.494	0.527	0.561	0.596	0.632	0.669	0.707	0.746	0.786
400	0.786	0.827	0.870	0.913	0.957	1.002	1.048	1.095	1.143	1.192	1.241
500	1.241	1.292	1.344	1.397	1.450	1.505	1.560	1.617	1.674	1.732	1.791
600	1.791	1.851	1.912	1.974	2.036	2.100	2.164	2.230	2.296	2.363	2.430
700	2.430	2.499	2.569	2.639	2.710	2.782	2.855	2.928	3.003	3.078	3.154
800	3.154	3.231	3.308	3.387	3.466	3.546	3.626	3.708	3.790	3.873	3.957
900	3.957	4.041	4.126	4.212	4.298	4.386	4.474	4.562	4.652	4.742	4.833
1000	4.833	4.924	5.016	5.109	5.202	5.297	5.391	5.487	5.583	5.680	5.777
1100	5.777	5.875	5.973	6.073	6.172	6.273	6.374	6.475	6.577	6.680	6.783
1200	6.783	6.887	6.991	7.096	7.202	7.308	7.414	7.521	7.628	7.736	7.845
1300	7.845	7.953	8.063	8.172	8.283	8.393	8.504	8.616	8.727	8.839	8.952
1400	8.952	9.065	9.178	9.291	9.405	9.519	9.634	9.748	9.863	9.979	10.094
1500	10.094	10.210	10.325	10.441	10.558	10.674	10.790	10.907	11.024	11.141	11.257
1600	11.257	11.374	11.491	11.608	11.725	11.842	11.959	12.076	12.193	12.310	12.426
1700	12.426	12.543	12.659	12.776	12.892	13.008	13.124	13.239	13.354	13.470	13.585
1800	13.585	13.699	13.814	—	—	—	—	—	—	—	—

Table XII

Output tolerances of standardized thermocouples, conforms to IEC-2, 1982.

Thermocouple	Tolerance class 1: Tolerance(±)	Tolerance class 1: Limits of validity (°C)	Tolerance class 2: Tolerance(±)	Tolerance class 2: Limits of validity (°C)	Tolerance class 3: Tolerance(±)	Tolerance class 3: Limits of validity (°C)
Cu–CuNi, code T	0.5%C or 0.004ϑ	−40 to +350	1°C or 0.0075ϑ	−40 to +350	1°C or 0.015ϑ	−200 to +40
NiCr–CuNi, code E	1.5°C or 0.004ϑ	−40 to +800	2.5°C or 0.0075ϑ	−40 to +900	2.5°C or 0.015ϑ	−200 to +40
Fe–CuNi, code J		−40 to +750		−40 to +750		
NiCr–NiAl, code K		−40 to +1000		−40 to +1200		−200 to +40
Pt10Rh–Pt, code S	1°C	0 to +1100	1.5°C or 0.0025ϑ	0 to +1600		
	[1+0.003 (ϑ −1100)]°C	+1100 to +1600				
Pt30Rh–Pt6Rh, code R	1°C	0 to +1100		0 to +1600		
	[1+0.003 (ϑ −1100)]°C	+1100 to +1600				
Pt30Rh–Pt6Rh, code B				+600 to +1700	4°C or 0.005ϑ	+600 to +1700

The greater value of tolerances applies.
Based on catalogue *Thermocouple & Resistance Thermometry Data* (TC Ltd, UK).
Class 1 conforms to ANSI, Special; Class 2 to ANSI, Standard.

Table XIII

Power series expansions and polynomials of output signals of standardized thermocouples for computer application; *E* in μV.

K

−270 to 0°C

$$E = \sum_{i=0}^{10} A_i \vartheta^i$$

$A_0 = 0$
$A_1 = 3.947\,543\,3139 \times 10^{1}$
$A_2 = 2.746\,525\,1138 \times 10^{-2}$
$A_3 = -1.656\,540\,6716 \times 10^{-4}$
$A_4 = -1.519\,091\,2392 \times 10^{-6}$
$A_5 = -2.488\,167\,0924 \times 10^{-8}$
$A_6 = -2.475\,791\,7816 \times 10^{-10}$
$A_7 = -1.558\,527\,6173 \times 10^{-12}$
$A_8 = -5.972\,992\,1255 \times 10^{-15}$
$A_9 = -1.268\,880\,1216 \times 10^{-17}$
$A_{10} = -1.138\,279\,7374 \times 10^{-20}$

0 to 1372°C

$$E = \sum_{i=0}^{8} A_i \vartheta^i + 125\exp\left[-\frac{1}{2}\left(\frac{\vartheta - 127}{65}\right)^2\right]$$

$A_0 = -1.853\,306\,3273 \times 10^{1}$
$A_1 = 3.891\,834\,4612 \times 10^{1}$
$A_2 = 1.664\,515\,4356 \times 10^{-2}$
$A_3 = -7.870\,237\,4448 \times 10^{-5}$
$A_4 = 2.283\,578\,5557 \times 10^{-7}$
$A_5 = -3.570\,023\,1258 \times 10^{-10}$
$A_6 = 2.993\,290\,9136 \times 10^{-13}$
$A_7 = -1.284\,984\,8798 \times 10^{-16}$
$A_8 = 2.223\,997\,4336 \times 10^{-20}$

T

−270 to 0°C

$$E = \sum_{i=0}^{14} A_i \vartheta^i$$

$A_0 = 0$
$A_1 = 3.874\,377\,3840 \times 10^{1}$
$A_2 = 4.412\,393\,2482 \times 10^{-2}$
$A_3 = 1.140\,523\,8498 \times 10^{-4}$
$A_4 = 1.997\,440\,6568 \times 10^{-5}$
$A_5 = 9.044\,540\,1187 \times 10^{-7}$
$A_6 = 2.276\,601\,8504 \times 10^{-8}$
$A_7 = 3.624\,740\,9380 \times 10^{-10}$
$A_8 = 3.864\,892\,4201 \times 10^{-12}$
$A_9 = 2.829\,867\,8519 \times 10^{-14}$
$A_{10} = 1.428\,138\,3349 \times 10^{-16}$
$A_{11} = 4.883\,325\,4364 \times 10^{-19}$
$A_{12} = 1.080\,347\,4683 \times 10^{-21}$
$A_{13} = 1.394\,929\,1026 \times 10^{-24}$
$A_{14} = 7.979\,589\,3156 \times 10^{-28}$

0 to 400°C

$$E = \sum_{i=0}^{8} A_i \vartheta^i$$

$A_0 = 0$
$A_1 = 3.874\,077\,3840 \times 10^{1}$
$A_2 = 3.319\,019\,8092 \times 10^{-2}$
$A_3 = 2.071\,418\,3645 \times 10^{-4}$
$A_4 = -2.194\,583\,4823 \times 10^{-6}$
$A_5 = 1.103\,190\,0550 \times 10^{-8}$
$A_6 = -3.092\,758\,1898 \times 10^{-11}$
$A_7 = 4.565\,333\,7165 \times 10^{-14}$
$A_8 = -2.761\,687\,8040 \times 10^{-17}$

J

−210 to 760°C

$$E = \sum_{i=0}^{7} A_i \vartheta^i$$

$A_0 = 0$
$A_1 = 5.037\,275\,3027 \times 10^{1}$
$A_2 = 3.042\,549\,1284 \times 10^{-2}$
$A_3 = -8.566\,975\,0464 \times 10^{-5}$
$A_4 = 1.334\,882\,5735 \times 10^{-7}$
$A_5 = -1.702\,240\,5966 \times 10^{-10}$
$A_6 = 1.941\,609\,1001 \times 10^{-13}$
$A_7 = -9.639\,184\,4859 \times 10^{-17}$

760 to 1200°C

$$E = \sum_{i=0}^{5} A_i \vartheta^i$$

$A_0 = 2.972\,175\,1778 \times 10^{5}$
$A_1 = -1.505\,963\,2873 \times 10^{3}$
$A_2 = 3.205\,106\,4215 \times 10^{0}$
$A_3 = -3.221\,017\,4230 \times 10^{-3}$
$A_4 = 1.594\,996\,8788 \times 10^{-6}$
$A_5 = -3.123\,980\,1752 \times 10^{-10}$

E

−270 to 0°C

$$E = \sum_{i=0}^{13} A_i \vartheta^i$$

$A_0 = 0$
$A_1 = 5.869\,585\,7799 \times 10^{1}$
$A_2 = 5.166\,751\,7705 \times 10^{-2}$
$A_3 = -4.465\,268\,3347 \times 10^{-4}$
$A_4 = -1.734\,627\,0905 \times 10^{-5}$
$A_5 = -4.871\,936\,8427 \times 10^{-7}$
$A_6 = -8.889\,655\,0447 \times 10^{-9}$
$A_7 = -1.093\,076\,7375 \times 10^{-10}$
$A_8 = -9.178\,453\,5039 \times 10^{-13}$
$A_9 = -5.257\,515\,8521 \times 10^{-15}$
$A_{10} = -2.016\,960\,1996 \times 10^{-17}$
$A_{11} = -4.950\,213\,8782 \times 10^{-20}$
$A_{12} = -7.017\,798\,0633 \times 10^{-23}$
$A_{13} = -4.367\,180\,8488 \times 10^{-26}$

0 to 1000°C

$$E = \sum_{i=0}^{9} A_i \vartheta^i$$

$A_0 = 0$
$A_1 = 5.869\,585\,7799 \times 10^{1}$
$A_2 = 4.311\,094\,5462 \times 10^{-2}$
$A_3 = 5.722\,035\,8202 \times 10^{-5}$
$A_4 = -5.402\,066\,8085 \times 10^{-7}$
$A_5 = 1.542\,592\,2111 \times 10^{-9}$
$A_6 = -2.485\,008\,9136 \times 10^{-12}$
$A_7 = 2.338\,972\,1459 \times 10^{-15}$
$A_8 = -1\,194\,629\,6815 \times 10^{-18}$
$A_9 = -2.556\,112\,7497 \times 10^{-22}$

Based on *Thermocouple and Resistance Thermometry Data* (TC Ltd, UK).

Table XIII *(continued)*

R

−50 to 630°C

$$E = \sum_{i=0}^{7} A_i\vartheta^i$$

$A_0 = 0$
$A_1 = 5.289\,139$
$A_2 = 1.391\,111 \times 10^{-2}$
$A_3 = -2.400\,524 \times 10^{-5}$
$A_4 = 3.620\,141 \times 10^{-8}$
$A_5 = -4.464\,502 \times 10^{-11}$
$A_6 = 3.849\,769 \times 10^{-14}$
$A_7 = -1.537\,264 \times 10^{-17}$

630 to 1064°C

$$E = \sum_{i=0}^{3} A_i\vartheta^i$$

$A_0 = -264.180$
$A_1 = 8.046\,868$
$A_2 = 2.989\,229 \times 10^{-3}$
$A_3 = -2.687\,606 \times 10$

1064 to 1665°C

$$E = \sum_{i=0}^{3} A_i \left(\frac{\vartheta - 1365}{300}\right)^i$$

$A_0 = 15\,540.414$
$A_1 = 4\,235.777$
$A_2 = 14.693$
$A_3 = -52.214$

1665 to 1767°C

$$E = \sum_{i=0}^{3} A_i \left(\frac{\vartheta - 1715}{50}\right)^i$$

$A_0 = 20416.695$
$A_1 = 668.509$
$A_2 = -12.301$
$A_3 = -2.786$

S

−50 to 630°C

$$E = \sum_{i=0}^{6} A_i\vartheta^i$$

$A_0 = 0$
$A_1 = 5.399\,578$
$A_2 = 1.251\,977 \times 10^{-2}$
$A_3 = -2.244\,822 \times 10^{-5}$
$A_4 = 2.845\,216 \times 10^{-8}$
$A_5 = -2.244\,058 \times 10^{-11}$
$A_6 = 8.505\,417 \times 10^{-15}$

630 to 1064°C

$$E = \sum_{i=0}^{2} A_i\vartheta^i$$

$A_0 = -238.245$
$A_1 = 8.237\,553$
$A_2 = 1.645\,391 \times 10^{-3}$

1064 to 1665°C

$$E = \sum_{i=0}^{3} A_i \left(\frac{\vartheta - 1365}{300}\right)^i$$

$A_0 = 13\,943.439$
$A_1 = 3\,639.869$
$A_2 = -5.028$
$A_3 = -42.451$

1665 to 1767°C

$$E = \sum_{i=0}^{3} A_i \left(\frac{\vartheta - 1715}{50}\right)^i$$

$A_0 = 18\,113.083$
$A_1 = 567.954$
$A_2 = -12.112$
$A_3 = -2.812$

B

300 to 1820°C

$$E = \sum_{i=0}^{8} A_i\vartheta^i$$

$A_0 = 0$
$A_1 = -2.467\,460\,1620 \times 10^{-1}$
$A_2 = 5.910\,211\,1169 \times 10^{-3}$
$A_3 = -1.430\,712\,3430 \times 10^{-6}$
$A_4 = 2.150\,914\,9750 \times 10^{-9}$
$A_5 = -3.175\,780\,0720 \times 10^{-12}$
$A_6 = 2.401\,036\,7459 \times 10^{-15}$
$A_7 = -9.092\,814\,8159 \times 10^{-19}$
$A_8 = 1.329\,950\,5137 \times 10^{-22}$

Table XIV
Reference calibration table of non-standardized thermocouples. Reference temperature 0°C; e.m.f. in mV.

	Thermocouple								
Temp-erature (°C)	W −55%Cu, 45% Ni	95% W, 5% Re −74% W, 26% Re	97% W, 3% Re −75% W, 25% Re	95% W, 5% Re −80% W, 30% Re	Ir −60% Jr, 40% Rn	W–Ir	W–Mo	Pallaplat	Ag–CuNi
0			0	0	0		0	0	0
200		3.089	2.602	2.871	0.841		−0.85	6.50	9.16
400		6.731	6.129	6.203	1.960		−1.14	15.05	20.56
600		10.606	10.085	9.605	3.173		−1.36	24.71	33.76
800		14.494	14.170	12.933	4.365		−1.54	35.08	
1000	14.500	18.257	18.226	16.125	5.495	14.25	−1.46	45.46	
1200	18.500	21.819	22.142	19.146	6.563	18.91	−0.98	55.39	
1400	22.500	25.148	25.875	21.971	7.590	23.70	0.06		
1600	26.500	28.236	29.403	24.588	8.610	28.62	1.42		
1800	30.400	31.078	32.702	26.992	9.656	33.69	2.88		
2000	34.100	33.660	35.707	29.181	10.753	38.88	4.10		
2200	37.300	35.932	38.289	31.138			4.94		
2400			40.223	32.853					
Source:	Zysk and Robertson, 1972 (Ref. to Ch. 4)	TC Ltd catalogue	TC Ltd catalogue	GOST 3044–77	Zysk and Robertson, 1972 (Ref. to Ch. 4)	Caldwell, 1962 (Ref. to Ch. 4)	Caldwell, 1962 (Ref. to Ch. 4)	Heraeus catalogue	Heraeus catalogue

Table XV
Characteristic data of standardized thermocouples wires.

Material	Diameter (mm)	Temperature limit for continuous use (°C)	Weight (kg/1000m)	Resistance at 20°C (Ω/m)	Average thermal resistance coefficient (1/°C)
Cu*	0.2	400	0.28	0.541	4.3×10^{-3} for the temperature range 20 to 600°C
	0.5	400	1.75	0.087	
	0.6	400	2.52	0.060	
	0.8	400	4.48	0.034	
	1.0	400	6.99	0.022	
	1.2	400	10.06	0.015	
	1.5	400	15.75	0.009	
	2.0	400	27.96	0.005	
	2.5	400	43.68	0.0035	
	3.0	600	62.91	0.0024	
	4.0	600	111.84	0.0013	
Fe*	0.2	400	0.25	3.821	$\sim 10 \times 10^{-3}$ for the temperature range 20 to 600°C
	0.5	400	1.54	0.612	
	0.6	400	2.25	0.424	
	0.8	400	4.00	0.239	
	1.0	600	6.16	0.152	
	1.2	600	9.00	0.106	
	1.5	600	13.87	0.067	
	2.0	700	24.66	0.038	
	2.5	700	38.53	0.024	
	3.0	800	55.48	0.017	
	4.0	800	98.64	0.009	
Constantan*	0.2	400	0.28	15.605	0.02×10^{-3} for the temperature range 20 to 600°C
	0.5	400	1.73	2.500	
	0.6	400	2.52	1.733	
	0.8	400	4.48	0.974	
	1.0	600	6.95	0.624	
	1.2	600	10.00	0.433	
	1.5	600	15.64	0.277	
	2.0	700	27.80	0.156	
	2.5	700	43.44	0.099	
	3.0	800	62.55	0.069	
	4.0	800	111.21	0.038	

continued

Table XV (*continued*)

Material	Diameter (mm)	Temperature limit for continuous use (°C)	Weight (kg/1000m)	Resistance at 20°C (Ω/m)	Average thermal resistance coefficient (1/°C)
NiCr*	0.2	900	0.27	23.089	0.25×10^{-3} for the temperature range 20 to 1000°C
	0.5	900	1.68	3.698	
	0.6	900	2.43	2.565	
	0.8	900	4.32	1.443	
	1.0	900	6.71	0.923	
	1.2	900	9.72	0.641	
	1.5	1000	15.11	0.410	
	2.0	1000	26.86	0.230	
	2.5	1000	41.96	0.147	
	3.0	1000	60.4	0.102	
	3.5	1100	82.26	0.075	
	5.0	1100	167.88	0.036	
NiAl*	0.2	800	0.27	9.550	1.8×10^{-3} for the temperature range 20 to 1000°C
	0.5	800	1.71	1.530	
	0.6	800	2.43	1.061	
	0.8	800	4.32	0.597	
	1.0	800	6.97	0.382	
	1.2	900	9.27	0.265	
	1.38	900	13.1	0.2	
	1.5	1000	15.37	0.169	
	2.0	1000	27.33	0.099	
	2.5	1000	42.70	0.061	
	3.0	1000	61.50	0.042	
	3.5	1100	83.70	0.031	
	5.0	1100	170.82	0.015	
Pt10Rh*	0.1	1300	0.155	24.750	1.4×10^{-3} for the temperature range 20 to 1600°C
	0.35	1300	1.92	2.09	
	0.5	1300	3.92	1.02	
Pt*	0.1	1300	0.168	13.550	3.1×10^{-3} for the temperature range 20 to 1600°C
	0.35	1300	2.06	1.11	
	0.5	1300	4.20	0.54	
Pt30Rh†	0.1	1400	0.14	25.42	
	0.35	1600	1.68	2.06	
	0.5	1800	3.44	1.01	
Pt6Rh†	0.1	1400	0.16	23.15	
	0.35	1600	1.95	1.87	
	0.5	1800	3.98	0.92	
Pt13Rh†	0.1	1300	0.15	29.07	
	0.35	1300	1.88	2.38	
	0.5	1300	3.83	1.17	

*Conforms to DIN 43712.
†Temperature Measurement Handbook (Omega Eng., Inc., USA)

Table XVI

Properties of hard and soft solders used for the measuring junctions of thermocouples.

Soft solders

Solder composition in %						Melting temperature (°C)
Pb	Sn	Ag	Sb max.	Cu max.	Bi max.	
~50	50	—	0.40	0.08	0.25	182 to 217
~50	40	—	0.40	0.08	0.25	181 to 250
~70	30	—	0.75	0.15	0.25	481 to 261
~97	—	2.5	0.50	0.08	0.25	300 to 303

Hard solders

Solder composition in %			Melting temperature (°C)
Ag	Cu	Zn	
20	45	35	777 to 816
45	30	23	677 to 744
63	20	15	694 to 718
80	16	4	738 to 794

Table XVII

Reference calibration table for Pt-100Ω resistor: resistance in Ω versus temperature, conforms to IEC 751.

°C	0	−1	−2	−3	−4	−5	−6	−7	−8	−9	−10
−200	18.49										
−190	22.80	22.37	21.94	21.51	21.08	20.65	20.22	19.79	19.36	18.93	18.49
−180	27.08	26.65	26.23	25.80	25.37	24.94	24.52	24.09	23.66	23.23	22.80
−170	31.32	30.90	30.47	30.05	29.63	29.20	28.78	28.35	27.93	27.50	27.08
−160	35.53	35.11	34.69	34.27	33.85	33.43	33.01	32.59	32.16	31.74	31.32
−150	39.71	39.30	38.88	38.46	38.04	37.63	37.21	36.79	36.37	35.95	35.53
−140	43.87	43.45	43.04	42.63	42.21	41.79	41.38	40.96	40.55	40.13	39.71
−130	48.00	47.59	47.18	46.76	46.35	45.94	45.52	45.11	44.70	44.28	43.87
−120	52.11	51.70	51.29	50.88	50.47	50.06	49.64	49.23	48.82	48.41	48.00
−110	56.19	55.78	55.38	54.97	54.56	54.15	53.74	53.33	52.92	52.52	52.11
−100	60.25	59.85	59.44	59.04	58.63	58.22	57.82	57.41	57.00	56.60	56.19
− 90	64.30	63.90	63.49	63.09	62.68	62.28	61.87	61.47	61.06	60.66	60.25
− 80	68.33	67.92	67.52	67.12	66.72	66.31	65.91	65.51	65.11	64.70	64.30
− 70	72.33	71.93	71.53	71.13	70.78	70.33	69.93	69.53	69.13	68.73	68.33
− 60	76.33	75.93	75.53	75.13	74.73	74.33	73.93	73.53	73.13	72.73	72.33
− 50	80.31	79.91	79.51	79.11	78.72	78.32	77.92	77.52	77.13	76.73	76.33
− 40	84.27	83.88	83.48	83.08	82.69	82.29	81.89	81.50	81.10	80.70	80.31
− 30	88.22	87.83	87.43	87.04	86.64	86.25	85.85	85.46	85.06	84.67	84.27
− 20	92.16	91.77	91.37	90.98	90.59	90.19	89.80	89.40	89.01	88.62	88.22
− 10	96.09	95.69	95.30	94.91	94.52	94.12	93.73	93.34	92.95	92.55	92.16
0	100.00	99.61	99.22	98.83	98.44	98.04	97.65	97.26	96.87	96.48	96.09

°C	0	1	2	3	4	5	6	7	8	9	10
0	100.00	100.39	100.78	101.17	101.56	101.95	102.34	102.73	103.12	103.51	103.90
10	103.90	104.29	104.68	105.07	105.46	105.85	106.24	106.63	107.02	107.40	107.79
20	107.79	108.18	108.57	108.98	109.35	109.73	110.12	110.51	110.90	111.28	111.67
30	111.67	112.06	112.45	112.83	113.22	113.61	113.99	114.38	114.77	115.15	115.54
40	115.54	115.93	116.31	116.70	117.08	117.47	117.85	118.24	118.62	119.01	119.40
50	119.40	119.78	120.16	120.55	120.93	121.32	121.70	122.09	122.47	122.86	123.24
60	123.24	123.62	124.01	124.33	124.77	125.16	125.54	125.92	126.31	126.69	127.07
70	127.07	127.45	127.84	128.22	128.60	128.98	129.37	129.75	130.13	130.51	130.89
80	130.89	131.27	131.66	132.04	132.42	132.80	133.18	133.56	133.94	134.32	134.70
90	134.70	135.08	135.46	135.84	136.22	136.60	136.98	137.36	137.74	138.12	138.50
100	138.50	138.88	139.26	139.64	140.02	140.39	140.77	141.15	141.53	141.91	142.29
110	142.29	142.66	143.04	143.42	143.80	144.17	144.55	144.93	145.31	145.68	146.06
120	146.06	146.44	146.81	147.19	147.57	147.94	148.32	148.70	149.07	149.45	149.82
130	149.82	150.20	150.57	150.95	151.33	151.70	152.08	152.45	152.83	153.20	153.58
140	153.58	153.95	154.32	154.70	155.07	155.45	155.82	156.19	156.57	156.94	157.31
150	157.31	157.69	158.06	158.43	158.81	159.18	159.55	159.93	160.30	160.67	161.04
160	161.04	161.42	161.79	162.16	162.53	162.90	163.27	163.65	164.02	164.39	164.76
170	164.76	165.13	165.50	165.87	166.24	166.61	166.98	167.35	167.72	168.09	168.46
180	168.46	168.83	169.20	169.57	169.94	170.31	170.68	171.05	171.42	171.79	172.16
190	172.16	172.53	172.90	173.26	173.63	174.00	174.37	174.74	175.10	175.47	175.84
200	175.84	176.21	176.57	176.94	177.31	177.68	178.04	178.41	178.78	179.14	179.51
210	179.51	179.88	180.24	180.61	180.97	181.34	181.71	182.07	182.44	182.80	183.17
220	183.17	183.53	183.90	184.26	184.63	184.99	185.36	185.72	186.09	186.45	186.82
230	186.82	187.18	187.54	187.91	188.27	188.63	189.00	189.36	189.72	190.09	190.45
240	190.45	190.81	191.18	191.54	191.90	192.26	192.63	192.99	193.35	193.71	194.07
250	194.07	194.44	194.80	195.16	195.52	195.88	196.24	196.60	196.96	197.33	197.69

continued

Table XVII (*continued*)

°C	0	1	2	3	4	5	6	7	8	9	10
260	197.69	198.05	198.41	198.77	199.13	199.49	199.85	200.21	200.57	200.93	201.29
270	201.29	201.65	202.01	202.36	202.72	203.08	203.44	203.80	204.16	204.52	204.88
280	204.38	205.23	205.59	205.95	206.31	206.67	207.02	207.38	207.74	208.10	208.45
290	208.45	208.81	209.17	209.52	209.88	210.24	210.59	210.95	211.31	211.66	212.02
300	212.02	212.37	212.73	213.09	213.44	213.80	214.15	214.51	214.86	215.22	215.57
310	216.57	215.93	216.28	216.64	216.99	217.35	217.70	218.05	218.41	218.76	219.12
320	219.12	219.47	219.82	220.18	220.53	220.88	221.24	221.59	221.94	222.29	222.65
330	222.65	223.00	223.35	223.70	224.06	224.41	224.76	225.11	225.46	225.81	226.17
340	226.17	226.52	226.87	227.22	227.57	227.92	228.27	228.62	228.97	229.32	229.67
350	229.67	230.02	230.37	230.72	231.07	231.42	231.77	232.12	232.47	232.82	233.17
360	233.17	233.52	233.87	234.22	234.56	234.91	235.26	235.61	235.96	236.31	236.65
370	236.65	237.00	237.35	237.70	238.04	238.39	238.74	239.09	239.43	239.78	240.13
380	240.13	240.47	240.82	241.17	241.51	241.86	242.20	242.55	242.90	243.24	243.59
390	243.69	243.93	244.28	244.62	244.97	245.31	245.66	246.00	246.35	246.69	247.04
400	247.04	247.38	247.73	248.07	248.41	248.76	249.10	249.45	249.79	250.13	250.48
410	250.43	250.82	251.16	251.50	251.85	252.19	252.53	252.83	253.22	253.56	253.90
420	253.90	254.24	254.59	254.93	255.27	255.61	255.95	256.29	256.64	256.98	257.32
430	257.32	257.68	258.00	258.34	258.68	259.02	259.36	259.70	260.04	260.38	260.72
440	260.72	261.06	261.40	261.74	262.08	262.42	262.76	263.10	263.43	263.77	264.11
450	264.11	264.45	264.79	265.13	265.47	265.80	266.14	266.48	266.82	267.15	267.49
460	267.49	267.83	268.17	268.50	268.84	269.18	269.51	269.85	270.19	270.52	270.86
470	270.86	271.20	271.53	271.87	272.20	272.54	272.88	273.21	273.55	273.88	274.22
480	274.22	274.55	274.89	275.22	275.56	275.89	276.23	276.56	276.89	277.23	277.56
490	277.56	277.90	278.23	278.58	278.90	279.23	279.56	279.90	280.23	280.56	280.90
500	280.90	281.23	281.56	281.89	282.23	282.56	282.89	283.22	283.55	283.89	284.22
510	284.22	284.65	284.88	285.21	285.54	285.87	286.21	286.54	286.87	287.20	287.53
520	287.53	287.86	288.19	288.52	288.85	289.18	289.51	289.84	290.17	290.50	290.83
530	290.83	291.16	291.49	291.81	292.14	292.47	292.80	293.13	293.46	293.79	294.11
540	294.11	294.44	294.77	295.10	295.43	295.75	296.08	296.41	296.74	297.06	297.39
550	297.39	297.72	298.04	298.37	298.70	299.02	299.35	299.68	300.00	300.33	300.65
560	300.65	300.98	301.31	301.63	301.96	302.28	302.61	302.93	303.26	303.58	303.91
570	303.91	304.23	304.56	304.88	305.20	305.53	305.86	306.18	306.50	306.82	307.15
580	307.15	307.47	307.79	308.12	308.44	308.76	309.09	309.41	309.73	310.05	310.38
590	310.38	310.70	311.02	311.34	311.67	311.99	312.31	312.63	312.95	313.27	313.59
600	313.59	313.92	314.24	314.56	314.88	315.20	315.52	315.84	316.16	316.48	316.80
610	316.80	317.12	317.44	317.76	318.08	318.40	318.72	319.04	319.36	319.68	319.99
620	319.99	320.31	320.63	320.95	321.27	321.59	321.91	322.22	322.54	322.86	323.18
630	323.18	323.49	323.81	324.13	324.45	324.76	325.08	325.40	325.72	326.03	326.35
640	326.35	326.66	326.98	327.30	327.61	327.93	328.25	328.56	328.88	329.19	329.51
650	329.51	329.82	330.14	330.45	330.77	331.08	331.40	331.71	332.03	332.34	332.66
660	332.66	332.97	333.28	333.60	333.91	334.23	334.54	334.85	335.17	335.48	335.79
670	335.79	336.11	336.42	336.73	337.04	337.36	337.67	337.98	338.29	338.61	338.92
680	338.92	339.23	339.54	339.85	340.16	340.48	340.79	341.10	341.41	341.72	342.03
690	342.03	342.34	342.65	342.96	343.27	343.53	343.89	344.20	344.51	344.82	345.13
700	345.13	345.44	345.75	346.06	346.37	346.68	346.99	347.30	347.60	347.91	348.22
710	348.22	348.53	348.84	349.15	349.45	349.76	350.07	350.38	350.69	350.99	351.30
720	351.30	351.61	351.91	352.22	352.53	352.83	353.14	353.45	353.75	354.06	354.37
730	354.37	354.07	354.98	355.28	355.59	355.90	356.20	356.51	356.81	357.12	357.42
740	357.42	357.73	358.03	358.34	358.64	358.95	359.25	359.55	359.86	360.16	360.47
750	360.47	360.77	361.07	361.38	361.68	361.98	362.29	362.59	362.89	363.19	363.50

continued

Table XVII (*continued*)

°C	0	1	2	3	4	5	6	7	8	9	10
760	363.50	363.80	364.10	364.40	364.71	365.01	365.31	365.61	365.91	366.22	366.52
770	366.52	366.82	367.12	367.42	367.72	368.02	368.32	368.63	368.93	369.23	369.53
780	369.53	369.83	370.13	370.43	370.73	371.03	371.33	371.63	371.93	372.22	372.52
790	372.52	372.82	373.12	373.42	373.72	374.02	374.32	374.61	374.91	375.21	375.51
800	375.51	375.81	376.10	376.40	376.70	377.00	377.29	377.69	377.89	378.19	378.48
810	378.48	378.78	379.06	379.37	379.67	379.97	380.26	380.56	380.85	381.15	381.45
820	381.45	381.74	382.04	382.33	382.63	382.92	383.22	383.51	383.81	384.10	384.40
830	384.40	384.69	384.98	385.28	385.57	385.87	386.16	386.45	386.75	387.04	387.34
840	387.34	387.63	387.92	388.21	388.51	388.80	389.09	389.39	389.68	389.97	390.26
850	390.26										

Table XVIII

Pt-100Ω resistor: permissible tolerances conforms to IEC 751.

Temperature (°C)	Tolerance			
	Class A		Class B	
	± °C	± Ω	± °C	± Ω
−200	0.55	0.24	1.3	0.56
−100	0.35	0.14	0.8	0.32
0	0.15	0.06	0.3	0.12
100	0.35	0.13	0.8	0.30
200	0.55	0.20	1.3	0.48
300	0.75	0.27	1.8	0.64
400	0.95	0.33	2.3	0.79
500	1.15	0.38	2.8	0.93
600	1.35	0.43	3.3	1.06
650	1.45	0.46	3.6	1.13
700	—	—	3.8	1.17
800	—	—	4.3	1.28
850	—	—	4.6	1.34

Table XIX
Reference calibration table for Ni-100Ω resistor: resistance in Ω versus temperature conforms to DIN 43760.

°C	0	−1	−2	−3	−4	−5	−6	−7	−8	−9	−10
−60	69.5	—	—	—	—	—	—	—	—	—	—
−50	74.3	73.8	73.2	72.8	72.3	71.9	71.4	70.9	70.5	70.0	69.5
−40	79.1	78.6	78.1	77.7	77.2	76.7	76.2	75.7	75.2	74.7	74.3
−30	84.1	83.6	83.1	82.6	82.1	81.6	81.1	80.6	80.1	79.6	79.1
−20	89.3	88.8	88.3	87.7	87.2	86.7	86.2	85.7	85.2	84.7	84.1
−10	94.6	94.0	93.5	93.0	92.5	91.9	91.4	90.9	90.3	89.8	89.3
0	100.0	99.5	98.9	98.4	97.8	97.3	96.7	96.2	95.7	95.1	94.6
°C	**0**	**1**	**2**	**3**	**4**	**5**	**6**	**7**	**8**	**9**	**10**
0	100.0	100.5	101.1	101.7	102.2	102.8	103.3	103.9	104.4	105.0	105.6
10	105.6	106.1	106.7	107.2	107.8	108.4	108.9	109.6	110.1	110.7	111.2
20	111.2	111.8	112.4	113.0	113.5	114.1	114.7	115.3	115.9	116.5	117.1
30	117.1	117.7	118.2	118.8	119.4	120.0	120.6	121.2	121.8	122.4	123.0
40	123.0	123.6	124.2	124.8	125.4	128.0	126.7	127.3	127.9	128.5	129.1
50	129.1	129.7	130.3	131.0	131.6	132.2	132.6	133.6	134.1	134.7	135.3
60	135.3	136.0	136.6	137.2	137.9	138.6	139.2	139.8	140.4	141.1	141.7
70	141.7	142.4	143.0	143.7	144.3	145.0	145.6	146.3	146.9	147.6	148.3
80	148.3	148.9	149.6	150.2	150.9	151.6	152.2	152.9	153.6	154.3	154.9
90	154.9	155.6	156.3	157.0	157.7	158.3	159.0	159.7	160.4	161.1	161.8
100	161.8	162.5	163.2	163.9	164.6	165.3	166.0	166.7	167.4	168.1	168.8
110	168.8	169.5	170.2	170.9	171.6	172.4	173.1	173.8	174.5	175.2	176.0
120	176.0	176.7	177.4	178.2	178.9	179.6	180.4	181.1	181.8	182.6	183.3
130	183.3	184.1	184.8	185.6	186.3	187.1	187.8	188.6	189.4	190.1	190.9
140	190.9	191.7	192.4	193.2	194.0	194.7	195.5	196.3	197.1	197.9	198.6
150	198.6	199.4	200.2	201.0	201.8	202.6	203.4	204.2	205.0	205.8	206.6
160	206.6	207.4	208.2	209.0	209.8	210.6	211.5	212.3	213.1	213.9	214.8
170	214.8	215.6	216.4	217.3	218.1	218.9	219.8	220.6	221.5	222.3	223.2
180	223.2	224.0	224.9	225.7	226.6	227.4	228.3	229.2	230.0	230.9	231.8
190	231.8	232.7	233.5	234.4	235.3	236.2	237.1	238.0	238.9	239.8	240.7
200	240.7	241.6	242.5	243.4	244.3	245.2	246.1	247.0	247.9	248.9	249.8
210	249.8	250.7	251.7	252.6	253.5	254.5	255.4	256.3	257.3	258.2	259.2
220	259.2	260.2	261.1	262.1	263.0	264.0	265.0	266.0	266.9	267.9	268.9
230	268.9	269.9	270.9	271.8	272.8	273.8	274.8	275.8	276.8	277.9	278.9
240	278.9	279.9	280.9	281.9	282.9	284.0	285.0	286.0	287.1	288.1	289.2
250	289.2	—	—	—	—	—	—	—	—	—	—

Table XX
Ni-100Ω resistor: permissible tolerances conform to DIN 43760.

Temperature (°C)	Tolerance	
	±Ω	±°C
−60	1.0	2.1
0	0.2	0.4
100	0.8	1.1
200	1.6	1.8
250	2.3	2.1

Table XXI
Reference calibration table for Cu-100Ω resistor: resistance in Ω versus temperature. Non-standardized by IEC.

°C	0	−1	−2	−3	−4	−5	−6	−7	−8	−9	−10
−50	78.70	—	—	—	—	—	—	—	—	—	—
−40	82.96	82.53	82.11	81.68	81.26	80.83	80.40	79.98	79.55	79.13	78.70
−30	87.22	86.79	86.37	85.94	85.52	85.09	84.66	84.24	83.81	83.39	82.96
−20	91.48	91.05	90.63	90.20	89.78	89.35	88.92	88.50	88.07	87.65	87.22
−10	95.74	95.31	94.89	94.46	94.04	93.61	93.18	92.76	92.33	91.91	91.48
0	100.00	99.57	99.15	98.72	98.30	97.87	97.44	97.02	96.59	96.17	95.74
°C	**0**	**1**	**2**	**3**	**4**	**5**	**6**	**7**	**8**	**9**	**10**
0	100.00	100.43	100.85	101.28	101.70	102.13	102.56	102.98	103.41	103.83	104.26
10	104.26	104.69	105.11	105.54	105.96	106.39	106.82	107.24	107.67	108.09	108.52
20	108.52	108.95	109.37	109.80	110.22	110.65	111.08	111.50	111.93	112.35	112.78
30	112.78	113.21	113.63	114.06	114.48	114.91	115.34	115.76	116.19	116.61	117.04
40	117.04	117.47	117.89	118.32	118.74	119.17	119.60	120.02	120.45	120.87	121.30
50	121.30	121.73	122.15	122.58	123.00	123.43	123.86	124.28	124.71	125.13	125.56
60	125.56	125.99	126.41	126.84	127.26	127.69	128.12	128.54	128.97	129.39	129.82
70	129.82	130.25	130.67	131.10	131.52	131.95	132.38	132.80	133.23	133.65	134.08
80	134.08	134.51	134.93	135.36	135.78	136.21	136.64	137.06	137.49	137.91	138.34
90	138.34	138.77	139.19	139.62	140.04	140.47	140.90	141.32	141.75	142.17	142.60
100	142.60	143.03	143.45	143.88	144.30	144.73	145.16	145.58	146.01	146.43	146.86
110	146.86	147.29	147.71	148.14	148.56	148.99	149.42	149.84	150.27	150.69	151.12
120	151.12	151.55	151.97	152.40	152.82	153.25	153.68	154.10	154.53	154.95	155.38
130	155.38	155.81	156.23	156.66	157.08	157.51	157.94	158.36	158.79	159.21	159.64
140	159.64	160.07	160.49	160.92	161.34	161.77	162.20	162.62	163.05	163.47	163.90
150	163.90	164.33	164.75	165.18	165.60	166.03	166.46	166.88	167.31	167.73	168.16
160	168.16	168.59	169.01	169.44	169.86	170.29	170.72	171.14	171.57	171.99	172.42
170	172.42	172.85	173.27	173.70	174.12	174.55	174.98	175.40	175.83	176.25	176.68
180	176.68	—	—	—	—	—	—	—	—	—	—

Table XXII
Emissivity of metals, measured perpendicularly to surface. Approximate values.

Material	Temperature ϑ (°C)	Total emissivity ϵ	Spectral emissivity ϵ_λ at $\lambda=0.65\ \mu m$	Specific total emissivity ϵ'	Specific spectral emissivity ϵ'_λ at $\lambda=0.65\ \mu m$
Aluminium	25	0.2	—	0.22	—
Aluminium, oxidized	—	0.1–0.4	0.3–0.5	—	—
Brass	0–500	0.03–0.4	—	—	—
Brass	20	—	—	0.035	—
Cast iron	200–1000	0.2–0.5	0.35–0.5	0.2	—
Cast iron, oxidized	—	0.5–0.95	0.6–0.95	—	—
Chrome	100	0.08	—	0.08	0.34
Chrome, oxidized	100	0.3–0.8	0.35–0.6	—	—
Chrome nickel alloy	0–1000	0.2–0.6	—	—	—
Cobalt	500	0.04–0.25	0.3–0.8	0.13	0.36
Copper	100	—	—	0.02	0.1
Copper	100–1000	0.03–0.2	0.1–0.2	—	—
Copper, oxidized	1000	0.4–0.8	0.4–0.8	—	—
Gold	100–1000	0.02–0.05	0.1–0.2	0.03	0.14
Iron and steel	100–1200	0.05–0.25	0.35	—	0.35
Iron and steel, oxidized	—	0.7–0.95	0.5–0.95	—	—
Lead	50–300	0.05–0.4	0.35–0.8	0.05	—
Molybdenum	100–500	0.03–0.3	0.4–0.6	0.13	0.37
Molybdenum	2000	0.2–0.5	0.5–0.8	—	—
Nickel	100–300	0.04–0.25	0.3–0.6	0.1	0.36
Nickel	1000	0.15–0.5	0.3–0.6	0.19	—
Platinum	1000	0.1–0.15	0.3–0.4	0.15	0.3
Rhodium	100–1500	0.05–0.1	0.15–0.3	—	—
Silver	100–900	0.01–0.04	0.04–0.13	—	0.07
Silver	500	—	—	0.035	—
Steel, stainless	100	0.1–0.3	0.3–0.45	0.08	—
Steel, oxidized	—	0.4–0.95	0.5–0.9	—	—
Tin	0–200	0.05–0.3	0.3–0.5	0.05	—
Tungsten	1500	0.2	0.45	0.23	0.43
Tungsten	3000	0.36	0.42	—	—
Zinc	20–400	0.02–0.3	0.25–0.45	0.05	—

Table XXIII

Total emissivity ϵ of some materials measured perpendicularly to surface. Approximate values.

Material	Temperature (°C)	Total emissivity
Al_2O_3, grain 1–2 μm	1000	0.25
Al_2O_3, grain 10–100 μm	1000	0.3–0.5
Asbestos	0–400	0.9
Asphalt	20	0.85
Brick	20	0.95
Ceramic	100	0.85–0.95
Chamotte	0–200	0.8
Concrete	0–200	0.95
Enamel	25	0.95
Graphite	—	0.75–0.95
Magnesite	100	0.7–0.8
Magnesite	1000	0.4–0.5
MgO_2	—	0.2
Paper	20	0.8–0.95
Plastics	20	0.7–0.9
Porcelain	1000	0.9
Quartz, cast	20	0.9
Roofing paper	20	0.9
Rubber	20	0.9
Slag	1000	0.7
Textiles	20	0.75–0.9
Varnish, aluminium	20	0.3–0.6
Varnish, black, flat	20	0.96–0.98
Varnish, white	20	0.9

Table XXIV

Pyrometric temperature measurement of plastics. Advised effective wavelengths, λ_e.

Plastic	$\lambda_e = 0.34$ μm	$\lambda_e = 6.85$ μm	$\lambda_e = 8.03$ μm
Cellulose acetate	x		
Polyester	x		x
Teflon			x
Polyimide		x	x
Polyurethane	x		x
PVC	x		x
Polyacrylate	x		x
Polycarbonate	x	x	x
Polyamide	x		x
Polypropylene	x		
Polyethylene	x	x	
Polystyrene	x		

Author and Organization Index

Subject Index